Deep Learning and Transformers' Methods Applied to Remotely Captured Data

Deep Learning and Transformers' Methods Applied to Remotely Captured Data

Editors

Moulay A. Akhloufi
Mozhdeh Shahbazi

MDPI

Basel • Beijing • Wuhan • Barcelona • Belgrade • Novi Sad • Cluj • Manchester

Editors

Moulay A. Akhloufi
Université de Moncton
Moncton, NB
Canada

Mozhdeh Shahbazi
University of Calgary
Calgary, AB
Canada

Editorial Office
MDPI
St. Alban-Anlage 66
4052 Basel, Switzerland

This is a reprint of articles from the Topic published online in the open access journals *Remote Sensing* (ISSN 2072-4292), *Applied Sciences* (ISSN 2076-3417), and *Sensors* (ISSN 1424-8220) (available at: https://www.mdpi.com/topics/remo_data_deep_lear).

For citation purposes, cite each article independently as indicated on the article page online and as indicated below:

Lastname, A.A.; Lastname, B.B. Article Title. *Journal Name* **Year**, *Volume Number*, Page Range.

ISBN 978-3-7258-0585-3 (Hbk)
ISBN 978-3-7258-0586-0 (PDF)
doi.org/10.3390/books978-3-7258-0586-0

Cover image courtesy of Moulay A. Akhloufi and Mozhdeh Shahbazi

Contents

About the Editors

Moulay A. Akhloufi

Moulay A. Akhloufi has a Bachelor of Science degree in Physics and a Bachelor of Engineering degree. He received his Master's degree and Ph.D. in Electrical Engineering from Ecole Polytechnique of Montreal and Laval University, respectively. He also holds an MBA from Laval University.

He is currently an Associate Professor in Computer Science, Head of the Perception, Robotics, and Intelligent Machines (PRIME) research group, and Director of the Center for Artificial Intelligence NB Power at Université de Moncton (Canada).

Before joining Université de Moncton in 2016, he worked in the computer vision industry with Matrox ES and in technology transfer in machine vision and robotics with CRVI inc. His research interests are in the areas of artificial intelligence, computer vision and intelligent robotic systems.

Professor Akhloufi is a senior member of the Institute of Electrical and Electronics Engineers (IEEE) and member of the Society of Photo-Optical Instrumentation Engineers (SPIE).

Mozhdeh Shahbazi

Mozhdeh Shahbazi (PhD, PEng) holds a BSc degree in survey engineering, MSc in photogrammetry engineering, and PhD in remote sensing. Her expertise is in the areas of computer vision, deep learning, photogrammetry, sensor integration and self-calibration, and robot vision. She joined the Department of Geomatics Engineering at the University of Calgary as a professor in 2016. Since 2018, she has been an adjunct professor at York University. In 2019, she took on the role of lead scientist at the Centre de Géomatique du Québec. During this period, she secured R&D funding of over CAD 5 million as the principal investigator and project lead of competitive grants using which she established two research laboratories and a technology access center and trained over 30 highly qualified personnel. She became a machine learning specialist at Genetec Inc in 2021. In this role, she designed algorithms and software related to intelligent surveillance. In 2022, she joined the government of Canada as a computer vision and photogrammetry scientist where she is developing automated techniques for geo-referencing and processing historical images to extract various statistics related to the natural resources of the country.

As part of her voluntary activities, she has been a secretary of the International Society of Photogrammetry and Remote Sensing, vice president of the Canadian Remote Sensing Society, associate editor of the *Canadian Journal of Remote Sensing*, and co-editor in chief of *Drone Systems and Applications*. She has been an active participant and organizer of STEM outreach programs such as NSERC Chairs for Women in Science and Engineering, Women in Data Science, and Cyber Mentor.

Preface

The fields of machine learning and deep learning have undergone remarkable advancements in recent years, driven by the convergence of increased processing power, cost-effective hardware, and vast datasets. In particular, the emergence of deep neural networks and transformers has revolutionized various domains, pushing the boundaries of what was previously deemed achievable. This rapid progress has not only spurred technological innovations but also led to new algorithms, methodologies, and unprecedented outcomes.

One area influenced by these advancements is processing remotely captured data, encompassing a wide array of imagery acquired from airborne and spaceborne platforms, underwater sources, mobile mapping technologies, and beyond. This reprint serves as an example of the transformative impact of deep learning and transformers in remote sensing applications.

Within these pages lie cutting-edge contributions from researchers utilizing deep learning techniques and transformers to analyze and interpret remotely captured data, covering diverse domains such as environmental studies, remote surveillance, change detection, and security. The breadth of the showcased research underscores the versatility and efficacy of these methodologies.

This compilation represents the leading edge of innovation in remote sensing, providing invaluable insights for both practitioners and scholars. Our goal is for it to serve as a foundational resource for researchers, scientists, and engineers keen on exploring the application of deep learning and transformers in remote sensing.

Moulay A. Akhloufi and Mozhdeh Shahbazi
Editors

remote sensing

MDPI

Article

Tree Species Classification in UAV Remote Sensing Images Based on Super-Resolution Reconstruction and Deep Learning

Yingkang Huang [1,2], Xiaorong Wen [1,2,*,†], Yuanyun Gao [3,†], Yanli Zhang [4] and Guozhong Lin [2]

1 Co-Innovation Center for Sustainable Forestry in Southern China, Nanjing Forestry University, Nanjing 210037, China; kerwin@njfu.edu.cn
2 College of Forestry, Nanjing Forestry University, Nanjing 210037, China
3 Ministry of Ecology and Environment, Nanjing Institute of Science, Nanjing 210037, China; gaoyuanyun@nies.org
4 Arthur Temple College of Forestry and Agriculture, Stephen F. Austin State University, Nacogdoches, TX 75962, USA
* Correspondence: wxr9872@njfu.edu.cn
† These authors contributed equally to this work.

Abstract: We studied the use of self-attention mechanism networks (SAN) and convolutional neural networks (CNNs) for forest tree species classification using unmanned aerial vehicle (UAV) remote sensing imagery in Dongtai Forest Farm, Jiangsu Province, China. We trained and validated representative CNN models, such as ResNet and ConvNeXt, as well as the SAN model, which incorporates Transformer models such as Swin Transformer and Vision Transformer (ViT). Our goal was to compare and evaluate the performance and accuracy of these networks when used in parallel. Due to various factors, such as noise, motion blur, and atmospheric scattering, the quality of low-altitude aerial images may be compromised, resulting in indistinct tree crown edges and deficient texture. To address these issues, we adopted Real-ESRGAN technology for image super-resolution reconstruction. Our results showed that the image dataset after reconstruction improved classification accuracy for both the CNN and Transformer models. The final classification accuracies, validated by ResNet, ConvNeXt, ViT, and Swin Transformer, were 96.71%, 98.70%, 97.88%, and 98.59%, respectively, with corresponding improvements of 1.39%, 1.53%, 0.47%, and 1.18%. Our study highlights the potential benefits of Transformer and CNN for forest tree species classification and the importance of addressing the image quality degradation issues in low-altitude aerial images.

Keywords: self-attention mechanism networks; convolutional neural networks; Real-ESRGAN; transformer; tree species classification

Citation: Huang, Y.; Wen, X.; Gao, Y.; Zhang, Y.; Lin, G. Tree Species Classification in UAV Remote Sensing Images Based on Super-Resolution Reconstruction and Deep Learning. *Remote Sens.* **2023**, *15*, 2942. https://doi.org/10.3390/rs15112942

Academic Editors: Moulay A. Akhloufi and Mozhdeh Shahbazi

Received: 4 May 2023
Revised: 31 May 2023
Accepted: 2 June 2023
Published: 5 June 2023

1. Introduction

Traditional tree species classification methods primarily rely on the expertise of forest workers who visually identify and judge trees based on features such as leaf shape, crown shape, and texture. These methods are often subjective and labor-intensive, requiring extensive fieldwork and manual identification. To improve the accuracy of tree species classification, LiDAR data are often combined with hyperspectral (HS) data [1–3]. However, the processing of airborne LiDAR data can be costly and complex, which makes this method unsuitable for large-scale forest classification [4]. Conventional research methods for tree species classification include manual feature extraction and classical machine learning algorithms, such as Support Vector Machines (SVMs) [5–7], Artificial Neural Networks (ANNs) [8,9], and Random Forest (RF) [10–12]. Burai, P. et al. [13] used airborne HS imagery and image classification methods (multi-label classification and SVM) combined with feature extraction to discriminate between species and clones of energy trees. They proposed an adaptive binary tree SVM classifier (ABTSVM) to improve the species-level classification accuracy. Rocha, S.J.S.S.D. et al. [14] used ANNs based on competition index and climatic and categorical variables to predict tree survival and mortality in the semideciduous seasonal forests of the Atlantic Forest biome, reaching a high classification performance. Freeman, E.A. et al. [15] proposed strategies to address the issues of varying sampling intensity across different strata and the imbalanced presence of target species in training data when using the RF model for species distribution modeling. However, traditional machine learning methods for tree species classification often rely on handcrafted features, including various vegetation indices and texture features. In contrast, CNNs offer a significant advantage over traditional approaches in extracting essential features from raw data, which enables their widespread application in tree species classification tasks.

Since LeCun, Y. and his colleagues pioneered the use of CNNs for image classification tasks in 1998 [16], and it has been demonstrated that CNNs have significant advantages in extracting low-level features and visual structures. More recently, Krizhevsky, A. and his team introduced a deep CNN architecture called AlexNet [17], which was trained on a large scale using GPUs and obtained breakthrough results on the ImageNet dataset. At the same time, the development of low-altitude UAVs has made the acquisition of high-resolution aerial images with richer texture information than satellite images easier [18]. This combination has resulted in significant advances in tree species classification. For example, Nezami, S. et al. [19] proposed a deep learning method based on 3D-CNN for the high-accuracy classification of three major tree species in a boreal forest using RGB and HS data layers. Kapil, R. et al. [20] proposed a RetinaNet-based method that reached an average accuracy of 98.95% in classifying different stages of bark beetle attacks on individual trees. Hu, M. et al. [21] used a transfer-learning-based approach that fused multiple deep learning models to solve tree species classification in complex backgrounds, attaining an overall average accuracy of 93.75%. Natesan, S. et al. [22] used DenseNet for classifying forest tree species at the individual tree level using high-resolution RGB images from UAVs. The validation results demonstrate an accuracy of over 84% in distinguishing coniferous tree species in eastern Canada. Ford, D. J. [23] delved into the use of high-resolution RGB imagery from UAVs for tree species classification in a tropical wet forest. The study compared three classifiers and found that U-Net obtained the highest overall accuracy of 71.2%, suggesting the suitability of CNNs for fine-grained species-level classification using UAV data. In addition, some researchers have used deep learning in combination with UAV-Borne LiDAR data for individual tree crown segmentation studies [24]. However, it is noted that canopy images are different from natural images, and mutual relationships between tree canopies can affect classification results in forests with medium to high canopy density. In addition, low-altitude drone canopy images exhibit intra-canopy heterogeneity at the level of individual tree crowns, while displaying repeated similarity at the overall level. Additionally, due to hardware limitations of the drone imaging equipment, the images may suffer from blurred tree crown boundaries and weak texture factors. Therefore, simply adopting a residual network model to identify tree species can lead to overlooking the importance of different channel feature maps for classification results and result in a limitation of accuracy. However, attention mechanism models have a natural long-range modeling ability that enables the full utilization of effective global information from shallow to deep layers. By taking advantage of this ability, researchers can more effectively classify tree species and overcome the unique challenges posed by canopy images.

The goal of this research article is to investigate tree species classification methods using both CNN models and Transformer models on UAV tree crown images. Our proposed approach involves several steps. Firstly, we utilized a sliding window cropping technique on low-altitude drone canopy images to obtain smaller image patches. Then, we applied Real-ESRGAN technology for super-resolution reconstruction to restore the blurry canopy images and enhance their spatial resolution. Next, we made use of both CNN models and Transformer models to extract features from the tree crown images, and performed a differential comparative analysis to evaluate the performance of these two approaches. Due to the relatively modest scale of the canopy sample set in this experiment, and taking into account the constraints imposed by computational resources, this study opted to train and validate the ResNet-50, ConvNeXt-T, Swin-T, and ViT-B models. These models boast fewer parameters and exhibit lower computational intricacy, thereby enhancing the efficacy of the training and inference processes. Through this efficient and accurate tree species classification method based on low-altitude aerial tree crown images, we aim to delve into the potential applications of artificial intelligence in forestry intelligence and information construction, and explore the possibilities of improving tree species classification accuracy using advanced deep learning techniques.

2. Study Area and Data Collection

2.1. Study Area

The study area was located in Dongtai Forest Farm, Yancheng City, Jiangsu Province, with geographical coordinates ranging from 120°47′11″E to 120°52′0″E and 32°53′30″N to 32°51′17″N, as shown in Figure 1. We used a Liortho high-resolution imaging system mounted on a Digital Green Earth octocopter UAV for data collection, with a flight altitude of 200 m and image resolution of 0.2 m. In the Dongtai Forest Farm study area, we picked out three representative plots as experimental areas. From these areas, we selected four predominant tree species, Poplar, Metasequoia, Bamboo, and Ginkgo, as the primary subjects of our research.

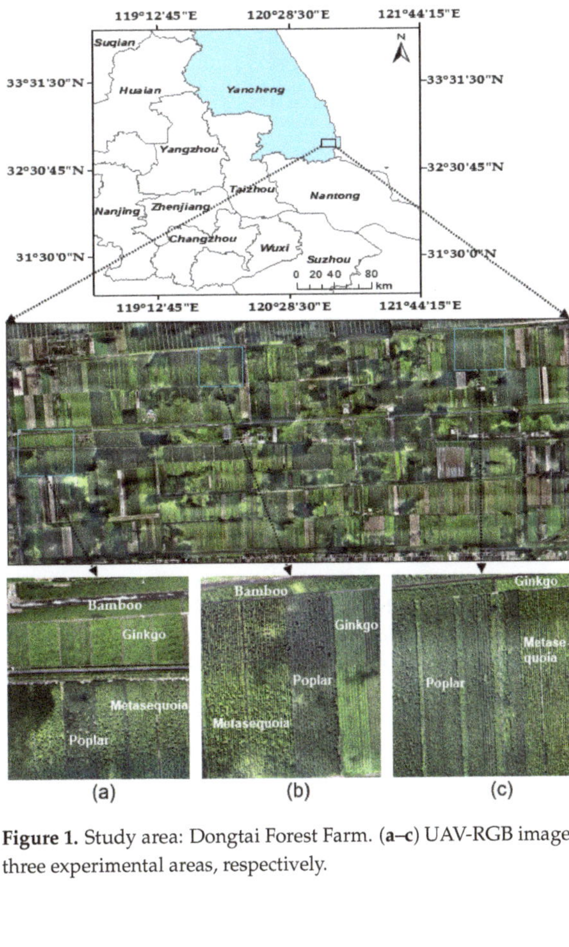

Figure 1. Study area: Dongtai Forest Farm. (**a–c**) UAV-RGB images of the main tree species for the three experimental areas, respectively.

2.2. Data Collection and Preprocessing

In this study, we selected four tree species, namely Metasequoia, Bamboo, Poplar, and Ginkgo, as the target recognition objects within the study area. To collect samples, we used sliding window cropping with a window size of 64 × 64 pixels, resulting in a total of 4338 samples. Specifically, the number of samples obtained for each tree species was as follows: 1101 for Bamboo, 880 for Ginkgo, 1105 for Metasequoia, and 1252 for Poplar. These samples were affected by various imaging factors, such as forest canopy closure, tree species, lighting, background, and shooting height, resulting in differences in intra-class and inter-class characteristics, as depicted in Figure 2.

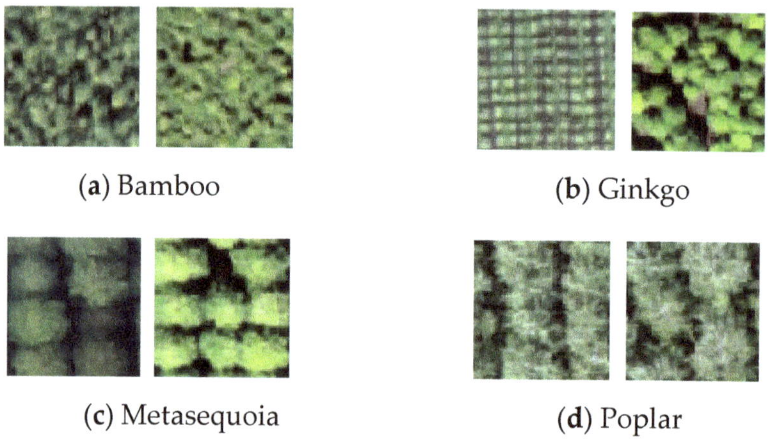

Figure 2. Sample tree species: (**a**) Bamboo; (**b**) Ginkgo; (**c**) Metasequoia; (**d**) Poplar.

Given that neural network models, especially Transformer models and their variants, can have millions or even billions of parameters, it is crucial to have sufficient samples for effective training. To increase the sample size, we performed data augmentation on the training and validation sample sets. The data augmentation techniques applied to the tree canopy image samples included random rotations by 90°, 180°, and 270°, horizontal flipping, and brightness adjustment as depicted in Figure 3. These transformations were implemented using the 'RandomRotation', 'RandomHorizontalFlip', and 'ColorJitter' functions provided by the Torchvision library. By incorporating these techniques into the data augmentation pipeline, variations were introduced to the dataset, allowing for enhanced training and improved model generalization. Random rotations provided diverse perspectives of the tree canopies, while horizontal flipping increased the dataset's diversity by mirroring the images. Additionally, brightness adjustment ensured robustness to varying lighting conditions, enabling the model to generalize better across different environments. These techniques collectively contributed to the overall robustness and performance of the tree canopy classification model.

Furthermore, we randomly divided the samples into training and validation sets in an 8:2 ratio, respectively. This ensured that both sets had a representative distribution of samples from each tree species, allowing for robust model evaluation and performance estimation. The augmented dataset with increased sample size and diversity, along with the appropriate training and validation set partitioning, facilitated the training of the neural network models for accurate tree species recognition in the study area.

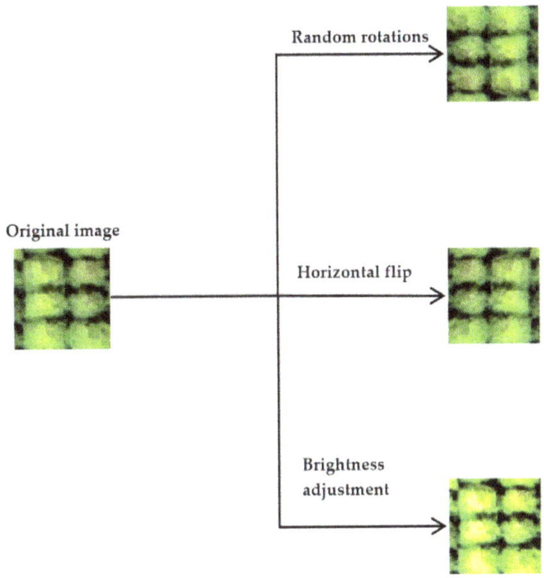

Figure 3. Schematic diagram of canopy sample data augmentation.

3. Methodology

3.1. Super-Resolution Reconstruction

Faced with the limitations of hardware equipment, such as high-denseness forest stands, aerial photography angles, and resolution, as well as complex geographic backgrounds, drone aerial images often suffer from degradation issues such as lost details, reduced brightness, and partially blurred tree crowns. To solve these problems, this paper uses Real-ESRGAN technology [25] to perform super-resolution reconstruction, denoising, and deblurring on the original tree crown images. Classical first-order degradation models, as expressed in Equation (1), often consider only one type of degradation operation, such as blur or noise, while ignoring the simultaneous presence of multiple degradation operations. However, in practical scenarios, images may undergo multiple degradations simultaneously. For example, during transmission, images may experience blur, resolution reduction, and the introduction of noise. In order to more accurately simulate the degradation process of real images, Real-ESRGAN proposes a high-order degradation model, as expressed in Equation (2), which utilizes multiple repeated degradation processes, with each degradation process representing a classical degradation model. Through ablation experiments, the second-order model has been proven to exhibit excellent performance and practicality, effectively meeting the requirements of most image processing tasks [25]. Real-ESRGAN's second-order degradation model, as shown in Figure 4, aims to simulate the degradation process of real images through several steps, including blur processing, downsampling, noise addition, and JPEG compression. Firstly, in the blur processing stage, isotropic and anisotropic Gaussian blur kernels are applied to the original image, resulting in the loss of details and clarity to mimic real-world blurring effects. Subsequently, the blurred image undergoes downsampling using randomly selected methods such as bilateral interpolation, bilinear interpolation, and regional interpolation, further reducing the image's resolution and diminishing details and clarity to simulate actual resolution reduction effects. Next, noise is added based on the image type (color or grayscale). For color images, both Gaussian noise and noise following the Poisson distribution are added to simulate environmental and sensor noise, while grayscale images only receive Gaussian noise. Finally, the image is subjected to JPEG compression based on a compression quality parameter ranging from 0 to 100, where lower compression quality leads to poorer image quality and more severe distortion.

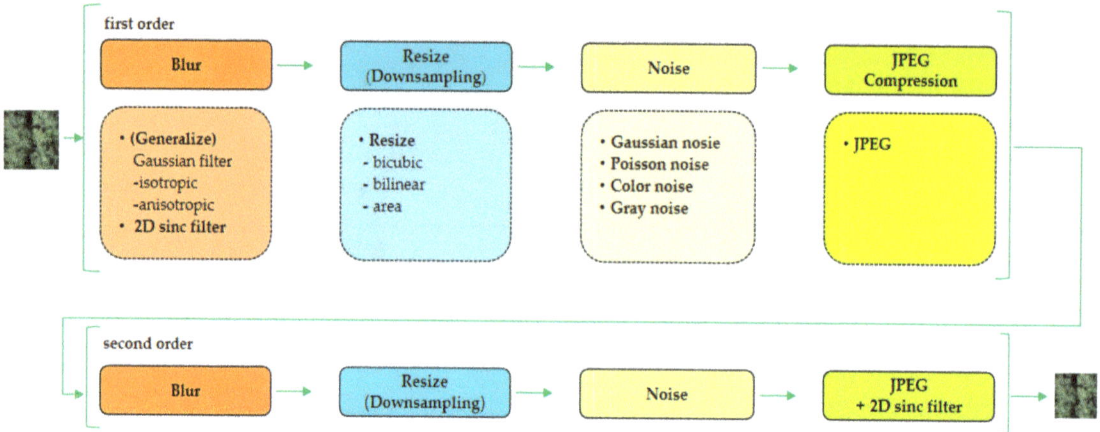

Figure 4. Real-ESRGAN second-order degradation model.

Both in the initial blur processing step and the final synthesis step, a sinc filter is used. The expression for the sinc filter is shown in Equation (3). The sinc filter exhibits high selectivity in the frequency domain, responding differently to signals of various frequencies. Consequently, the sinc filter can smooth and blur specific frequency details in the image, thereby reducing its details and clarity. Additionally, the sinc filter possesses inverse filtering properties, allowing it to repair blurred or degraded images during the restoration process, mitigating degradation effects and improving image quality. It is worth noting that in the final synthesis step, the order of applying the sinc filter and JPEG compression is randomly exchanged to cover a broader range of degradation scenarios.

In summary, Real-ESRGAN's second-order degradation model adopts various degradation operations, such as blur processing, downsampling, noise addition, and JPEG compression, to simulate the degradation process of real images. These processed degraded images lose their detail and clarity, exhibiting visual effects such as blurring, reduced resolution, noise addition, and distortion, thereby providing challenging inputs for the subsequent Real-ESRGAN super-resolution reconstruction process.

$$x = D(y) = [(y \otimes k) \downarrow_r + n]_{JPEG} \quad (1)$$

$D(\cdot)$ denotes the degradation process, y denotes the input image, k denotes the blur function, $\downarrow_r$ denotes the downsampling factor, n denotes the noise, and $[\,]_{JPEG}$ denotes the compression of the obtained result in JPEG format.

$$x = D^n(y) = (D_n \circ \cdots \circ D_2 \circ D_1)(y) \quad (2)$$

The above equation is actually a multiple repetition operation of the first-order degradation, where each D represents the execution of one first-order degradation.

$$k(i,j) = \frac{\omega_c}{2\pi\sqrt{i^2+j^2}} J_1(\omega_c\sqrt{i^2+j^2}) \quad (3)$$

(i,j) is the kernel coordinate, ω_c is the cutoff frequency, and J_1 is a first-order Bessel function of the first type.

Real-ESRGAN builds upon the generator structure of ESRGAN [26] as a foundation and further enhances and refines it through design improvements, comprising numerous Residual-in-Residual Dense Blocks (RRDB) for enhanced performance. These processed images, obtained through the preceding pre-processing steps elucidated earlier, are subsequently channeled into the generator of ESRGAN, as demonstrated in Figure 5 underneath. Primarily, pixel shuffling is implemented to diminish the spatial dimensions and augment the channel properties. Then, the resultant outcome is fed into the principal architecture of ESRGAN for super-resolution reconstruction.

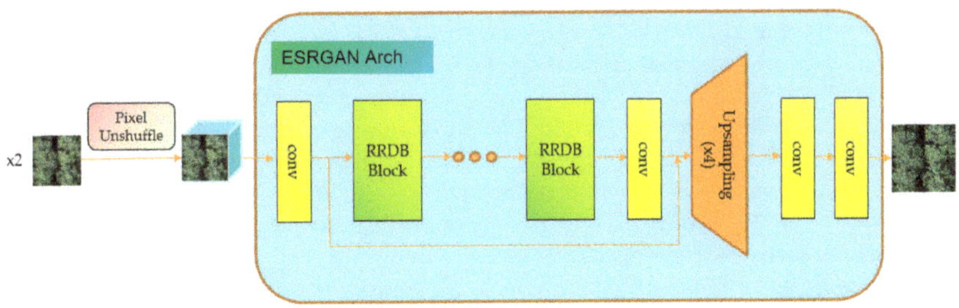

Figure 5. ESRGAN architecture. We used a pixel-unshuffle operation to diminish the spatial dimensions and re-arrange information to the channel dimension for scale factors of x2.

Moreover, owing to Real-ESRGAN's aspiration to confront a considerably wider range of degradations than ESRGAN, the original VGG-style discriminator design in ESRGAN is no longer suitable. Instead, Real-ESRGAN introduces a U-Net framework with skip connections for the discriminator, inspired by the referenced research endeavors [27,28]. Finally, the generated images are mixed with the input images and fed into the discriminator for discrimination, which uses spectrally normalized U-Net to mitigate excessive sharpness and artifacts introduced by GAN training. The original tree species canopy images were restored by Real-ESRGAN super-resolution reconstruction as shown in Figure 6 below, and the canopy texture details were restored and processed to facilitate the neural network model to extract clear canopy edge contour features and detailed texture features.

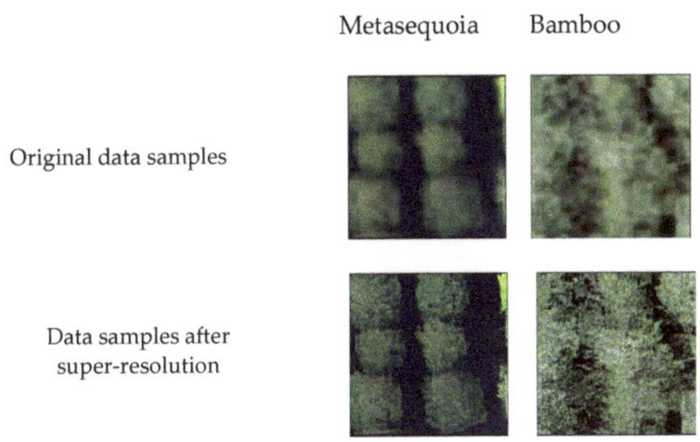

Figure 6. Comparison of original canopy images and super-resolution reconstructed canopy images.

3.2. *Model Training*

3.2.1. ResNet Model

Residual blocks proposed by ResNet [29], as shown in Figure 7, can effectively solve the problem of network degradation in deep networks. Two varieties of shortcut connections are adopted within the layers of ResNet. Identity shortcuts are utilized when the dimensions of the input and output are equal, while projection shortcuts are used to align dimensions [29]. In this paper, we use ResNet-50 as a representative of such models. The decision to adopt ResNet-50 was grounded on empirical observations and experiments conducted by the researchers who proposed the ResNet architecture. Their findings revealed that surpassing a certain depth threshold in the network resulted in diminishing performance improvements or even a decline in accuracy, primarily due to the issue of vanishing gradients [29]. ResNet-50 successfully strikes an excellent balance between model depth and performance. As the name suggests, ResNet-50 consists of 50 layers, as shown in Figure 8, divided into five stages. Stage 0 can be regarded as the preprocessing of input data, while stages 1 to 4 are each composed of 3, 4, 6, and 3 bottleneck blocks, respectively. (1) In Stage 0, the first step was to convert the canopy image, which has a size of 224 × 224, into a digital matrix with dimensions [224, 224, 3]. Subsequently, we used a convolutional layer with a 7 × 7 kernel, stride of 2, and 64 output channels, resulting in a feature map measuring 112 × 112 × 64. Furthermore, a 3 × 3 max pooling layer with a stride of 2 was applied to reduce the feature map's size, yielding a 56 × 56 × 64 representation. (2) Moving forward, the 56 × 56 × 64 feature map underwent processing in Stage 1. At this stage, we adopted three bottleneck blocks to facilitate the integration process. Each block consisted of a sequence of convolutional operations, including 1 × 1 convolutions with 64 output channels, 3 × 3 convolutions with 64 output channels, and another 1 × 1 convolution with 256 output channels. These operations reshaped the feature map, resulting in a 56 × 56 × 256 representation. (3) Continuing the progression, the transformed 56 × 56 × 256 feature map proceeded to Stage 2, housing four bottleneck blocks. Each block implemented a series of convolutions to reshape the feature map from 56 × 56 × 256 to 28 × 28 × 512. (4) Continuing the sequence, the 28 × 28 × 512 feature map was inputted into Stage 3, containing six bottleneck blocks. Through a cascade of convolutions, the feature map underwent size transformation from 28 × 28 × 512 to 14 × 14 × 1024. (5) Subsequently, the modified 14 × 14 × 1024 feature map advanced to Stage 4, incorporating three bottleneck blocks. These convolutions modified the feature map's dimensions from 14 × 14 × 1024 to 7 × 7 × 2048. (6) Next, a global average pooling operation was applied to the 7 × 7 × 2048 feature map, computing the average value of each channel. Consequently, a feature map of size 1 × 1 × 2048 was obtained. (7) Finally, the 1 × 1 × 2048 feature map was flattened into a one-dimensional vector and subjected to processing through a fully connected layer for classification purposes. Given that this experiment encompasses four tree species, the output provides probability values for the four categories. The detailed architecture specifications of ResNet-50 are described in Table 1.

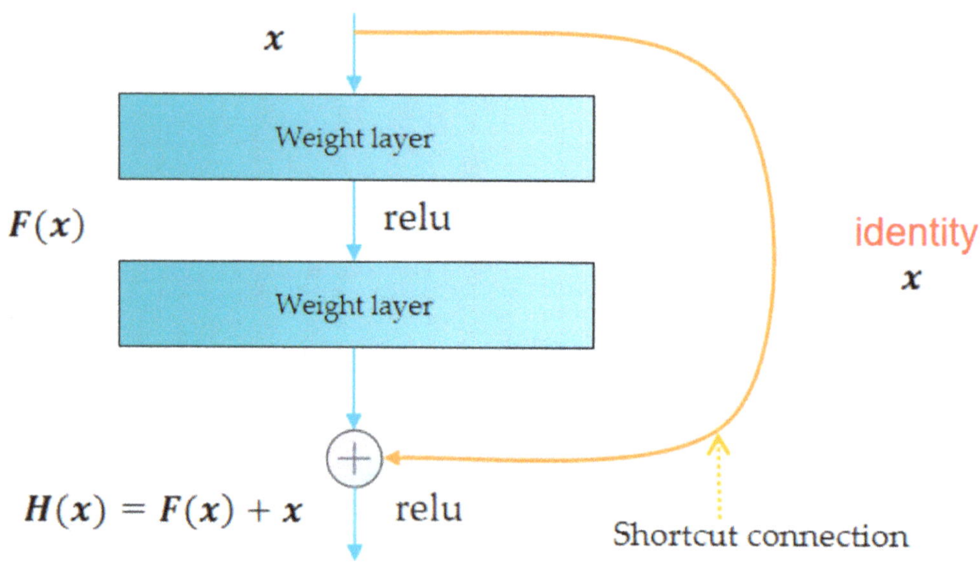

Figure 7. Residual block structure.

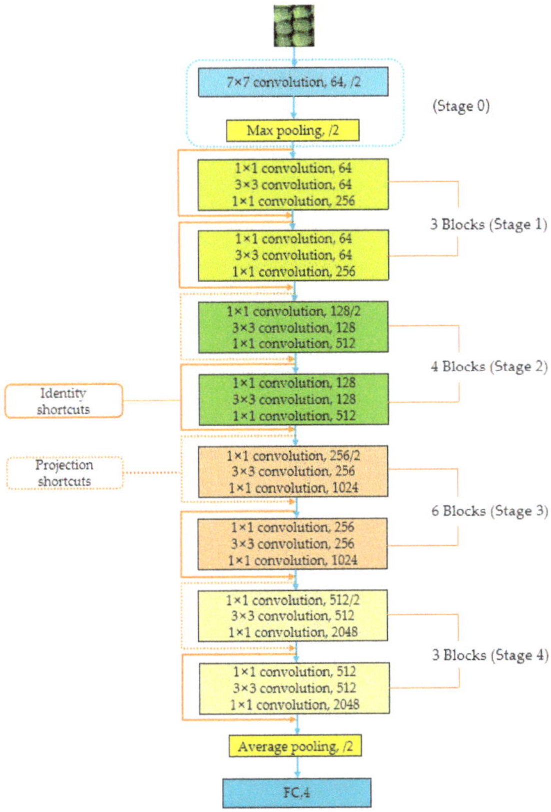

Figure 8. A depiction of the ResNet-50 network architecture. '/2' means stride is set to 2.

Table 1. Detailed architecture specifications for ResNet-50. '4-d fc' denotes a fully connected layer with 4 dimensions.

Stage	Output_Size	ResNet-50
Stage0	112 × 112	7 × 7, 64, stride 2
Stage1	56 × 56	3 × 3 max pooling, stride 2
		$\begin{bmatrix} 1\times1, 64 \\ 3\times3, 64 \\ 1\times1, 256 \end{bmatrix} \times 3$
Stage2	28 × 28	$\begin{bmatrix} 1\times1, 128 \\ 3\times3, 128 \\ 1\times1, 512 \end{bmatrix} \times 4$
Stage3	14 × 14	$\begin{bmatrix} 1\times1, 256 \\ 3\times3, 256 \\ 1\times1, 1024 \end{bmatrix} \times 6$
Stage4	7 × 7	$\begin{bmatrix} 1\times1, 512 \\ 3\times3, 512 \\ 1\times1, 2028 \end{bmatrix} \times 3$
	1 × 1	average pooling, 4-d fc, softmax

3.2.2. ConvNeXt Model

The ConvNeXt model [30] improves upon the ResNet architecture by incorporating ideas from the Transformer network [31], as depicted in Figure 9. Specifically, ConvNeXt implements techniques such as mimicking depthwise convolutions to form convolution blocks and referencing the design of the inverted bottleneck structure, resulting in enhanced performance. In this study, we made use of the ConvNeXt-T model. The process of classifying tree species with ConvNeXt-T involved the following steps: (1) To begin, we started with an input canopy image of size 224 × 224 × 3. This image was passed through a convolutional layer with a kernel size of 4 and a stride of 4. Afterward, a Layer Norm (LN) was applied to normalize the feature map, resulting in a 56 × 56 × 96-sized feature map. The LN operation plays a crucial role in enhancing network stability and generalization by standardizing the feature map. (2) Moving on, the feature map underwent four stages, each containing a ConvNeXt block. In each stage, there were 3, 3, 9, and 3 ConvNeXt blocks, respectively. These ConvNeXt blocks were composed of a 7 × 7 depthwise convolution with a stride of 1 and a padding of 3, followed by an LN. In addition, two 1 × 1 Conv2d layers with a stride of 1 and a GeLU activation function [32] were adopted. The output channels of the ConvNeXt blocks were 96, 192, 384, and 768, respectively. During the second to fourth stages, a downsampling operation was performed on the feature map. This downsampling operation included applying an LN and using a Conv2d layer with a stride of 2. This operation reduces the size of the feature map to half that of the previous layer. As a result, the feature map ended up containing 768 channels after passing through all four stages. (3) Next, the feature map was passed through a global average pooling layer, and a feature vector with a size of 1 × 1 × 768 was obtained. (4) The final step involved passing the feature vector through a fully connected layer to convert the 768-dimensional vector into an output vector of size 4. This output vector represents the probability values of the four tree species categories. Table 2 shows the detailed structural specifications of the ConvNeXt-T.

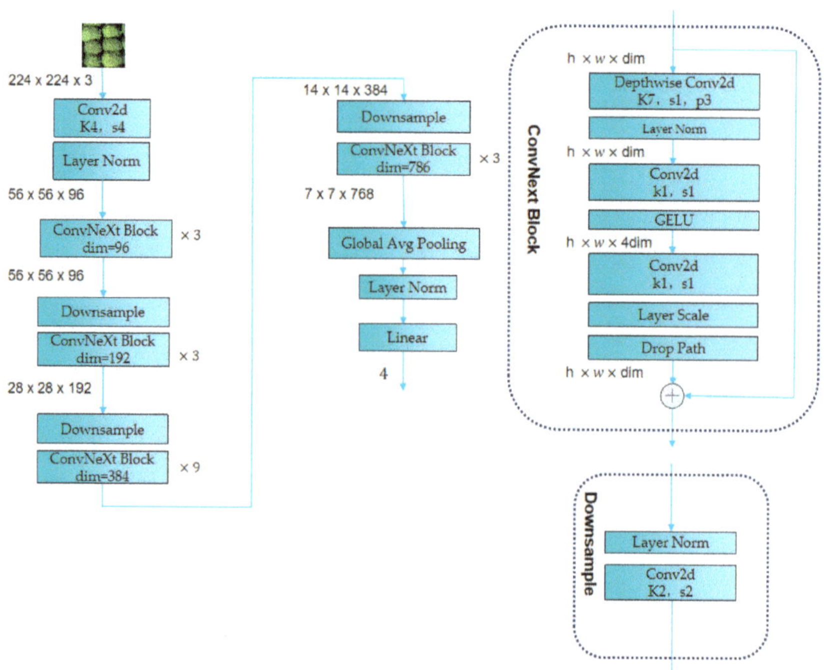

Figure 9. ConvNeXt-T Network Architecture.

Table 2. Detailed architecture specifications for ConvNeXt-T. 'd' is short for depthwise convolution.

Layer_Name	Output_Size	ConvNeXt-T
Conv1	56×56	4×4, 96, stride 4
Conv2_x	56×56	$\begin{bmatrix} d7 \times 7, 96 \\ 1 \times 1, 384 \\ 1 \times 1, 96 \end{bmatrix} \times 3$
Conv3_x	28×28	$\begin{bmatrix} d7 \times 7, 192 \\ 1 \times 1, 768 \\ 1 \times 1, 192 \end{bmatrix} \times 3$
Conv4_x	14×14	$\begin{bmatrix} d7 \times 7, 384 \\ 1 \times 1, 1536 \\ 1 \times 1, 384 \end{bmatrix} \times 9$
Conv5_x	7×7	$\begin{bmatrix} d7 \times 7, 768 \\ 1 \times 1, 3072 \\ 1 \times 1, 768 \end{bmatrix} \times 3$
	1×1	global average pooling, 4-d fc

3.2.3. ViT Model

The ViT model [33] is a specially designed Transformer model for image classification. It is composed of three modules: Embedding, Transformer Encoder, and Multi-Layer Perceptron (MLP) Head. The Transformer Encoder comprises LN, Multi-head Attention, Dropout, and MLP Block, and for this paper, we used the ViT-B model, as shown in Figure 10. Here is how the image classification process worked using ViT-B: (1) Initially, we input a tree canopy image with dimensions of $224 \times 224 \times 3$. It underwent convolution using a 16×16 kernel and a stride of 16, resulting in a feature map sized $14 \times 14 \times 768$. (2) Next, we flattened the feature map in both the height and width directions, transforming its size to 196×768. Subsequently, we concatenated a class token and applied positional encoding to the feature map, yielding a transformed size of 197×768. (3) Following this step, we applied Dropout and passed the input through 12 stacked Encoder Blocks. The output from the Encoder was then processed with LN, maintaining the feature map size at 197×768. We proceeded to extract the output corresponding to the class token and slice it, obtaining a vector of size 1×768. This vector was then fed into the MLP Head. (4) Finally, the feature vectors were fed into a fully connected layer with four neurons for classification, where each neuron represented a tree species category, resulting in the final classification results.

3.2.4. Swin Transformer Model

Due to the high computational cost and memory consumption of the self-attention mechanism in ViT models when processing high-resolution image tasks, the Swin Transformer model was proposed [34], which adopts a hierarchical structure as shown in Figure 11. The model comprises four key components: Patch Partition, Linear Embedding, Swin Transformer Block, and Patch Merging. Two successive Swin Transformer Blocks illustrated in Figure 12, incorporate Window-based Multi-head Self-attention (W-MSA) and Shifted Window-based Multi-head Self-attention (SW-MSA) to address memory consumption challenges while maintaining efficient performance. Additionally, Patch Merging adopts pooling-like operations to progressively reduce the feature map size and merge image blocks, constructing a hierarchical feature map in deeper layers. LN layers are used before each MSA module and each MLP, and residual connections are used after each MSA and MLP. These characteristics make it a versatile backbone for image classification and dense recognition tasks. In this study, we adopted the Swin-T model as a representative example of such models. (1) Initially, the input was a tree crown image with dimensions 224 × 224 × 3, which underwent Patch Partition. This process involved dividing the image into fixed-sized blocks using a 4 × 4 convolutional kernel. The resulting feature map dimensions were 56 × 56 × 48. (2) Next, the Linear Embedding layer was subsequently applied to each channel of the pixel data, resulting in a linear transformation that changed the feature map dimensions to 56 × 56 × 96. (3) Moving forward, we proceeded to a series of Swin Transformer blocks. These blocks consisted of two variations: one utilizing the W-MSA structure and the other adopting the SW-MSA structure. Consequently, the Swin Transformer blocks appeared in even numbers, with 2 blocks in each of the first, second, and fourth stages, and 6 blocks in the third stage. The Swin Transformer block incorporated window partitioning and window reverse operations, maintaining the output feature map size at 56 × 56 × 96. Therefore, the input and output sizes of the Swin Transformer block remained unchanged. (4) Subsequently, Patch Merging was performed to reduce the spatial dimensions by half and double the channel count. This process was repeated across the four stages, eventually transforming the feature map dimensions to 7 × 7 × 768. (5) Finally, global average pooling was utilized to reduce the spatial dimension to 1, resulting in a feature vector of size 1 × 1 × 768. A linear classifier with four neurons was used to map the output vector to probability values corresponding to the four tree species categories, yielding the final prediction. The detailed Swin-T architecture specification is described in Table 3.

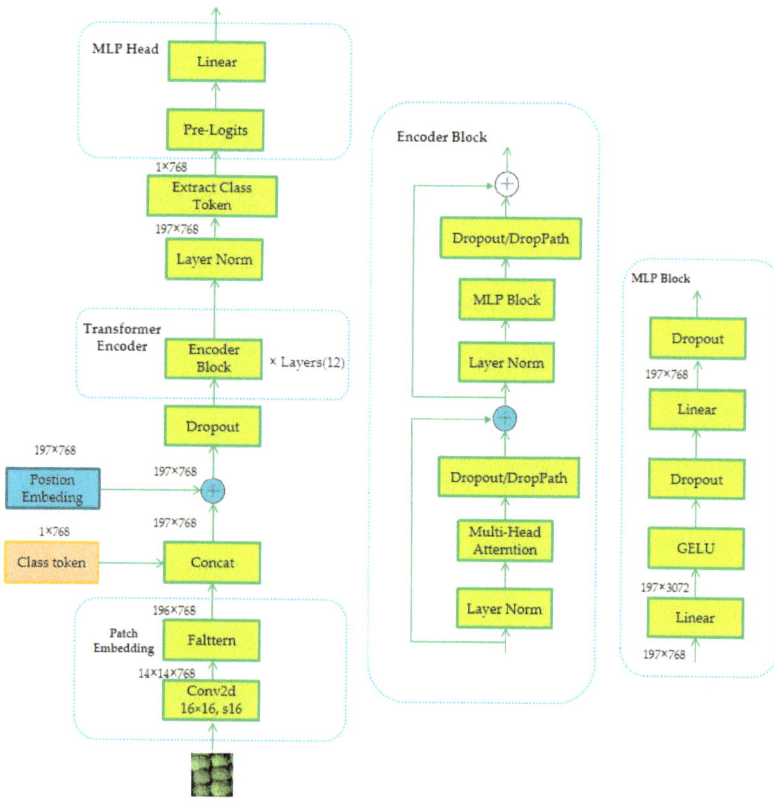

Figure 10. ViT-B Network Architecture.

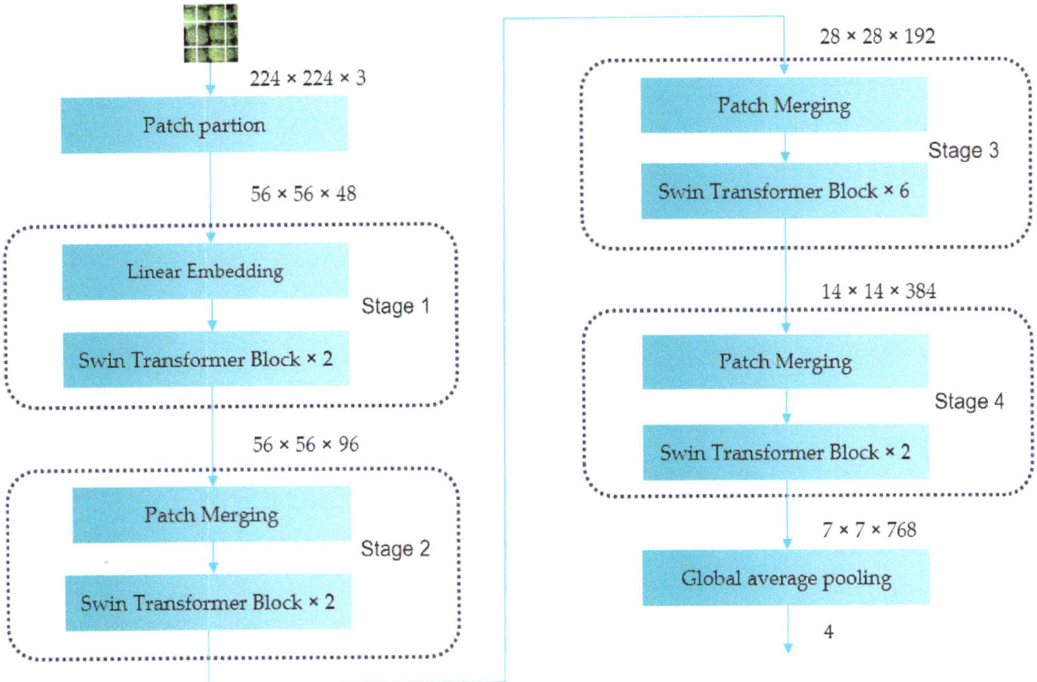

Figure 11. Swin-T Network Architecture.

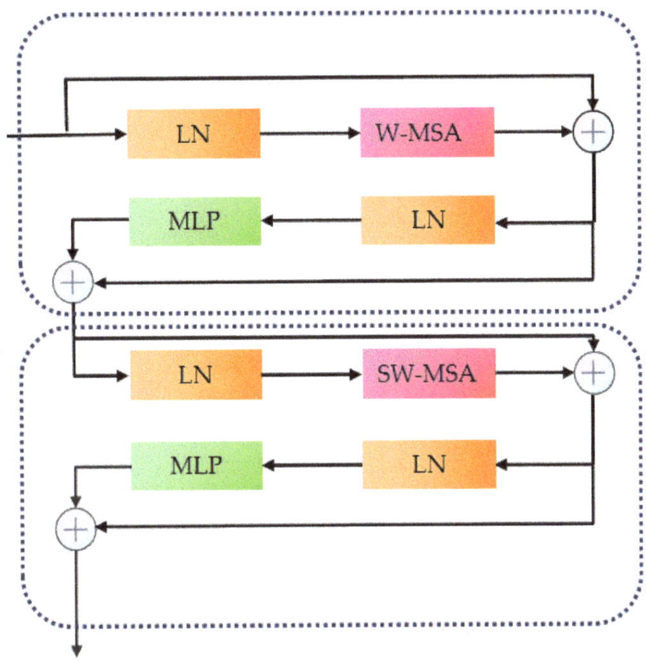

Figure 12. Two Successive Swin Transformer Blocks.

Table 3. Detailed architecture specifications for Swin-T. win. sz. indicates the size of the window used; dim indicates the channel depth of the feature map; head indicates the number of heads in a multi-headed attention module.

Stage_Name	Output_Size	Swin-T
Stage1	56 × 56	concat 4 × 4, 96, LN
		[win. sz. 7 × 7, dim 96, head 3] × 2
Stage2	28 × 28	concat 2 × 2, 192, LN
		[win. sz. 7 × 7, dim 192, head 6] × 2
Stage3	14 × 14	concat 2 × 2, 384, LN
		[win. sz. 7 × 7, dim 384, head 12] × 6
Stage4	7 × 7	concat 2 × 2, 768, LN
		[win. sz. 7 × 7, dim 768, head 24] × 2
	1 × 1	global average pooling, 4-d fc

3.3. Experimental Environment

The experiments were conducted on a computer with the following specifications: Windows 10 operating system, AMD Ryzen 5 5600X 6-Core Processor CPU, and NVIDIA GeForce RTX 3070Ti GPU with 8GB of memory. The deep learning platform used for training and evaluation was PyTorch 1.12.0, along with cudatoolkit11.3 for GPU acceleration. For raster and vector data processing, Arcgis10.8 was utilized, and Matplotlib 3.5.2 was used for data visualization. Python 3.9.13 was the programming language used for implementation and analysis.

4. Results

4.1. Comparison of CNN and Transformer Models for Tree Species Classification

The main objective of this investigation was to evaluate the impact of different models on tree species classification using canopy images. To this end, we selected two representative pre-trained models from both the CNN and Transformer models, specifically ResNet-50, ConvNeXt-T, ViT-B, and Swin-T. For these four models, we set the image input size to 224 × 224 pixels. During training, we used a batch size of 16 and the AdamW optimization algorithm, and trained for a total of 150 epochs. The initial learning rate was set to 4×10^{-4}, and the weight decay factor was set to 5×10^{-2}. We also used the LambdaLR strategy for learning rate adjustment.

To assess the performance of the models, we monitored the changes in the loss function and recognition accuracy of the original image data samples on the training set and validation set across 150 epochs. These changes are visualized in Figure 13 to understand the training progress and stability of each model. It is observed that for all models, the accuracy and loss tend to stabilize after around 120 to 130 epochs, with recognition classification accuracy exceeding 95%, indicating strong stability and high accuracy in the tree species classification task.

Furthermore, we evaluated the overall classification accuracy (OA), Kappa coefficient, and confusion matrix of each model, as shown in Figure 14. These evaluation metrics provide quantitative measures of the performance of the models in the tree species classification task. By analyzing these results, we can gain a comprehensive understanding of the performance of each model in tree species classification using canopy images.

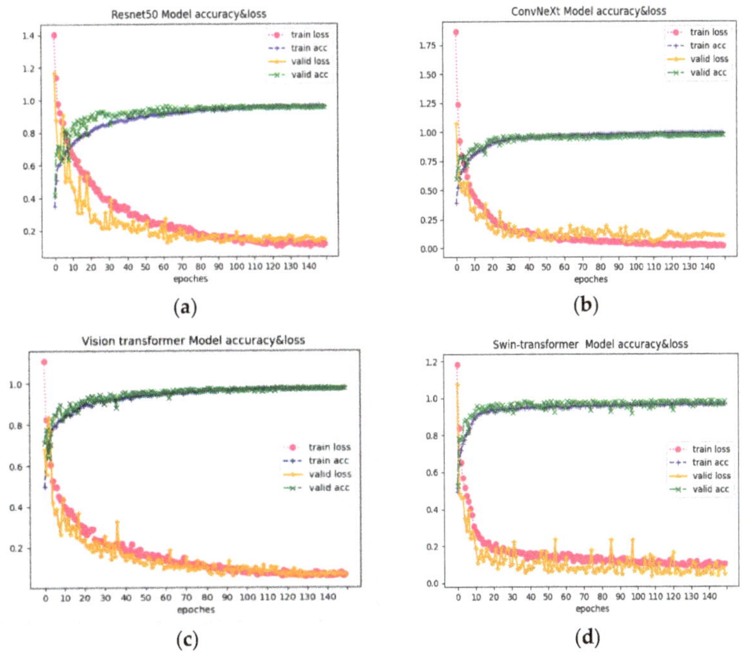

Figure 13. (**a**–**d**) are plots of the accuracy and loss rates of the ResNet-50, ConvNeXt-T, ViT-B, and Swin-T models trained and validated on the original dataset, respectively.

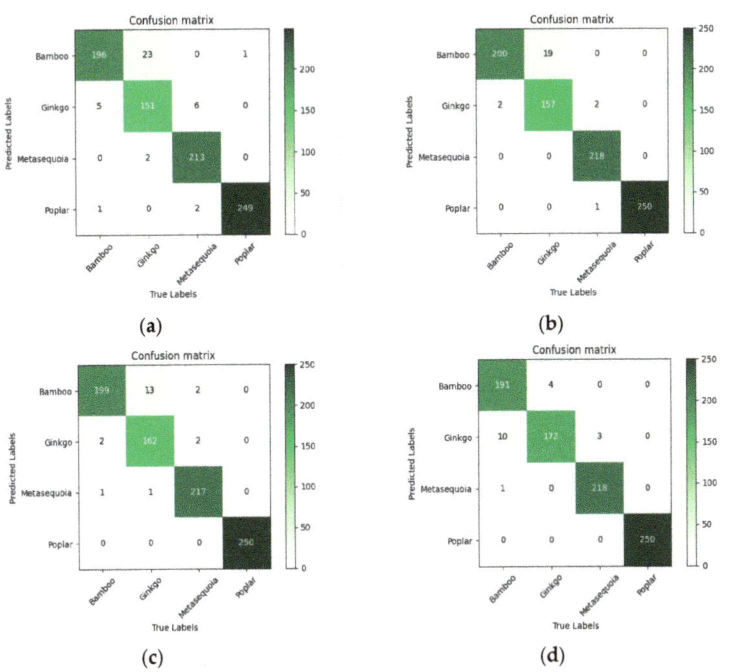

Figure 14. (**a**–**d**) are the confusion matrix plots for the ResNet-50, ConvNeXt-T, ViT-B, and Swin-T models validated on the original dataset, respectively.

The ResNet-50 model attained an OA of 95.32% with a Kappa coefficient of 0.9368. The ConvNeXt-T model reached an OA of 97.17% with a Kappa coefficient of 0.9621. The ViT-B and Swin-T models reached even higher OAs of 97.41% (with a Kappa coefficient of 0.9668) and 97.43% (with a Kappa coefficient of 0.9671), respectively. This study's finding showed that during training and validation, the Transformer models ViT-B and Swin-T outperformed the CNN model represented by ResNet-50 and ConvNeXt-T in terms of OA and Kappa coefficient.

This superiority of the Transformer models could be attributed to their multi-head attention mechanism, which allows them to capture richer feature information more effectively. The multi-head attention mechanism enables the models to attend to different regions of the input image simultaneously, capturing both local and global contextual information. In contrast, the CNN models, ResNet-50 and ConvNeXt-T, were found to be sensitive to factors such as blurry edges and weak texture in the original canopy images, which could impact the matching accuracy of feature points during feature extraction.

In summary, the research results show that the ViT-B and Swin-T models perform better than the ResNet-50 and ConvNeXt-T models in terms of OA and Kappa coefficient, due to the influence of factors such as blurry edges and weak textures in the original canopy images. This suggests that the ViT-B and Swin-T models have the potential to be effective in tree species classification tasks. However, an accurate judgment of model performance requires a comprehensive consideration of multiple evaluation metrics, as well as the needs of practical application scenarios, dataset characteristics, and model strengths and weaknesses. Furthermore, further research and empirical analysis can validate the performance of these models on different datasets and tasks.

4.2. Super-Resolution Reconstruction for Improved Tree Species Classification

Real-ESRGAN, a proven super-resolution image restoration algorithm [25], was utilized in this study to reconstruct the original image dataset. The reconstructed dataset was then used as input for training and validation in four different models. To ensure a fair comparison of the performance of the models, consistent hyperparameter tuning methods were applied during training.

Figure 15 provides a visual representation of the changes in the loss function and recognition accuracy for the training and validation sets of the reconstructed dataset for each of the four models. The figure allows for a detailed analysis of the training progress and stability of each model. It is observed that after approximately 90 to 100 iterations, the accuracy and loss of each model tend to stabilize, indicating convergence of the training process. This implies that the dataset reconstructed and repaired through super-resolution exhibits a faster convergence of the model with fewer training iterations compared to the original dataset, and demonstrates improved stability on the training set and validation set.

Furthermore, Figure 16 presents the OA, Kappa coefficient, and confusion matrix of each model. These metrics provide a comprehensive assessment of the performance of each model in terms of accuracy, agreement, and confusion among different classes. The detailed analysis of these metrics can provide insights into the effectiveness of each model in accurately classifying tree species based on the reconstructed dataset.

After comparing the model parameters reached and tree species classification results in Table 4, the following conclusions can be drawn: the ResNet-50 model reached an OA of 96.71% and a Kappa coefficient of 0.9558; the ConvNeXt-T model reached an OA of 98.70% and a Kappa coefficient of 0.9826; the ViT-B model reached an OA of 97.88% and a Kappa coefficient of 0.9716; and the Swin-T model attained an OA of 98.59% and a Kappa coefficient of 0.9810. Compared with the data samples that were not reconstructed using Real-ESRGAN, the recognition accuracy of each model increased by 1.39%, 1.53%, 0.47%, and 1.16%, respectively. Among them, the ConvNeXt-T model reached the best result. Therefore, we can conclude that the original image data, which may contain factors such as blurry edges and weak texture in the canopy, can benefit from reconstruction using

Real-ESRGAN. This reconstruction method can help improve the accuracy of tree species classification recognition to a certain extent.

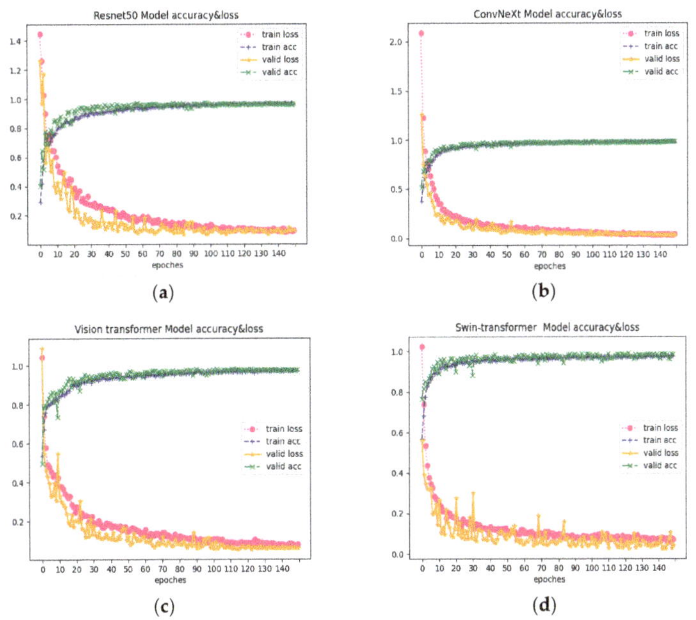

Figure 15. (**a**–**d**) depict the accuracy and loss rate of ResNet-50, ConvNeXt-T, ViT-B, and Swin-T models trained and validated on super-resolution reconstructed dataset, respectively.

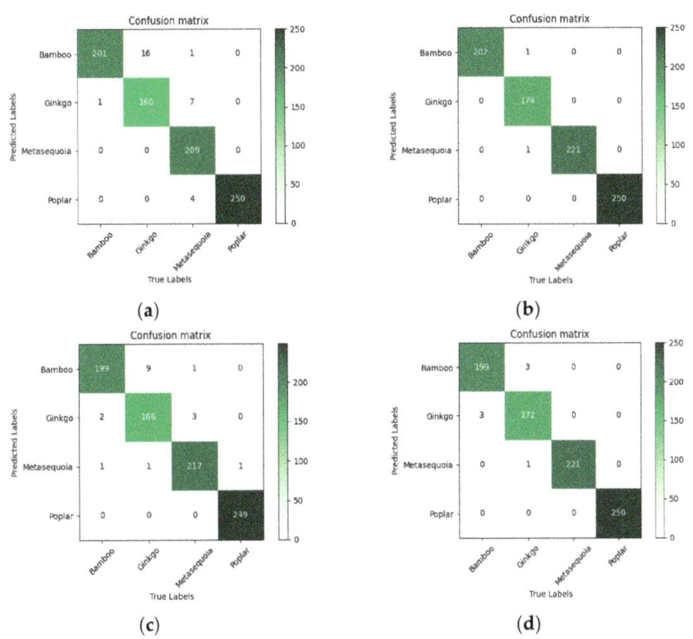

Figure 16. (**a**–**d**) depict the confusion matrix plots of ResNet-50, ConvNeXt-T, ViT-B, and Swin-T models validated on super-resolution reconstructed dataset, respectively.

Table 4. Comparison of classification results of different model tree species.

Model	Model Size/Mb Size	Average Accuracy of Model Validation for the Original Dataset/%	Average Accuracy of Model Validation after Super-Resolution Reconstruction Restoration of the Dataset/%
ResNet-50	25.55	95.32	96.71
ConvNeXt-T	28.58	97.17	98.70
ViT-B	86.41	97.41	97.88
Swin-T	28.26	97.43	98.59

4.3. Distribution Map of Tree Species in Dongtai Forest Plot

In this study, we selected a typical plot from the study area of Dongtai Forest Farm with UAV remote sensing images as a test sample set. To match the sample size of the original dataset, we used the same sliding window method to crop this sample. Then, we used the ConvNeXt-T model to test this sample set and created a tree species distribution map based on the model's predictions, as shown in Figure 17. The plot of this forest farm is divided as follows: the left plot is mainly planted with Metasequoia, the middle plot is mainly planted with Poplar, and the right plot is mainly distributed with Ginkgo. In addition, Bamboo mainly grows in the plot below.

Figure 17. Tree species distribution map of Dongtai Forestry Sample Site.

This study examined the impact of different models on tree species classification in crown images. The study selected two representative pre-training models from the CNN and Transformer models, including ResNet-50, ConvNeXt-T, ViT-B, and Swin-T. The experimental results show that compared to traditional CNN models, Transformer models are more stable in feature extraction, and have better classification accuracy and stability. Additionally, this study used the Real-ESRGAN algorithm to perform super-resolution reconstruction and repair on the original image dataset, resulting in an improvement in the accuracy of tree species classification as demonstrated in the results. Finally, the study presents a distribution map of tree species in Dongtai Forest Farm, demonstrating the practical application of the Real-ESRGAN algorithm and serving as a reference for further research.

5. Discussion

5.1. Performance of CNN and Transformer in Classifying Tree Species Using the Original Dataset

For the application of tree species classification in low-altitude remote sensing images obtained from UAV, this paper further evaluated the classification accuracy performance of four models, namely, ResNet-50 and ConvNeXt-T as representatives of CNN models, and ViT-B and Swin-T as representatives of Transformer models, using the original canopy image dataset. Transformer has demonstrated its exceptional ability to capture global information, thereby bolstering a wide range of vision-related tasks such as image classification, object detection, and particularly semantic segmentation [33,34]. CNN and Transformer use object-based classification to achieve end-to-end tree species classification and avoid the non-transferability of manual feature extraction. The experimental results reveal that, as depicted in Figure 18, all four models exhibit classification validation accuracies exceeding 95%. Notably, the Swin Transformer reaches the highest classification accuracy, demonstrating an OA of 97.43% and a Kappa coefficient of 0.9671. Conversely, the CNN models, particularly the traditional CNN model, are more susceptible to the challenges posed by low-spatial-height aerial images, including detail loss, brightness reduction, blurred canopy edges, and weak image texture. These issues, coupled with small inter-class differences and significant intra-class differences, adversely impact the feature point matching accuracy of the CNN model, whereas the Transformer model is comparatively less affected.

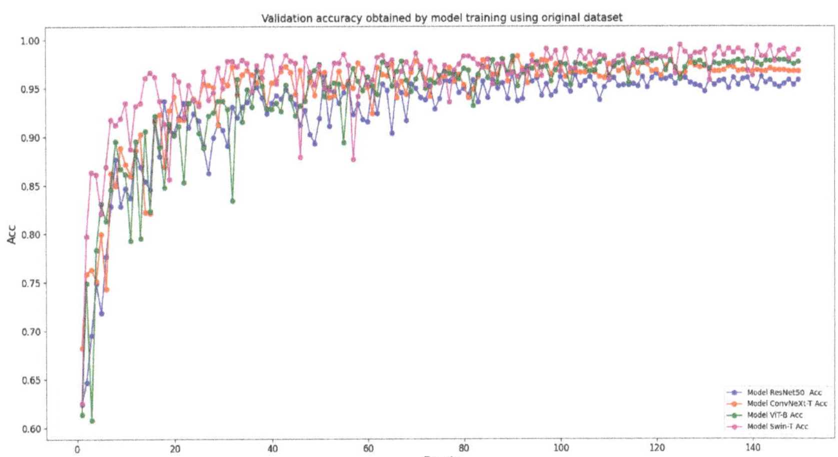

Figure 18. Validation accuracy obtained by training the model using the original dataset.

5.2. Performance of CNN and Transformer in Tree Species Classification Using Super-Resolution Reconstructed Dataset

To address these issues, this paper introduced the Real-ESRGAN super-resolution reconstruction technique to recover low-quality tree canopy images captured by UAVs.

The recovery process improved the validation accuracy of the four models. For example, the OA of the ConvNeXt-T model increased by 1.53% and the Kappa coefficient increased by 0.0205. The validation accuracy comparison derived from the model using the original dataset and the Real-ESRGAN processed dataset for training is depicted in Figure 19. Although the Real-ESRGAN super-resolution reconstruction technique has some limitations and shortcomings, it can be further improved in future research by introducing more fuzzy kernels and enhancing the image super-resolution algorithm model. These findings suggest that models trained on datasets restored and reconstructed by super-resolution may achieve stability faster while reaching higher accuracy on both training and validation sets compared to models trained on the original dataset. This phenomenon may be due to the fact that the restored and reconstructed datasets provide higher-quality images, which help the models to quickly acquire features related to tree species classification. However, further empirical evidence and validation are needed to confirm the correctness of this inference.

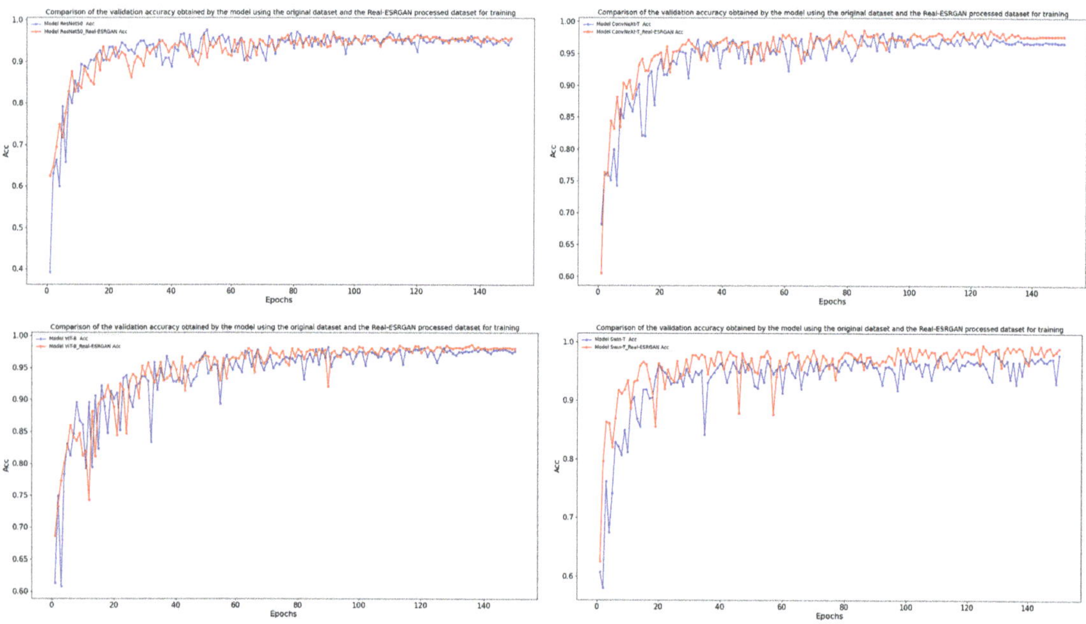

Figure 19. Comparison of the validation accuracy obtained by the model using the original dataset and the Real-ESRGAN processed dataset for training.

6. Conclusions

The assessment of tree species classification in Dongtai Forest utilizing RGB images captured by UAV has yielded promising outcomes through deep-learning-based approaches. Four models, including CNN models (ResNet-50 and ConvNeXt-T) and Transformer models (ViT-B and Swin-T), were trained and validated using UAV RGB tree crown images, achieving classification accuracies surpassing 95%. CNN models have been extensively used in forest resource surveys for tree species classification tasks, demonstrating exceptional classification accuracy [22,35]. Transformer models have also started finding applications in plant classification using UAV imagery [36] and exhibit significant potential for future advancements in forest surveys. However, the limited spatial resolution of aerial images introduces degradation challenges, such as detail loss, decreased brightness, blurred tree crown edges, and weak image texture. These issues negatively impact the feature point matching accuracy of CNN models and the capture of crucial information in the images. In contrast, Transformer models, with their inherent attention mechanisms, effectively leverage contextual information and global correlations in the images, resulting in comparatively less susceptibility to such issues. To address these challenges, Real-ESRGAN technology was adopted to perform super-resolution reconstruction and restoration on the original tree crown image dataset, leading to improved classification accuracy across all four models. This study confirms and underscores the observed enhancement in classification accuracy when using neural network models trained on images reconstructed through super-resolution. Super-resolution reconstruction techniques facilitate the restoration of low-quality images by recovering details, enhancing brightness, improving tree crown edge clarity, and augmenting image texture. These reconstructed images provide higher-quality information, enabling CNN and Transformer models to more accurately learn and extract features relevant to tree species classification. Consequently, when trained on these repaired and reconstructed image datasets, the four models exhibit improved stability and accuracy on validation sets.

Author Contributions: Conceptualization, Y.H. and X.W.; methodology, Y.H.; software, Y.H.; validation, Y.H., X.W. and Y.G.; formal analysis, Y.H.; investigation, Y.H.; resources, Y.H.; data curation, Y.H. and G.L.; writing—original draft preparation, Y.H.; writing—review and editing, Y.H., X.W., Y.G. and Y.Z.; visualization, Y.H.; supervision, X.W. and Y.G.; project administration, X.W.; funding acquisition, X.W. All authors have read and agreed to the published version of the manuscript.

Funding: This research was funded by the Forestry Innovation Foundation of Guangdong Province (No. 2021KJCX001) and the Priority Academic Program Development of Jiangsu Higher Education Institutions (PAPD). Additionally, financial support was provided by the Nanjing Institute of Environmental Sciences, MEE(GYZX230202).

Data Availability Statement: Data are available upon request from the section editors.

Acknowledgments: Y.H. is thankful for the patience and support from Jie Yang, who cooperated with him to write the paper in English and correct it and for the helpful discussion with Li Yang from the College of Forest, Nanjing Forestry University of China.

Conflicts of Interest: The authors declare no conflict of interest.

References

1. Zhao, D.; Pang, Y.; Liu, L.; Li, Z. Individual tree classification using airborne LiDAR and hyperspectral data in a natural mixed forest of northeast China. *Forests* **2020**, *11*, 303. [CrossRef]
2. Marrs, J.; Ni-Meister, W. Machine learning techniques for tree species classification using co-registered LiDAR and hyperspectral data. *Remote Sens.* **2019**, *11*, 819. [CrossRef]
3. Ballanti, L.; Blesius, L.; Hines, E.; Kruse, B. Tree species classification using hyperspectral imagery: A comparison of two classifiers. *Remote Sens.* **2016**, *8*, 445. [CrossRef]
4. Sun, Y.; Xin, Q.; Huang, J.; Huang, B.; Zhang, H. Characterizing tree species of a tropical wetland in southern china at the individual tree level based on convolutional neural network. *IEEE J. Sel. Top. Appl. Earth Obs. Remote Sens.* **2019**, *12*, 4415–4425. [CrossRef]
5. Heikkinen, V.; Tokola, T.; Parkkinen, J.; Korpela, I.; Jaaskelainen, T. Simulated multispectral imagery for tree species classification using support vector machines. *IEEE Trans. Geosci. Remote Sens.* **2009**, *48*, 1355–1364. [CrossRef]
6. Zhang, Z.; Liu, X. Support vector machines for tree species identification using LiDAR-derived structure and intensity variables. *Geocarto Int.* **2013**, *28*, 364–378. [CrossRef]
7. Ab Majid, I.; Abd Latif, Z.; Adnan, N.A. Tree species classification using worldview-3 data. In Proceedings of the 2016 7th IEEE Control and System Graduate Research Colloquium (ICSGRC), Shah Alam, Malaysia, 8 August 2016; pp. 73–76. [CrossRef]
8. Bondarenko, A.; Aleksejeva, L.; Jumutc, V.; Borisov, A. Classification tree extraction from trained artificial neural networks. *Procedia Comput. Sci.* **2017**, *104*, 556–563. [CrossRef]
9. Raczko, E.; Zagajewski, B. Tree species classification of the UNESCO man and the biosphere Karkonoski National Park (Poland) using artificial neural networks and APEX hyperspectral images. *Remote Sens.* **2018**, *10*, 1111. [CrossRef]
10. Karlson, M.; Ostwald, M.; Reese, H.; Bazié, H.R.; Tankoano, B. Assessing the potential of multi-seasonal WorldView-2 imagery for mapping West African agroforestry tree species. *Int. J. Appl. Earth Obs. Geoinf.* **2016**, *50*, 80–88. [CrossRef]
11. Immitzer, M.; Atzberger, C.; Koukal, T. Tree species classification with random forest using very high spatial resolution 8-band WorldView-2 satellite data. *Remote Sens.* **2012**, *4*, 2661–2693. [CrossRef]
12. Hologa, R.; Scheffczyk, K.; Dreiser, C.; Gärtner, S. Tree species classification in a temperate mixed mountain forest landscape using random forest and multiple datasets. *Remote Sens.* **2021**, *13*, 4657. [CrossRef]

13. Burai, P.; Beko, L.; Lenart, C.; Tomor, T. Classification of energy tree species using support vector machines. In Proceedings of the 2014 6th Workshop on Hyperspectral Image and Signal Processing: Evolution in Remote Sensing (WHISPERS), Lausanne, Switzerland, 24–27 June 2014; pp. 1–4. [CrossRef]
14. da Rocha, S.J.S.S.; Torres, C.M.M.E.; Jacovine, L.A.G.; Leite, H.G.; Gelcer, E.M.; Neves, K.M.; Schettini, B.L.S.; Villanova, P.H.; da Silva, L.F.; Reis, L.P. Artificial neural networks: Modeling tree survival and mortality in the Atlantic Forest biome in Brazil. *Sci. Total Environ.* **2018**, *645*, 655–661. [CrossRef] [PubMed]
15. Freeman, E.A.; Moisen, G.G.; Frescino, T.S. Evaluating effectiveness of down-sampling for stratified designs and unbalanced prevalence in Random Forest models of tree species distributions in Nevada. *Ecol. Model.* **2012**, *233*, 1–10. [CrossRef]
16. LeCun, Y.; Bottou, L.; Bengio, Y.; Haffner, P. Gradient-based learning applied to document recognition. *Proc. IEEE* **1998**, *86*, 2278–2324. [CrossRef]
17. Krizhevsky, A.; Sutskever, I.; Hinton, G.E. Imagenet classification with deep convolutional neural networks. *Commun. ACM* **2017**, *60*, 84–90. [CrossRef]
18. Ling, C.; Jia, J.; Haiqing, W. An overview of applying high resolution remote sensing to natural resources survey. *Remote Sens. Nat. Resour.* **2019**, *31*, 1–7.
19. Nezami, S.; Khoramshahi, E.; Nevalainen, O.; Pölönen, I.; Honkavaara, E. Tree species classification of drone hyperspectral and RGB imagery with deep learning convolutional neural networks. *Remote Sens.* **2020**, *12*, 1070. [CrossRef]
20. Kapil, R.; Marvasti-Zadeh, S.M.; Goodsman, D.; Ray, N.; Erbilgin, N. Classification of Bark Beetle-Induced Forest Tree Mortality using Deep Learning. *arXiv* **2022**, arXiv:2207.07241. [CrossRef]
21. Hu, M.; Fen, H.; Yang, Y.; Xia, K.; Ren, L. Tree species identification based on the fusion of multiple deep learning models transfer learning. In Proceedings of the 2018 Chinese Automation Congress (CAC), Xi'an, China, 30 November–2 December 2018; pp. 2135–2140. [CrossRef]
22. Natesan, S.; Armenakis, C.; Vepakomma, U. Individual tree species identification using Dense Convolutional Network (DenseNet) on multitemporal RGB images from UAV. *J. Unmanned Veh. Syst.* **2020**, *8*, 310–333. [CrossRef]
23. Ford, D.J. UAV Imagery for Tree Species Classification in Hawai'i: A Comparison of MLC, RF, and CNN Supervised Classification. Ph.D. Thesis, University of Hawai'i at Manoa, Honolulu, HI, USA, 2020.
24. Chen, X.; Jiang, K.; Zhu, Y.; Wang, X.; Yun, T. Individual tree crown segmentation directly from UAV-borne LiDAR data using the PointNet of deep learning. *Forests* **2021**, *12*, 131. [CrossRef]
25. Wang, X.; Xie, L.; Dong, C.; Shan, Y. Real-esrgan: Training real-world blind super-resolution with pure synthetic data. In Proceedings of the IEEE/CVF International Conference on Computer Vision, Virtual, 11–17 October 2021; pp. 1905–1914.
26. Wang, X.; Yu, K.; Wu, S.; Gu, J.; Liu, Y.; Dong, C.; Qiao, Y.; Loy, C.C. ESRGAN: Enhanced Super-Resolution Generative Adversarial Networks. In Proceedings of the European Conference on Computer Vision (ECCV) 2018, Munich, Germany, 8–14 September 2018; pp. 63–79. [CrossRef]
27. Schonfeld, E.; Schiele, B.; Khoreva, A. A u-net based discriminator for generative adversarial networks. In Proceedings of the IEEE/CVF Conference on Computer Vision and Pattern Recognition, Seattle, WA, USA, 13–19 June 2020; pp. 8207–8216.
28. Yan, Y.; Liu, C.; Chen, C.; Sun, X.; Jin, L.; Peng, X.; Zhou, X. Fine-grained attention and feature-sharing generative adversarial networks for single image super-resolution. *IEEE Trans. Multimed.* **2021**, *24*, 1473–1487. [CrossRef]
29. He, K.; Zhang, X.; Ren, S.; Sun, J. Deep residual learning for image recognition. In Proceedings of the IEEE Conference on Computer Vision and Pattern Recognition, Las Vegas, Nevada, USA, 27–30 June 2016; pp. 770–778. [CrossRef]
30. Liu, Z.; Mao, H.; Wu, C.Y.; Feichtenhofer, C.; Darrell, T.; Xie, S. A ConvNet for the 2020s. *arXiv* **2022**, arXiv:2201.03545. [CrossRef]
31. Vaswani, A.; Shazeer, N.; Parmar, N.; Uszkoreit, J.; Jones, L.; Gomez, A.N.; Kaiser, Ł.; Polosukhin, I. Attention is all you need. *arXiv* **2017**, arXiv:1706.03762. [CrossRef]
32. Hendrycks, D.; Gimpel, K. Gaussian error linear units (gelus). *arXiv* **2016**, arXiv:1606.08415. [CrossRef]
33. Dosovitskiy, A.; Beyer, L.; Kolesnikov, A.; Weissenborn, D.; Zhai, X.; Unterthiner, T.; Dehghani, M.; Minderer, M.; Heigold, G.; Gelly, S. An image is worth 16x16 words: Transformers for image recognition at scale. *arXiv* **2020**, arXiv:2010.11929. [CrossRef]
34. Liu, Z.; Lin, Y.; Cao, Y.; Hu, H.; Wei, Y.; Zhang, Z.; Lin, S.; Guo, B. Swin transformer: Hierarchical vision transformer using shifted windows. In Proceedings of the IEEE/CVF International Conference on Computer Vision, Virtual, 11–17 October 2021; pp. 10012–10022. [CrossRef]
35. Egli, S.; Höpke, M. CNN-based tree species classification using high resolution RGB image data from automated UAV observations. *Remote Sens.* **2020**, *12*, 3892. [CrossRef]
36. Reedha, R.; Dericquebourg, E.; Canals, R.; Hafiane, A. Transformer neural network for weed and crop classification of high resolution UAV images. *Remote Sens.* **2022**, *14*, 592. [CrossRef]

remote sensing

MDPI

Article

AF-OSD: An Anchor-Free Oriented Ship Detector Based on Multi-Scale Dense-Point Rotation Gaussian Heatmap

Zizheng Hua [1], Gaofeng Pan [2], Kun Gao [1,*], Hengchao Li [3] and Su Chen [1]

[1] School of Optics and Photonic, Beijing Institute of Technology, Beijing 100811, China
[2] School of Cyberspace Science and Technology, Beijing Institute of Technology, Beijing 100811, China
[3] School of Information Science and Technology, Southwest Jiaotong University, Chengdu 610032, China
* Correspondence: gaokun@bit.edu.cn; Tel.: +86-139-1101-2489

Abstract: Due to the complexity of airborne remote sensing scenes, strong background and noise interference, positive and negative sample imbalance, and multiple ship scales, ship detection is a critical and challenging task in remote sensing. This work proposes an end-to-end anchor-free oriented ship detector (AF-OSD) framework based on a multi-scale dense-point rotation Gaussian heatmap (MDP-RGH) to tackle these aforementioned challenges. First, to solve the sample imbalance problem and suppress the interference of negative samples such as background and noise, the oriented ship is modeled via the proposed MDP-RGH according to its shape and direction to generate ship labels with more accurate information, while the imbalance between positive and negative samples is adaptively learned for the ships with different scales. Then, the AF-OSD based on MDP-RGH is further devised to detect the multi-scale oriented ship, which is the accurate identification and information extraction for multi-scale vessels. Finally, a multi-task object size adaptive loss function is designed to guide the training process, improving its detection quality and performance for multi-scale oriented ships. Simulation results show that extensive experiments on HRSC2016 and DOTA ship datasets reveal that the proposed method achieves significantly outperforms the compared state-of-the-art methods.

Keywords: anchor-free; arbitrarily oriented ship detection; deep learning; Gaussian heatmap; multi-scale

Citation: Hua, Z.; Pan, G.; Gao, K.; Li, H.; Chen, S. AF-OSD: An Anchor-Free Oriented Ship Detector Based on Multi-Scale Dense-Point Rotation Gaussian Heatmap. *Remote Sens.* **2023**, *15*, 1120. https://doi.org/10.3390/rs15041120

Academic Editors: Moulay A. Akhloufi and Mozhdeh Shahbazi

Received: 23 December 2022
Revised: 15 February 2023
Accepted: 16 February 2023
Published: 18 February 2023

1. Introduction

Ship detection is a highly regarded topic in remote sensing. The precise identification of ships through remote sensing images is a crucial task in the field of target recognition. Typically, ships are located offshore or near coasts, leading to a lot of similarities in the imagery. Despite this, accurately determining the size, position, and orientation of ships remains a daunting challenge due to the intricate nature of remote sensing scenarios and the varying sizes of ships. With the development of imaging hardware, remote sensing images have higher resolution, so the method of ship detection based on remote sensing images has been widely studied in various fields of marine supervision [1–3], port traffic flow [4,5], ship reconnaissance and statistics [6,7], etc.

For traditional ship detection methods [1,8], the common problems are poor system robustness, high leakage rate in complex scenarios, and low detection accuracy. With the development of deep learning in various applications, deep learning-based methods [9–11] can provide automated, high-precision, and high-accuracy results for target detection from remote sensing images. Due to the specificity of remote sensing, there are still some open problems: (1) the bounding box may contain much background in the selected area, and cannot accurately represent the position and direction of ships [12]; (2) false detection because of the small and densely distributed ships [6]; (3) difficulty of recognizing multi-scale ships [13]. Aiming at these problems, some anchor-based methods have been developed [2,13,14]. However, to ensure satisfactory detection accuracy, these methods usually require the manual design of pre-selected boxes according to the actual scene, hindering their applications in practice because of a large number of hyper-parameters and high computational complexity.

To solve the problems of anchor-based detection methods, anchor-free methods have been further proposed [6,15–17], which do not require the anchor parameters and can directly predict the class and location information of objects. Thus, these methods can avoid hyper-parameters and reduce computational processes, gradually attracting the extensive attention of researchers. The CenterNet models [6,9,16] have been built based on the center point of the object as a positive sample point to represent an oriented object, as shown in Figure 1a. However, if a target is represented by only one point, or if the shape characteristics of the oriented target are not considered. These methods fail to balance the number of positive and negative samples well, which also causes some positive locations to be misjudged as negative samples.

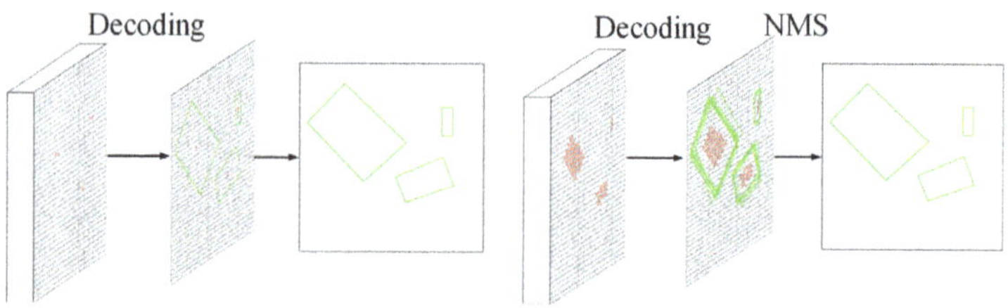

(**a**) AF-center methods (**b**) AF-multi-reference points

Figure 1. Comparison of target detection algorithms.

Aiming at the misjudgment problem of balancing the number of positive and negative samples, dense points-based methods are further proposed to increase the number of positive samples to achieve better detection performance. Zinelli et al. [18] proposed a dense point method to represent and predict the oriented objects, thus optimizing the imbalance of positive and negative samples to some extent. Nevertheless, the number of dense points is different with different scales of objects, as shown in Figure 1b. Specifically, the small objects usually contain fewer dense points than the large ones, and the contribution of the small objects to the loss function is small and easily ignored by the optimizer during training. Moreover, different locations of positive sample points exhibit different effects on the prediction results of the same object. The richer the object features extracted from the sample points close to the object center, the greater the impact on the prediction. However, these methods can not suppress background and noise interference well, which may result in the same confidence level for the feature points at the target centroid and edges, thus generating worse prediction results after non-maximum suppression (NMS), as shown in Figure 2. Additionally, these methods do not take the shape of the target into account, which may cause some negative locations to be misallocated as positive samples, such as the two ends of a ship.

To tackle the misallocation problem, some methods shrink the object bounding box to obtain the core region and take the core region as a positive location [18,19], while the other regions of the object bounding box are the transition from the positive location to the negative location. In this way, the problem of mislabeling can be alleviated to a certain extent. However, it fails to reflect the shape and direction characteristics of the oriented ships. Gaussian rotation heatmaps [6,20] have been used as the supervisory information to distinguish the positive and negative positions of oriented ships. Ref. [20] proposed a target detection method by adopting a Gaussian rotation heatmap as prior information and an adaptive weight-adjustment mechanism (OWAM) algorithm to weight the positive and negative samples at different positions. While these methods address the issue of mislabeling and incorporate the shape and direction of oriented ships, they only assign positive and negative labels through a continuous two-dimensional function represented by a Gaussian heatmap. This approach has a crucial limitation, as the Gaussian heatmap cannot be utilized as the confidence output in the prediction stage. Additionally, the contribution of positive samples from different locations to the object is neglected, leading to the possibility of positive samples at the edge of the target location being overlooked and increasing the impact of noise on accurate target detection.

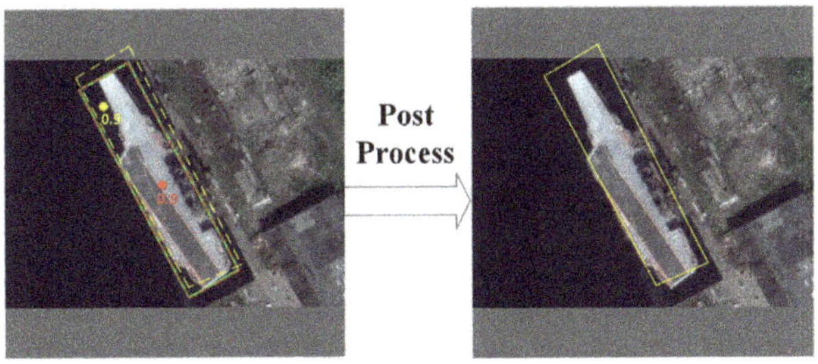

Figure 2. Diagram of prediction results of different reference points of the same rotation ship.

We aim to solve the imbalance problem caused by the number of positive and negative samples, better suppress background and noise interference, and simultaneously improve the robustness for remote sensing multi-scale ships and the training model's capability. Therefore, the multi-scale dense-point rotation Gaussian heatmap (MDP-RGH) method is proposed to overcome background and noise interference while enhancing the robustness of multi-scale ship detection and the training model's capabilities. The MDP-RGH is a discrete two-dimensional function based on a Gaussian heatmap that operates on dense points, allowing for the modeling of multi-scale oriented ships based on their shape and direction. The positive samples are weighted using the MDP-RGH to ensure rotational Gaussian distribution. The Gaussian heatmap confidence is used to predict whether a target is a positive or negative sample, thereby improving detection and reducing network computation. After that, a new anchor-free oriented ship detector network (AF-OSD) is constructed using the MDP-RGH method to detect multi-scale oriented ships. Additionally, a multi-task object size adaptive loss (OSALoss) function is designed to address the training imbalance caused by varying ship sizes. The weight of this function is determined by both the object area and the density of dense points, leading to improved ship detection accuracy. The contributions of this work can be summarized as follows.

1. An oriented ship model based on MDP-RGH is proposed, which can balance the number of positive and negative samples, suppress the interference of negative samples such as background and noise in the image, and improve the training accuracy.
2. An AF-OSD based on MDP-RGH is designed to achieve a better prediction for oriented ships with multi-scale attributes.
3. A multi-task OSALoss function is constructed to further overcome the training imbalance problem caused by different ship sizes to improve the detection quality and performance of the whole model for multi-scale ships.

The rest of the paper is organized as follows. In Section 2, the ship model based on MDP-RGH is elaborated in detail. Section 3 describes the AF-OSD based on the MDP-RGH ship model, which shows the detail of the network. Section 4 reports the hyper-parameter settings. Section 5 shows the experimental results and analysis. Section 6 discusses the experiment. Finally, Section 7 is the conclusion.

The variables themselves indicate a certain class of meaning, e.g., (x, y) for coordinate variables, $\mathbf{G}$ for ground truth, $\mathbf{F}$ for output convolutional features, and $\mathbf{M}$ for masks. The subscript of a variable indicates the qualification of the variable, indicating that the variable belongs to the range indicated by the subscript, and the next-level subscript is the qualification of the previous-level subscript.

2. The Oriented Ship Model Based on MDP-RGH

To solve the problem of an unbalanced number of positive and negative samples, predict the target using Gaussian heatmap confidence, and account for the contribution of positive samples at different locations, a new model of the ship, MDP-RGH, is proposed to suppress the background and noise interference, and effectively describe the characteristics of the shape and direction of the multi-scale ship. Specifically, there are three steps to complete the model of an oriented ship: (1) dividing the image region; (2) obtaining multi-scale dense points by down-sampling the image; (3) weighting the dense points with a rotation Gaussian Heatmap.

To address the issue of the imbalanced number of positive and negative samples, a new ship model called MDP-RGH is proposed. This model incorporates Gaussian heatmap confidence in target prediction and considers the contribution of positive samples from different locations. MDP-RGH aims to reduce background and noise interference and effectively describe the shape and direction of multi-scale ships, as shown in Figure 3. The process of creating the oriented ship model involves three steps: (1) dividing the image region; (2) obtaining multi-scale dense points through down-sampling the image; (3) weighting the dense points with a rotation Gaussian heatmap.

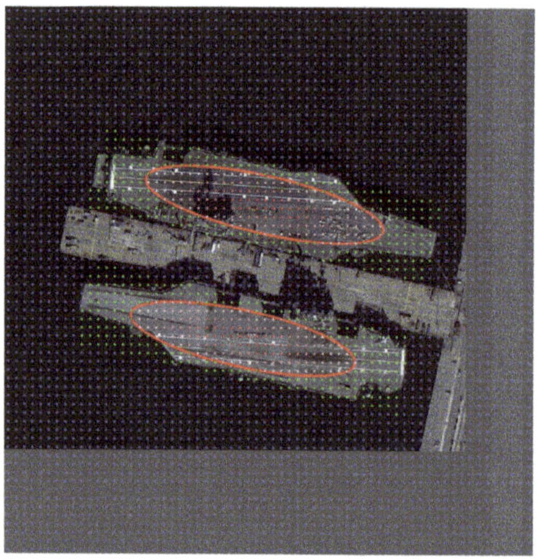

Figure 3. Ship representation format based on a dense-point Gaussian heatmap. The target in the red circle are delineated as the object regions.

2.1. *Dividing the Image Region*

As the shape of the ship is similar to a shuttle [6], if all the areas in the object bounding box are divided into object regions, part of the background pixels will be included in the object region.

To fit the shape and direction characteristics of the ship and coordinate with the subsequent rotation Gaussian heatmap, an image region division method is designed, as shown in Figure 3. Specifically, the shrink rotation ellipse regions are delineated as the object regions. Other regions in the object-bounding box are regarded as ignorable regions, while the regions not in any object-bounding box are regarded as the background regions. In this case, we only need to determine the object regions and the ignorable regions. More specifically, here are three steps to divide the image region.

2.1.1. Transformation the Coordinate for the Oriented Ship

To facilitate the subsequent calculation, we transform the rotating ship into an upright ship. Therefore, as shown in Figure 4, coordinate transformation is performed. Specifically, we transform the points $\begin{bmatrix} x_{\text{pixel}} \\ y_{\text{pixel}} \end{bmatrix}$ in the original pixel coordinate system **XOY** to the new coordinate system **X′O′Y′** with the center point of the oriented ship being set as the origin, the long axis of the ship (the line between the center points of the bow and stern) is set as the Y' axis, and the short axis is set as the X' axis. The coordinates of pixel points in the coordinate system **X′O′Y′** can be expressed as

$$\begin{bmatrix} x'_{\text{pixel}} \\ y'_{\text{pixel}} \end{bmatrix} = \begin{bmatrix} \cos\alpha & \sin\alpha \\ -\sin\alpha & \cos\alpha \end{bmatrix} \left(\begin{bmatrix} x_{\text{pixel}} \\ y_{\text{pixel}} \end{bmatrix} - \begin{bmatrix} x_c \\ y_c \end{bmatrix} \right), \tag{1}$$

where $\begin{bmatrix} x_c \\ y_c \end{bmatrix}$ means the coordinates of the center point of the oriented ship in the coordinate system **XOY**. α represents the counterclockwise angle between the positive half-axis of the oriented ship (the direction from the center point of the ship to the center point of the bow is positive) and the positive Y axis of the pixel coordinate system **XOY**.

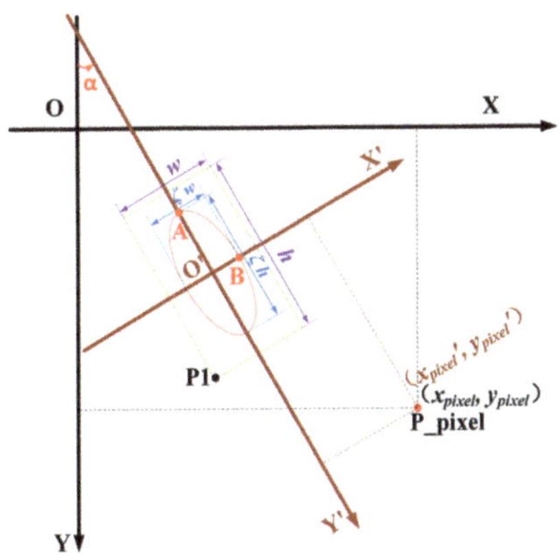

Figure 4. Coordinate transformation diagram. The green rectangular box is the boundary box of the ship. The red ellipse is a rotation ellipse with ξh as its long axis and ξw as its short axis after scaling the boundary frame of a ship according to the scale factor $\xi \in (0, 1]$, where h and w are the width and height of the boundary frame of a ship. A and B are the two vertices of the scaled rotation ellipse.

2.1.2. Creating a Shrink Rotation Ellipse Equation for the Oriented Ship

We create a shrink rotate ellipse equation for the oriented ship to determine which region the pixels belong to. The ellipse of the shrink rotation ship bounding box in the coordinate system **X′O′Y′** is a standard ellipse. Specifically, the ellipse equation is written as

$$f_{\text{ellipse}}\left(\begin{bmatrix} x' \\ y' \end{bmatrix} \right) = \frac{(x')^2}{\left(\frac{\xi w}{2}\right)^2} + \frac{(y')^2}{\left(\frac{\xi h}{2}\right)^2} = 1. \tag{2}$$

2.1.3. Identifying the Region to Which the Pixels in the Image Belong

After the above two steps, we can determine whether the pixel point $\begin{bmatrix} x_{\text{pixel}} \\ y_{\text{pixel}} \end{bmatrix}$ belongs to the object region $\text{area}_{\text{ship}}$ or the ignorable region $\text{area}_{\text{ignore}}$ as

$$\begin{cases} \begin{bmatrix} x_{\text{pixel}} \\ y_{\text{pixel}} \end{bmatrix} \in \text{area}_{\text{ship}}, & f_{\text{ellipse}}\left(\begin{bmatrix} x'_{\text{pixel}} \\ y'_{\text{pixel}} \end{bmatrix}\right) \leq 1 \text{ and} \left|x'_{\text{pixel}}\right| \leq \frac{w}{2} \\ \begin{bmatrix} x_{\text{pixel}} \\ y_{\text{pixel}} \end{bmatrix} \in \text{area}_{\text{ignore}}, & f_{\text{ellipse}}\left(\begin{bmatrix} x'_{\text{pixel}} \\ y'_{\text{pixel}} \end{bmatrix}\right) > 1 \text{ and} \left|y'_{\text{pixel}}\right| \leq \frac{h}{2}, \left|x'_{\text{pixel}}\right| \leq \frac{w}{2}. \end{cases} \quad (3)$$

Particularly, a pixel point belongs to the background region when it does not belong to any ship's object or ignorable region.

2.2. Multi-Scale Dense Points by Down-Sampling

Considering the multi-scale attributes of different ships, as shown in Figure 5, a multi-scale dense-point sampling method is further proposed to balance the number of positive samples of ships of different sizes. In particular, dense points can also reduce the amount of model computation since not all points need to be involved.

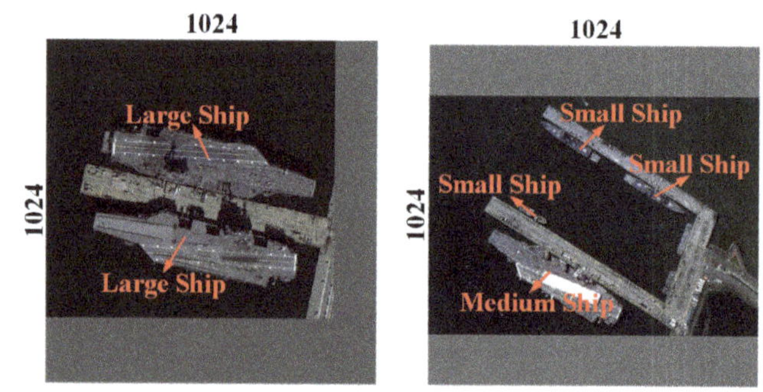

(**a**) Large ships (**b**) Medium and small ships

Figure 5. Some examples in the DOTA dataset.

Specifically, according to the different scales of the ships, the image is down-sampled at different multiples to obtain different low-resolution images. We conduct s (s = 4, 8, 16) down-samplings of the image with three scales. Then, the low-resolution image is mapped to the original image to obtain a three-scale dense point matrix. The coordinate calculation formula of dense points can be described as

$$\begin{cases} x_{\text{denps}_{ij}} = (i-1)(s) + \frac{s}{2} \\ y_{\text{denps}_{ij}} = (j-1)(s) + \frac{s}{2}, \end{cases} \quad (4)$$

where $x_{\text{denps}_{ij}}$ and $y_{\text{denps}_{ij}}$ denote the horizontal and vertical coordinates of the j-th row and i-th column sample point $\text{denp}_{s_{ij}}$ in the original image in the dense point matrix after s samplings, respectively.

Thus, small-scale dense points are used to represent large ships and large-scale dense points are used to represent small ships, which could balance the number of positive samples of the large, medium, and small ships to a certain extent. Among them, large-scale, medium-scale, and small-scale ships can be clustered with the k-means clustering algorithm. To better delineate the effective area of the image, dense points in the background region are defined as negative sample points, dense points in the ignorable region are ignorable dense points, and dense points in the object region are positive sample points, as shown in Figure 6a–c, respectively. Particularly, the ignorable dense points do not participate in calculating the loss function when training the model.

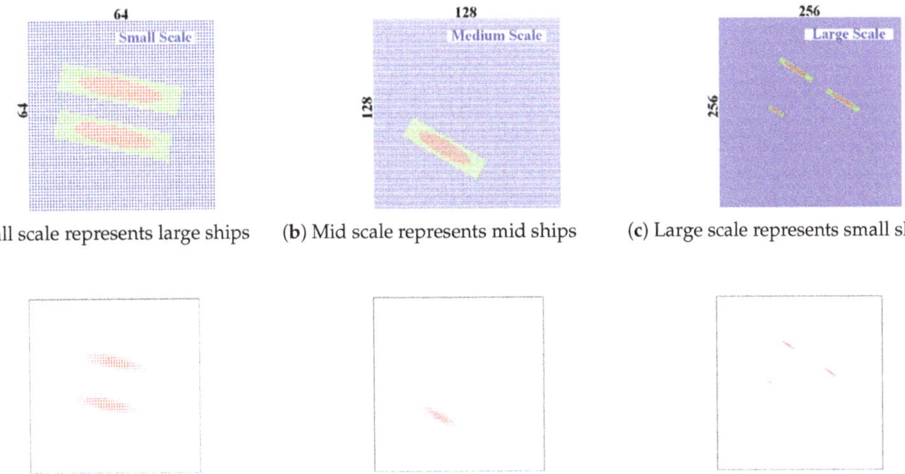

(**a**) Small scale represents large ships (**b**) Mid scale represents mid ships (**c**) Large scale represents small ships

(**d**) Large ship Gaussian Heatmap (**e**) Medium ship Gaussian Heatmap (**f**) Small ship Gaussian Heatmap

Figure 6. Schematic diagram of multi-scale dense-point sampling, in which blue points represent negative sample points, green points represent ignorable sample points, and red points represent positive sample points. Corresponding to (**a–c**) are two-dimensional rotation Gaussian heatmaps of large, medium, and small-sized ships, where the initial Gaussian heatmap value $g_{init} = 0.3$ and the scaling factor $\xi = 0.7$.

2.3. Weighting the Dense Points with Rotation Gaussian Heatmap

It has been proved that the closer the sample point is to the center of the ship, the richer the features of the ship are extracted during model inference, obtaining more accurate detection results. Therefore, the Gaussian weighting on the dense points with the object region is performed, where the dense points with different positions mean the various degrees of importance of the ship. Furthermore, the multi-scale dense-point rotation Gaussian heatmap is obtained, as shown in Figure 6.

Specifically, the calculation method of the rotation Gaussian heatmap value of the dense points in the ship region at each scale can be represented as

$$g\left(x_{\text{denp}_{s_{ij}}}, y_{\text{denp}_{s_{ij}}}\right) = g'\left(x'_{\text{denp}_{s_{ij}}}, y'_{\text{denp}_{s_{ij}}}\right) = \text{Exp}\left(-\left(\frac{\left(x'_{\text{denp}_{s_{ij}}}\right)^2}{2\sigma_w^2} + \frac{\left(y'_{\text{denp}_{s_{ij}}}\right)^2}{2\sigma_h^2}\right)\right), \tag{5}$$

where $g(\cdot)$ means the rotation Gaussian heatmap function; $g'(\cdot)$ means the general Gaussian heatmap function; $x'_{\text{denp}_{s_{ij}}}$ and $y'_{\text{denp}_{s_{ij}}}$ denote the horizontal and vertical coordinates of the dense point in the coordinate system $\mathbf{X'O'Y'}$, respectively; $\text{Exp}(\cdot)$ is the exponential function; σ_w and σ_h are the parameters related to the width and height of the ship, respectively.

To determine the values of σ_w and σ_h, a hyper-parameter $g_{init} \in (0,1)$ is introduced, which represents the value of the rotation Gaussian heatmap when the dense point is located at the boundary of the shrink rotation ellipse. Take two points on the scaled rotating ellipse boundary, and then the values of σ_w and σ_h can be obtained by combining the initial Gaussian heatmap value g_{init}, scale factor ξ, (1), (2), and (5). For example, we take the two vertices of this scaled rotation ellipse (e.g., the two points A and B in Figure 4) to determine the values of σ_w and σ_h. At this point, the horizontal and vertical coordinates of these two points in the coordinate system $\mathbf{X'O'Y'}$ can be calculated by (1) and (2). Bringing g_{init}, A, and B into (5) yields

$$\begin{cases} g_{init} = \mathrm{Exp}\left(-\frac{(\xi h/2)^2}{2\sigma_h^2}\right) \\ g_{init} = \mathrm{Exp}\left(-\frac{(\xi w/2)^2}{2\sigma_w^2}\right), \end{cases} \tag{6}$$

then the values of σ_w and σ_h can be obtained.

In particular, the rotation Gaussian heatmap value of the dense points located in the ignorable region or background region is 0. After that, the Gaussian rotation heatmap of the ships with three scales can be generated in Figure 6d–f, respectively.

Based on the above designs, an oriented ship model based on MDP-RGH can be obtained, which can better balance the influence of positive and negative samples of the training dataset, and reduce the influence of noise and background on ship recognition. The positive samples near the edge of the ship are made to conform to the rotated Gaussian distribution by avoiding noise interference. The closer the positive sample is to the edge, the lower its value, so our positive sample is soft. A higher positive sample score means that the positive sample is more representative of the ship. Thus, the MDP-RGH improves the robustness for different size targets and the calculation speed of the proposed model. Next, we will introduce the oriented ship detection algorithm base on MDP-RGH.

3. Oriented Ship Detection Algorithm Based on MDP-RGH

Based on the proposed MDP-RGH method for training dataset improvement, an end-to-end AF-OSD is devised to detect the multi-scale oriented ship, which is accurate identification and information extraction of multi-scale ships, as shown in Figure 7. Precisely, the AF-OSD mainly consists of three parts: (1) label assignment based on MDP-RGH; (2) the AF-OSD; (3) multitask OSALoss.

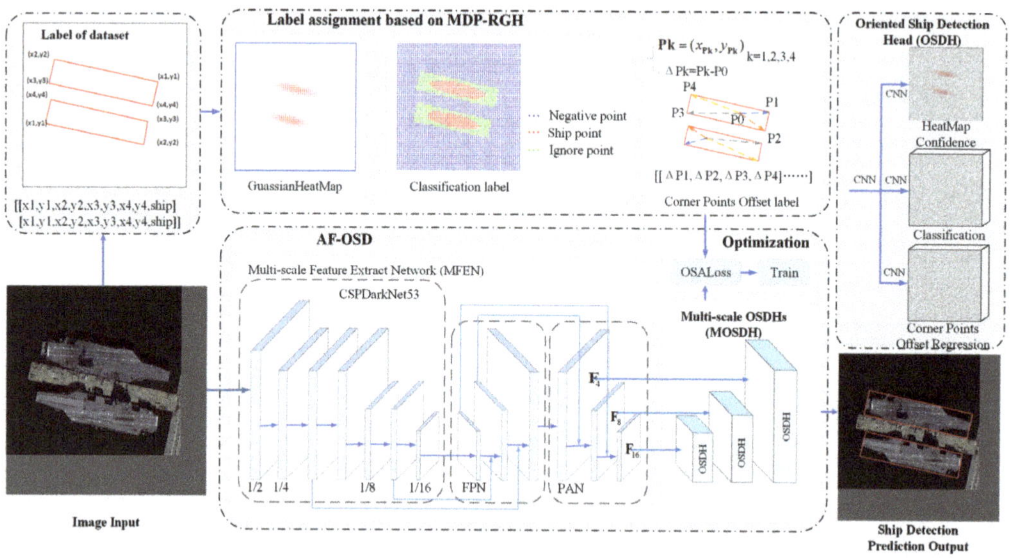

Figure 7. Overview diagram of AF-OSD based on MDP-RGH. (1) Label assignment based on MDP-RGH; (2) the AF-OSD; (3) OSALoss.

3.1. *Label Assignment Based on MDP-RGH*

According to Section 2, based on the MDP-RGH model, the label of supervising the training process of the AF-OSD can be obtained. The label allocation strategy processes the original dataset labels into three parts: (1) multi-scale dense-point rotation Gaussian heatmap confidence label, which is the weight of dense points; (2) multi-scale dense point category label, namely, regions divided by the image; (3) multi-scale dense point ship bounding box corner offsets label, namely, the offsets of the four corner points of the ship bounding box relative to the dense points in the corresponding object region.

3.1.1. Positive Sample Confidence Labels for Dense Points Based on MDP-RGH

The confidence labels of dense points based on MDP-RGH indicate the probability that the dense points belong to a ship, which can be expressed as $\mathbf{G}_{\mathrm{conf}_s} \in \mathbf{R}^{\left(\frac{H}{s}\right)\times\left(\frac{W}{s}\right)\times 1}$, where $s = 4, 8, 16$ represents the sampling scales with three values over the image and each element is a scalar.

Specifically, for the dense point $\mathrm{denp}_{s_{ij}}$ falling into the object region, its Gaussian heatmap confidence label value can be calculated by (5), while its value is 0 for the dense point outside the object region, which can be represented as

$$\mathbf{G}_{\mathrm{conf}_{s_{ij}}} = \begin{cases} g\left(x_{\mathrm{denp}_{s_{ij}}}, y_{\mathrm{denp}_{s_{ij}}}\right), & \mathrm{denp}_{s_{ij}} \in \mathrm{area}_{\mathrm{ship}} \\ 0, & \text{otherwise}, \end{cases} \tag{7}$$

where $\mathbf{G}_{\mathrm{conf}_{s_{ij}}}$ means the label value of the Gaussian heatmap of the dense point $\mathrm{denp}_{s_{ij}}$.

3.1.2. The Classification Label for Dense Points

The classification label $\mathbf{G}_{\mathrm{cls}_s} \in \mathbf{R}^{\left(\frac{H}{s}\right)\times\left(\frac{W}{s}\right)\times N_c}$ of dense points is a three-dimensional matrix, which can be written as

$$\mathbf{G}_{\mathrm{cls}_s} = \left[\mathbf{G}^{(1)}_{\mathrm{cls}_{s_{ij}}}, \ldots, \mathbf{G}^{(c)}_{\mathrm{cls}_{s_{ij}}}, \ldots, \mathbf{G}^{(N_c)}_{\mathrm{cls}_{s_{ij}}}\right], \tag{8}$$

where each element is a vector with size N_c (here N_c represents the number of categories, including the background category). $\mathbf{G}^{(c)}_{\mathrm{cls}_{s_{ij}}}$ denotes the score of dense point $\mathrm{denp}_{s_{ij}}$ belonging to category c, which can be determined by

$$\mathbf{G}^{(c)}_{\mathrm{cls}_{s_{ij}}} = \begin{cases} 1, & \mathrm{denp}_{s_{ij}} \in c\text{−th} \\ 0, & \text{otherwise}, \end{cases} \tag{9}$$

where the dense point belongs to the c-th category. The c-th component of the above vector is set to 1, and the others are set to 0.

3.1.3. The Corner Points offsets Label of the Ship Bounding Box of the Dense Points

We built corner point offset labels of the ship bounding box for regression when training. The label $\mathbf{G}_{\mathrm{Poffsets}_s} \in \mathbf{R}^{\left(\frac{H}{s}\right)\times\left(\frac{W}{s}\right)\times 8}$ of dense points is a two-dimensional matrix, where each element is a vector with size 8. The label represents the horizontal and vertical coordinates of the four corner points offsets of the ship bounding box of dense point $\mathrm{denp}_{s_{ij}}$, which can be written as

$$\mathbf{G}_{\mathrm{Poffsets}_{s_{ij}}} = \left[\mathbf{G}^{(1)}_{\mathrm{Poffsets}_{s_{ij}}}, \ldots, \mathbf{G}^{(8)}_{\mathrm{Poffsets}_{s_{ij}}}\right]. \tag{10}$$

If $\text{denp}_{s_{ij}}$ does not belong to the object region of any oriented ship, all components of $\mathbf{G}_{\text{Poffsets}_{s_{ij}}}$ are 0. If $\text{denp}_{s_{ij}}$ belongs to the object region of an oriented ship, the component of $\mathbf{G}_{\text{Poffsets}_{s_{ij}}}$ can be calculated as

$$\begin{cases} \mathbf{G}^{(2k-1)}_{\text{Poffsets}_{s_{ij}}} = \left(x_{\mathbf{Pk}} - x_{\text{denp}_{s_{ij}}} \right) / \text{scale}_s \\ \mathbf{G}^{(2k)}_{\text{Poffsets}_{s_{ij}}} = \left(y_{\mathbf{Pk}} - y_{\text{denp}_{s_{ij}}} \right) / \text{scale}_s, \end{cases} \tag{11}$$

where k ($k = 1,2,3,4$) represents the k-th corner point of the ship bounding box, $x_{\mathbf{Pk}}$ and $y_{\mathbf{Pk}}$ are the horizontal and vertical coordinates of the k-th corner point of the oriented ship bounding box, respectively, and scale_s is the normalized parameter of different scales.

3.2. AF-OSD

Normally, the AF-OSD consists of two parts: (1) a multi-scale feature extraction network (MFEN) and (2) a multi-scale oriented ship detection head (MOSDH). The MFEN extracts the features of the oriented ship in the input image, while the MOSDH applies the extracted features to identify and detect the oriented ship.

3.2.1. MFEN

PANet [21] is selected for the feature extraction structure of the MFEN for the fact that the size of the ship object has a large scale range. To reduce the problem of information loss arising from long transmission distances, the bottom-up path augmentation of the FPN [22] structure is applied, which provides a shorter path for high-level information transmission and effectively reduces information loss, as shown in Figure 8.

For the backbone of MFEN, ship detection is based on remote sensing data, visual range, and complex environments to fully extract the attribute information of ships. Hence, a relatively large feature extraction backbone network is needed to design. Because the cross-stage partial network (CSPNet) [23] solves the problem of repeated gradient information in the optimization process of large CNN frameworks, CSPDarknet53 is chosen as the backbone network, which divides the feature map of the base layer into two parts and then merges them through a cross-stage hierarchy.

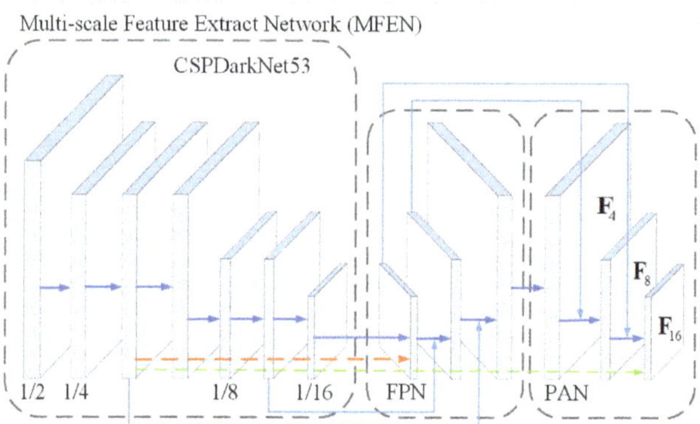

Figure 8. Multi-scale feature extraction network. The red dotted line represents the transmission path from the FPN low-level feature layer to the high-level feature layer. The dotted green line indicates the path from the PAN low-level feature layer to the high-level feature layer.

Based on the designed MFEN, the feature extraction of the input image can be realized. Specifically, the image $I \in \mathbb{R}^{H \times W \times 3}$ (H and W are the height and width, respectively) is input into the MFEN E_{MFEN}. Then, the multi-scale deep feature $\mathbf{F}_s \in \mathbb{R}^{(H/s)\times(W/s)\times C}$ can be extracted as

$$\mathbf{F}_s = E_{\text{MFEN}}(\mathbf{I}, s; \theta_{\text{E}}), \tag{12}$$

where C is the channel number, and θ_{E} is the weight parameters of the designed MFEN E_{MFEN}. The multi-scale deep feature $\mathbf{F}_s$ is fed into the subsequent MOSDH to identify and detect the ship.

3.2.2. MOSDH

The MOSDH is used to identify and detect the oriented ships through the features extracted by the MFEN. The output structure of MOSDH refers to the decoupling output mode of YoloX [24]. Corresponding to the three feature layers of PANet structure, the MOSDH contains three scales of oriented ship detection heads (OSDHs), where the small-scale detection head predicts large ships, the medium-scale detection head predicts medium ships, and the large-scale detection head predicts small ships.

Moreover, each OSDH has three decoupling output branches, as shown in Figure 9: (a) Gaussian heatmap confidence output branch of dense points: to predict the score that each dense point to be a positive sample (belonging to a ship); (b) Gaussian heatmap classification output branch of dense points: to predict the category of each dense point; (c) ship bounding box corner points offsets output branch of dense points: to predict the offsets of the four corner points of the ship (which ship the dense point belongs to) relative to the dense point.

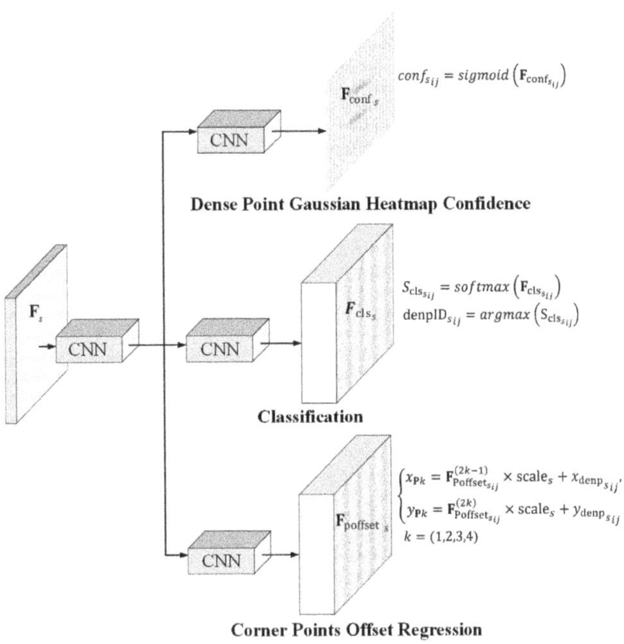

Figure 9. Structure of OSDH.

(a) The Dense Point Gaussian Heatmap Confidence Output Branch

To suppress background and noise interference, the confidence output of the Gaussian heatmap of dense points is designed. The multi-scale deep feature $\mathbf{F}_s$ is fed into the dense point Gaussian heatmap confidence output branch of MOSDH $H_{\text{MOSDH-conf}_s}$, and the Gaussian heatmap confidence output feature $\mathbf{F}_{\text{conf}_s} \in R^{(H/s)\times(W/s)\times 1}$ can be obtained as

$$\mathbf{F}_{\text{conf}_s} = H_{\text{MOSDH-conf}_s}(\mathbf{F}_s, s; \theta_{\mathbf{H}}), \tag{13}$$

where $\theta_{\mathbf{H}}$ is the weight parameters of the MOSDH. Each feature point of the Gaussian heatmap confidence output feature layer predicts the score of the dense point corresponding to this feature point belonging to the ship object region ($\text{area}_{\text{ship}}$). When the score is greater than the set threshold $T_{\text{conf}} \in (0, 1)$, the dense point is judged to be a dense positive point (belonging to the ship object region). The judgment method can be expressed as

$$\begin{cases} \text{denp}_{s_{ij}} \in \text{area}_{\text{ship}}, & sigmoid(\mathbf{F}_{\text{conf}_{s_{ij}}}) \geq T_{\text{conf}} \\ \text{denp}_{s_{ij}} \notin \text{area}_{\text{ship}}, & \text{otherwise}, \end{cases} \tag{14}$$

where $\mathbf{F}_{\text{conf}_{s_{ij}}}$ means the eigenvalue of the j-th row and the i-th column of the Gaussian heatmap confidence output feature layer, the activation function of the confidence output uses the $sigmoid(\cdot)$ function, and T_{conf} represents the confidence threshold.

Suppose that the dense point is determined to be a positive dense point. In that case, the category of the oriented object to which the dense point belongs and the coordinates of the four corner points of the oriented ship can be determined. Otherwise, it is directly judged that the dense matter belongs to the background (dense negative point), and no further discrimination and calculation will be carried out.

(b) Dense Point Classification Output Branch

We design the dense point classification output branch to determine the category of dense points. When a dense point is determined as a positive sample point, the dense point classification output branch $H_{\text{MOSDH}-\text{cls}_s}$ is used to predict the category of the dense point.

Specifically, the multi-scale deep feature $\mathbf{F}_s$ is input into the classification output branch of the MOSDH $H_{\text{MOSDH-cls}_s}$, and the classification output feature $\mathbf{F}_{\text{cls}_s} \in R^{(H/s)\times(W/s)\times N_c}$ is obtained as

$$\mathbf{F}_{\text{cls}_s} = H_{\text{MOSDH-cls}_s}(\mathbf{F}_s, s; \theta_{\mathbf{H}}). \tag{15}$$

Each feature point with size N_c of the feature layer of the category output predicts the score of each category that the corresponding dense point is, which can be written as

$$\text{denpID}_{s_{ij}} = argmax\left(softmax\left(\mathbf{F}_{\text{cls}_{s_{ij}}}\right)\right), \tag{16}$$

where $\text{denpID}_{s_{ij}}$ denotes the category number of the dense point $\text{denp}_{s_{ij}}$, and $\mathbf{F}_{\text{cls}_{s_{ij}}}$ represents the N_c dimension feature vector of the j-th row and the i-th column of the output feature layer of the category.

Therefore, the output feature vector of the category is first converted into the score of each type by the $softmax(\cdot)$ function, and then the category ID number of the highest category score is calculated by $argmax(\cdot)$ function to obtain the category of the oriented object.

(c) The Corner Point offsets Output Branch

To determine the position of the bounding box, the corner points to offset the output branch are further built. When a dense matter is judged as a dense positive point, the corner point offset output branch of the dense points predicts the offsets of the four corner points of the ship bounding box relative to the dense point.

Specifically, $\mathbf{F}_s$ is input into the corner point offset output branch of the dense points of the MOSDH $H_{\text{MOSDH-Poffsets}_s}$, and the output of offsets feature $\mathbf{F}_{\text{Poffsets}_s} \in R^{(H/s)\times(W/s)\times 8}$ is calculated as

$$\mathbf{F}_{\text{Poffsets}_s} = H_{\text{MOSDH-Poffsets}_s}(\mathbf{F}_s, s, \theta_{\mathbf{H}}). \tag{17}$$

Each feature point with size 8 of the output feature layer predicts the offsets of the four corner points of the ship bounding box relative to the dense point corresponding to the feature point. Therefore, the coordinates of the four corner points are the offsets plus the coordinate of the dense point, which can be expressed as

$$\begin{cases} x_{\mathbf{Pk}} & = x_{\Delta\mathbf{Pk}} + x_{\mathrm{denp}_{s_{ij}}} = \mathbf{F}^{(2k-1)}_{\mathrm{Poffsets}_{s_{ij}}} \times \mathrm{scale}_s + x_{\mathrm{denp}_{s_{ij}}} \\ y_{\mathbf{Pk}} & = y_{\Delta\mathbf{Pk}} + y_{\mathrm{denp}_{s_{ij}}} = \mathbf{F}^{(2k)}_{\mathrm{Poffsets}_{s_{ij}}} \times \mathrm{scale}_s + y_{\mathrm{denp}_{s_{ij}}} \end{cases}, \quad (18)$$

where $x_{\mathbf{Pk}}$ and $y_{\mathbf{Pk}}$ are the horizontal and vertical coordinates of the k-th corner point **Pk** of the ship bounding box, respectively, $x_{\Delta\mathbf{Pk}}$ and $y_{\Delta\mathbf{Pk}}$ represent the offsets of the k-th corner point **Pk** of the ship bounding box relative to the dense point $\mathrm{denp}_{s_{ij}}$, respectively, and $\mathbf{F}^{(2k-1)}_{\mathrm{Poffsets}_{s_{ij}}}$ is the $(2k-1)$-th value of the j-th row and i-th column feature vector of the output feature.

Multiple dense points may simultaneously predict the same ship object, or output layers of different scales may predict the same object. Therefore, after decoding all prediction boxes, it is necessary to use the NMS technology to filter out the redundant oriented object bounding boxes, obtaining the final detection result of the oriented ship.

3.3. Multi-Task Object Size Adaptive Loss (MOSALoss)

The multi-task OSALoss function, as the target function of the optimizer in the training stage, guides the optimization of the weight parameters of the proposed AF-OSD.

3.3.1. OSALoss Weight

The multi-scale structure of the AF-OSD network has been designed to solve the unbalanced training of different sizes of ships due to the other numbers of positive dense points. Specifically, three detection heads are designed to detect oriented ships of various sizes. However, since the object size is approximately continuous from small to large, the predicted objects of a certain output scale are large or small. The loss function model [25] focuses more on a small target with only one best anchor as a positive sample paired with it. Different from [25], we design the loss function OSALoss suitable for the MDP-RGH-based multi-scale directional ship detection in this paper.

To further solve the training imbalance caused by different object sizes, increasing the number of output characteristic layers at different scales will not only fail to completely solve this problem but also increase the amount of calculation, as shown in Figure 10. As such, an OSALoss weight (OSAWeight) is further designed, which can be adaptive to ship size.

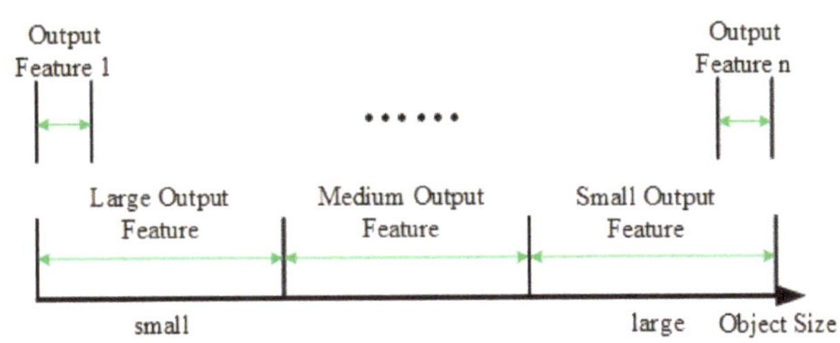

Figure 10. The relationship between the scale size of the target and the model output.

Specifically, if the dense point $\mathrm{denp}_{s_{ij}}$ belongs to the object region of a ship Ships_{s_n}, the weight of the sample point $W_{\text{adap-denp}_{s_{ij}}}$ is obtained as

$$\begin{cases} W_{\text{adap-denp}_{s_{ij}}} = \frac{d_{\text{denp}_s}}{\sqrt{A_{\text{Ships}_{sn}}}}, & \text{denp}_{s_{ij}} \in \text{area}_{\text{ship}} \\ W_{\text{adap-denp}_{s_{ij}}} = 1, & \text{otherwise,} \end{cases} \tag{19}$$

where Ships_{s_n} represents the n-th ship on the down-sampling scale of s times, $A_{\text{Ships}_{sn}}$ is the area of the ship, and d_{denp_s} is the distance between two adjacent dense points. According to (19), the weight of this dense point is inversely proportional to the square root of the area of the ship in a dense point matrix of the same scale. When the shipping area is large, the weight is small, while the weight is significant when the shipping area is small.

The difference in object size directly affects the difference in the number of positive sample points, which affects the loss related to positive sample points, leading to the problem of training imbalance. Therefore, the OSAWeight will be introduced into the loss function related to positive sample points to eliminate the influence caused by different object sizes.

3.3.2. Multi-Task OSALoss

In order to solve the situation of easy sample imbalance during training, the multi-task OSALoss $Loss_{\text{adap-multi}}$ consists of the Gaussian heatmap confidence loss $\mathbf{GHC}Loss_s = Loss(\mathbf{F}_{\text{conf}_s}, \mathbf{G}_{\text{conf}_s})$ of dense points, the category prediction loss $\mathbf{CP}Loss_s = Loss(\mathbf{F}_{\text{cls}_s}, \mathbf{G}_{\text{cls}_s})$ of dense points, and the corner points offsets prediction loss $\mathbf{OP}Loss_s = Loss(\mathbf{F}_{\text{Poffsets}_s}, \mathbf{G}_{\text{Poffsets}_s})$ of the ship bounding box. Then, the proposed multi-task OSALoss can be written as

$$Loss_{\text{adap-multi}} = \frac{1}{3} \sum_{s \in (4,8,16)} (\mathbf{GHC}Loss_s + \gamma \mathbf{CP}Loss_s + \lambda \mathbf{OP}Loss_s), \tag{20}$$

where $\mathbf{G}_{\text{conf}_s}$, $\mathbf{G}_{\text{cls}_s}$, and $\mathbf{G}_{\text{Poffsets}_s}$ represent the confidence ground truth of Gaussian heatmap, the classification ground truth of dense points and the corner offsets ground truth of ship bounding box, respectively. γ and λ are the weight parameters, which are adaptive with the training.

Next, we will introduce the Gaussian heatmap confidence loss, the category prediction loss, and the corner points offset prediction loss in detail.

(a) The Gaussian Heatmap Confidence Loss:

Adaptive wing loss [26] is used for the $\mathbf{GHC}Loss_s$, which can be represented as

$$\mathbf{GHC}Loss_s = AWingLoss(\mathbf{F}_{\text{conf}_s}, \mathbf{G}_{\text{conf}_s}). \tag{21}$$

Since the $\mathbf{GHC}Loss_s$ adopts the relatively mature adaptive wing loss, we do not introduce the OSAWeight into this loss function.

(b) The Category Prediction Loss:

The cross-entropy function is used for the $\mathbf{CP}Loss_s$. Particularly, the $\mathbf{CP}Loss_s$ is composed of positive $\mathbf{CP}Loss_s$ and negative $\mathbf{CP}Loss_s$, which can be written as

$$\mathbf{CP}Loss_s = \mathbf{CP}Loss_s^{\text{Pos}} + \mathbf{CP}Loss_s^{\text{Neg}}, \tag{22}$$

where the positive category prediction loss $\mathbf{CP}Loss_s^{\text{Pos}}$ can be given as

$$\begin{aligned} \mathbf{CP}Loss_s^{\text{Pos}} = & \sum_{j=1}^{H/s} \sum_{i=1}^{W/s} W_{\text{adap-denp}_{sij}} \times \mathbf{M}_{\text{obj}_{sij}} \times \sum_{c=1}^{N_c} \Big(\mathbf{G}^c_{\text{cls}_{sij}} \log\left(\mathbf{S}^c_{\text{cls}_{sij}}\right) \\ & + \left(1 - \mathbf{G}^c_{\text{cls}_{sij}}\right) \log\left(1 - \mathbf{S}^c_{\text{cls}_{sij}}\right) \Big), \end{aligned} \tag{23}$$

where $\mathbf{M}_{\text{obj}_{sij}}$ is the positive dense point mask. $\mathbf{S}^c_{\text{cls}_{sij}}$ is the score when the dense point denp_{sij} is the c-th category. $\mathbf{G}^c_{\text{cls}_{sij}}$ is the ground truth of whether the dense point denp_{sij} is the c-th category.

$\mathbf{M}_{\text{obj}_{sij}}$ can be written as

$$\begin{cases} \mathbf{M}_{\text{obj}_{sij}} = 1, & \text{denp}_{s_{ij}} \in \text{area}_{\text{ship}} \\ \mathbf{M}_{\text{obj}_{sij}} = 0, & \text{otherwises} \end{cases} \tag{24}$$

and $\mathbf{S}^c_{\text{cls}_{sij}}$ can be expressed as

$$\mathbf{S}_{\text{cls}_{s_{ij}}} = softmax\left(\mathbf{F}_{\text{cls}_{s_{ij}}}\right). \tag{25}$$

For negative dense point category prediction loss, to further solve the imbalance between the number of positive and negative samples, the hard negative example mining technique is applied. Specifically, the negative dense category prediction loss for each dense point can be written as

$$Loss_{\text{neg}}\left(\mathbf{F}_{\text{cls}_{s_{ij}}}, \mathbf{G}_{\text{cls}_{s_{ij}}}\right) = \sum_{c=1}^{N_c}\left(\mathbf{G}^c_{\text{cls}_{sij}} \log\left(\mathbf{S}^c_{\text{cls}_{sij}}\right) + \left(1 - \mathbf{G}^c_{\text{cls}_{sij}}\right)\log\left(1 - \mathbf{S}^c_{\text{cls}_{sij}}\right)\right) \times \mathbf{M}_{\text{bg}_{sij}}, \tag{26}$$

where $\mathbf{M}_{\text{bg}_{sij}}$ is a negative sample mask, which can be written as

$$\begin{cases} \mathbf{M}_{\text{bg}_{sij}} = 0, & \text{denp}_{s_{ij}} \in \text{area}_{\text{ship}} \\ \mathbf{M}_{\text{bg}_{sij}} = 1, & \text{otherwise.} \end{cases} \tag{27}$$

Then, the negative sample category prediction loss of dense points is sorted from the largest to the smallest, and the ranked negative sample category prediction loss $Loss_{\text{neg}_{\text{sorted}}}\left(\mathbf{F}_{\text{cls}_{s_l}}, \mathbf{G}_{\text{cls}_{s_l}}\right)$ can be obtained, where l denotes the serial number of the l-th dense point after the negative dense category prediction loss ranking of dense points. Finally, only the sum of the negative sample category prediction loss of the top L is taken as the negative sample category prediction loss. The category prediction loss $\mathbf{CP}Loss^{\text{Neg}}_{s}$ can be represented as

$$\mathbf{CP}Loss^{\text{Neg}}_{s} = \sum_{l=1}^{L} Loss_{\text{neg}_{\text{sorted}}}\left(\mathbf{F}_{\text{cls}_{s_l}}, \mathbf{G}_{\text{cls}_{s_l}}\right). \tag{28}$$

In particular, if the number of negative samples is greater than twice the number of positive samples, L takes two times the number of positive samples. Otherwise, L is equal to the number of positive samples.

(c) The Corner Points Offsets Prediction Loss:

The ship bounding box corner point offset prediction loss uses the object size adaptive weighted Smooth-L1 error loss. Specifically, we first use the Smooth-L1 loss to predict each dense point of the ship bounding box corner point offsets as

$$Loss\left(\mathbf{F}_{\text{cls}_{s_{ij}}}, \mathbf{G}_{\text{cls}_{s_{ij}}}\right) = \begin{cases} \sum_{m=1}^{8} 0.5 \times (\mathbf{F}^m_{\text{Poffsets}_{sij}} - \mathbf{G}^m_{\text{Poffsets}_{s_{ij}}})^2, if \left|\mathbf{F}^m_{\text{Poffsets}_{sij}} - \mathbf{G}^m_{\text{Poffsets}_{s_{ij}}}\right| < 1 \\ \sum_{m=1}^{8}\left(\left|\mathbf{F}^m_{\text{Poffsets}_{sij}} - \mathbf{G}^m_{\text{Poffsets}_{s_{ij}}}\right| - 0.5\right), \text{otherwise.} \end{cases} \tag{29}$$

Then, the target-scale adaptive weighted summation of the predicted losses for all positive dense points ship target frame corner point offsets is performed. The total loss $\mathbf{OP}Loss_s$ can be written as

$$\mathbf{OP}Loss_s = \sum_{j=1}^{H/s}\sum_{i=1}^{W/s} W_{\text{adap-denp}_{sij}} \times \mathbf{M}_{\text{obj}_{sij}} \times Loss\left(\mathbf{F}_{\text{cls}_{s_{ij}}}, \mathbf{G}_{\text{cls}_{s_{ij}}}\right). \tag{30}$$

Using the above label assignment strategy, network design, and loss function, we can train the deep learning network of the AF-OSD through the Adam optimization method [27], as summarized in Algorithm 1.

Algorithm 1: Training AF-OSD based on Adam optimization

INPUT: Image data set $\mathbb{I} = \left\{\mathbf{I}^{(1)}, \mathbf{I}^{(2)}, \ldots\right\}$, and the corresponding labels $\mathbb{Y} = \left\{\mathbf{Y}^{(1)}, \mathbf{Y}^{(2)}, \ldots\right\}$
OUTPUT: The trained AF-OSD with fixed neural network parameters $\theta_{\mathbf{E}}$ and $\theta_{\mathbf{H}}$
Construct AF-OSD
Initialize $\theta_{\mathbf{E}}, \theta_{\mathbf{H}}$ **randomly**
for k to *iterations* **do**
    Select randomly $minibatch\ \mathcal{I}_k \subset \mathbb{I}, \mathcal{Y}_k \subset \mathbb{Y}$
    Label assign as Section 3.1 : $\mathcal{G}_k \leftarrow \mathcal{Y}_k$
    $\mathcal{F}_k = AF\text{-}OSD(\mathcal{I}_k; \theta_{\mathbf{E}k}, \theta_{\mathbf{H}k})$
    $\mathcal{L}_k = Loss_{adap\text{-}multi}(\mathcal{F}_k, \mathcal{G}_k)$
    $\theta_{\mathbf{E}k+1}, \theta_{\mathbf{H}k+1} \leftarrow Optimizer_{Adam}(\mathcal{L}_k; \theta_{\mathbf{E}k}, \theta_{\mathbf{H}k})$
end for
Freeze $\theta_{\mathbf{E}}, \theta_{\mathbf{H}}$

4. Experimental Conditions

In this section, experiments on two public-oriented ship datasets are conducted to quantitatively and qualitatively evaluate the proposed AF-OSD. A series of experiments are conducted using the DOTA ship dataset [28] and HRSC2016 dataset [29]. Next, we will introduce the experimental conditions, including experimental platforms, datasets, evaluation metrics, and implementation details.

4.1. Experimental Platforms

All the experiments are implemented on a desktop computer with an Intel Core i7-9700 CPU, 32 GB of memory, and a single NVIDIA GeForce RTX 3090 with 24 GB GPU memory.

4.2. Dataset

4.2.1. DOTA Ship Dataset

DOTA is a large-scale aerial image dataset containing 2806 images, with 15 categories labeled with oriented boxes and image sizes ranging from 800 × 800 to 1600 × 1600. We select 434 images containing ships (including 37,028 ships), where 90% of them are randomly selected as the training set, and the other 10% are as the validation set. The size distribution of ships is not uniform, and there are few large-scale ships. So, we enhance and expand the large ships in the dataset (rotation and flip enhancement), and the size distribution of the original dataset and the enhanced dataset is shown in Figure 11a, and the rotation angle distribution is shown in Figure 11b.

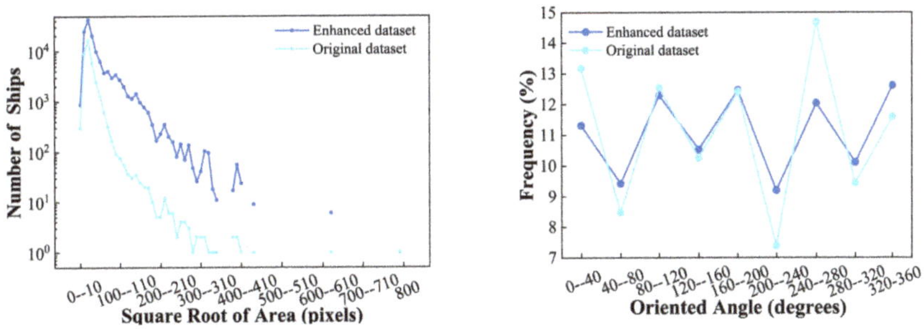

(**a**) Size distribution (**b**) Rotation angle

Figure 11. Size and rotation angle of ship targets in the DOTA dataset.

4.2.2. HRSC2016 Dataset

HRSC2016 is a remote sensing image dataset containing ship targets labeled as arbitrary orientations. It consists of 1061 images (436 images are the training set, 181 images are the validation set, and 444 images are the test set), whose spatial sizes range from 300 × 300 to 1500 × 900. We use the training set to train the model and the validation set to test the model. The size and rotation angle distributions of ships are illustrated in Figure 12, and the ship sizes and angles are more uniformly distributed. Therefore, we do not target special scale ships when we perform enhancement expansion (flip enhancement).

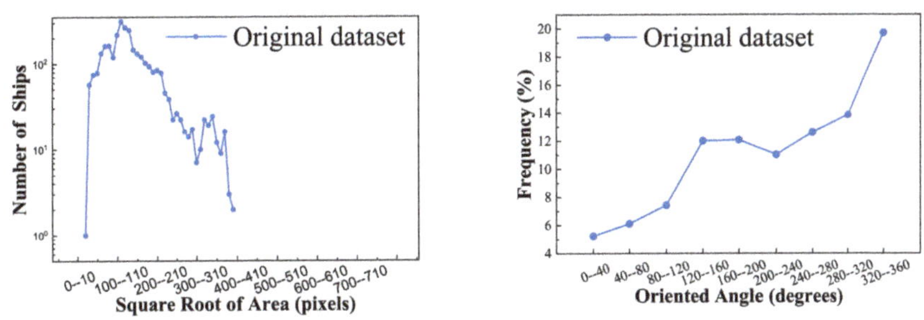

(**a**) Size distribution (**b**) Directional distribution

Figure 12. Size and directional distributions of ship targets in the HRSC2016 dataset.

4.3. Evaluation Metrics

In this article, the widely used metrics in object detection are adopted to measure the detection performance, i.e., precision, recall, and average precision (AP) (IOU threshold set to 0.5). Moreover, we use the number of parameters (Params) and average running time to evaluate the complexity and speed of the model, respectively.

4.4. Implementation Details

Due to the large size of the original input image, we cut it before network training. Specifically, the input resolution is set to 800 × 800. As such, the original image and the corresponding labels are first preprocessed, where the cut image is uniform 800 × 800 in size (if the image is smaller than 800 × 800, the padding operation is applied). To avoid the situation that an object is cut into two halves and disappears, an overlapping area of 200 pixels is set in the experiments. Of course, there will be some cases where only a part of the object is in the cropped image.

If this happens, we will judge the ratio of the area of the target on the cropped image to the area of the target itself. If this ratio is greater than the set threshold (taken as 0.6), the label information of the object will be kept. The judgment method of whether to keep or discard the target on the cropped picture is represented as

$$\begin{cases} \text{Keep}, & \frac{\mathbf{A}(\text{box}_{\text{Object}} \cap \text{box}_{\text{Crop}_{\text{image}}})}{\mathbf{A}(\text{box}_{\text{Object}})} > \text{Threshold} \\ \text{Discard}, & \text{otherwise}, \end{cases} \tag{31}$$

where $\mathbf{A}(\cdot)$ denotes the function of the area and $box_{Object} \cap box_{Crop_{image}}$ means the intersection of the target box and the cut picture box to be calculated. If it is determined that the target needs to be retained, its coordinates on the cut image are calculated as

$$p^{O*} = p_i^{O*} = p_i^O - p_1^{\text{Crop}}, i \in 1,2,3,4 \tag{32}$$

where p^{O*} is the new coordinate of the target frame in the cropped picture, p_i^O denotes the i-th coordinate point of the target frame in the original picture, and p_1^{Crop} denotes the first coordinate point of the cropped picture frame in the original pictures. The image cutting schematic is shown in Figure 13.

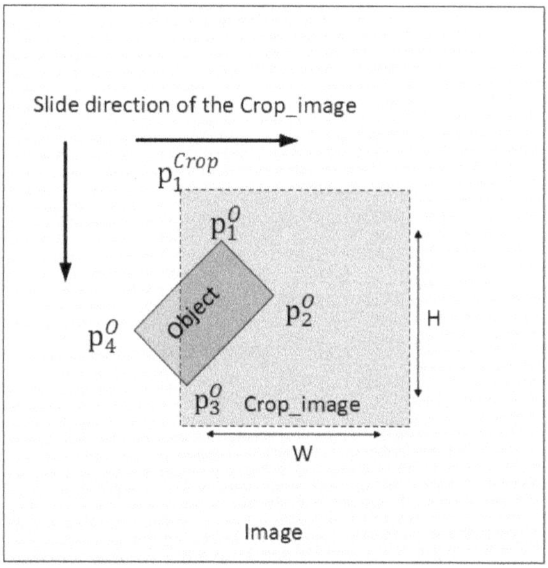

Figure 13. Image cutting schematic.

During the training process, the Adam optimizer is used to optimize the weight parameters of our AF-OSD, and the initial learning rate is set to 0.001 with an exponential decay of 5%. The batch size is set to 10, and all networks are trained for 200 epochs.

During the testing process, since the image size of HRSC2016 is not very large, before input to the model, we resize the image and then pad its size to 800 × 800. For the test images of DOTA, we used the same processing method as the training data. The confidence threshold $score_t$ is set to 0.1 and the NMS threshold is set to 0.6.

5. Experimental Results

To verify the rationality and advantages of the proposed AF-OSD, some comparison experiments are implemented to study the performance differences between the proposed oriented ship detector and other state-of-the-art (SOTA) oriented-object detection methods. Moreover, some ablation experiments are conducted to analyze the effects of each part on the performance of the proposed oriented ship detector.

5.1. Ablation Experiments

To testify to the necessity of the oriented ship detector proposed, three sets of ablation comparison experiments are designed: (1) comparing the performance of a multi-scale network structure and a single-scale network structure for detecting oriented ships; (2) comparing the performance of a network without MDP-RGH confidence output branch and a network with MDP-RGH confidence output branch for detecting oriented ships; (3) comparing the performance of the AF-OSD trained with the traditional multi-task loss function and the AF-OSD trained with multi-task OSALoss function.

All the ablation comparisons of DNN networks are trained using the HRSC2016 training set, parameter settings, and training algorithms mentioned in the previous section. They are then tested in the HRSC2016 validation set for measuring their detection performance.

5.1.1. Ablation Experiment on Different Scales

To verify the effect of multi-scale detection head on model enhancement, three single-scale models and multi-scale models were compared in terms of detection accuracy on the HRSC2106 dataset. Single-scale oriented ship detection network contains different detection scales, including 50×50, 100×100, and 200×200, and the multi-scale oriented ship detection network has three detection scales output.

As shown in Table 1, the comparison results show that the detection accuracy of the multi-scale oriented ship detection network is better than that of any single-scale detection network. Since it is difficult for a single-scale detection head to cover the remote sensing oriented ship with a large scale range, the multi-scale detection head can solve this problem well. So, compared with the AF-OSD network that does not use the multi-scale detection head structure, the performance of the whole model can be improved by 1.31–2.69% by introducing this structure.

Table 1. Ablation study of the multi-scale structure of the AF-OSD. The ✓ means network with the structure. The ✕ means network without the structure. The bold results mean the best performance.

Single Head	Multi Head	Recall	Precision	AP
50×50	✕	0.9020	0.8761	0.8789
100×100	✕	0.9020	0.7948	0.8700
200×200	✕	0.9039	0.8180	0.8833
✕	✓	**0.9068**	**0.8963**	**0.8969**

5.1.2. Ablation Experiment on MDP-RGH

To ensure the MDP-RGH confidence is favorable for the system, the comparison of the detection accuracy of the oriented ship detection network without the MDP-RGH confidence output branch and the oriented ship detection network with this branch is designed.

As shown in Table 2, the comparison results show that the detection accuracy of the oriented ship detection network with the MDP-RGH confidence branch is better. Compared with the AF-OSD without the MDP-RGH structure, the performance of the whole framework can have achieved 4.75% AP improvement by introducing this structure on HRSC2016.

Table 2. Ablation study of the MDP-RGH confidence output branch of the AF-OSD.

FPN	PAN	MDP-RGH	Recall	Precision	AP
✓	✓	✕	0.8854	0.8615	0.8494
✓	✓	✓	**0.9068**	**0.8963**	**0.8969**

5.1.3. Ablation Experiment on Multi-Task OSALoss

To demonstrate the effectiveness of the OSAloss designed in this paper, the detection accuracy between the AF-OSD trained with the traditional multi-task loss (TMLoss) function and the network trained with the multi-task OSALoss (MOSALoss) function on HRSC2016 is compared.

As shown in Table 3, comparison results show that our AF-OSD trained with the MOSALoss function has better detection accuracy. The OSALoss weight in the MOSALoss function is proportional to the distance between the dense points and inversely proportional to the square root of the ship area. Therefore, the MOSALoss function can be adjusted adaptively with the different ship sizes. So, compared with the AF-OSD network trained with TMLoss, the performance of the model which trained through MOSAloss can be improved by 4.56%. Therefore, the AF-OSD network with the MOSALoss function can effectively solve the training imbalance caused by the different scales of oriented ships.

Table 3. Ablation study of the multi-task OSALoss function.

TMLoss	MOSALOSS	Recall	Precision	AP
✓	×	0.8595	0.9082	0.8513
×	✓	**0.9068**	**0.8963**	**0.8969**

Based on the ablation experiments, the three parts of the AF-OSD are fully proven effective in remote sensing ship detection.

5.2. Comparisons with SOTA

To demonstrate the effectiveness of the method in this paper, the AF-RPN [18] is used as a baseline method, and the anchor-based methods RRPN [30], ROI-Transformer [31], ref. [20], and R^3Det [32] are compared with our method. The anchor-free methods IENet [33], GGHL [20], and GRS-DET [6] are also used for the comparison. The baseline method shrinks the object bounding box to obtain the core region and takes all the core region points as positive. The proposed method utilized the MDP-RGH to model a ship compared to the baseline method. We can compare the benefits of MDP-RGH through experiments.

Repeatability experiments on the DOTA dataset by the proposed method in this paper yield better results on the DOTA test dataset, as shown in Figure 14. The first row in the figure is the ground truth of the dataset. From Figure 14a,b, we can find that the proposed method has a high recognition rate for small targets in remote sensing images in the DOTA dataset. From Figure 14a,c, it can be seen that our method can achieve a good result both for large and dense small-oriented ships at the same time. Moreover, AF-OSD can accurately identify the small targets in the ground truth that are not completely labeled in the ground truth, as shown in Figure 14a. However, the missed detection phenomenon of the method proposed is significantly less than the ground truth. Meanwhile, it can be seen that the bounding box accuracy of the detection algorithm in this paper is high, as shown in Figure 15, in which the green bounding box is our method and the red bounding box is the ground truth.

5.2.1. Results on the DOTA Ships Dataset

Table 4 further shows the quantitative comparison results between the proposed AF-OSD and the SOTA-oriented ship detection algorithms on the DOTA ship dataset. Among these methods, the R^3Det method has the highest accuracy among anchor-based methods. Compared with the highest accuracy of the anchor-based method R^3Det, our method has improved by 0.32%. Moreover, among anchor-free methods, the experimental results show that the GRS-DET achieves about 2.36% AP improvement compared to the baseline method because of the Gaussian mask and selective concatenation module (SCM) structure. Compared to the GRS-DET, the proposed MDP-RGH can better suppress the interference of negative samples such as background and noise in the image, and the OSALoss function can further overcome the training imbalance problem caused by the different ship sizes, so our method outperforms the GRS-DET with 3.39% AP improvement on the DOTA ship dataset. The positive and negative sample distribution of GGHL is based on Gaussian prior and is adjusted during the training process, making it applicable to remote sensing targets of any shape, but it will have some impact on regular symmetrical targets (the distribution of positive samples in symmetrical rules will change to some extent after adjustment), resulting in its AP value being 2% lower than the proposed method on the DOTA ship targets.

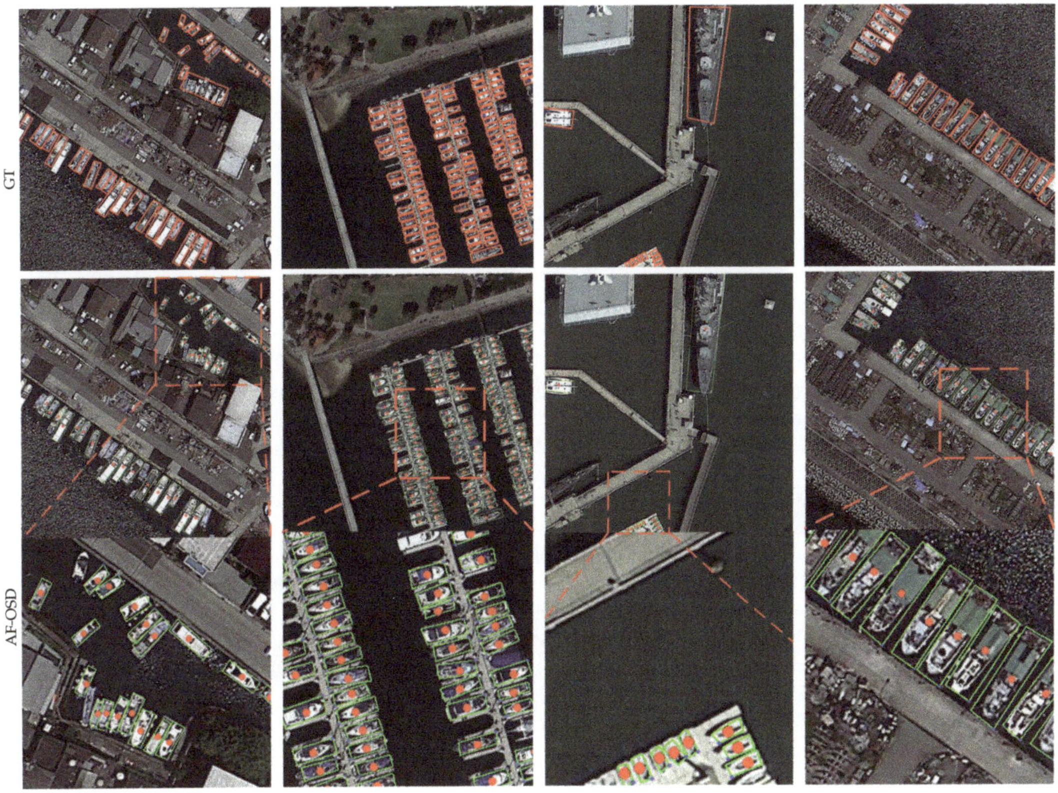

(**a**) Image1 (**b**) Image2 (**c**) Image3 (**d**) Image4

Figure 14. Detection comparisons of different test images on the DOTA ships dataset (red box: ground truth bounding box; red point: ship central points of proposed method; green box: proposed method bounding box).

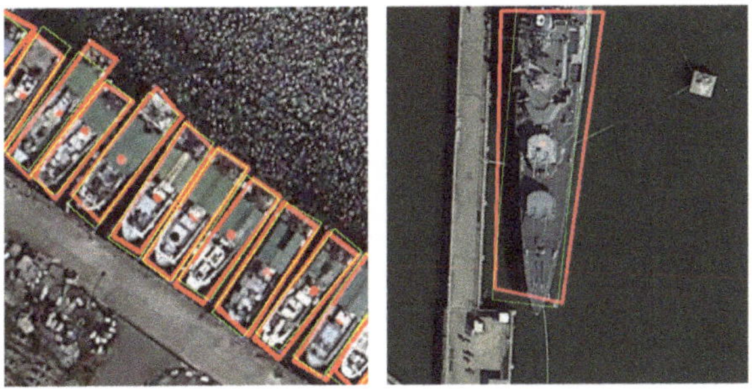

(**a**) image3 Local area (**b**) image4 Local area

Figure 15. Comparison of AF-OSD bounding box on DOTA dataset.

Table 4. Comparison results on the DOTA ship dataset.

Methods	Anchor-Free	Backbone	AP
RRPN	×	ResNet-101	0.4454
ROI-Transfomer	×	ResNet-101	0.8359
R^3Det	×	ResNet-152	0.8784
IENet	✓	ResNet-101	0.7161
AF-RPN	✓	CSPDarkNet53	0.8241
GRS-DET	✓	ResNet-101	0.8477
GGHL	✓	DarkNet53	0.8616
AF-OSD (Ours)	✓	CSPDarkNet53	**0.8816**

5.2.2. Results on the HRSC2016 Dataset

For the HRSC2016 test dataset, we directly resize and pad the image to 800 × 800 pixels, and then input it into the model for processing. Based on the qualitative comparison of the detection effectiveness of the proposed AF-OSD and the baseline method (AF-RPN) under the HRSC2016 dataset, it can be seen that our method tends to use the point near the center of the ship to predict the ship's bounding box. The point marked red means it had the highest score to predict the ship. This indicates that the proposed MDP-RGH can overcome the interference of negative samples such as background and noise in the image by using the center point to detect the ship. Thus, a better result can be realized by our method, as shown in Figure 16. From Figure 16b, it can be obtained that the proposed method effectively reduces the false detection rate of the ships. Moreover, AF-OSD can accurately identify the small targets in ground truth that are not completely labeled in the ground truth, as shown in Figure 16a,d.

Table 5 further shows the quantitative comparison results between the proposed AF-OSD and the SOTA-oriented ship detection algorithm on the HRSC2016 dataset. Among the anchor-based methods, the R^3Det method has the highest accuracy. Compared with R^3Det, the accuracy can be improved by 0.43% by our method. Moreover, among anchor-free methods, the experimental results show that the GRS-DET [6] achieves about 4.63% AP improvement compared to the baseline method. Due to the scale of ships in the HRSC2016 dataset being more uniform, our method, with the help of the MDP-RGH and the OSALoss function, outperforms the GRS-DET with 0.12% AP improvement on the HRSC2016 dataset.

Table 5. Comparison results on the HRSC2016 dataset.

Methods	RRPN	ROI-Transformer	R^3Det	IENet	AF-RPN	GRS-DET	AF-OSD (Ours)
AP	0.7905	0.8620	0.8926	0.7501	0.8494	0.8957	**0.8969**

The results proved that the proposed AF-OSD based on MDP-RGH can achieve SOTA through qualitative and quantitative comparative experiments.

5.3. Network Complexity Analysis

Based on the experimental process analysis, considering the network model's complexity and the computational effort, the proposed AF-OSD based on MDP-RGH in this paper is compared with the anchor-based and anchor-free models, as shown in Table 6. The comparison of results is based on the amount of computation and computing time within the corresponding methods' articles.

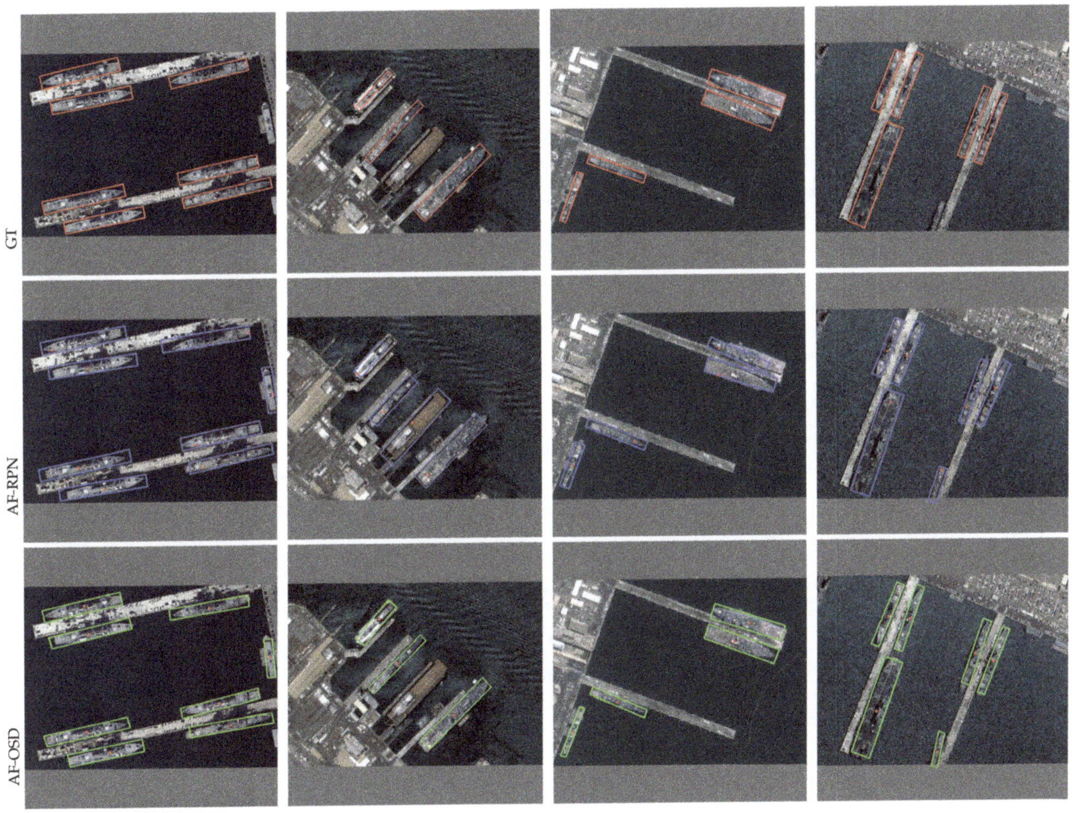

(**a**) Image1 (**b**) Image2 (**c**) Image3 (**d**) Image4

Figure 16. Detection comparisons of different test images on HRSC2016 (red box: ground truth bounding box; blue box: baseline method bounding box; green box: proposed method bounding box).

Table 6. Comparison of different networks on HRSC2016.

Methods	AP	Params	GPU	Speed
RRPN	0.7905	**181 MB**	GTX 1080TI	0.2857 s
ROI-Transformer	0.8620	273 MB	RTX 3090	0.1282 s
R^3Det	0.8926	227 MB	RTX 3090	0.0950 s
IENet	0.7501	212 MB	GTX 1080TI	0.0592 s
AF-RPN	0.8494	189 MB	RTX 3090	**0.0109 s**
GRS-DET	0.8957	200 MB	GTX 1080TI	0.0729 s
AF-OSD (Ours)	**0.8969**	192 MB	RTX 3090	0.0162 s

The network parameters of our proposed AF-OSD are significantly less, and its parameters are 35 MB less than those in the R^3Det method. Moreover, compared to the anchor-free method, the parameters of our method are 3 MB more than the baseline, 20 MB less than IENet, and 8 MB less than GRS-DET. Based on the analysis of the toxic network model, the computational complexity of this paper is 136 G FLOPS.

6. Discussion

In this paper, in order to solve the sample imbalance problem and suppress the interference of negative samples such as background and noise, the oriented ship is modeled via the proposed MDP-RGH according to its shape and direction to generate ship labels with more accurate information, which can predict the target for the contribution of positive samples at different positions to judge the target. Additionally, we designed an end-to-end anchor-free oriented ship detector (AF-OSD) net based on MDP-RG and validate its detection performance.

For the necessity study of network modules, the ablation experiment includes three parts: multi-scale structure, MDP-RGH-based label confidence, and MOSA loss. It has been proved that the framework with a multi-scale structure can extract the features of the oriented ship with different scales better. Multi-scale-based detection head setup enhances the robustness of the model. Therefore, the design of the multi-scale structure of the oriented ship detection network is reasonable. We use MDP-RGH to weight the positive samples so that the samples are rotationally Gaussian distributed and have values closer to the edge. The positive samples of our method are soft. As the result of the ablation experiment, the output can use the Gaussian heatmap confidence to predict whether the target is a negative sample or a positive one. A higher positive sample score means that the sample is more representative of the ship. Therefore, the MDP-RGH confidence can better suppress the interference of negative samples such as background and noise in the image, so the detection performance has improved. The improved loss function comparison experiments can prove that the weight can better solve the loss imbalance caused by the large and small ships. The larger the target, the higher the number of positive sample densities included; the smaller the target, the lower the number of positive sample densities included. The ablation experiment shows that the weight can adaptively solve the situation of sample imbalance that easily occurs during training.

Based on the test results on the DOTA ship dataset and HRSC2016 dataset, one can reach that the proposed AF-OSD has the best target recognition effect in remote sensing images with multi-scale ships, complex scenes, and positive and negative sample imbalance. The AF-OSD has features of high accuracy, few parameters of the network model, and high robustness. Due to the complexity of airborne remote sensing scenes, strong background and noise interference, positive and negative sample imbalance, and multiple ship scales, ship detection is a key and challenging task in remote sensing. The proposed method in this paper can better solve the above challenges. In summary, the proposed AF-OSD has a low number of parameters and higher model accuracy, detection accuracy, and computational speed.

As our main research object is ship targets. Most ships in the HRSC2016 and DOTA ship datasets have the characteristic of being narrow at both ends and wide in the middle, and their external contours are similar to rectangles. Therefore, the proposed method in this paper can be extended to the detection of remote sensing targets with similar external contours as rectangles (such as vehicles and stadium-like remote sensing targets) in any orientation, and even be possibly applied to symmetrical targets. However, AF-OSD is not an optimal solution for irregular and asymmetrical remote sensing targets (such as port-like targets). The application of the method in this paper to the detection of irregular targets is somewhat affected by the poor matching of the Gaussian heatmap. Therefore, the detection of full targets will be developed in the subsequent work. Moreover, to improve recognition speed and increase detection accuracy by decoupling the scale and task, the channel and spatial attention mechanism need to be considered in the design. However, the results showed that the inclusion of the attention mechanism actually led to a decrease in the detection accuracy of the network, which means that aspect requires further research and study in future work.

7. Conclusions

This paper presents the AF-OSD (anchor-free oriented ship detector) based on the MDP-RGH (multi-scale dense-point rotation Gaussian heatmap), which is designed for detecting arbitrarily oriented ships. The AF-OSD adopts a PANet structure for multi-scale feature extraction and a decoupled detection head for predicting the rotate Gaussian heatmap confidence, class, and location information of the oriented ship. The AF-OSD is capable of handling ships of different sizes, whose bounding boxes can be of any shape that fits the target. Experiments on two commonly used oriented ship datasets demonstrate the superiority of the AF-OSD over other state-of-the-art methods. These results showcase the high robustness of the AF-OSD in remote sensing ship identification applications and its ability to balance the number of positive and negative samples of various scale targets unsupervised while reducing the impact of noise and background on target identification. As a result, the proposed model has achieved state-of-the-art performance.

Author Contributions: Conceptualization, Z.H.; Methodology, Z.H.; Software, Z.H.; Formal analysis, G.P., H.L. and S.C.; Investigation, H.L.; Writing—original draft, Z.H.; Writing—review & editing, Z.H., G.P., K.G., H.L. and S.C.; Visualization, S.C.; Supervision, G.P., K.G., H.L. and S.C.; Project administration, G.P. and K.G.; Funding acquisition, K.G. All authors have read and agreed to the published version of the manuscript.

Funding: This research was funded by National Natural Science Foundation of China, grant number U2241275 and 61827814; Beijing Natural Science Foundation, grant number Z190018; China High-resolution Earth Observation System Project, grant number 52-L10D01-0613-20/22.

Data Availability Statement: Data available in a publicly accessible repository. The data presented in this study are openly available in DOTA dataset at [https://doi.org/10.1109/CVPR.2018.00418], reference number [28], and HRSC dataset at [https://doi.org/10.5220/0006120603240331], reference number [29].

Conflicts of Interest: The authors declare no conflict of interest. The funders had no role in the design of the study; in the collection, analyses, or interpretation of data; in the writing of the manuscript; or in the decision to publish the results.

References

1. Yang, G.; Li, B.; Ji, S.; Gao, F.; Xu, Q. Ship Detection From Optical Satellite Images Based on Sea Surface Analysis. *IEEE Geosci. Remote Sens. Lett.* **2014**, *11*, 641–645. [CrossRef]
2. Zou, Z.; Shi, Z. Ship Detection in Spaceborne Optical Image With SVD Networks. *IEEE Trans. Geosci. Remote Sens.* **2016**, *54*, 5832–5845. [CrossRef]
3. Dou, Z.; Gao, K.; Zhang, X.; Wang, H.; Wang, J. Improving Performance and Adaptivity of Anchor-Based Detector Using Differentiable Anchoring With Efficient Target Generation. *IEEE Trans. Image Process* **2021**, *30*, 712–724. [CrossRef]
4. Hu, Z.; Gao, K.; Zhang, X.; Wang, J.; Wang, H.; Han, J. Probability Differential-Based Class Label Noise Purification for Object Detection in Aerial Images. *IEEE Geosci. Remote Sens. Lett.* **2022**, *19*, 6509705. [CrossRef]
5. Yang, Y.; Tang, X.; Cheung, Y.M.; Zhang, X.; Liu, F.; Ma, J.; Jiao, L. AR2Det: An Accurate and Real-Time Rotational One-Stage Ship Detector in Remote Sensing Images. *IEEE Trans. Geosci. Remote Sens.* **2022**, *60*, 5605414. [CrossRef]
6. Zhang, X.; Wang, G.; Zhu, P.; Zhang, T.; Li, C.; Jiao, L. GRS-Det: An Anchor-Free Rotation Ship Detector Based on Gaussian-Mask in Remote Sensing Images. *IEEE Trans. Geosci. Remote Sens.* **2021**, *59*, 3518–3531. [CrossRef]
7. Ren, Z.; Tang, Y.; He, Z.; Tian, L.; Yang, Y.; Zhang, W. Ship Detection in High-Resolution Optical Remote Sensing Images Aided by Saliency Information. *IEEE Trans. Geosci. Remote Sens.* **2022**, *60*, 5623616. [CrossRef]
8. Liu, G.; Zhang, Y.; Zheng, X.; Sun, X.; Fu, K.; Wang, H. A New Method on Inshore Ship Detection in High-Resolution Satellite Images Using Shape and Context Information. *IEEE Geosci. Remote Sens. Lett.* **2014**, *11*, 617–621. [CrossRef]
9. Guo, W.; Chen, H.; Zhang, Z.; Zhang, Y.; Yu, W. Direct Oriented Ship Localization Regression in Remote Sensing Imagery with Curriculum Learning. In Proceedings of the 2021 IEEE International Geoscience and Remote Sensing Symposium IGARSS, Brussels, Belgium, 11–16 July 2021; pp. 2584–2587. [CrossRef]
10. Ren, S.; He, K.; Girshick, R.; Sun, J. Faster R-CNN: Towards Real-Time Object Detection with Region Proposal Networks. *IEEE Trans. Pattern Anal. Mach. Intell.* **2017**, *39*, 1137–1149. [CrossRef]
11. Redmon, J.; Divvala, S.; Girshick, R.; Farhadi, A. You Only Look Once: Unified, Real-Time Object Detection. In Proceedings of the 2016 Conference on Computer Vision and Pattern Recognition (CVPR), Las Vegas, NV, USA, 27–30 June 2016; pp. 779–788. [CrossRef]
12. Liu, Z.; Wang, H.; Weng, L.; Yang, Y. Ship Rotated Bounding Box Space for Ship Extraction From High-Resolution Optical Satellite Images With Complex Backgrounds. *IEEE Geosci. Remote Sens. Lett.* **2016**, *13*, 1074–1078. [CrossRef]
13. Yang, X.; Sun, H.; Fu, K.; Yang, J.; Sun, X.; Yan, M.; Guo, Z. Automatic Ship Detection in Remote Sensing Images from Google Earth of Complex Scenes Based on Multiscale Rotation Dense Feature Pyramid Networks. *Remote Sens.* **2018**, *10*, 132. [CrossRef]
14. Lin, H.; Shi, Z.; Zou, Z. Fully Convolutional Network With Task Partitioning for Inshore Ship Detection in Optical Remote Sensing Images. *IEEE Geosci. Remote Sens. Lett.* **2017**, *14*, 1665–1669. [CrossRef]

15. Pan, X.; Ren, Y.; Sheng, K.; Dong, W.; Yuan, H.; Guo, X.; Ma, C.; Xu, C. Dynamic Refinement Network for Oriented and Densely Packed Object Detection. In Proceedings of the IEEE/CVF Conference on Computer Vision and Pattern Recognition, Seattle, WA, USA, 14–19 June 2020; pp. 11204–11213. [CrossRef]
16. Yi, J.; Wu, P.; Liu, B.; Huang, Q.; Qu, H.; Metaxas, D. Oriented Object Detection in Aerial Images with Box Boundary-Aware Vectors. In Proceedings of the IEEE/CVF Winter Conference on Applications of Computer Vision, Virtual, 5–9 January 2021; pp. 2149–2158. [CrossRef]
17. Yang, F.; Xu, Q.; Li, B. Ship Detection From Optical Satellite Images Based on Saliency Segmentation and Structure-LBP Feature. *IEEE Geosci. Remote Sens. Lett.* **2017**, *14*, 602–606. [CrossRef]
18. Zinelli, A.; Musto, L.; Pizzati, F. A Deep-Learning Approach for Parking Slot Detection on Surround-View Images. In Proceedings of the 2019 IEEE Intelligent Vehicles Symposium (IV), Paris, France, 9–12 June 2019; pp. 683–688. [CrossRef]
19. Zhong, Z.; Sun, L.; Huo, Q. An Anchor-Free Region Proposal Network for Faster R-CNN-based Text Detection Approaches. *Int. J. Doc. Anal. Recognit.* **2019**, *22*, 315–327. [CrossRef]
20. Huang, Z.; Li, W.; Xia, X.G.; Tao, R. A General Gaussian Heatmap Label Assignment for Arbitrary-Oriented Object Detection. *IEEE Trans. Image Process* **2022**, *31*, 1895–1910. [CrossRef]
21. Liu, S.; Qi, L.; Qin, H.; Shi, J.; Jia, J. Path Aggregation Network for Instance Segmentation. In Proceedings of the IEEE Conference on Computer Vision and Pattern Recognition, Salt Lake City, UT, USA, 18–22 June 2018; pp. 8759–8768. [CrossRef]
22. Lin, T.Y.; Dollár, P.; Girshick, R.; He, K.; Hariharan, B.; Belongie, S. Feature Pyramid Networks for Object Detection. In Proceedings of the 2017 IEEE Conference on Computer Vision and Pattern Recognition (CVPR), Honolulu, HI, USA, 21–26 July 2017; pp. 936–944. [CrossRef]
23. Wang, C.Y.; Mark Liao, H.Y.; Wu, Y.H.; Chen, P.Y.; Hsieh, J.W.; Yeh, I.H. CSPNet: A New Backbone that can Enhance Learning Capability of CNN. In Proceedings of the 2020 IEEE/CVF Conference on Computer Vision and Pattern Recognition Workshops (CVPRW), Seattle, WA, USA, 14–19 June 2020; pp. 1571–1580. [CrossRef]
24. Ge, Z.; Liu, S.; Wang, F.; Li, Z.; Sun, J. YOLOX: Exceeding YOLO Series in 2021. *arXiv* **2021**, arXiv:2107.08430.
25. Wang, P.; Sun, X.; Diao, W.; Fu, K. FMSSD: Feature-Merged Single-Shot Detection for Multiscale Objects in Large-Scale Remote Sensing Imagery. *IEEE Trans. Geosci. Remote Sens.* **2020**, *58*, 3377–3390. [CrossRef]
26. Wang, X.; Bo, L.; Fuxin, L. Adaptive Wing Loss for Robust Face Alignment via Heatmap Regression. In Proceedings of the 2019 IEEE/CVF International Conference on Computer Vision (ICCV), Seoul, Republic of Korea, 27 October 2019–2 November 2019; pp. 6970–6980. [CrossRef]
27. Kingma, D.P.; Ba, J. Adam: A Method for Stochastic Optimization. In Proceedings of the International Conference on Learning Representations (ICLR), San Diego, CA, USA, 7–9 May 2015.
28. Xia, G.S.; Bai, X.; Ding, J.; Zhu, Z.; Belongie, S.; Luo, J.; Datcu, M.; Pelillo, M.; Zhang, L. DOTA: A Large-Scale Dataset for Object Detection in Aerial Images. In Proceedings of the 2018 IEEE/CVF Conference on Computer Vision and Pattern Recognition, Salt Lake City, UT, USA, 18–23 June 2018; pp. 3974–3983. [CrossRef]
29. Liu, Z.; Yuan, L.; Weng, L.; Yang, Y. A High Resolution Optical Satellite Image Dataset for Ship Recognition and Some New Baselines. In Proceedings of the 6th International Conference on Pattern Recognition Applications and Methods, Porto, Portugal, 20 October 2017; pp. 324–331. [CrossRef]
30. Ma, J.; Shao, W.; Ye, H.; Wang, L.; Wang, H.; Zheng, Y.; Xue, X. Arbitrary-Oriented Scene Text Detection via Rotation Proposals. *IEEE Trans. Multimed.* **2018**, *20*, 3111–3122. [CrossRef]
31. Ding, J.; Xue, N.; Long, Y.; Xia, G.S.; Lu, Q. Learning RoI Transformer for Oriented Object Detection in Aerial Images. In Proceedings of the 2019 IEEE/CVF Conference on Computer Vision and Pattern Recognition (CVPR), Long Beach, CA, USA, 15–20 June 2019; pp. 2844–2853. [CrossRef]
32. Yang, X.; Yan, J.; Feng, Z.; He, T. R3Det: Refined Single-Stage Detector with Feature Refinement for Rotating Object. *Proc. AAAI Conf. Artif. Intell.* **2021**, *35*, 3163–3171. [CrossRef]
33. Lin, Y.; Feng, P.; Guan, J.; Wang, W.; Chambers, J. IENet: Interacting Embranchment One Stage Anchor Free Detector for Orientation Aerial Object Detection. *arXiv* **2019**, arXiv:1912.00969.

applied sciences

MDPI

Article

ReBiDet: An Enhanced Ship Detection Model Utilizing ReDet and Bi-Directional Feature Fusion

Zexin Yan [1], Zhongbo Li [1,*], Yongqiang Xie [1,*], Chengyang Li [1,2], Shaonan Li [1] and Fangwei Sun [1]

[1] Institute of System Engineering, Academy of Military Sciences, Beijing 100000, China; yanzexin@outlook.com (Z.Y.); chengyang_li@stu.pku.edu.cn (C.L.); sfw1208@outlook.com (F.S.)
[2] School of Computer Science, Peking University, Beijing 100871, China
* Correspondence: zbli2021@outlook.com (Z.L.); yqxie2021@outlook.com (Y.X.)

Abstract: To enhance ship detection accuracy in the presence of complex scenes and significant variations in object scales, this study introduces three enhancements to ReDet, resulting in a more powerful ship detection model called rotation-equivariant bidirectional feature fusion detector (ReBiDet). Firstly, the feature pyramid network (FPN) structure in ReDet is substituted with a rotation-equivariant bidirectional feature fusion feature pyramid network (ReBiFPN) to effectively capture and enrich multiscale feature information. Secondly, K-means clustering is utilized to group the aspect ratios of ground truth boxes in the dataset and adjust the anchor size settings accordingly. Lastly, the difficult positive reinforcement learning (DPRL) sampler is employed instead of the random sampler to address the scale imbalance issue between objects and backgrounds in the dataset, enabling the model to prioritize challenging positive examples. Through numerous experiments conducted on the HRSC2016 and DOTA remote sensing image datasets, the effectiveness of the proposed improvements in handling complex environments and small object detection tasks is validated. The ReBiDet model demonstrates state-of-the-art performance in remote sensing object detection tasks. Compared to the ReDet model and other advanced models, our ReBiDet achieves mAP improvements of 3.20, 0.42, and 1.16 on HRSC2016, DOTA-v1.0, and DOTA-v1.5, respectively, with only a slight increase of 0.82 million computational parameters.

Keywords: artificial intelligence; deep learning; remote sensing images; ship detection; bi-directional feature fusion; feature pyramid network; anchor size; K-means; sampler

Citation: Yan, Z.; Li, Z.; Xie, Y.; Li, C.; Li, S.; Sun, F. ReBiDet: An Enhanced Ship Detection Model Utilizing ReDet and Bi-Directional Feature Fusion. *Appl. Sci.* **2023**, *13*, 7080. https://doi.org/10.3390/app13127080

Academic Editor: Luis Javier Garcia Villalba

Received: 14 April 2023
Revised: 7 June 2023
Accepted: 9 June 2023
Published: 13 June 2023

1. Introduction

With the advancement of space remote sensing technology, the observation capability of optical remote sensing satellites has significantly improved, leading to an increase in the spatial resolution of remote sensing images. Currently, the spatial resolution can reach 0.3 m per pixel [1], providing high-quality data for various scientific research applications. Computer vision technology plays a crucial role in extracting information from these images for tasks such as geological mapping, terrain measurement, and land cover change detection [2]. Object detection and recognition in remote sensing images are particularly important, especially in identifying ships, which hold strategic value in both economic and military domains. The accurate detection and recognition of ships in port areas and key waterways are vital for civilian and military purposes. However, the presence of complex backgrounds and variations in object scales in remote sensing images [3] pose significant challenges to achieving satisfactory ship detection accuracy.

1.1. Difficulties in Optical Remote Sensing Image Object Detection

Remote sensing images exhibit complexity, covering a wide range with large image scales. While they provide valuable visual features for detection models, they also introduce complex and irrelevant background information that hinders accurate detection. Firstly, background interference arises from similar objects in the scene, leading to potential misjudgment. For example, small islands with similar colors and thin buildings on land can be mistakenly identified as ships. Secondly, self-interference occurs due to the diverse textures, similar colors, and wake features of ships, affecting the final recognition results. Thirdly, the random orientation of ships in remote sensing images poses another difficulty. Unlike natural scene images captured from a horizontal perspective, satellite images are taken from a bird's-eye view, resulting in ships appearing at any position with any orientation, making detection challenging.

Furthermore, vehicles, boats, helicopters, airplanes, and large structures such as stadiums appear much smaller in remote sensing images compared to natural scene images captured from a horizontal viewpoint. Consequently, existing visual models often struggle to achieve accurate detection for these objects. For instance, in the optical remote sensing image dataset DOTA [4], small objects, such as cars, occupy approximately 30×50 pixels, large trucks occupy around 35×90 pixels, and the smallest yacht of interest is merely about 35×60 pixels in size.

1.2. The Significant Advantages of Deep Learning Techniques

Traditional ship detection methods [5–8] based on artificial feature modeling rely on manually designed algorithms to extract shallow features. These algorithms often focus on specific scenes and exhibit limited generalization abilities. They still face challenges in terms of accuracy, computational efficiency, and robustness in complex environments, making them less applicable [9–11].

In recent years, significant progress has been made in the field of object detection in natural images, largely attributed to the rapid development of deep learning technology [12]. The introduction of deep learning models has greatly improved the accuracy of object localization and recognition in images. Convolutional neural networks (CNNs) have emerged as powerful tools for feature extraction, enabling automatic learning and extraction of features from images. The renowned Faster R-CNN model [13] has become a representative model in natural image object detection, opening possibilities for applying deep learning-based object detection models to remote sensing images.

1.3. Related Works

After years of research, several mature methods have been developed for ship detection. Early studies predominantly focused on ship detection methods based on artificial feature modeling. However, since 2012, deep learning-based detection methods have emerged as a transformative force in the field of ship detection. Recent years have witnessed the proposal of numerous methods for ship detection in remote sensing images. Leveraging the powerful feature expression capabilities of convolutional neural networks (CNNs), these deep learning-based object detection methods can extract higher-level semantic features of ships, delivering significantly improved performance in complex remote sensing scenarios compared to traditional approaches. Nevertheless, most of these methods originated from object detection models designed for natural images captured from horizontal viewpoints and are limited in their ability to detect objects with arbitrary orientations. Given that ships possess relatively large aspect ratios and can exhibit random orientations in images, employing horizontal detection boxes would result in significant background information being included in the detection boxes. This issue becomes particularly pronounced when detecting densely distributed ships in a harbor, where interference from neighboring objects leads to poor detection performance.

To address these challenges, researchers have proposed models suitable for detecting objects with arbitrary orientations. For instance, the rotational R-CNN (R2CNN) model [14], initially developed for scene text detection, has shown excellent performance in detecting slanted text and has been adapted for ship detection. Liu et al. [15] enhanced this approach by introducing a rotation region of interest (RoI) pooling layer and a rotation bounding box regression module based on R2CNN. This improvement enables accurate feature extraction and localization, resulting in highly efficient and accurate ship detection. The rotated region proposal network (RRPN) [16] was designed as an enhancement of R2CNN to improve the quality of candidate regions generated by the region proposal network (RPN). RRPN generates 54 rotating anchor boxes with 3 scales, 3 aspect ratios, and 6 directions at each point in the feature map. Yang et al. [17] incorporated the dense feature pyramid network (FPN) based on RRPN to integrate low-level positional information and high-level semantic information, effectively detecting densely distributed ships using prow direction prediction. They also reduced redundant detection areas using rotational non-maximum suppression (NMS) to obtain more accurate object positions. Liu et al. [16] proposed the one-stage model CHPDet, which utilizes an orientation-invariant model to extract features. It selects the peak value of the feature map as the center point and then regresses the offset, size, and head point on the corresponding feature map of each center point, achieving impressive results in terms of speed and accuracy. Ding et al. [18] introduced a module called RoI Transformer based on the Faster R-CNN [13]. This module predicts a rough rotating RoI in the first stage based on the horizontal RoI generated by RPN. In the second stage, RoI Align [19] extracts the feature map of the RoI and refines the prediction result of the first stage to obtain a more accurate rotating RoI. Compared to RRPN, RoI Transformer significantly reduces the number of rotating anchors, leading to improvements in computational efficiency and detection accuracy. Oriented R-CNN [20], building upon the Faster R-CNN, introduced the midpoint offset representation and expanded the output parameters of the RPN regression branch from four to six to achieve skewed candidate boxes, resulting in substantial improvements in detection accuracy and computational efficiency.

Most object detection models are based on traditional CNNs, which possess translational equivariance but lack rotational equivariance. To enhance the adaptability of CNNs to rotational changes, numerous rotation-based object detection models have made significant efforts in data augmentation, rotation anchor boxes, or generating rotation RoIs to enhance the feature extraction of rotational objects in remote sensing images. Although these models have achieved some performance improvements, the issue of CNNs lacking rotational equivariance remains unaddressed.

To tackle this problem, ReDet [21] introduces the concept of E(2)-Equivariant Steerable CNNs [22] (E2-CNNs) into object detection by building upon the RoI Transformer [23]. ReDet adopts the *e2cnn* extension [22] for PyTorch [24] and employs it to reconfigure the architecture of ResNet-50 [25], renaming it ReResNet50. Additionally, the RiRoI Align module is redesigned to align channels and spatial dimensions, enabling the extraction of rotation-invariant features and further enhancing detection accuracy.

1.4. Goal of the Research

Although the methods mentioned above demonstrated good performance in ship detection, they still encounter several challenges. This study primarily focuses on two key issues:

(1) High-resolution optical remote sensing images contain rich pixel information of ships and other objects, but they also introduce a considerable amount of irrelevant, redundant, and potentially interfering noise. In specific scenarios, such as variations in lighting conditions and the limitations of satellite-borne sensors, the color contrast between ships and the background in some images may be low, leading to visual similarity to the human eye. Moreover, in harbor images, the shape of the dock can resemble that of large ships moored nearby. These factors contribute to the potential confusion of ships with complex backgrounds by detection models, resulting in missed detections and false alarms.

(2) Ships are typically sparsely and unevenly distributed in optical remote sensing images. Inappropriate sampling strategies may lead to the insufficient learning of ship features by the detection model. For example, the random sampling strategy commonly employed by many models, including ReDet, may exclude certain potential high-quality positive samples from the training process.

The ReDet model fails to address the aforementioned challenges. The upsampling operation in its FPN module may lead to the loss of crucial information, resulting in incomplete feature fusion and ultimately affecting the accuracy of the detection results. Additionally, the random sampler used in the RPN stage may lead to a small number of certain classes or samples due to class imbalance or sample imbalance, impacting the effectiveness of model training.

The objective of this study is to propose a solution based on ReDet to tackle the aforementioned problems, specifically focusing on ship object detection in remote sensing images. We introduce the ReBiDet model, which encompasses the following contributions: Firstly, we design a feature pyramid module called ReBiFPN based on rotational equivariance theory. ReBiFPN ensures balanced output levels and replaces the FPN module in the ReDet model. By integrating high-resolution features from higher layers with detailed information from lower layers, it enhances the detection performance of ships at multiple scales. Secondly, we propose the positive reinforcement learning sampler and an anchor optimization module based on K-means clustering. These components are utilized to balance the difficulty and proportion of positive samples during the training process, as well as optimizing the sizes and aspect ratios of anchors, further enhancing the model's detection performance. Thirdly, we validate and compare the generalization performance of ReBiDet by selecting images from different optical remote sensing datasets. Finally, following the practices of other researchers in the field of optical remote sensing image object detection, we conduct experiments on the HRSC2016 and DOTA datasets to demonstrate the effectiveness of the proposed modules and the performance advantages of ReBiDet.

2. The Proposed Method

This section presents the architecture of ReBiDet. Figure 1 illustrates the five modules of ReBiDet: the feature extraction module ReResNet [21], the feature pyramid construction module ReBiFPN, the bounding box generation module RPN, the RoIAlign module for extracting feature maps of horizontal bounding boxes and performing preliminary classification and rotation bounding box generation, and the RiRoIAlign [21] module for extracting feature maps of rotation bounding boxes and performing classification refinement and rotation bounding box parameter refinement.

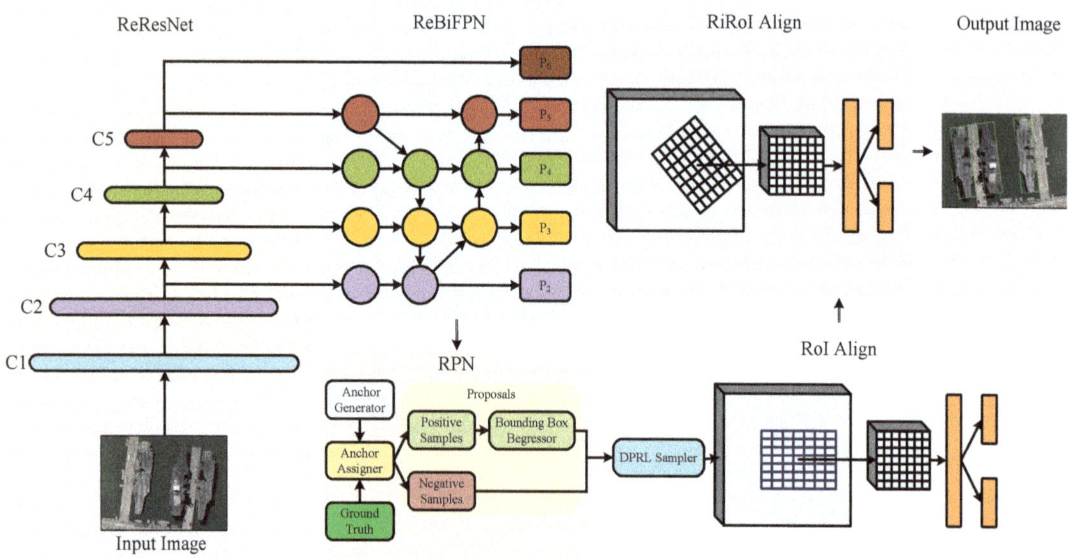

Figure 1. Architecture of rotation-equivariant bidirectional feature fusion detector (ReBiDet).

ReBiDet is an improved version of ReDet that specifically addresses the characteristics of optical remote sensing images. The objective is to tackle the challenge of detecting ships in complex scenes with scale differences and enhance the detection accuracy of ships in remote sensing images using specialized modules. Three changes were made to the original ReDet architecture. Firstly, we introduced ReBiFPN, constructed based on *e2cnn* [22], to enable the network to capture multi-scale feature information from both bottom–up and top–down paths, facilitating more comprehensive fusion of low-level positional information and high-level semantic information. Secondly, we utilized the K-means [26] algorithm to cluster the aspect ratios of ground truth boxes in the dataset and adjusted the size of the anchor boxes generated by the anchor generator accordingly. Finally, in the RoIAlign stage, we employed the difficult positive reinforcement learning (DPRL) sampler to address class imbalance in the dataset and make the model more sensitive to challenging positive samples. Experimental results demonstrate the effectiveness of these improvements in enhancing the detection performance of small objects in optical remote sensing images.

2.1. Rotation-Equivariant Networks

Convolutional neural networks (CNNs) typically employ fixed-size, two-dimensional convolution kernels to extract features from images. During convolution, these kernels apply the same weights to every position in the image, thereby allowing the detection of the same feature, even if the image is horizontally or vertically shifted. This property is known as translation invariance. However, when an image undergoes a rotation, the pixel arrangement of all objects in the image changes, causing the features that can be extracted by the CNN to also change and potentially become distorted. This lack of rotation equivariance in CNNs poses challenges in accurately recognizing rotated objects.

The absence of rotation equivariance implies that as an object in the input image changes its orientation, the features extracted by the CNN also change, adversely affecting the accuracy of the rotated object detection. This effect is particularly pronounced for elongated objects, such as buses and ships.

Traditional CNNs are designed based on the assumption of translation invariance in input data. However, for objects with rotational symmetry, the extracted features may change, resulting in the loss of important information and poor training outcomes. E2-CNNs provide a solution to this problem. Their core is rooted in group theory and convolution operations, enabling them to extract features from an image in multiple directions simultaneously, thereby ensuring rotation equivariance throughout the convolution process. This property ensures that objects yield the same feature output regardless of their orientation or angle. Since rotation involves a continuous operation that entails substantial floating-point computation and information storage, it is impractical and inefficient to include the entire 2D rotation group in the computation. Thus, optimization becomes necessary in this context.

The backbone network employed in our model, ReResNet [21], is constructed based on E2-CNNs. Specifically, the 2D rotation group is discretized initially, establishing a discretized 2D rotation group with eight discrete parameters, representing eight rotation angles. This discretization significantly reduces computational complexity and memory resource consumption. Subsequently, each convolutional kernel extracts features based on this discretized 2D rotation group, resulting in feature maps (K, N, H, W) as depicted in Figure 2, where K denotes the number of convolutional kernels, N represents the number of directions (eight), and H and W denote the height and width of the feature maps, respectively.

2.2. Rotation-Equivariant Bidirectional FPN

In the feature extraction network, lower-level features contain more detailed and positional information but have lower resolution. On the other hand, higher-level features have higher resolution and more semantic information but lack detail. In ReDet's ReFPN module, inspired by the traditional FPN [27], a top–down feature fusion approach is adopted as shown in Figure 3. However, the traditional FPN has a relatively simple feature fusion method, performing simple feature upsampling from high to low levels. To provide a more intuitive understanding, we simplify the convolution layer and upsampling layer, representing them with Equation (1):

$$P_i = \sum_{n=i}^{l_{max}} C_n (i = 2, 3, 4, 5), \tag{1}$$

here, C_n denotes the feature map extracted by the backbone network, P_i represents the feature map output by the FPN module, and i_{max} is the total number of layers inputted into FPN from the backbone network. The output P_i fuses only the feature maps from layer i to i_{max}, ignoring the importance of feature maps below layer i and not fully utilizing the feature information from different levels, which may lead to information loss. Typically, top-level feature maps with large strides are used to detect large objects, while bottom-level feature maps with small strides are used to detect small objects [28]. However, in the case of P_3, if only the features of C_5, C_4, and C_3 from top to bottom are fused, the importance of C_2, dedicated to detecting small objects, is ignored. Additionally, both P_5 and P_6 are extracted from the C_5 layer inputted into the backbone network, but P_6 only uses 1×1 convolution downsampling with stride 2, which may result in the loss of significant feature information. In summary, the FPN design in the ReDet model suffers from unbalanced feature fusion, which may affect the detection performance of small objects in the overall model.

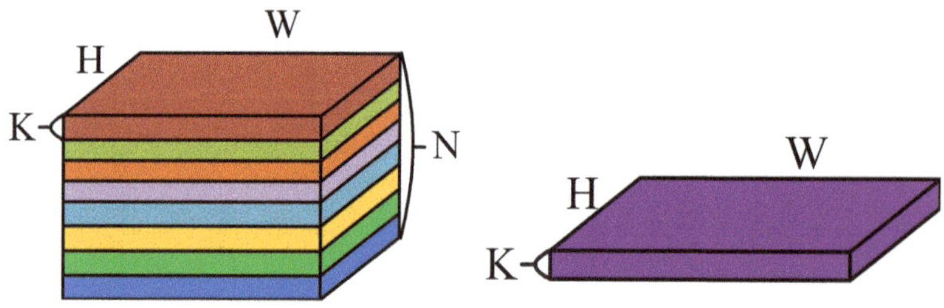

(**a**) (**b**)

Figure 2. Difference between the feature maps extracted by ReResNet and ResNet networks. (**a**) Feature maps extracted by the ReResNet network, denoted as (K, N, H, W), where N is an additional dimension compared to traditional ResNet feature maps shown in (**b**). Here, N = 8 indicates that discrete rotation group convolution can extract feature maps at 8 different rotation angles.

To address these issues, we introduce the rotation-equivariant bidirectional feature fusion feature pyramid network module, named ReBiFPN, based on the idea of PaNet [29]. Figure 4 illustrates the ReBiFPN module. ReBiFPN balances the output levels by integrating the feature maps of each resolution. It combines the high-resolution of the high-level features with the high-detail information of the low-level features, enhancing the network's ability to extract and integrate multi-scale features through bidirectional path feature fusion. This approach captures richer multi-scale features and improves the detection performance of small objects. For simplicity, we represent the core idea of ReBiFPN with Equation (2):

$$P_i = \sum_{n=l_{min}}^{l_{max}} C_n (i = 2, 3, 4, 5), \tag{2}$$

where C_n represents the n_{th} feature map extracted by ReResNet, P_i denotes the feature map output by the ReBiFPN module, l_{max} is the total number of layers inputted into ReBiFPN from ReResNet, and l_{min} is the number of the lowest layer inputted into ReBiFPN from ReResNet. Each level of the output P_i balances the features fused from $\{C_2, C_3, C_4, C_5\}$. The P_6 layer is excluded from the feature fusion because experimental results indicate that downsampling directly from the C_5 layer to obtain P_6 improves the detection accuracy slightly compared to downsampling from the P_5 layer. This improvement may be due to the weakening of object features after multiple convolution operations. A 3×3 convolution is added before all the final output feature maps $\{P_2, P_3, P_4, P_5\}$ to reduce the aliasing effect of upsampling.

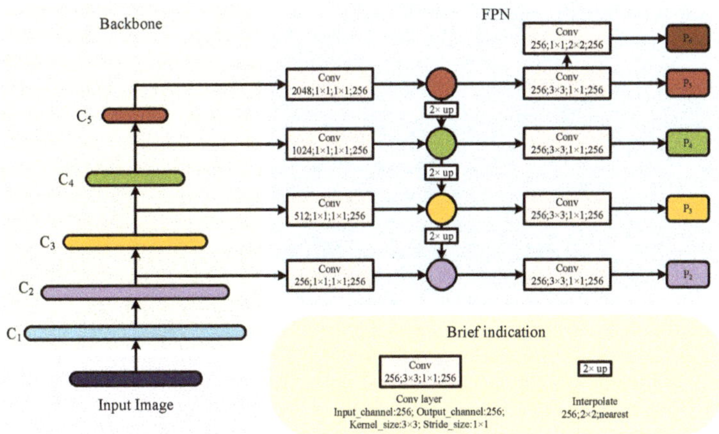

Figure 3. Traditional feature pyramid network (FPN) structure schematic.

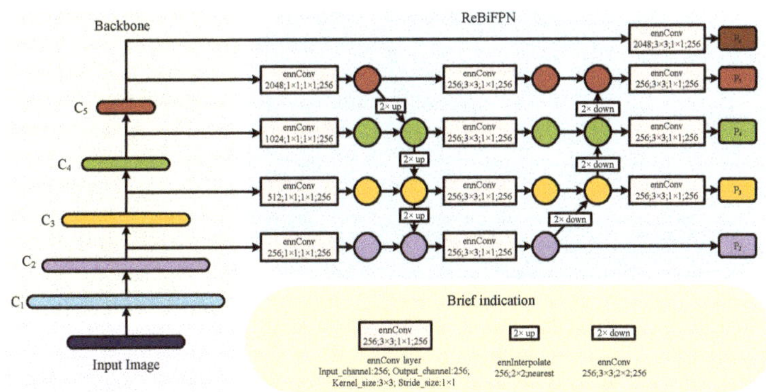

Figure 4. Structure diagram of rotation-equivariant bidirectional feature fusion feature pyramid network (ReBiFPN), where all convolution and interpolation operations are based on E(2)-equivariant steerable CNNs [22] (E2-CNNs).

Our proposed ReBiFPN achieves the following objectives:

(1) Multi-scale information fusion: By leveraging the structure of the feature pyramid, we extract features at different scales. The fusion of features from various scales enhances the network's receptive field, enabling it to detect ships of different scales. This capability allows the network to better adapt to variations in ship size and shape, thereby improving the robustness of ship detection.

(2) Bi-directional feature propagation: The use of a bi-directional feature propagation mechanism facilitates the exchange and interaction of information between different levels of the feature pyramid. This bidirectional propagation enables high-level semantic information to propagate to lower levels, enriching the semantic representation of lower-level features and enhancing their expressive capacity. Simultaneously, lower-level features can also propagate to higher levels, providing more accurate positional information, which aids in precise target localization.

(3) Multi-level feature fusion: We adopt a multi-level feature fusion strategy that progressively merges features from different layers. This approach allows the network to fuse semantic and positional information at different levels, resulting in more comprehensive

and accurate feature representations. Through multi-level feature fusion, the finer details and contextual information of ships can be better captured, thereby improving the accuracy of ship detection.

From a theoretical standpoint, our proposed ReBiFPN demonstrates superior adaptability to scale variations compared to FPN. It effectively extracts richer semantic information and accurately localizes targets, ultimately leading to improved precision in object detection. Subsequent experiments provide further validation of these findings.

2.3. Anchor Improvement

ReBiDet is a two-stage object detection model that follows an anchor-based approach, where anchors are predefined boxes placed at various positions in an image [13]. The selection of anchor sizes and aspect ratios should be based on the distribution of object annotation boxes in the dataset, which can be considered as prior knowledge [30]. Improper anchor settings can hinder the network from effectively learning features of certain objects or result in the misidentification of adjacent objects. Therefore, it is crucial to carefully consider the size and shape distribution of objects in the training dataset and experiment with different numbers and ratios of anchors to find the most suitable settings for improving the network's detection performance. To accomplish this, we employ the K-means algorithm to group the aspect ratios of ship bounding boxes in the HRSC2016 dataset into multiple clusters and determine anchor sizes based on the average aspect ratio of each cluster.

The K-means algorithm, presented in Algorithm 1, is an unsupervised learning technique. It starts by taking a set of aspect ratio data from the HRSC2016 dataset, randomly selecting K (set to 5 in our case) initial cluster centers. Each data point is then assigned to the nearest cluster center, and the cluster centers are iteratively recalculated and adjusted until they converge or reach a preset number of iterations. The objective is to maximize similarity within each cluster while minimizing similarity between clusters.

Algorithm 1 K-means used to cluster the aspect ratios of ground truth boxes.

Input: Set of bounding box aspect ratios $R = r_1, r_2, \ldots, r_n$, number of clusters K
Output: Set of K cluster centers $m_1, m_2, \ldots, m_K$ and cluster labels for each bounding box aspect

Randomly initialize K centroids $m_1, m_2, \ldots, m_K$ from R
repeat
    for $i = 1$ to n **do**
        Assign r_i to the nearest centroid m_j, j in $[1, K]$
    for $j = 1$ to K **do**
        Recompute m_j as the mean of all r_i assigned to centroid m_j
until convergence or maximum number of iterations is reached
return Set of K cluster centers $m_1, m_2, \ldots, m_K$ and cluster labels for each bounding box aspect

Figure 5 illustrates the aspect ratio data of bounding boxes in the HRSC2016 dataset, grouped into five categories. The deep blue dots represent the cluster centroids of each aspect ratio group, which are {3.9, 5.8, 6.2, 6.3, 7.3}. Although these centroids span from 3.9 to 7.3, they are concentrated around 6. This suggests that the aspect ratios of objects in the HRSC2016 dataset tend to cluster around the value of 6. After conducting multiple experiments, aspect ratios of 7.3 and above did not yield satisfactory results in practical detection. Analysis indicates that such aspect ratios are too large. Although they appear in the clustering results, many ground truth boxes in the dataset are not strictly horizontal or vertical, but rather arbitrary. Large aspect ratios result in low intersection over union (IoU) values between anchor boxes and ground truth boxes, thus failing to enhance the model's performance and introducing numerous negative samples that adversely impact performance. Consequently, we set the ratios of the RPN anchor generator as {1/6, 1/4, 1/2, 1, 2, 4, 6}.

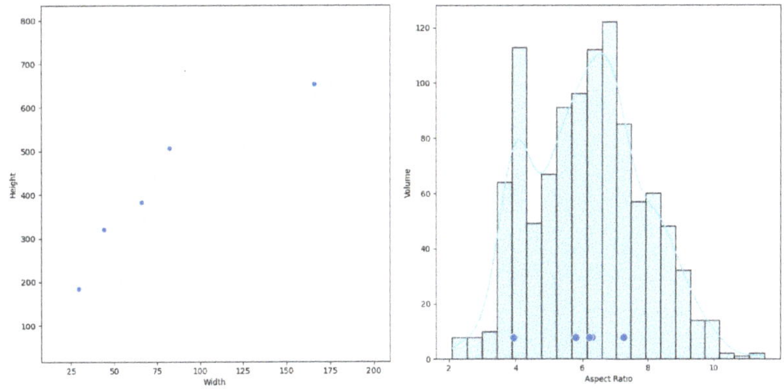

(a) (b)

Figure 5. Aspect ratio distribution and K-means clustering results of labeled boxes in the HRSC2016 dataset. (**a**) Scatter plot of the distribution of the length and width values of all labeled boxes. (**b**) Histogram of the aspect ratio distribution of all labeled boxes.

2.4. *Difficult Positive Reinforcement Learning Sampler*

The RPN module generates anchor boxes of varying sizes and aspect ratios on the feature map. It assigns positive and negative samples to these anchors based on the IoU threshold with the ground truth boxes. Specifically, an anchor box with an IoU greater than 0.7 is labeled as a positive sample, while an IoU less than 0.3 results in a negative sample. Any anchors falling between these thresholds are ignored. However, training the model with all candidate boxes becomes impractical due to their large number. To expedite the training process, ReDet adopts a random sampler that randomly selects a small subset of candidate boxes for training [21]. Nevertheless, the random sampler has notable drawbacks. The ratio of positive to negative samples is often imbalanced, and the random sampling strategy may excessively emphasize easy samples, impeding the model's ability to learn from challenging samples such as small objects and overlapping objects. This not only affects the learning efficiency but also hampers the final detection accuracy.

Inspired by the concept of online hard example mining (OHEM) sampler [31], we propose selecting the candidate boxes with the highest loss for positive and negative samples during training to address the aforementioned issues with the random sampler. However, in practice, the distribution of negative samples is exceptionally complex, as depicted in Figure 6. During the sample selection stage in remote sensing ship detection, some anchor boxes may contain ship features but fall below the preset IoU threshold, while others may contain ships that were not annotated by the dataset's original author. Despite these cases being classified as negative samples with high classification loss, they actually contain object features and belong to the category of hard samples. Prioritizing such negative samples with high classification loss during training could significantly affect the detection accuracy as confirmed by our experiments. As it is not possible to entirely avoid incorrect answers, we focus on learning the correct answers presents a more viable solution. Furthermore, since the number of negative samples typically exceeds the number of positive samples, randomly sampling negative samples may have a higher likelihood of selecting samples that only contain background and lack ship instances. Based on these observations, we propose the difficult positive reinforcement learning (DPRL) sampler to address this problem. The DPRL sampler concentrates on learning challenging positive samples to enhance the model's performance, while continuing to employ the traditional random sampling approach for negative samples.

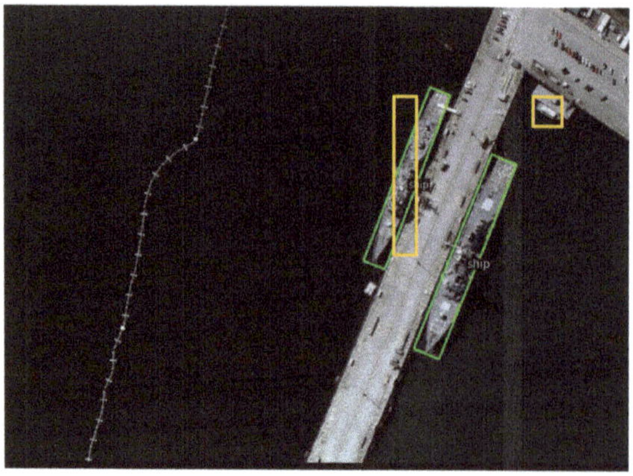

Figure 6. Illustration of the intersection over union (IoU) between anchor boxes and ground truth boxes. The green box represents the ground truth box, while the yellow box represents the anchor box generated by the model with a 1:6 ratio.

Algorithm 2 outlines the calculation process of the DPRL sampler, which can be summarized as follows. For each batch of training samples $D_{batchsize}$, the forward propagation is initially performed to generate a set of candidate boxes using RPN, followed by loss calculation. The DPRL sampler maintains positive and negative samples in L_{pos} and L_{neg}, respectively. The samples in L_{pos} are then sorted in descending order based on the loss value, and the top $pos_{expected}$ samples are selected. Simultaneously, $neg_{expected}$ negative samples are randomly chosen from L_{neg}. These selected samples are combined with L_{pos} to form a new sample set L, which is used for backward propagation and updating the neural network parameters. This method intelligently selects the most challenging positive samples for training, thereby mitigating the issue of neglecting difficult samples and effectively enhancing the training efficiency of the neural network.

Algorithm 2 Difficult positive reinforcement learning (DPRL) sampler.

Input: image set $D_{batchsize}$, samples expected $num_{expected}$, positive sample ratio pos_{ratio}
Output: The collection of selected samples L

$pos_{expected} = num_{expected} \times pos_{ratio}$
$neg_{expected} = num_{expected} \times (1 - pos_{ratio})$
Clear d_{pos} and d_{neg}
for j in range[len($D_{minibatch}$)] **do**
  RPN generate a set of positive samples $mini_{pos}$
  RPN generate a set of negative samples $mini_{neg}$
  $d_{pos} = d_{pos} + mini_{pos}$
  $d_{neg} = d_{neg} + mini_{neg}$
if $d_{pos} \leqslant pos_{expected}$ **then**
  $L_{pos} = d_{pos}$
else
  calculate the loss value for each sample in d_{pos}
  sort all elements in d_{pos} in descending order of their classification loss value
  extract the top $pos_{expected}$ samples, and form a new positive sample set L_{pos}
if $dpos \leqslant neg_{expected}$ **then**
  $L_{neg} = d_{neg}$
else
  randomly shuffle the order of each sample in d_{neg}
  extract $neg_{expected}$ samples, and form a new negative sample set L_{neg}
merge L_{pos} and L_{neg} into L
return L

3. Experimental Results

3.1. Datasets

HRSC2016 [32] was released by Northwestern Polytechnic University in 2016 and is a unique dataset focusing on ship detection in remote sensing images. It consists of a total of 1680 images captured from 6 renowned ports available on Google Earth. Among these images, 619 do not contain any ships and serve as background images. The image sizes range from 300×300 to 1500×900. The dataset is divided into training, validation, and test sets, with 436, 181, and 444 images, respectively. The annotation format for ship bounding boxes is oriented bounding boxes (OBB) rather than horizontal bounding boxes (HBB), and a total of 2976 ships are annotated. HRSC2016 has become one of the most widely used benchmark datasets in remote sensing detection. The dataset presents several challenges for ship detection: (1) The majority of ships are docked closely together, exhibiting a dense arrangement of ships side by side. (2) Some ships are integrated with long piers or located in shipyards or ship repair yards, where the similarity between the ships and the nearby textures is high, making their features less distinguishable. (3) Certain land buildings bear resemblance to luxury yachts, leading to potential misclassification. (4) The scale of ships varies greatly, and an image may contain an aircraft carrier as well as a very small civilian ship simultaneously. The dataset provides annotations for ship classification at three levels; however, due to their low frequency, detecting and classifying ship subcategories is challenging. Therefore, our RePaDet model focuses only on training and inferring the first-level ship category, following the approach of most researchers.

DOTA [4,33], publicly released by Wuhan University in November 2017, has become the most commonly used public dataset for object detection in remote sensing. DOTA-v1.0 [4] consists of 2806 images obtained from various platforms, including Google Earth, JL-1, and GF-2 satellites. The dataset contains a total of 188,282 annotated objects belonging to 15 different categories, as represented by the abbreviations in the tables presented later: airplane (PL), baseball diamond (BD), bridge (BR), ground track field (GTF), small vehicle (SV), large vehicle (LV), ship (SH), tennis court (TC), basketball court (BC), storage tank (ST), soccer-ball field (SBF), roundabout (RA), harbor (HA), swimming pool (SP), and helicopter (HC). Unlike HRSC2016, the annotation boxes in DOTA-v1.0 are quadrilaterals defined by four points $\{x_1, y_1; x_2, y_2; x_3, y_3; x_4, y_4\}$. DOTA-v1.5 [33], an extension of DOTA-v1.0, increases the number of bounding boxes to 402,089. It includes additional annotations for small objects (around 10 pixels or less) that were missed in DOTA-v1.0 and introduces a new category, container crane (CC), making it a more challenging dataset. In addition to the challenges faced by the HRSC2016 dataset, DOTA presents additional difficulties, including (1) a wide range of image pixel sizes, from 800×800 to 4000×4000, making it challenging to train on GPUs such as GTX 3090ti, and (2) small object sizes, with 98% of the objects in the dataset being smaller than 300 pixels, and 57% smaller than 50 pixels [4], resulting in significant scale differences between tiny and large objects, thereby complicating detection. It is worth noting that the DOTA dataset does not release annotations for the test set; instead, researchers are required to upload their detection results for evaluation, which limits comprehensive analysis of experimental results.

3.2. Setup

The experimental setup employed a workstation equipped with an AMD Threadripper PRO 5995WX processor, 2 NVIDIA GTX3090Ti graphics cards, and 128 GB of memory. The workstation ran on the Ubuntu20.04 operating system and was installed with various software, including CUDA 11.8, Python 3.8, PyTorch 1.11.0 [24], and torchvision 0.12.0. The MMRotate 0.3.2 [34] framework was used for the entire experimental process and result generation. Both the hardware and software versions were relatively new and exhibited strong compatibility, allowing for easy replication in various environments.

During training, data augmentation techniques, such as horizontal and vertical flips, were applied, with a batch size of 2. The entire network was optimized using the stochastic gradient descent algorithm, with a momentum of 0.9 and a weight decay of 0.0001. The IoU threshold for positive samples in the RPN was set to 0.7. The horizontal NMS threshold was set to 0.7, and the oriented NMS threshold was set to 0.1.

For the HRSC2016 dataset, the longer side of the images was adjusted to 1280 pixels, while maintaining the aspect ratio. If the shorter side was larger than 1280 pixels, it was reduced to 1280 pixels as well. During training, the blank regions were filled with black, and the model was trained for 12 epochs. The initial learning rate was set to 0.01, which was divided by 10 at epoch 9 and epoch 11.

For the DOTA dataset, the original images were cropped to a size of 1024 × 1024, with a stride of 824 pixels, resulting in overlapping patches by 200 pixels. RePaDet was trained for 12 epochs with an initial learning rate of 0.01, which was also divided by 10 at epoch 9 and 11. Figure 7 illustrates the loss curve of our model during training on the DOTA dataset, with 3200 iterations per epoch. A significant decrease in loss was observed at the 8th epoch (25,600 iterations). To ensure a fair comparison with other models, multi-scale training was conducted by randomly cropping and rotating the dataset at three scales {0.5, 1.0, 1.5} for both training and testing.

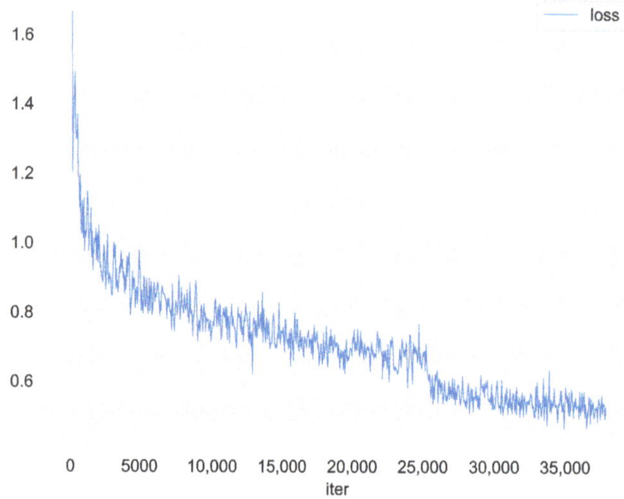

Figure 7. Train loss curve of ReBiDet on DOTA.

3.3. Results and Analysis

3.3.1. Analysis Based on HRSC2016 Dataset

Firstly, we conducted ablation experiments on the HRSC2016 dataset to evaluate the aforementioned modules. The average precision (AP) was used as the main evaluation metric, calculated using the 11-interpolated precision method proposed in VOC2007 [35]. AP50 represents the AP value at an IoU threshold of 0.5. As shown in Equation (3),

$$AP50 = \frac{1}{11} \sum_{r \in \{0,0.1,\ldots,1\}} \max_{\widetilde{r}:\widetilde{r} \geqslant r} p(\widetilde{r}), \tag{3}$$

where r is the recall rate and $p(\widetilde{r})$ is the precision at recall rate $\widetilde{r}$. AP75 represents the AP value at an IoU threshold of 0.75. Its calculation formula is the same as that of AP50, except that the IoU threshold is changed to 0.75. AP50 and AP75 are commonly used to evaluate model performance [36]. The mean average precision (mAP) is the average AP calculated at different IoU thresholds ranging from 50% to 95% with a step size of 5%. Its calculation formula is given in Equation (4):

$$mAP = \frac{1}{|T|} \sum_{t \in T} AP_t \tag{4}$$

here, T represents the set of IoU thresholds, and AP_t is the average precision at IoU threshold t. This calculation method was initially used in the COCO detection challenge (bounding box) and is now commonly used in object detection evaluation [36].

Table 1 presents the performance of each module in ReDet. ReDet represents the experimental results reported by the authors, while ReDet* represents our best performance achieved in the MMRotate framework [34] for comparison. The three added modules showed varying degrees of improvement compared to the baseline. Although AP50 showed a small improvement and even a 0.1% drop after the DPRL module was added, mAP significantly improved. This suggests that the accuracy of the detection results under high IoU thresholds is greatly improved. Our ReBiDet model achieved a 1.27% improvement in mAP compared to the baseline, demonstrating the effectiveness of our proposed method.

Table 1. The ablation study results on the HRSC2016 dataset. Noted that mean average precision (mAP) refers to the average precision over all possible IoU thresholds. AP50 refers to the average precision at an IoU threshold of 0.5, and AP75 refers to the average precision at an IoU threshold of 0.75.

Method	ReBiFPN	DPRL Sampler	Anchor Improvement	mAP	AP50	AP75
ReDet	-	-	-	70.41 [1]	90.46 [1]	89.46 [1]
ReDet *	-	-	-	72.34 [2]	90.30 [2]	88.50 [2]
	✓	-	-	72.80	90.40	89.20
	-	✓	-	72.86	90.20	89.10
	-	-	✓	72.41	90.40	89.10
ReBiDet (ours)	✓	✓	✓	73.61	90.50	89.80

* Represents the detection results achieved by reproducing this model on the experimental platform we built. [1] The detection results published by the original paper authors. [2] The best result we obtained by reproducing ReDet. However, there are some differences from the results of the original authors, which cannot be ignored.

We further analyzed the detection results of the ReBiDet model on the HRSC2016 dataset and generated a confusion matrix as shown in Figure 8. The test set contained 1228 annotated ground truth boxes. Our model correctly detected 1178 ships, missed 50 ships, and misclassified 106 background images as ships. We were surprised by the misclassification of the background images, but upon visualizing the results, we discovered that incomplete annotations in the HRSC2016 dataset might be the main reason for this. The original author did not fully annotate all objects that should have been labeled as ships in this dataset. Incomplete annotations in the test set can negatively impact model performance during testing. Figure 9 illustrates that the model detected some ships that were not fully labeled during testing. However, due to the incomplete annotation information, the validation program could not accurately determine if these ships were correctly detected, leading to their misclassification as false positives and a decrease in the model's mAP.

Considering these findings, we thoroughly examined the HRSC2016 test set and manually re-annotated a portion of the missing targets, adding 80 new target annotations to the existing 1228 annotations. We then re-evaluated the detection performance of ReBiDet on the newly annotated test set. To ensure fairness, all compared models were baseline models from the MMRotate framework, replicated by the authors based on the original papers and the HRSC2016 dataset. Among them, S^2A-Net-TO [18] was the best model obtained after adjusting the hyperparameters by us. The final experimental results are presented in Table 2. On the new test set, ReBiDet showed slight improvements in mAP, AP50, and AP75. In terms of target detection statistics, at an IoU threshold of 0.5, ReBiDet correctly detected 1256 targets, missed 52 targets, and misdetected 29 targets. Its precision, recall, and accuracy were significantly better than those of ReDet and other detection models.

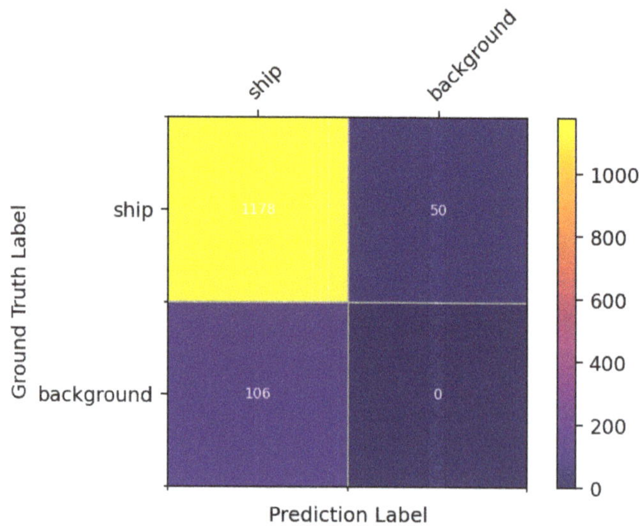

Figure 8. Confusion matrix of the detection results of the ReBiDet model on the HRSC2016 dataset. The numbers in the figure refer to the number of targets.

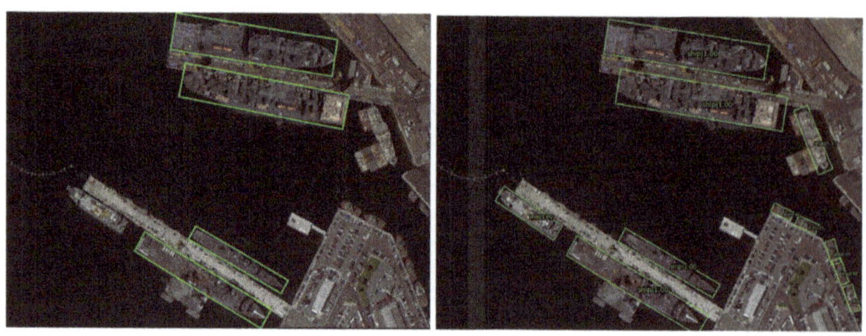

(**a**) (**b**)

Figure 9. Comparison between the ground truth label and detection result of an image in the test set of HRSC2016. (**a**) Image annotated with the ground truth label. (**b**) Our detection result on the same image.

Table 2. Comparison of detection results on the re-annotated HRSC2016 test set. All models compared are baseline models from the MMRotate framework.

Method	mAP	AP50	AP75	TP	FN	FP	TN	Precision	Recall	Accuracy
RetinaNet-O [37]	52.51	86.80	59.20	997	311	67	0	93.70	76.22	72.51
RetinaNet-O [37] + KLD [38]	55.43	88.70	62.00	1054	254	36	0	96.70	80.58	78.42
S^2A-Net [18]	55.31	89.80	58.90	1185	123	118	0	90.94	90.60	83.10
S^2A-Net-TO [18]	61.00	90.20	71.10	1185	123	63	0	94.95	90.60	86.43
ReDet [21]	72.46	90.70	89.70	1243	65	44	0	96.58	95.03	91.94
ReBiDet (ours)	73.66	90.80	90.00	1256	52	29	0	97.74	96.02	93.94

Note. The model configurations and parameters of RetinaNet OBB (RetinaNet-O) [37], RetinaNet-O [37]+ KLD [38], S^2A-Net [18], and ReDet [21] were all trained by the MMRotate team based on HRSC2016. We did not make any modifications to them. S^2A-Net-TO is a model that we optimized based on the MMRotate framework.

In terms of Precision, ReBiDet outperformed S^2A-Net by 6.8% at an IoU threshold of 0.5, but the difference in AP50 was only 1%. This can be attributed to the lower precision of ReBiDet at certain recall rates, affecting its overall AP50. Compared to RetinaNet + KLD, ReBiDet had only a 1.04% higher precision, but its precision at certain recall rates was much higher, resulting in a 2.1% higher AP50 than RetinaNet+KLD. This demonstrates that AP and precision are not directly related [36]. AP is a more comprehensive indicator, as it considers multiple factors and evaluates the model's performance under different conditions [36]. Compared to the baseline model ReDet, ReBiDet consistently outperformed in AP50, AP75, and mAP, indicating its superior detection accuracy under high IoU threshold conditions. This further validates the superiority of our proposed ReBiDet model.

3.3.2. Analysis Based on DOTA-v1.0

Earlier, we noted that the authors of DOTA-v1.0 did not provide ground truth annotation files for the test set. Instead, participants were required to submit their test results to a designated server, receiving mAP and AP values for each class as feedback. However, these metrics were insufficient for a comprehensive performance analysis. In our experiments, we trained ReBiDet and ReDet models on the training set and evaluated them on the validation set, which was not used for training. The MMRotate framework was employed for training and evaluation. The ReDet model's learning rate was set to 0.01, consistent with ReBiDet. Other settings for ReDet remained at their default values. To facilitate a more intuitive comparison of detection performance, we generated confusion matrices for both models. Due to the large number of objects per class in the test set, we normalized the confusion matrix rows to present the correct and false recognition rates as percentages. The diagonal line in the matrix represents the proportion of true positive (TP) detections, while the rightmost column represents the number of false negative (FN) detections. Other areas denote false positive (FP) detections. Figure 10 illustrates that ReBiDet achieved a 92% TP rate for the ship class, surpassing ReDet by 1%, while also exhibiting a 7% FN rate, 1% lower than ReDet. For other classes, ReBiDet demonstrated either higher or equivalent TP numbers compared to ReDet, along with generally significantly lower FN numbers. This further confirms the effectiveness of our proposed modules.

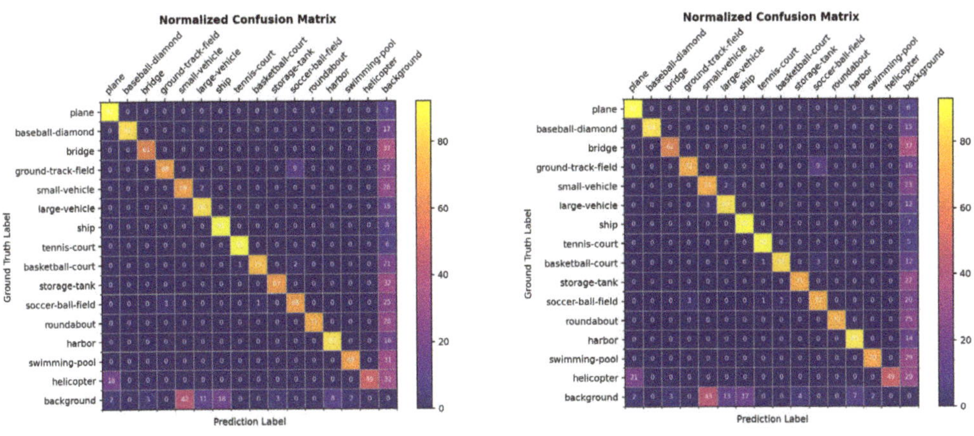

(a) (b)

Figure 10. Normalized confusion matrix on the DOTA dataset. (**a**) The detection results of ReDet reproduced by us. (**b**) The detection results of ReBiDet. The unit of the numbers in the figure is percentage (%), representing the percentage of ground truth label targets.

Figure 11 compares the ground truth labels of some validation set images with the detection results of the ReBiDet model trained solely on the training set. The images demonstrate the superior ability of ReBiDet to detect objects not annotated in the original dataset. Figure 11a,b showcase the strong performance of ReBiDet in detecting densely packed vessels, while Figure 11c,d exemplify its recognition capability for vessels in motion. Figure 11e,f,k,l demonstrate the effectiveness of ReBiDet in detecting small objects with indistinct features. Furthermore, Figure 11g–j demonstrate the robust detection performance of ReBiDet, even in complex scenes with cloud and mist interference.

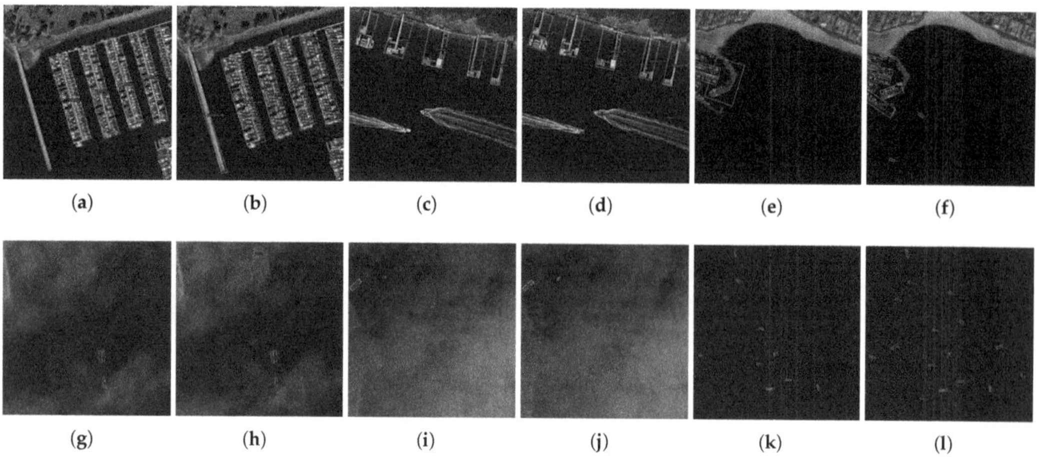

Figure 11. Comparison between the ground truth label and detection result of images in the validation set of DOTA-v1.0. (**a**,**c**,**e**,**g**,**i**,**k**) are image annotated with the ground truth label. (**b**,**d**,**f**,**h**,**j**,**l**) are our detection result on the same image.

Additionally, when comparing ReBiDet with other models on a single RTX4090 computing platform, ReBiDet exhibited a slightly higher computational parameter count while achieving superior performance. Table 3 presents the comparison of ReBiDet with other models in terms of computational parameters, inference speed, and other dimensions. Within the MMRotate framework, ReBiDet had 32.56 million computational parameters, merely 0.92 M more than the 31.64 M of ReDet. This less than 1 M increase resulted in an additional 2.5 ms per image during model inference, corresponding to a decrease of 1.4 img/s in FPS. Thus, ReBiDet incurred a small computational cost, while significantly improving detection accuracy. The table reveals that computational parameters and inference speed are not completely correlated, with models having similar parameters but different inference speeds, and vice versa. Specifically, although ReBiDet had a relatively small parameter count, its inference speed was not fast enough. This may be attributed to the backbone network ReResNet50 and ReBiFPN, which are constructed based on E2CNN. These networks utilize rotation-equivariant convolutional layers, involving additional operations, such as rotation, interpolation, and grouping, which can reduce the calculation speed. Additionally, unlike traditional ResNet employing PyTorch's built-in convolutional layers, ReResNet50 and ReBiFPN require calling the *e2cnn* library for rotation-equivariant network calculations, potentially affecting the computational efficiency.

Table 3. Comparison of ReBiDet with other models on computational parameters and inference speed.

Method	Backbone	SH	mAP	Params	FPS	TPI (ms)
RoI Transformer * [23]	ResNet-101	83.59 [1]	69.56 [1]	74.12	30.00	33.30
R^3Det [39]	ResNet-152	78.21 [1]	73.74 [1]	76.54	20.40	49.10
S^2A-Net [18]	ResNet-50	87.25 [1]	74.12 [1]	87.25	32.10	31.20
Oriented R-CNN [20]	ResNet-50	88.20 [1]	75.87 [1]	41.14	39.30	25.40
Oriented R-CNN [20]	ResNet-101	87.52 [1]	76.28 [1]	60.13	32.00	31.30
ReDet [21]	ReResNet-50	88.04 [1]	76.25 [1]	31.64	24.70	40.50
ReBiDet	ReResNet-50	88.41	76.50	32.56	23.30	43.00

* indicates multi-scale training and testing. [1] The detection results published by the original paper authors. Note. SH refers to the AP of the ship category. Params refers to the computational parameter quantity of the model, in million units. FPS refers to the number of images detected per second by the model during inference. TPI is an abbreviation for times per image.

It is important to note that ReBiDet also exhibits excellent generalization performance. Figure 12 presents the detection results when applying the ReBiDet model trained on the DOTA dataset to image from the HRSC2016 dataset. The ReBiDet trained on the DOTA dataset outperformed the one trained on HRSC2016, showing improved alignment between the labeled detection boxes and the actual targets. ReBiDet detected more ships, including three additional small-sized vessels. This improvement can be attributed to the HRSC2016 training dataset's incomplete annotation, which hinders model training completeness and leads to subpar detection performance. Additionally, the ReBiDet model underwent specific optimizations for ship detection tasks, particularly enhancing feature extraction capabilities through the ReBiFPN module. These improvements contribute to better generalization performance following thorough training.

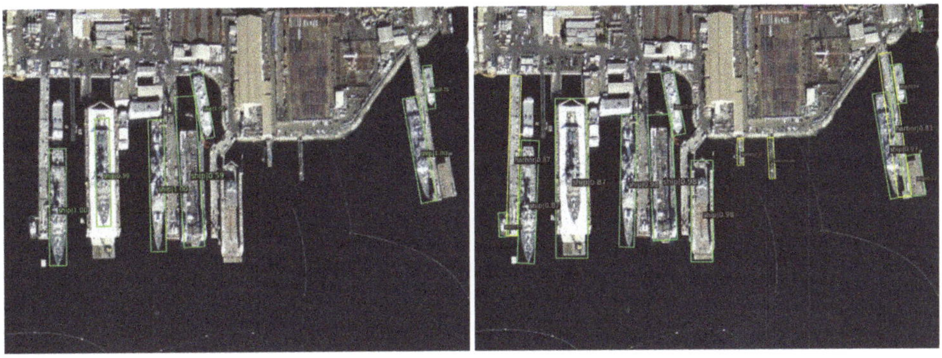

(**a**) (**b**)

Figure 12. Illustration of the generalization ability of the ReBiDet model trained on the DOTA dataset, using an original image from the HRSC2016 test set. Subfigure (**a**) displays the detection results obtained by the ReBiDet model trained on the HRSC2016 dataset. Subfigure (**b**) illustrates the detection results achieved by the ReBiDet model trained on the DOTA dataset.

To further assess the models' generalization capability, we selected an image from the FAIR1M [40] dataset. To ensure fairness, all compared models were baseline models from the MMRotate framework, reconstructed by the authors based on the original papers and the DOTA dataset. Figure 13 reveals that only ReBiDet and RoI Transformer [23] detected the small vessel on the left side of the image. However, RoI Transformer mistakenly identified three land objects as ships. Although R^3Det [39] detected more ships, most of the annotated detection boxes had confidence scores below 0.5. S^2A-Net [18] exhibited a similar situation, with confidence scores around 0.5 for the detected large ships, indicating uncertainty in the results and potentially affecting the AP metric. Oriented R-CNN [20] falsely detected a land object, and ReDet [21] failed to detect an adequate number of ships. For this particular image, ReBiDet demonstrated superior generalization capability.

(**a**) (**b**) (**c**)

Figure 13. *Cont.*

(d) (e) (f)

Figure 13. Comparison of the generalization capability of ReBiDet and other five models trained on the DOTA dataset, using an original image from the FAIR1M [40] dataset. Subfigures (**a–c**) depict the detection results of RoI Transformer [23], R^3Det [39], and S^2A-Net [18], respectively, while subfigures (**d–f**) represent the detection results of Oriented R-CNN [20], ReDet [21], and ReBiDet, respectively. With the exception of ReBiDet, all other models were replicated by the authors of the MMRotate framework based on the original papers and the DOTA dataset.

3.3.3. Results

This section primarily presents the performance test results of ReBiDet. We follow the practices commonly employed by researchers in the field of optical remote sensing image object detection. We compare the ReBiDet model with state-of-the-art models reported in recent years on two benchmark datasets, namely HRSC2016 and DOTA, to evaluate and compare its performance. The selected models are from reputable journals and conferences, and we directly quote their performance metrics without modification. The backbone networks used by these models are denoted as R50, R101, R152, H-104, and DLA34, representing ResNet-50, ResNet-101, ResNet-152, 104-layer hourglass network, and 34-layer deep layer aggregation network, respectively.

Detailed comparison results for the HRSC2016, DOTA-v1.0, and DOTA-v1.5 datasets are reported in Tables 4–6. On the HRSC2016 dataset, ReBiDet outperforms all existing models, demonstrating outstanding performance compared to the state-of-the-art approaches. In the single-scale detection task on DOTA-v1.0, ReBiDet surpasses all compared methods with a mAP of 76.50%, which is 0.22 higher than the baseline model. In most fine-grained categories, ReBiDet ranks within the top three, except for the ship category, where it achieves the highest AP. In the multi-scale detection task, ReBiDet exhibits various degrees of improvement compared to the single-scale detection results across all categories. ReBiDet also demonstrates improvement in the PL, BD, GTF, SV, SH, TC, ST, SBF, and HA categories compared to the baseline model, with first-place rankings in GTF, TC, ST, SBF, and HA categories. On the DOTA-v1.5 dataset, ReBiDet showcases superior performance compared to the baseline method. This advantage can be attributed to the presence of numerous small objects, approximately 10 pixels or smaller, in DOTA-v1.5, which ReBiDet excels at learning and detecting. Furthermore, even when compared to RTMDet (a research collaboration among Shanghai AI Laboratory, Nanyang Technological University, Tianjin University, Shanghai Jiaotong University, and Northeastern University) [41], ReBiDet exhibits certain advantages in terms of performance and computational parameters. These results indicates the significant capability of ReBiDet in detecting large and elongated objects with ship features. Notably, despite utilizing a 50-layer backbone network, our model outperforms all compared methods with backbone networks of 101 layers or more. Figure 14 presents some results for the DOTA dataset.

Table 4. Comparison results on the HRSC2016 dataset.

Method	Backbone	mAP	AP50	AP75
RRPN [16]	R50-FPN	-	79.08	-
R^2PN [42]	R50-FPN	-	79.60	-
RRD [43]	VGG-16	-	84.30	-
RoI Transformer [23]	R101-FPN	-	86.20	-
Gliding Vertex [44]	R101-FPN	-	88.20	-
R^3Det [39]	R101-FPN	-	89.26	-
CSL [45]	R152-FPN	-	89.62	-
S^2A-Net [18]	ResNet50	-	89.75	-
ReDet [21]	ReR50-ReFPN	70.41	90.46	89.46
ReBiDet (Ours)	ReR50-ReBiFPN	73.61	90.50	89.80

Table 5. Comparisons with state-of-the-art methods on DOTA-v1.0 OBB task.

Method	Backbone	PL	BD	BR	GTF	SV	LV	SH	TC	BC	ST	SBF	RA	HA	SP	HC	mAP
single-scale:																	
FR-O [4]	R101	79.42	77.13	17.70	64.05	35.30	38.02	37.16	89.41	69.64	59.28	50.30	52.91	47.89	47.40	46.30	54.13
ICN [46]	R101-FPN	81.36	74.30	47.70	70.32	64.89	67.82	69.98	90.76	79.06	78.20	53.64	62.90	67.02	64.17	50.23	68.16
CADNet [47]	R101-FPN	87.80	82.40	49.40	73.50	71.10	63.50	76.60	90.90	79.20	73.30	48.40	60.90	62.00	67.00	62.20	69.90
DRN [48]	H-104	88.91	80.22	43.52	63.35	73.48	70.69	84.94	90.14	83.85	84.11	50.12	58.41	67.62	68.60	52.50	70.70
CenterMap [49]	R50-FPN	88.88	81.24	53.15	60.65	78.62	66.55	78.10	88.83	77.80	83.61	49.36	66.19	72.10	72.36	58.70	71.74
SCRDet [50]	R101-FPN	89.98	80.65	52.09	68.36	68.36	60.32	72.41	90.85	87.94	86.86	65.02	66.68	66.25	68.24	65.21	72.61
R^3Det [39]	R152-FPN	89.49	81.17	50.53	66.10	70.92	78.66	78.21	90.81	85.26	84.23	61.81	63.77	68.16	69.83	67.17	73.74
S^2A-Net [18]	R50-FPN	89.11	82.84	48.37	71.11	78.11	78.39	87.25	90.83	84.90	85.64	60.36	62.60	65.26	69.13	57.94	74.12
Oriented R-CNN [20]	R50-FPN	89.46	82.12	54.78	70.86	78.93	83.00	88.20	90.90	87.50	84.68	63.97	67.69	74.94	68.84	52.28	75.87
ReDet [21]	ReR50-ReFPN	88.79	82.64	53.97	74.00	78.13	84.06	88.04	90.89	87.78	85.75	61.76	60.39	75.96	68.07	63.59	76.25
Oriented R-CNN [20]	R101-FPN	88.86	83.48	55.27	76.92	74.27	82.10	87.52	90.90	85.56	85.33	65.51	66.82	74.36	70.15	57.28	76.28
ReBiDet (Ours)	ReR50-ReBiFPN	89.36	83.77	53.04	71.55	79.01	83.56	88.41	90.89	87.31	86.00	65.45	61.79	76.73	70.30	60.36	76.50
multi-scale:																	
RoI Transformer * [23]	R101-FPN	88.64	78.52	43.44	75.92	68.81	73.68	83.59	90.74	77.27	81.46	58.39	53.54	62.83	58.93	47.67	69.56
O^2-Dnet * [51]	H104	89.30	83.30	50.10	72.10	71.10	75.60	78.70	90.90	79.90	82.90	60.20	60.00	64.60	68.90	65.70	72.80
DRN * [48]	H104	89.71	82.34	47.22	64.10	76.22	74.43	85.84	90.57	86.18	84.89	57.65	61.93	69.30	69.63	58.48	73.23
Gliding Vertex * [44]	R101-FPN	89.64	85.00	52.26	77.34	73.01	73.14	86.82	90.74	79.02	86.81	59.55	70.91	72.94	70.86	57.32	75.02
BBAVectors * [52]	R101	88.63	84.06	52.13	69.56	78.26	80.40	88.06	90.87	87.23	86.39	56.11	65.62	67.10	72.08	63.96	75.36
CenterMap * [49]	R101-FPN	89.83	84.41	54.60	70.25	77.66	78.32	87.19	90.66	84.89	85.27	56.46	69.23	74.13	71.56	66.06	76.03
CSL * [45]	R152-FPN	90.25	85.53	54.64	75.31	70.44	73.51	77.62	90.84	86.15	86.69	69.60	68.04	73.83	71.10	68.93	76.17
SCRDet++ * [53]	R152-FPN	88.68	85.22	54.70	73.71	71.92	84.14	79.39	90.82	87.04	86.02	67.90	60.86	74.52	70.76	72.66	76.56
S^2A-Net * [18]	R50-FPN	88.89	83.60	57.74	81.95	79.94	83.19	89.11	90.78	84.87	87.81	70.30	68.25	78.30	77.01	69.58	79.42
ReDet * [21]	ReR50-ReFPN	88.81	82.48	60.83	80.82	78.34	86.06	88.31	90.87	88.77	87.03	68.65	66.90	79.26	79.71	74.67	80.10
ReBiDet * (Ours)	ReR50-ReBiFPN	89.39	85.19	60.44	82.17	79.48	85.74	88.54	90.90	88.62	87.85	72.24	66.32	79.32	78.06	73.52	80.52

* indicates multi-scale training and testing. Note. For the convenience of reading and comparison, we bolded and marked the red, yellow and green numbers in the column as the 1st, 2nd, and 3rd largest values in that column, respectively.

Table 6. Comparisons with state-of-the-art methods on DOTA-v1.5 OBB task.

Method	PL	BD	BR	GTF	SV	LV	SH	TC	BC	ST	SBF	RA	HA	SP	HC	CC	mAP
single-scale:																	
RetinaNet-O [37]	71.43	77.64	42.12	64.65	44.53	56.79	73.31	90.84	76.02	59.96	46.95	69.24	59.65	64.52	48.06	0.83	59.16
FR-O [4]	71.89	74.47	44.45	59.87	51.28	68.98	79.37	90.78	77.38	67.50	47.75	69.72	61.22	65.28	60.47	1.54	62.00
Mask R-CNN [19]	76.84	73.51	49.90	57.80	51.31	71.34	79.75	90.46	74.21	66.07	46.21	70.61	63.07	64.46	57.81	9.42	62.67
HTC [54]	77.80	73.67	51.40	63.99	51.54	73.31	80.31	90.48	75.12	67.34	48.51	70.63	64.84	64.48	55.87	5.15	63.40
CMR [54]	77.77	74.62	51.09	63.44	51.64	72.90	79.99	90.35	74.90	67.58	49.54	72.85	64.19	64.88	55.87	3.02	63.41
DAFNe [55]	-	-	-	-	-	-	-	-	-	-	-	-	-	-	-	-	64.76
FR-O [4] + RT [23]	71.92	76.07	51.87	69.24	52.05	75.18	80.72	90.53	78.58	68.26	49.18	71.74	67.51	65.53	62.16	9.99	65.03
ReDet [21]	79.20	82.81	51.92	71.41	52.38	75.73	80.92	90.83	75.81	68.64	49.29	72.03	73.36	70.55	63.33	11.53	66.86
ReBiDet (Ours)	80.54	82.90	53.62	74.55	52.55	79.65	87.53	90.84	84.57	72.93	65.02	73.05	75.87	65.56	65.18	7.32	69.48
multi-scale:																	
DAFNe * [55]	-	-	-	-	-	-	-	-	-	-	-	-	-	-	-	-	71.99
OWSR * [56]	-	-	-	-	-	-	-	-	-	-	-	-	-	-	-	-	74.90
RTMDet-R-tiny [41]	88.14	83.09	51.80	77.54	65.99	82.22	89.81	90.88	80.54	81.34	64.64	71.51	77.13	76.32	72.11	46.67	74.98
RTMDet-R-s [41]	88.14	85.82	52.90	82.09	65.58	81.83	89.78	90.82	83.31	82.47	68.51	70.93	78.00	75.77	73.09	47.32	76.02
RTMDet-R-m [41]	89.07	86.71	52.57	82.47	66.13	82.55	89.77	90.88	84.39	83.34	69.51	73.03	77.82	75.98	80.21	42.00	76.65
ReDet * [21]	88.51	86.45	61.23	81.20	67.60	83.65	90.00	90.86	84.30	75.33	71.49	72.06	78.32	74.73	76.10	46.98	76.80
ReBiDet * (Ours)	86.23	85.89	61.99	82.41	67.86	83.94	89.78	90.88	86.37	83.70	72.12	77.58	78.38	73.24	75.01	52.05	77.96

* indicates multi-scale training and testing. Note. The results of RetinaNet OBB (RetinaNet-O) [37], Faster R-CNN OBB (FR-O) [4], Mask R-CNN [19] and Hybrid Task Cascade (HTC) [54] are re-implemented [33] version for DOTA, and used by some scholars [21,41]. For the convenience of reading and comparison, we bolded and marked the red, yellow and green numbers in the column as the 1st, 2nd, and 3rd largest values in that column, respectively.

Figure 14. Examples of detection results on the DOTA dataset using ReBiDet.

4. Discussion

The experimental results validate the effectiveness of our proposed ReBiDet model in ship detection tasks, indicating the efficacy of our designed methods and modules. ReBiFPN efficiently performs cross-fusion between deep features, which possess large receptive fields and strong semantic representation ability, and shallow features that lack spatial geometric feature details and have low resolution. This approach exhibits significant performance advantages compared to traditional feature pyramid networks. Each layer acquires more comprehensive and rich feature information, thereby facilitating downstream task execution and considerably improving the model's detection ability. The anchor improvement method aggregates features from annotated boxes in the dataset, adjusting the parameters of the anchor generator accordingly, thereby enhancing the model's detection capability for specific objects. The DPRL sampler module enables the model to focus on learning difficult positive samples, while avoiding the interference of false difficult negative samples. In summary, these three proposed methods or modules hold value in improving the ship detection capability in optical remote sensing images.

Nevertheless, it is important to acknowledge that any method or research inherently possesses certain limitations. Our research primarily falls within the qualitative category, and there is still ample room for exploration in the quantitative aspects, which represents our future direction. The ReBiFPN module, inspired by PANet [29], indeed yielded positive results. The EfficientDet research [57] demonstrates that stacking multiple bi-directional feature fusion paths further enhances the detection accuracy on the COCO dataset [58]. Moreover, the detection accuracy varies with the number of stacked layers. Hence, in our field of optical remote sensing detection, it is crucial to investigate whether a similar effect exists, which will be the focus of our forthcoming quantitative research. Regarding the anchor improvement method, we have only roughly aggregated the features of ships in the existing dataset and adjusted the parameters of the anchor generator, achieving a qualitative effect. To put it figuratively, this can be likened to "casting a wide net to catch more fish" (a common idiom). Additionally, the quantitative research lies in how to set the anchor generator reasonably and accurately to strike a balance between accuracy and computational complexity. After introducing the OHEM sampler module, we observed a decrease in detection accuracy instead of an increase. Through repeated experiments and in-depth analysis, we proposed the DPRL sampler module to enhance the learning of difficult positive samples, thereby beneficially improving the model's detection accuracy. As for low-loss positive samples, it raises the question of whether it is the correct choice to abandon direct learning. In terms of negative samples, ships that are only partially anchored and interference from land structures, roads, and vegetation are considered difficult negative samples. Although random sampling is a feasible approach, it remains uncertain whether more reasonable methods exist to address these challenges. We firmly believe that such methods do indeed exist.

For ship detection in optical remote sensing images, our research is still in its early stages, with quantitative research serving as the focal point of our future work. Concerning the issue of inaccurate detection accuracy evaluation due to incomplete object annotations in HRSC2016, we intend to manually re-annotate all ship targets in the HRSC2016 dataset in our next study. This will enable us to conduct a comprehensive comparison of various evaluation indicators based on the model's performance in existing detection models, ultimately leading to breakthroughs in improving detection performance. Furthermore, as illustrated in Figure 15, ReBiDet encounters an issue of inaccurate detection box positioning when detecting elongated objects such as ships. This issue primarily stems from inadequate training of the detection box during the alignment regression stage. We plan to explore potential solutions to this problem, including, but not limited to, modifying the loss function and the bounding box alignment regression module.

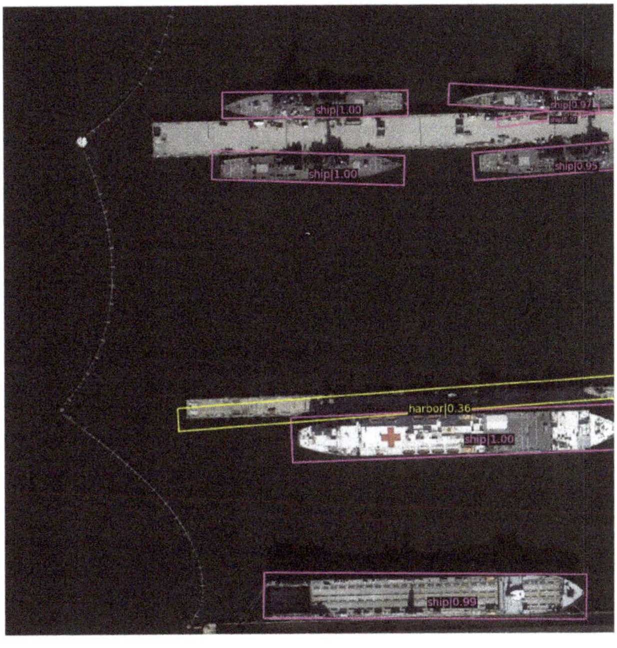

Figure 15. Illustration of inaccurate detection box positioning by ReBiDet.

5. Conclusions

This study is dedicated to addressing the ship detection problem in optical remote sensing images. The proposed ReBiDet, an enhanced version of ReDet, incorporates three improvements: ReBiFPN, DPRL sampler, and anchor improvement. These enhancements respectively bolster the network's ability to capture multi-scale features, learn difficult positive samples, and adapt to ship target aspect ratios. As a result, ReBiDet provides an effective and practically valuable solution for ship detection. The experimental results on HRSC2016 and DOTA datasets underscore the model's exceptional performance. Future research can concentrate on enhancing the model's performance by exploring more advanced network structures, loss functions, and optimization methods. Additionally, researchers can delve into additional optimization schemes for ship detection in remote sensing images to enhance accuracy and robustness. Moreover, it would be intriguing to explore the model's generalization capability on other datasets, such as FAIR1M [40].

Author Contributions: Methodology, Z.Y.; Software, Z.Y.; Supervision, Z.L. and Y.X.; Writing—Original Draft, Z.Y. and S.L.; Writing—Review and Editing, Z.Y., C.L. and Z.L.;Visualization, F.S. All authors have read and agreed to the published version of the manuscript.

Funding: This research received no external funding.

Data Availability Statement: The HRSC2016 dataset used in this study was publicly released by Northwestern Polytechnical University in 2016 and includes complete sets of training, validation, and testing images and annotation files. It can be downloaded at the following address: https://aistudio.baidu.com/aistudio/datasetdetail/54106 (accessed on 11 June 2023). The DOTA dataset was publicly released by Wuhan University in November 2017 and includes complete sets of training, validation, and testing images, as well as annotation files for the training and validation sets. It can be downloaded at this address: https://captain-whu.github.io/DOTA/dataset.html (accessed on 11 June 2023). Since the downloaded package does not include annotation files for the testing set, users need to submit their model's inference results to the Wuhan University official Evaluation Server to obtain the accuracy of the inference results, at this address: https://captain-whu.github.io/DOTA/evaluation.html (accessed on 11 June 2023). Finally, we published some of the raw experiment results for this study at https://github.com/YanZeGit/ReBiDet (accessed on 11 June 2023). After this article is published, our code and other related materials will also be released at this address.

Acknowledgments: This work was supported by the Institute of Systems Engineering, Academy of Military Sciences, and We would like to express our gratitude to Yongqiang Xie and Zhongbo Li, for their support and guidance throughout the project. Their contributions have been invaluable to the success of this study.

Conflicts of Interest: The authors declare no conflict of interest.

Abbreviations

The following abbreviations are used in this manuscript:

ReBiDet	Rotation-equivariant Bidirectional Feature Fusion Detector
CNN	Convolutional Neural Networks
R2CNN	Rotational R-CNN
E2-CNNs	E(2)-Equivariant Steerable CNNs
RoI	Region of Interest
RRPN	Rotated Region Proposal Network
RPN	Region Proposal Network
NMS	Non-Maximum Suppression
IoU	Intersection over Union
FPN	feature pyramid network
ReFPN	Rotation-equivariant Feature Pyramid Network
ReBiFPN	Rotation-equivariant Bidirectional Feature Fusion Feature Pyramid Network
OHEM	Online Hard Example Mining

DPRL	Difficult Positive Reinforcement Learning
mAP	mean Average Precision
OBB	Oriented Bounding Boxes
TP	True Positive
FN	False Negative
FP	False Positive

References

1. Feng, X.; Zhang, W.; Su, X.; Xu, Z. Optical Remote Sensing Image Denoising and Super-Resolution Reconstructing Using Optimized Generative Network in Wavelet Transform Domain. *Remote Sens.* **2021**, *13*, 1858. [CrossRef]
2. Zhou, H.; Tian, C.; Zhang, Z.; Li, C.; Ding, Y.; Xie, Y.; Li, Z. Position-Aware Relation Learning for RGB-Thermal Salient Object Detection. *IEEE Trans. Image Process.* **2023**, *32*, 2593–2607. [CrossRef] [PubMed]
3. Li, Q.; Mou, L.; Liu, Q.; Wang, Y.; Zhu, X.X. HSF-Net: Multiscale Deep Feature Embedding for Ship Detection in Optical Remote Sensing Imagery. *IEEE Trans. Geosci. Remote Sens.* **2018**, *56*, 7147–7161. [CrossRef]
4. Xia, G.S.; Bai, X.; Ding, J.; Zhu, Z.; Belongie, S.; Luo, J.; Datcu, M.; Pelillo, M.; Zhang, L. DOTA: A Large-Scale Dataset for Object Detection in Aerial Images. In Proceedings of the 2018 IEEE/CVF Conference on Computer Vision and Pattern Recognition, Salt Lake City, UT, USA, 18–23 June 2018; pp. 3974–3983. [CrossRef]
5. Shuai, T.; Sun, K.; Wu, X.; Zhang, X.; Shi, B. A ship target automatic detection method for high-resolution remote sensing. In Proceedings of the 2016 IEEE International Geoscience and Remote Sensing Symposium (IGARSS), Beijing, China, 10–15 July 2016; pp. 1258–1261. [CrossRef]
6. Diao, W.; Sun, X.; Zheng, X.; Dou, F.; Wang, H.; Fu, K. Efficient Saliency-Based Object Detection in Remote Sensing Images Using Deep Belief Networks. *IEEE Geosci. Remote Sens. Lett.* **2016**, *13*, 137–141. [CrossRef]
7. Proia, N.; Pagé, V. Characterization of a Bayesian Ship Detection Method in Optical Satellite Images. *IEEE Geosci. Remote Sens. Lett.* **2010**, *7*, 226–230. [CrossRef]
8. Ding, Z.; Yu, Y.; Wang, B.; Zhang, L. An approach for visual attention based on biquaternion and its application for ship detection in multispectral imagery. *Neurocomputing* **2012**, *76*, 9–17. [CrossRef]
9. Li, B.; Xie, X.; Wei, X.; Tang, W. Ship detection and classification from optical remote sensing images: A survey. *Chin. J. Aeronaut.* **2021**, *34*, 145–163. [CrossRef]
10. Yin, Y.; Cheng, X.; Shi, F.; Liu, X.J.; Huo, H.; Chen, S. High-order Spatial Interactions Enhanced Lightweight Model for Optical Remote Sensing Image-based Small Ship Detection. *arXiv* **2023**, arXiv:abs/2304.03812.
11. Ren, Z.; Tang, Y.; He, Z.; Tian, L.; Yang, Y.; Zhang, W. Ship Detection in High-Resolution Optical Remote Sensing Images Aided by Saliency Information. *IEEE Trans. Geosci. Remote Sens.* **2022**, *60*, 5623616. [CrossRef]
12. Li, C.; Zhou, H.; Liu, Y.; Yang, C.; Xie, Y.; Li, Z.; Zhu, L. Detection-Friendly Dehazing: Object Detection in Real-World Hazy Scenes. *IEEE Trans. Pattern Anal. Mach. Intell.* **2023**, *45*, 8284–8295. [CrossRef]
13. Ren, S.; He, K.; Girshick, R.; Sun, J. Faster R-CNN: Towards Real-Time Object Detection with Region Proposal Networks. *IEEE Trans. Pattern Anal. Mach. Intell.* **2017**, *39*, 1137–1149. [CrossRef] [PubMed]
14. Jiang, Y.; Zhu, X.; Wang, X.; Yang, S.; Li, W.; Wang, H.; Fu, P.; Luo, Z. R2CNN: Rotational Region CNN for Orientation Robust Scene Text Detection. *arXiv* **2017**, arXiv:abs/1706.09579.
15. Liu, Z.; Hu, J.; Weng, L.; Yang, Y. Rotated region based CNN for ship detection. In Proceedings of the 2017 IEEE International Conference on Image Processing (ICIP), Beijing, China, 17–20 September 2017; pp. 900–904. [CrossRef]
16. Liu, W.; Ma, L.; Chen, H. Arbitrary-Oriented Ship Detection Framework in Optical Remote-Sensing Images. *IEEE Geosci. Remote Sens. Lett.* **2018**, *15*, 937–941. [CrossRef]
17. Yang, X.; Sun, H.; Sun, X.; Yan, M.; Guo, Z.; Fu, K. Position Detection and Direction Prediction for Arbitrary-Oriented Ships via Multitask Rotation Region Convolutional Neural Network. *IEEE Access* **2018**, *6*, 50839–50849. [CrossRef]
18. Han, J.; Ding, J.; Li, J.; Xia, G.S. Align Deep Features for Oriented Object Detection. *IEEE Trans. Geosci. Remote Sens.* **2022**, *60*, 5602511. [CrossRef]
19. He, K.; Gkioxari, G.; Dollár, P.; Girshick, R. Mask R-CNN. *IEEE Trans. Pattern Anal. Mach. Intell.* **2020**, *42*, 386–397. [CrossRef]
20. Xie, X.; Cheng, G.; Wang, J.; Yao, X.; Han, J. Oriented R-CNN for Object Detection. In Proceedings of the 2021 IEEE/CVF International Conference on Computer Vision (ICCV), Montreal, QC, Canada, 10–17 October 2021; pp. 3500–3509. [CrossRef]
21. Han, J.; Ding, J.; Xue, N.; Xia, G. ReDet: A Rotation-equivariant Detector for Aerial Object Detection. In Proceedings of the 2021 IEEE/CVF Conference on Computer Vision and Pattern Recognition (CVPR), Nashville, TN, USA, 20–25 June 2021; pp. 2785–2794. [CrossRef]
22. Weiler, M.; Cesa, G. General E(2)-Equivariant Steerable CNNs. In Proceedings of the Thirty-Third Conference on Neural Information Processing Systems (NeurIPS 2019), Vancouver, BC, Canada, 8–14 December 2019. Available online: https://proceedings.neurips.cc//paper/2019/hash/45d6637b718d0f24a237069fe41b0db4-Abstract.html (accessed on 11 June 2023).
23. Ding, J.; Xue, N.; Long, Y.; Xia, G.; Lu, Q. Learning RoI Transformer for Oriented Object Detection in Aerial Images. In Proceedings of the 2019 IEEE/CVF Conference on Computer Vision and Pattern Recognition (CVPR), Long Beach, CA, USA, 15–20 June 2019; pp. 2844–2853. [CrossRef]

24. Paszke, A.; Gross, S.; Massa, F.; Lerer, A.; Bradbury, J.; Chanan, G.; Killeen, T.; Lin, Z.; Gimelshein, N.; Antiga, L.; et al. PyTorch: An Imperative Style, High-Performance Deep Learning Library. In Proceedings of the Thirty-Third Conference on Neural Information Processing Systems (NeurIPS 2019), Vancouver, BC, Canada, 8–14 December 2019. Available online: https://proceedings.neurips.cc/paper/2019/hash/bdbca288fee7f92f2bfa9f7012727740-Abstract.html (accessed on 11 June 2023).
25. He, K.; Zhang, X.; Ren, S.; Sun, J. Deep Residual Learning for Image Recognition. In Proceedings of the 2016 IEEE Conference on Computer Vision and Pattern Recognition (CVPR), Las Vegas, NV, USA, 27–30 June 2016; pp. 770–778. [CrossRef]
26. Arthur, D.; Vassilvitskii, S. K-means++ the advantages of careful seeding. In Proceedings of the Eighteenth Annual ACM-SIAM Symposium on Discrete Algorithms, New Orleans, LA, USA, 7–9 January 2007; pp. 1027–1035.
27. Lin, T.Y.; Dollár, P.; Girshick, R.; He, K.; Hariharan, B.; Belongie, S. Feature Pyramid Networks for Object Detection. In Proceedings of the 2017 IEEE Conference on Computer Vision and Pattern Recognition (CVPR), Honolulu, HI, USA, 21–26 July 2017; pp. 936–944. [CrossRef]
28. Gong, Y.; Yu, X.; Ding, Y.; Peng, X.; Zhao, J.; Han, Z. Effective Fusion Factor in FPN for Tiny Object Detection. In Proceedings of the 2021 IEEE Winter Conference on Applications of Computer Vision (WACV), Waikoloa, HI, USA, 3–8 January 2021; pp. 1159–1167. [CrossRef]
29. Liu, S.; Qi, L.; Qin, H.; Shi, J.; Jia, J. Path Aggregation Network for Instance Segmentation. In Proceedings of the 2018 IEEE/CVF Conference on Computer Vision and Pattern Recognition, Salt Lake City, UT, USA, 18–23 June 2018; pp. 8759–8768. [CrossRef]
30. Zhong, Y.; Wang, J.; Peng, J.; Zhang, L. Anchor Box Optimization for Object Detection. In Proceedings of the 2020 IEEE Winter Conference on Applications of Computer Vision (WACV), Snowmass, CO, USA, 1–5 March 2020; pp. 1275–1283. [CrossRef]
31. Shrivastava, A.; Gupta, A.; Girshick, R. Training Region-Based Object Detectors with Online Hard Example Mining. In Proceedings of the 2016 IEEE Conference on Computer Vision and Pattern Recognition (CVPR), Las Vegas, NV, USA, 27–30 June 2016; pp. 761–769. [CrossRef]
32. Liu, Z.; Yuan, L.; Weng, L.; Yang, Y. A High Resolution Optical Satellite Image Dataset for Ship Recognition and Some New Baselines. In Proceedings of the 6th International Conference on Pattern Recognition Applications and Methods–ICPRAM, Porto, Portugal, 24–26 February 2017; pp. 324–331. [CrossRef]
33. Ding, J.; Xue, N.; Xia, G.S.; Bai, X.; Yang, W.; Yang, M.Y.; Belongie, S.; Luo, J.; Datcu, M.; Pelillo, M.; et al. Object Detection in Aerial Images: A Large-Scale Benchmark and Challenges. *IEEE Trans. Pattern Anal. Mach. Intell.* **2022**, *44*, 7778–7796. [CrossRef] [PubMed]
34. Zhou, Y.; Yang, X.; Zhang, G.; Wang, J.; Liu, Y.; Hou, L.; Jiang, X.; Liu, X.; Yan, J.; Lyu, C.; et al. MMRotate: A Rotated Object Detection Benchmark using PyTorch. In Proceedings of the 30th ACM International Conference on Multimedia, Lisboa, Portugal, 10–14 October 2022; pp. 7331–7334. [CrossRef]
35. Everingham, M.; Van Gool, L.; Williams, C.K.I.; Winn, J.; Zisserman, A. The PASCAL Visual Object Classes Challenge 2007 (VOC2007) Results. Available online: http://host.robots.ox.ac.uk/pascal/VOC/voc2007/ (accessed on 11 June 2023).
36. Padilla, R.; Netto, S.L.; Silva, E.A.B.d. A Survey on Performance Metrics for Object-Detection Algorithms. In Proceedings of the 2020 International Conference on Systems, Signals and Image Processing (IWSSIP), Niteroi, Brazil, 1–3 July 2020; pp. 237–242. [CrossRef]
37. Lin, T.Y.; Goyal, P.; Girshick, R.; He, K.; Dollár, P. Focal Loss for Dense Object Detection. In Proceedings of the 2017 IEEE International Conference on Computer Vision (ICCV), Venice, Italy, 22–29 October 2017; pp. 2999–3007. [CrossRef]
38. Yang, X.; Yang, X.; Yang, J.; Ming, Q.; Wang, W.; Tian, Q.; Yan, J. Learning High-Precision Bounding Box for Rotated Object Detection via Kullback-Leibler Divergence. In Proceedings of the Thirty-Fifth Conference on Neural Information Processing Systems (NeurIPS 2021), Virtual, 6–14 December 2021. Available online: https://proceedings.neurips.cc/paper/2021/hash/98f13708210194c475687be6106a3b84-Abstract.html (accessed on 11 June 2023).
39. Yang, X.; Liu, Q.; Yan, J.; Li, A. R3Det: Refined Single-Stage Detector with Feature Refinement for Rotating Object. In Proceedings of the AAAI Conference on Artificial Intelligence, Virtual, 8–9 February 2021; pp. 3163–3171. [CrossRef]
40. Sun, X.; Wang, P.; Yan, Z.; Xu, F.; Wang, R.; Diao, W.; Chen, J.; Li, J.; Feng, Y.; Xu, T.; et al. FAIR1M: A benchmark dataset for fine-grained object recognition in high-resolution remote sensing imagery. *ISPRS J. Photogramm. Remote Sens.* **2022**, *184*, 116–130. [CrossRef]
41. Lyu, C.; Zhang, W.; Huang, H.; Zhou, Y.; Wang, Y.; Liu, Y.; Zhang, S.; Chen, K. RTMDet: An Empirical Study of Designing Real-Time Object Detectors. *arXiv* **2022**, arXiv:abs/2212.07784.
42. Zhang, Z.; Guo, W.; Zhu, S.; Yu, W. Toward Arbitrary-Oriented Ship Detection With Rotated Region Proposal and Discrimination Networks. *IEEE Geosci. Remote Sens. Lett.* **2018**, *15*, 1745–1749. [CrossRef]
43. Liao, M.; Zhu, Z.; Shi, B.; Xia, G.S.; Bai, X. Rotation-Sensitive Regression for Oriented Scene Text Detection. In Proceedings of the 2018 IEEE/CVF Conference on Computer Vision and Pattern Recognition, Salt Lake City, UT, USA, 18–23 June 2018; pp. 5909–5918. [CrossRef]
44. Xu, Y.; Fu, M.; Wang, Q.; Wang, Y.; Chen, K.; Xia, G.S.; Bai, X. Gliding Vertex on the Horizontal Bounding Box for Multi-Oriented Object Detection. *IEEE Trans. Pattern Anal. Mach. Intell.* **2021**, *43*, 1452–1459. [CrossRef]
45. Yang, X.; Yan, J. Arbitrary-Oriented Object Detection with Circular Smooth Label. In Proceedings of the Computer Vision–ECCV 2020: 16th European Conference on Computer Vision, Glasgow, UK, 23–28 August 2020; pp. 677–694.

46. Azimi, S.M.; Vig, E.; Bahmanyar, R.; Körner, M.; Reinartz, P. Towards Multi-class Object Detection in Unconstrained Remote Sensing Imagery. In Proceedings of the Computer Vision–ACCV 2018: 14th Asian Conference on Computer Vision, Perth, Australia, 2–6 December 2018; pp. 150–165. [CrossRef]
47. Zhang, G.; Lu, S.; Zhang, W. CAD-Net: A Context-Aware Detection Network for Objects in Remote Sensing Imagery. *IEEE Trans. Geosci. Remote Sens.* **2019**, *57*, 10015–10024. [CrossRef]
48. Pan, X.; Ren, Y.; Sheng, K.; Dong, W.; Yuan, H.; Guo, X.; Ma, C.; Xu, C. Dynamic Refinement Network for Oriented and Densely Packed Object Detection. In Proceedings of the 2020 IEEE/CVF Conference on Computer Vision and Pattern Recognition (CVPR), Seattle, WA, USA, 13–19 June 2020; pp. 11204–11213. [CrossRef]
49. Wang, J.; Yang, W.; Li, H.C.; Zhang, H.; Xia, G.S. Learning Center Probability Map for Detecting Objects in Aerial Images. *IEEE Trans. Geosci. Remote Sens.* **2021**, *59*, 4307–4323. [CrossRef]
50. Yang, X.; Yang, J.; Yan, J.; Zhang, Y.; Zhang, T.; Guo, Z.; Sun, X.; Fu, K. SCRDet: Towards More Robust Detection for Small, Cluttered and Rotated Objects. In Proceedings of the 2019 IEEE/CVF International Conference on Computer Vision (ICCV), Seoul, Republic of Korea, 27 October–2 November 2019; pp. 8231–8240. [CrossRef]
51. Wei, H.; Zhang, Y.; Chang, Z.; Li, H.; Wang, H.; Sun, X. Oriented objects as pairs of middle lines. *ISPRS J. Photogramm. Remote Sens.* **2020**, *169*, 268–279. [CrossRef]
52. Yi, J.; Wu, P.; Liu, B.; Huang, Q.; Qu, H.; Metaxas, D.N. Oriented Object Detection in Aerial Images with Box Boundary-Aware Vectors. In Proceedings of the 2021 IEEE Winter Conference on Applications of Computer Vision (WACV), Waikoloa, HI, USA, 3–8 January 2021; pp. 2149–2158. [CrossRef]
53. Yang, X.; Yan, J.; Yang, X.; Tang, J.; Liao, W.; He, T. SCRDet++: Detecting Small, Cluttered and Rotated Objects via Instance-Level Feature Denoising and Rotation Loss Smoothing. *IEEE Trans. Pattern Anal. Mach. Intell.* **2022**, *45*, 2384–2399. [CrossRef] [PubMed]
54. Chen, K.; Pang, J.; Wang, J.; Xiong, Y.; Li, X.; Sun, S.; Feng, W.; Liu, Z.; Shi, J.; Ouyang, W.; et al. Hybrid Task Cascade for Instance Segmentation. In Proceedings of the 2019 IEEE/CVF Conference on Computer Vision and Pattern Recognition (CVPR), Long Beach, CA, USA, 15–20 June 2019; pp. 4969–4978. [CrossRef]
55. Lang, S.; Ventola, F.G.; Kersting, K. DAFNe: A One-Stage Anchor-Free Deep Model for Oriented Object Detection. *arXiv* **2021**, arXiv:abs/2109.06148.
56. Li, C.; Xu, C.; Cui, Z.; Wang, D.; Jie, Z.; Zhang, T.; Yang, J. Learning Object-Wise Semantic Representation for Detection in Remote Sensing Imagery. In Proceedings of the IEEE Conference on Computer Vision and Pattern Recognition Workshops, CVPR Workshops 2019, Long Beach, CA, USA, 16–20 June 2019; pp. 20–27. Available online: http://openaccess.thecvf.com/content_CVPRW_2019/html/DOAI/Li_Learning_Object-Wise_Semantic_Representation_for_Detection_in_Remote_Sensing_Imagery_CVPRW_2019_paper.html (accessed on 11 June 2023).
57. Tan, M.; Pang, R.; Le, Q.V. EfficientDet: Scalable and Efficient Object Detection. In Proceedings of the 2020 IEEE/CVF Conference on Computer Vision and Pattern Recognition (CVPR), Seattle, WA, USA, 13–19 June 2020; pp. 10778–10787. [CrossRef]
58. Lin, T.Y.; Maire, M.; Belongie, S.; Hays, J.; Perona, P.; Ramanan, D.; Dollár, P.; Zitnick, C.L. Microsoft COCO: Common Objects in Context. In Proceedings of the Computer Vision–ECCV 2014: 13th European Conference on Computer Vision, Zurich, Switzerland, 6–12 September 2014; pp. 740–755. [CrossRef]

remote sensing

MDPI

Article

Daytime Sea Fog Identification Based on Multi-Satellite Information and the ECA-TransUnet Model

He Lu [1,2], Yi Ma [3], Shichao Zhang [1], Xiang Yu [1] and Jiahua Zhang [1,2,*]

1 Remote Sensing Information and Digital Earth Center, College of Computer Science and Technology, Qingdao University, Qingdao 266071, China; 2021023975@qdu.edu.cn (H.L.); 2021010034@qdu.edu.cn (S.Z.); 2020010031@qdu.edu.cn (X.Y.)
2 Key Laboratory of Digital Earth Science, Aerospace Information Research Institute, Chinese Academy of Sciences, Beijing 100094, China
3 First Institute of Oceanography, State Oceanic Administration, Qingdao 266061, China; mayimail@fio.org.cn
* Correspondence: zhangjh@radi.ac.cn

Abstract: Sea fog is a weather hazard along the coast and over the ocean that seriously threatens maritime activities. In the deep learning approach, it is difficult for convolutional neural networks (CNNs) to fully consider global context information in sea fog research due to their own limitations, and the recognition of sea fog edges is relatively vague. To solve the above problems, this paper puts forward an ECA-TransUnet model for daytime sea fog recognition, which consists of a combination of a CNN and a transformer. By designing a two-branch feed-forward network (FFN) module and introducing an efficient channel attention (ECA) module, the model can effectively take into account long-range pixel interactions and feature channel information to capture the global contextual information of sea fog data. Meanwhile, to solve the problem of insufficient existing sea fog detection datasets, we investigated sea fog events occurring in the Yellow Sea and Bohai Sea and their territorial waters, extracted remote sensing images from Moderate Resolution Imaging Spectroradiometer (MODIS) data at corresponding times, and combined data from the Cloud-Aerosol Lidar and Infrared Pathfinder Satellite Observation (CALIPSO), cloud and sea fog texture features, and waveband feature information to produce a manually annotated sea fog dataset. Our experiments showed that the proposed model achieves 94.5% accuracy and an 85.8% F1 score. Compared with the existing models relying only on CNNs such as UNet, FCN8s, and DeeplabV3+, it achieves state-of-the-art performance in sea fog recognition.

Keywords: sea fog recognition; deep learning; transformer; convolutional neural network (CNN); efficient channel attention (ECA)

Citation: Lu, H.; Ma, Y.; Zhang, S.; Yu, X.; Zhang, J. Daytime Sea Fog Identification Based on Multi-Satellite Information and the ECA-TransUnet Model. *Remote Sens.* **2023**, *15*, 3949. https://doi.org/10.3390/rs15163949

Academic Editor: Jing Wei

Received: 26 June 2023
Revised: 16 July 2023
Accepted: 3 August 2023
Published: 9 August 2023

1. Introduction

Sea fog is a water vapor condensation phenomenon that results from the massive accumulation of water droplets or ice crystals in the lower atmosphere over the ocean [1]. The horizontal visibility of the atmosphere in sea fog is less than 1 km [2]. Sea fog poses significant risks to aviation and maritime transportation because of its low visibility, leading to accidents and resulting in serious safety concerns and economic losses [3,4]. The Yellow Sea and Bohai Sea are critical maritime transportation and development zones in China, and they are also prone to frequent occurrences of sea fog. [5]. According to statistics, more than 50% of maritime accidents that occurred in the Yellow Sea from 1981 to 2010 were due to sea fog [6]. Therefore, effective monitoring of sea fog is crucial for the safety of human life, property, and social activities.

Satellite remote sensing technology is the primary tool for sea fog monitoring. It provides more real-time, accurate, and large-scale sea fog information compared with ground-based observatories [7–9]. Sea fog and clouds have a small brightness temperature

(TB) difference in the 3.7 μm and 11 μm channels compared to each other [10–12], respectively; therefore, the dual-channel difference (DCD) [13] method is often employed for distinguishing between sea spray and clouds. For example, Ellrod et al. [14] tried to perform night-time fog detection based on the DCD method using a geostationary operational environmental satellite and experimentally demonstrated that the DCD method can effectively detect sea fog in the absence of cloud occlusion. Zhang et al. [15] identified sea fog by conducting a comparative analysis of the relative frequency of BT between sea fog and low clouds. Han et al. [16] used the reflectivity difference between fog and other objects at 0.63 μm and the DCD method for daytime and night-time sea fog monitoring. In addition, the thresholding method is typically used for sea fog monitoring. Ryu et al. [17] proposed a sea fog detection algorithm using Himawari-8 satellite data, based on the reflectance of visible (VIS) and near-infrared (NIR) bands, which can be applied to optical satellites without shortwave infrared (SWIR) bands. Wu et al. [18] constructed an automatic sea fog monitoring algorithm based on Moderate Resolution Imaging Spectroradiometer (MODIS) data using several variables such as the normalized snow index (NDSI), the normalized difference of near-infrared water vapor index (NWVI), etc. These investigations described earlier proved that the algorithm can effectively identify sea fog under various weather conditions. The detection of sea fog using the threshold method is stable and straightforward, but simple combinations of thresholds are not able to handle sea fog monitoring in complex scenarios, and using multiple combinations of thresholds requires more statistical work and experimental data.

The deep learning approach has the ability to autonomously learn feature information from data [19] and has been increasingly utilized into the process of remote sensing data analysis in recent years [20–25]. For example, convolutional neural networks (CNNs) were used to classify terrestrial fog in Anhui Province, China [26]. Jeon et al. [27] used a migration learning model (CNN-TL) to identify sea fog while verifying the effect of different combinations of bands on sea fog identification. Zhu et al. [28] applied UNet [29], a semantic segmentation model in the field of deep learning, to sea fog monitoring and combined MODIS data to identify the specific range. The purpose of semantic segmentation is to obtain the classification of each pixel, which provides a new method for remote sensing classification. Zhou et al. [30] proposed a two-branch sea fog detection network (DB-SFNet) to achieve comprehensive and accurate sea fog monitoring by combining sea fog events recorded by the Geostationary Ocean Color Imager (GOCI). To address the scarcity of sea fog samples, Huang et al. [31] performed sea fog sample augmentation by using generative adversarial networks (GANs) [32]. Currently, most of the deep learning applications for sea fog monitoring studies are based on CNNs. The capability of CNNs is unquestionable, but they also have their own limitations and cannot model long-range dependencies well [33,34]. In sea fog images, clouds and fog have high similarity and mask each other, and more global context information needs to be considered.

Recently, the ViT model [35] was proposed, so transformers started to be applied more in the realm of computer vision, which provides a new research idea for image processing. In remote sensing images, different semantic classes may have similar size and spectral features, which are not easily distinguishable, so remote sensing images need to focus on more global contextual information. Transformers have also been applied in remote sensing images [36–39]. It is difficult for CNNs to model contextual information, and transformers based on a self-attentive mechanism can better model the global context, and a combination of the two can bring out greater capabilities, based on which TransUnet was first proposed to be applied in the field of medical image segmentation while having great potential, but the segmentation capability on remote sensing sea fog recognition has not been proven [40].

In this study, we selected MODIS data for sea fog studies. Cloud-Aerosol Lidar and Infrared Pathfinder Satellite Observation (CALIPSO) is used to differentiate between clouds and sea fog by obtaining atmospheric vertical profile information and is also applied to sea fog verification [41–43]. Therefore, in the process of creating sea fog samples, we utilize CALIPSO data to assist in addition to combining different cloud and sea fog texture features [27] and MODIS band features. Furthermore, we propose an ECA-TransUnet model for daytime sea fog monitoring. Through enhancement of the transformer and the incorporation of the attention mechanism, the model was enabled to achieve superior performance for sea fog detection. Finally, we used fifth generation of global climate reanalysis data (ERA5) to analyze the meteorological conditions for sea fog generation.

The overall structure of this paper consists of seven sections. The Section 2 presents the study area and the data used. The Section 3 introduces the ECA-TransUnet model. The Section 4 shows the data pre-processing and experimental setup, and the Section 5 compares the experimental results of different models and also analyzes the meteorological conditions that produce sea fog. The Section 6 is a discussion, and the Section 7 concludes the study.

2. Materials

2.1. Study Area

In this study, we selected the Yellow and Bohai seas and their territorial coastal waters close to China as our main study area. These seas experience cold, dry winters and warm, humid summers due to their proximity to the northwest Pacific Ocean [44]. Affected by the warm and humid air, sea fog is mainly concentrated in April-July of the year, mostly in advective fog [45]. The specific study area is shown in Figure 1.

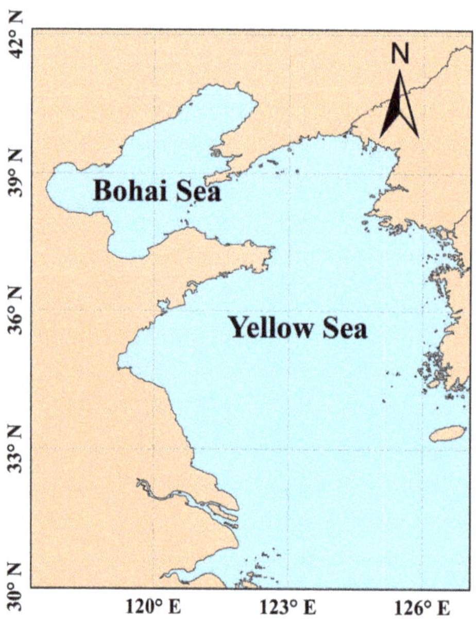

Figure 1. Study area.

2.2. Data

2.2.1. MODIS Data

Aqua/MODIS L1B data were selected as the primary data for sea fog monitoring. MODIS has 36 bands with a spectral range of 0.4–14.4 μm, providing comprehensive and detailed data support for our study due to its long runtime and wide band. The MODIS and CALIPSO observation intervals differ by approximately 1.5 min, which is relatively consistent in time and space. [46]. The MYD021KM data products we used were downloaded from Google Earth Engine (https://ladsweb.modaps.eosdis.nasa.gov/, accessed on 14 June 2022). We selected 30 MODIS images from 2013 to 2018.

2.2.2. CALIPSO Data

Cloud-Aerosol Lidar with Orthogonal Polarization (CALIOP) is the first satellite-based polarimetric lidar capable of observing the state of the global atmosphere for observing aerosols and micron-sized cloud particles and obtaining information on the vertical profile configuration of the atmosphere. In the process of sea fog sample classification, we chose Vertical Feature Mask (VFM) data, a secondary data product of CALIPSO, to assist in the work. This data allows for obtaining a vertical profile of the atmosphere with a horizontal resolution of 333 m. The profile includes target object types such as cloud, aerosol, surface, and clear sky and their corresponding altitudes. A certain distinction between clouds and fog can be made using this data. The CALIOP VFM product used in this article was downloaded from https://subset.larc.nasa.gov/calipso/login.php, accessed on 26 July 2022.

2.2.3. Fifth Generation of Global Climate Reanalysis Data

We utilized the fifth generation of global climate reanalysis data (ERA5) from the European Centre for Medium-Range Weather Forecasts (ECMWF) to analyze the meteorological conditions in the area where sea fog occurs in terms of wind direction, wind speed, pressure, and temperature differences to analyze the causes of sea fog. The selected meteorological data consists primarily of a 10 m u-wind component, a 10 m v-wind component, mean sea level pressure, surface air temperature (SAT), and sea surface temperature (SST). The ERA-5 product used in this paper was downloaded from: https://cds.climate.copernicus.eu/cdsapp#!/dataset/reanalysis-era5-single-levels?tab=form, accessed on 20 December 2022.

3. Method

The work of this study is divided into two parts: the production of the sea fog dataset and model training. The overall workflow is shown in Figure 2. First, MODIS images with sea fog events were selected and combined with sea fog texture features, MODIS waveband features, and CALIPSO radar data for manual interpretation to generate sea fog samples. We trained the model by inputting the generated sea fog data, and the performance of each model was analyzed through model comparison, while the model results were analyzed by ERA5 meteorological data for sea fog meteorology.

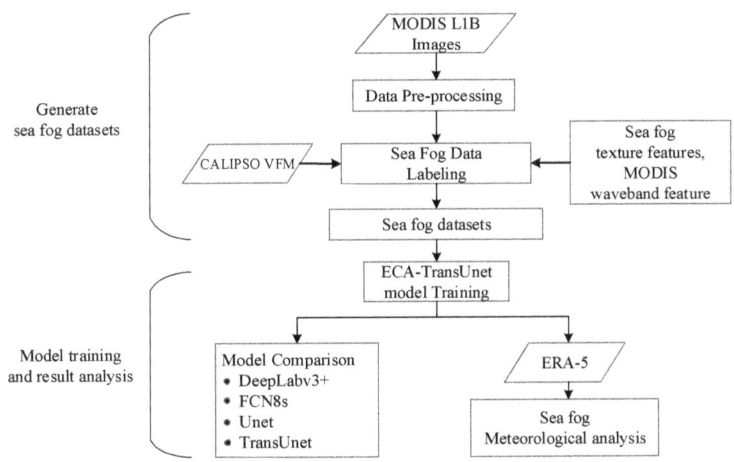

Figure 2. General workflow of this study.

In this section, we first introduce the general structure of the proposed ECA-TransUnet model. This is followed by an explanation of the improvement of the FFN in the transformer module, and a detailed depiction of the ECA module is introduced in the decoder section.

3.1. ECA-TransUnet Network Model

The overall network architecture of the ECA-TransUnet model is depicted in Figure 3. The model structure is segmented into two parts: an encoder and a decoder. In the encoder part, a CNN and transformer are used in series to extract image features. After the CNN extracts the image feature information, it will pass the feature information into the transformer module. However, the transformer's global self-attention mechanism ignores position information, so the position embedding operation is performed before going on. In this way, the image is converted to sequence information processing, which enables better long-range modeling. To further improve the model, some enhancements were made to the FFN part of the transformer module. These enhancements include the use of two parallel branches to filter the feature information and suppress less informative features in the feature transmission, allowing more useful information to be passed into the network hierarchy. The ECA module is added to each layer of the decoder, for most of the existing attention mechanisms are to achieve better performance of the network through relatively complex structures, while the ECA module involves only a few parameters. This allows the performance of the network to improve while simultaneously diminishing the complexity of the network and focusing more on the channel information of the characteristics.

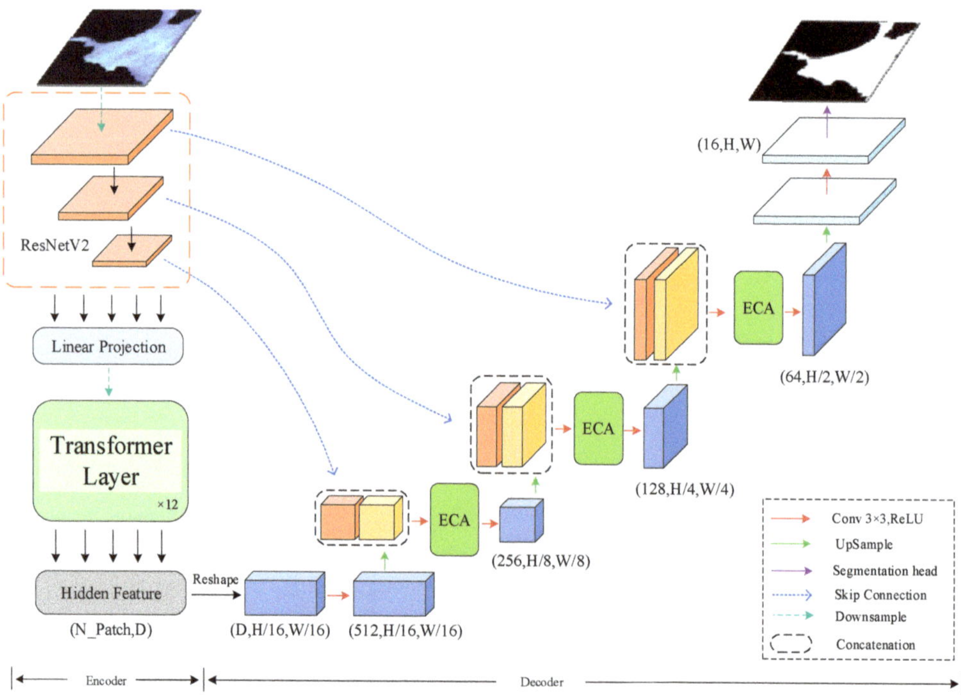

Figure 3. ECA-TransUnet model network structure.

3.2. FFN Improvement

The transformer module introduced into the encoder part implements the extraction of global context information. It mainly consists of two main parts: the multi-headed attention (MSA) mechanism and the FFN [47]. Figure 4b shows the standard FFN module structure, which consists of two linear transformation layers, and the GeLu activation function is added between the two linear layers, with an overall single-branch structure. Figure 4a indicates the improved FFN module, which changes from a single branch to a two-branch structure, where one branch is nonlinearly activated by the GeLu activation function after passing through the linear transformation layer, which can retain better feature information.

The other branch is passed through the linear layer, and then the two parallel branches are multiplied element by element to further enhance the superior feature information, and the final output is passed through the dropout layer. The design of the two-branch structure allows the model to focus on the details of each layer and other complementary layers, which can retain and enhance the superior feature information.

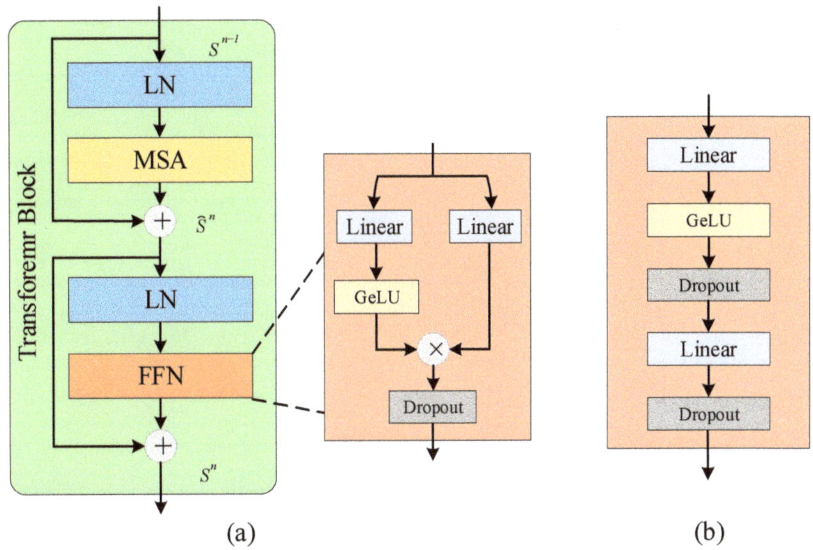

Figure 4. (**a**) Transformer module of the ECA-TransUnet network model and improved FFN structure. (**b**) Standard FFN module structure.

3.3. ECA Model

The proposed ECA attention mechanism is based on the SE module [48]. The ECA attention mechanism first performs a global average pooling operation on the input feature maps, and then uses a 1-dimensional convolution of convolution kernel size k to achieve local cross-channel interaction with shared weights. The weights ω of each channel are obtained after the sigmoid activation function, as shown in (1). The size k of the convolution kernel can be chosen adaptively according to the given number of channels in the nonlinear mapping. After obtaining the dependencies between channels, the weights are multiplied with the corresponding elements of the original input feature map, the calculated weights are added to the feature map, and, finally, the feature map is output.

$$\omega = \sigma(\mathrm{C1D_k}(y)) \quad (1)$$

where ω is the channel weight, σ is the sigmoid activation function, C1D denotes 1D convolution, and k is the number of parameter information. Compared to the SE module, the ECA module reduces the side effects associated with predicting channel attention by avoiding the operation of dimensionality reduction. Second, unlike the original approach of capturing the dependencies of all channels, ECA employs appropriate cross-channel interactions to improve the efficiency of the network, achieving better performance with a small number of parameters.

In the decoder part of the network, the ECA module [49] is added after feature fusion. The overall structure of this attention mechanism is shown in Figure 5.

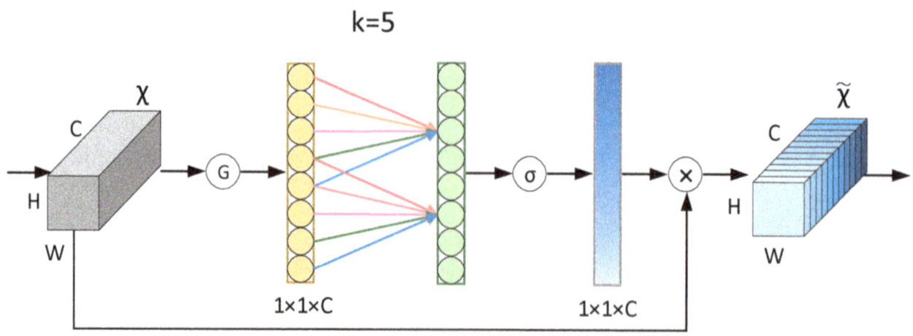

Figure 5. Structure of the efficient channel attention module; k is proportional to the number of channels.

In the whole network model architecture, the improvement of the FFN allows the network to prioritize the important feature information during feature transmission. Additionally, the inclusion of an ECA module after feature fusion in the decoder helps the network to better focus on channel information, leading to improved overall performance.

Precision Evaluation

In this study, the statistical prediction results per unit pixel are obtained for true positive samples (TPs), false negative samples (FNs), false positive samples (FPs), and true negative samples (TNs) based on the confusion matrix. Four metrics, accuracy, recall, intersection ratio (IoU), and F1-score, were selected to evaluate the performance of the model in sea fog identification [50]. The sensitivity and specificity of the model were measured comprehensively.

$$\text{Accuracy} = \frac{\text{TP} + \text{TN}}{\text{TP} + \text{FP} + \text{FN} + \text{TN}} \tag{2}$$

$$\text{Recall} = \frac{\text{TP}}{\text{TP} + \text{FN}} \tag{3}$$

$$\text{IoU} = \frac{\text{TP}}{\text{TP} + \text{FP} + \text{FN}} \tag{4}$$

$$\text{F1} = \frac{2\text{TP}}{2\text{TP} + \text{FP} + \text{FN}} \tag{5}$$

where TP: sea fog present and detected, FN: sea fog is present but not detected, FP: sea fog is not present but detected, and TN: sea fog does not exist and is not detected.

4. Data Pre-Processing and Experiment

4.1. Data Pre-Processing

We used MODIS L1B data in conducting the sea fog monitoring study. MODIS data of large volume and a broad spectral range are applied in several fields of remote sensing. Among the established sea fog samples, the sea fog data were selected from a total of 30 MODIS sea fog images of the Yellow Sea, Bohai Sea, and the territorial waters of China from 2013 to 2018.

During MODIS data preprocessing, the data were subjected to radiometric calibration, solar zenith angle correction, geometric correction, bowtie processing, band synthesis, and land-ocean masking operations. Bands 1, 17, and 32 of MODIS were adopted for the sea fog samples based on previous studies [51]. Table 1 shows the details of the three bands.

Table 1. MODIS L1B band information.

Band	Central Wavelength (nm)	Spectral Range (nm)	Main Applications	Resolution (m)
1	645	620–670	Land/Cloud Border	250
17	905	890–920	Atmospheric water vapor	1000
32	12.02	11.77–12.27	Cloud Temperature	1000

In the process of sea fog labeling, CALIPSO's VFM product is used to assist in a distinction between clouds and sea fog. The MODIS image from 29 March 2018 was used as an example. Figure 6 shows the MODIS image from 26 March 2018, the yellow line in Figure 6 is the trajectory line of CALIPSO, and Figure 7 shows the vertical section of this trajectory line. In order to facilitate the analysis, we divided the track line of CALIPSO in Figure 6 (yellow line in Figure 6) into two segments, AB and CD, which correspond to the areas labeled AB and CD in Figure 7.

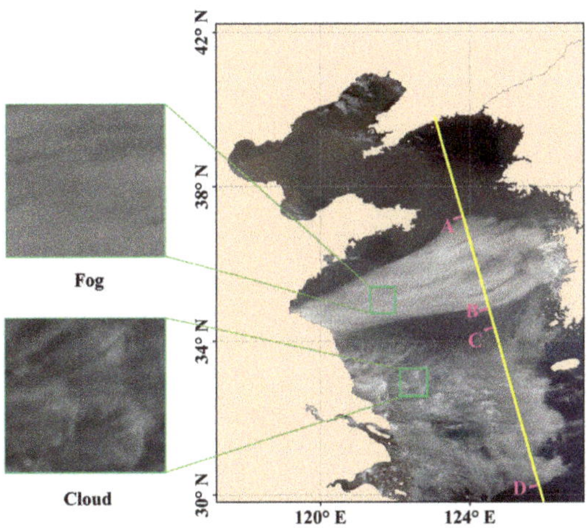

Figure 6. True color red–green–blue (RGB) MODIS image from 29 March 2018 at 05:00 UTC; the yellow line represents the trajectory line of CALIPSO. Segment AB is a sea fog area and segment CD is a mixture of sea fog and clouds.

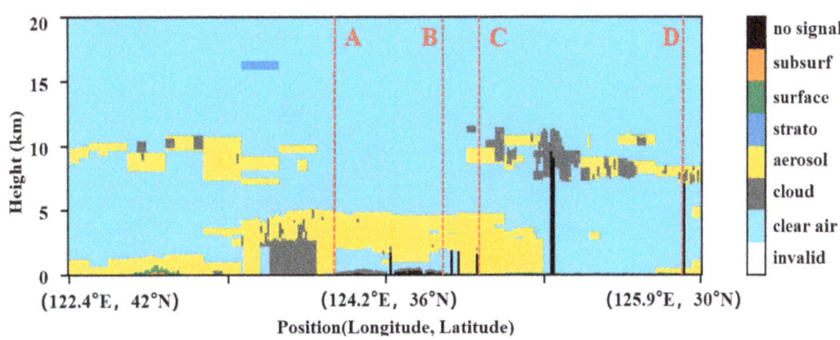

Figure 7. Vertical profile of CALIPSO. Segment AB is a sea fog area, segment CD is a mixture of sea fog and clouds.

Relying on subjective human judgment alone, areas AB and CD are likely to be labeled as sea fog based on the MODIS imagery in Figure 6. However, the vertical profile from CALIPSO in Figure 7 reveals that there are clouds present in the CD area at an altitude of approximately 10 km above sea level. Therefore, the AB area should be labeled as sea fog and the CD area should be labeled as a mixture of sea fog and clouds.

In addition to using CALIPSO data to assist in the production of the sea fog samples, certain distinctions can be made by the textural features of sea spray and clouds [27]. As seen in the comparison of the two areas on the left side of Figure 6, the texture of sea mist tends to be finer and smoother compared to the rough texture of clouds.

We combined various approaches (CALIPSO data, sea fog texture features, and MODIS waveband combinations) to assist in sea fog identification in the process of producing sea fog sample datasets to minimize the subjectivity of manual labeling. Ultimately, the sea fog dataset was divided into two categories: sea fog and others (land, clouds, and sea).

4.2. Experimental Setup

To facilitate the experimentation process, the sea fog images and labels are randomly cropped to a size of 256 × 256. Moreover, data augmentation [52] is a method to increase the number and diversity of training samples by transforming the original data, which aims to change the appearance, shape, or characteristics of the data so as to make the model more robust and generalizable. We used random rotation, horizontal and vertical flipping, and cropping operations to further expand this data based on the existing sea spray data, and then input it into the model for training.

In the experiments, we set batch_size = 8 and the number of iterations to 80, and the initial learning rate was set to 0.0001. During training of the model, the RMSprop optimizer was used, while the learning rate was dynamically adjusted during the training process with the exponential decay strategy, and the learning rate calculation formula for exponential decay is defined as follows:

$$\text{ExponentialLR} = \text{Base}_{\text{lr}} \times \text{decay\_rate}^{\frac{\text{global\_step}}{\text{decay\_steps}}} \tag{6}$$

where Base_lr is the initial learning rate, decay_rate is the decay coefficient, global_step is the number of training iterations, and decay_steps is the decay rate, i.e., how many rounds to update the parameters.

Considering the characteristics of the unbalanced sea fog samples, the area occupied by the ground and the sea surface is larger, while the sea fog is relatively smaller. In order to improve the accuracy and stability of the model, the focal loss function was used in the process of network training [53]; it is able to solve the problem of category imbalance by decreasing the weight of easily categorized samples during the training process. The focal loss function is defined as follows:

$$\text{FL}(p_t) = -\alpha_t(1-p_t)^{\gamma}\log(p_t) \tag{7}$$

$$p_t = \begin{cases} p & \text{if } y = 1 \\ 1-p & \text{otherwise} \end{cases} \tag{8}$$

where the focusing parameter γ smoothly adjusts the rate at which the easy examples reduce their weights, p_t responds to the proximity of the actual category to the predicted category, and $p \in [0, 1]$ is the estimated probability of the model for $y = 1$.

Eventually, when the model stopped training, the accuracy of the model stabilized at 98.3%.

5. Results

5.1. Model Comparison and Evaluation

For the presentation of the model results, we selected three days with sea fog events in the Yellow and Bohai seas: 3 June 2013, 12 July 2013, and 26 March 2018. In addition to our proposed ECA-TransUnet model, it was also compared with some other CNN models, including the DeepLabV3+ [54], FCN8s [55], UNet [29], and TransUnet [40] models. Figure 8 illustrates the detailed outcomes of sea fog identification for each model.

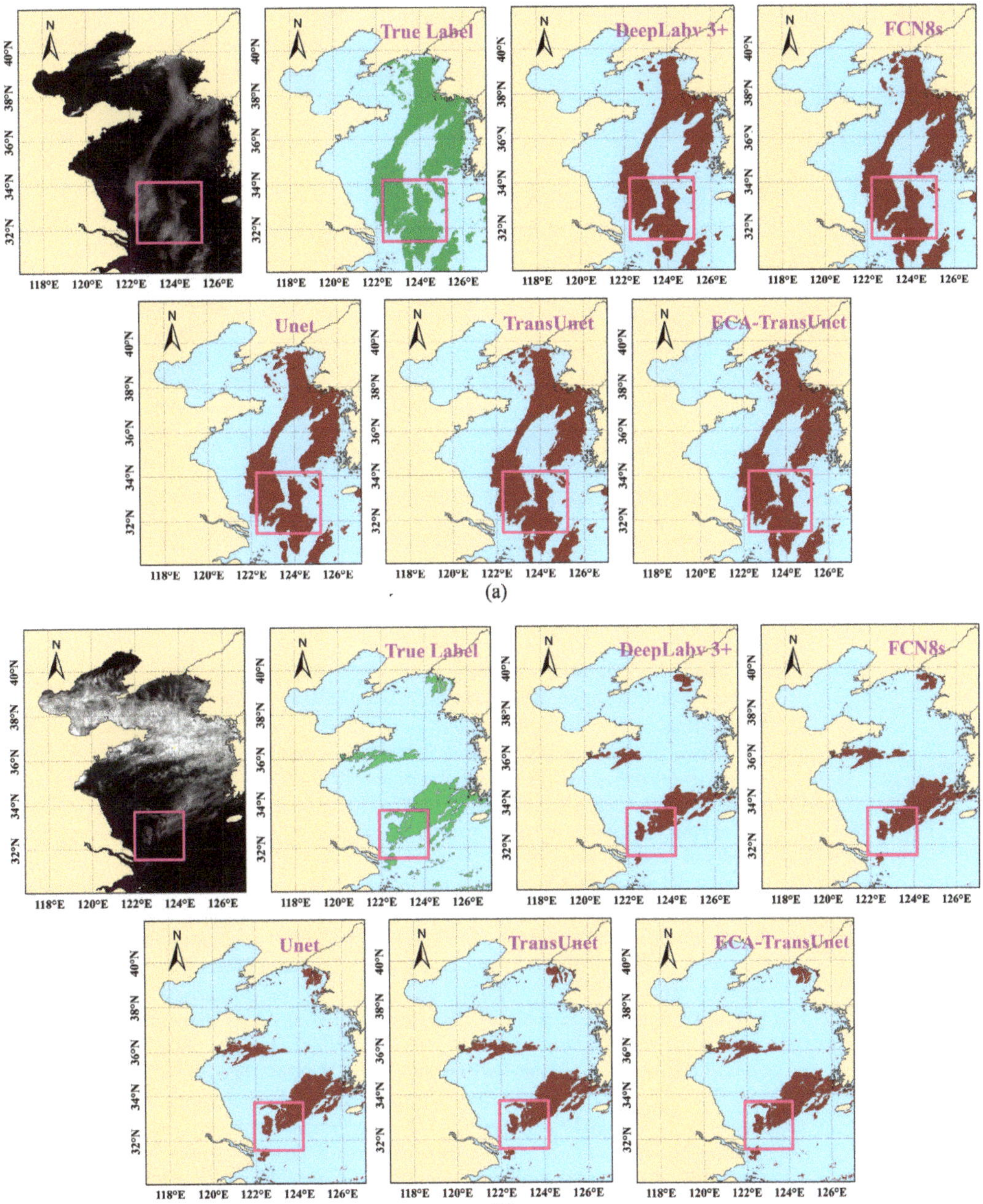

Figure 8. *Cont.*

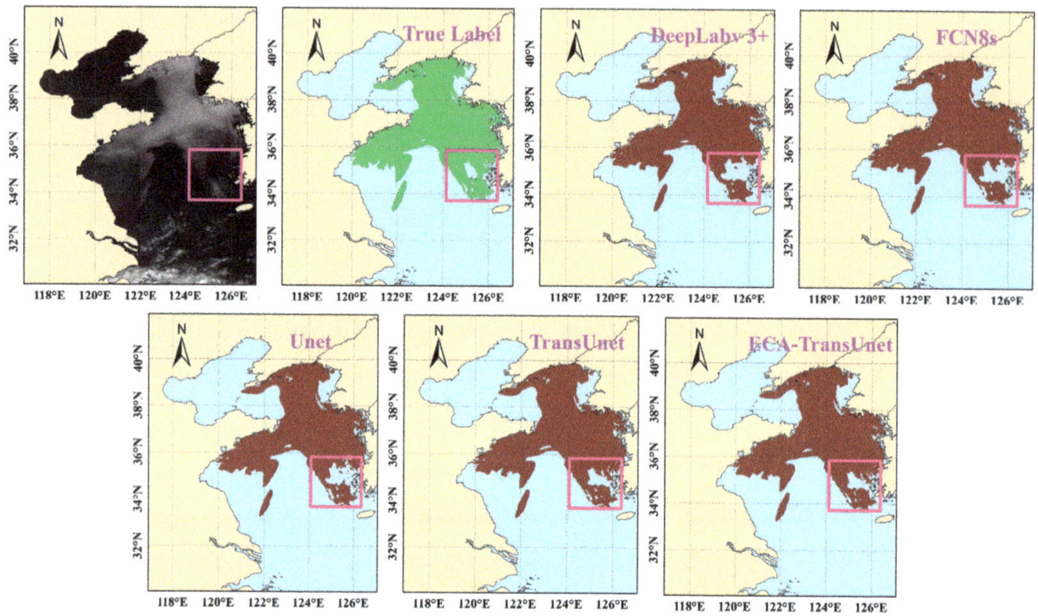

(c)

Figure 8. Comparison of the model results for different sea fog date events. From left to right, the RGB images of MODIS sea fog, sea fog labels, and the results of each model are shown. (**a**) 3 June 2013 at 05:00 UTC, (**b**) 12 July 2013 at 05:00 UTC, and (**c**) 26 March 2018 at 05:00 UTC.

As seen in Figure 8a, the ECA-TransUnet model performs better in identifying small, scattered sea fog areas than other models, and the overall sea fog areas identified are approximately the same as the labeled sea fog areas. Figure 8b shows that the ECA-TransUnet model is relatively more accurate in separating sea fog and clouds. Across the three groups of sea fog events, the DeepLabV3+ model presents relatively poor prediction results for the overall sea fog, and the ECA-TransUnet model can identify the sea fog area better than the traditional CNN model. In order to depict the sea spray identification results of each model more distinctly, we have labeled some areas of Figure 8 (red boxes), which will be magnified and exhibited in the follow-up. The ECA-TransUnet model shows some ambiguity for some specific thinner sea fog edge recognition, but the overall sea fog detection results are closest to the true labels. It is worth noting that although we use various auxiliary methods for sea fog labeling, the manual labeling process still has some errors, particularly the distinction between sea fog and low clouds. This is a challenging task in sea fog monitoring since the texture features of sea fog and low clouds can be similar, and also, the two can transform into each other under certain conditions.

There are some variations in the ability of the different models to recognize sea fog, which is reflected in their ability to distinguish between clouds and fog as well as their recognition of thinner sea fog edges. The ECA-TransUnet model performs better overall in these tasks. To more clearly compare the results of the different models, we present the sea fog results of each model overlaid with the sea fog labels for the event on 3 June 2013 as an example in Figure 9.

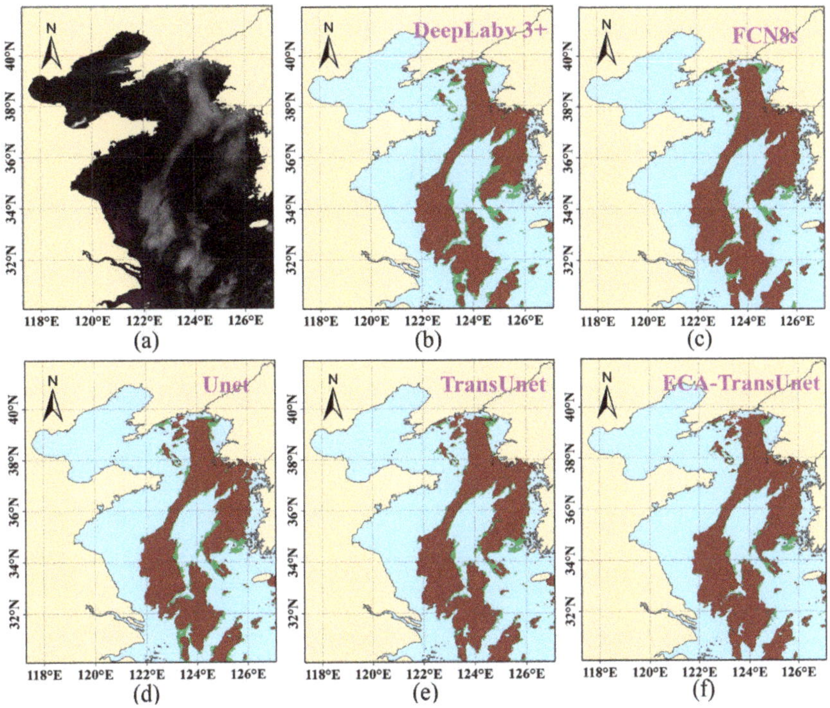

Figure 9. The different model sea fog recognition results are shown overlaid with real tags: red is the recognition result of the model and green is the sea fog range of the real tags. (**a**) The RGB images of MODIS sea fog, (**b**) DeeplabV3+, (**c**) FCN8s, (**d**) UNet, (**e**) TransUnet, and (**f**) ECA-TransUnet.

Figure 9 illustrates that the overall sea fog results of the ECA-TransUnet model (Figure 9f in red) match more closely with the labeled green areas and can identify some irregular sea fog edges more accurately. In addition to comparing the sea fog results of the different models visually, we also show the differences of each model for sea fog recognition specifically through the evaluation metrics of the models. We evaluated a test set of sea fog data.

Table 2 illustrates the evaluation metrics of each model for sea fog recognition in the test set. The combination of a transformer and a convolutional neural network shows better parameter metrics than most traditional convolutional neural network models. The ECA-TransUnet model improves by 1.49% and 3.95% on Acc and the IoU, respectively, relative to the TransUnet model. On the IoU, ECA-TransUnet is 8.81% higher relative to DeepLabV3+. It is clear from the evaluation metrics that our proposed ECA-TransUnet model has better experimental results in the identification of sea fog relative to several other models.

Table 2. Comparison of the evaluation indexes of different models.

	Acc	**Recall**	**IoU**	**F1**
DeeplabV3+	0.9151	0.7353	0.7276	0.7946
FCN8s	0.9244	0.745	0.73	0.802
Unet	0.9279	0.783	0.7533	0.8152
TransUnet	0.9302	0.8019	0.7762	0.8374
ECA-TransUnet	0.9451	0.8231	0.8157	0.8587

In order to observe the effect of each model on sea fog recognition more distinctly, we marked some areas with rectangular boxes for the three sea fog events in Figure 8 and then zoomed in on the marked areas to show the specific results as shown in Figure 10.

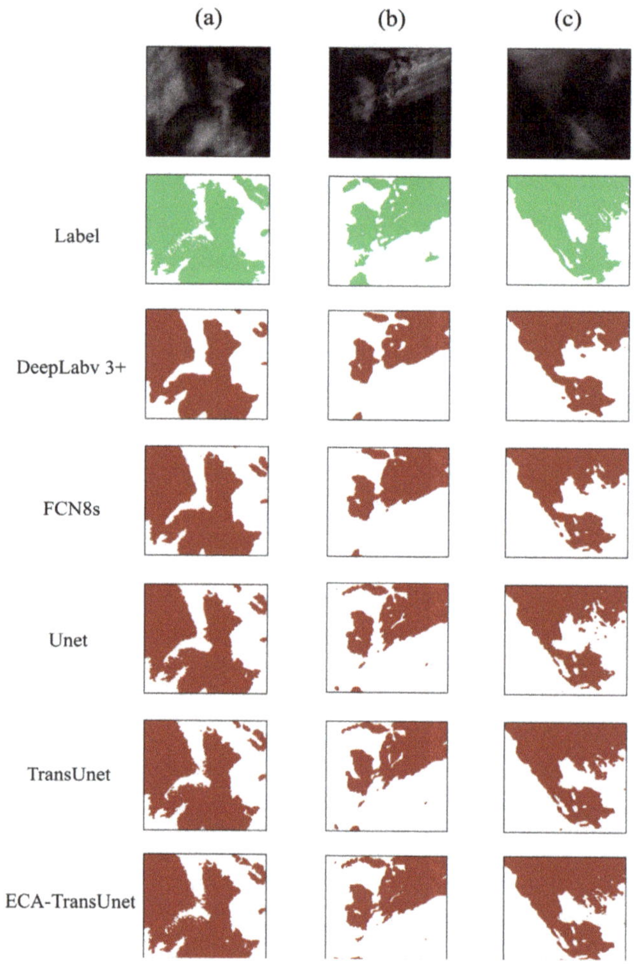

Figure 10. Marking the sea fog area for enlarged display: (**a**) 3 June 2013 at 05:00 UTC, (**b**) 12 July 2013 at 05:00 UTC, and (**c**) 26 March 2018 at 05:00 UTC.

Figure 10a,b exhibits the outcomes of each model in terms of sea fog recognition, and it is apparent that ECA-TransUnet demonstrates superior performance in capturing intricate sea fog details compared to other models. Furthermore, the overall extent of sea fog recognized by ECA-TransUnet closely aligns with the ground truth label. In Figure 10c, ECA-TransUnet overlooks some of the thinner sea fog areas in the sea fog identification results, but the overall sea fog recognition performance is unmatched among other models.

5.2. The Meteorological Conditions of Sea Fog Generation Analysis

Meteorological data from ERA5, including the 10 m u-wind component, 10 m v-wind component, mean sea level pressure, SAT and SST, are used to analyze the meteorological conditions in the area of sea fog occurrence and validate the sea fog results identified by the ECA-TransUnet model. We selected two sea fog events for the analysis.

Firstly, the analysis of the sea fog event in the Yellow and Bohai seas on 25 July 2016, is shown in Figure 11. The red area in Figure 11b is the sea fog area identified by the ECA-TransUnet model, and Figure 11c shows the superimposed sea surface mean pressure and wind direction corresponding to the time of sea fog. The warm and humid south wind was transported to the distant sea by the influence of the high-pressure area in the southeastern part of the Korean Peninsula. The difference between the surface air temperature and the sea surface temperature (SAT-SST) exceeded 0.5 °C, which is consistent with the sea fog generation meteorological conditions. Under this meteorological condition, the ECA-TransUnet model identifies the overall sea fog range.

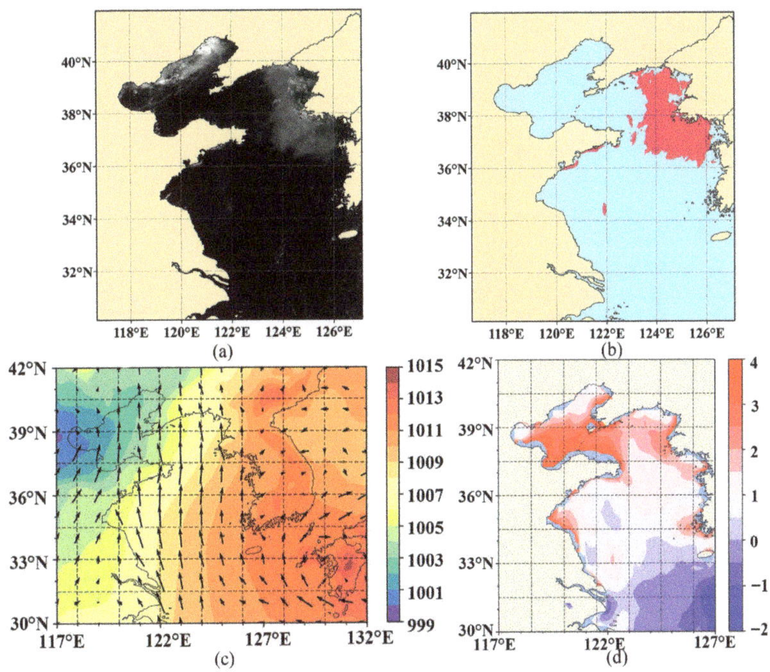

Figure 11. Analysis of the sea fog event on 25 July 2016. (**a**) RGB image from MODIS at 05:00 UTC on 25 July 2016. (**b**) Results of the ECA-TransUnet model for identifying sea fog under this sea fog event, with the red area as the sea fog region. (**c**) Superimposed display of mean sea level pressure and 10 m wind vectors. (**d**) The difference between SAT and SST is shown.

The other sea fog event was dated 26 March 2018. Figure 12c shows the superposition of the mean sea surface pressure and wind direction corresponding to the time of sea fog. It can be seen that there is a high-pressure area at sea level in the southern part of the Korean Peninsula, and the southeastern winds move far offshore after passing through the sea fog region, and the difference between the surface air temperature and the sea surface temperature (SAT-SST) is greater than 0.5 °C due to the influence of the sea surface pressure and wind direction, which is consistent with the meteorological conditions for sea fog generation. The sea fog region identified by the ECA-TransUnet model has some correctness for the overall sea fog region except for a small area in the western part of the Korean Peninsula.

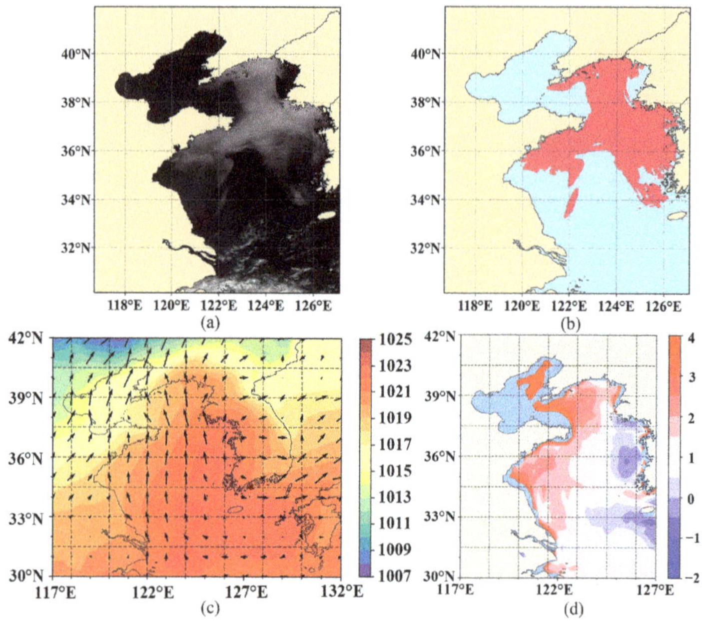

Figure 12. Analysis of the 26 March 2018 sea fog event. (**a**) RGB image of MODIS at 05:00 UTC on 26 March 2018. (**b**) Results of the ECA-TransUnet model for identifying sea fog under this sea fog event, with the red area as the sea fog region. (**c**) Superimposed display of mean sea level pressure and 10 m wind vectors. (**d**) The difference between SAT and SST is shown.

6. Discussion

We have combined the TransUnet network with ECA modules and improved the FFN module to propose an ECA-TransUnet model for daytime sea fog recognition. The combination of a transformer and a CNN can make up for the deficiency of CNNs in establishing long-range dependencies and improve the ability of the model to capture global contextual information. In order to verify the effectiveness of the improvement of the model (ECA-TransUnet), we conducted ablation experiments and proposed two variants of the model: TransUnet-block1: in the decoder part, the ECA module is added after each layer of feature fusion, and the rest remains unchanged, and TransUnet-block2: only the FFN module is improved.

As shown in Table 3, ECA-TransUnet has better evaluation metrics compared with several other models, and it can be initially verified that our improvement of the model (ECA-TransUnet) is effective and can improve the ability to identify sea fog.

Table 3. Model ablation results.

	Recall	IoU	F1
TransUnet	0.8019	0.7662	0.8374
TransUnet-block1	0.8029	0.7860	0.8302
TransUnet-block2	0.8146	0.8059	0.8425
ECA-TransUnet	0.8231	0.8157	0.8587

Sea fog monitoring by deep learning methods has improved automation. However, it is worth noting that in the process of producing sea fog datasets, manual labeling inevitably leads to certain errors, especially the distinction between sea fog and low clouds, which were grouped together in previous studies [56,57], and how to effectively distinguish between sea fog and low clouds is a problem that needs to be noted in subsequent research. In addition, land and ocean masks were performed during the data processing, and the problem of sea fog identification under remote sensing intricate images should be considered in future research.

7. Conclusions

In this paper, we used MODIS remote sensing images and combined various approaches (CALIPSO data, cloud and sea fog texture features, and MODIS waveband features) to produce sea fog samples with improved accuracy in manual labeling and proposed an ECA-TransUnet model for monitoring sea fog in the Yellow and Bohai seas during the daytime.

We designed a new FFN module in the ECA-TransUnet network to make the model more focused on the better-performing feature information through a dual branch structure and added an ECA module in the decoder part of the model to further consider channel information. To address the extreme class imbalance in the sea fog sample data, we employed the focal loss function to improve the accuracy of the model in identifying daytime sea fog. Our proposed ECA-TransUnet model demonstrated improved performance in sea fog identification compared to some traditional CNN models, with an accuracy of 94.5%. In addition, we also analyzed the meteorological conditions of sea fog generation using ERA5 weather data. The range of sea fog identified by ECA-TransUnet is consistent with the conditions of sea fog generation, which indicates that the proposed model is feasible for daytime sea fog monitoring.

Author Contributions: H.L.: data curation, conceptualization, writing—original draft preparation, software, methodology, validation, and formal analysis; J.Z.: conceptualization, funding acquisition, writing—review and editing, and supervision; Y.M.: software, visualization, and writing—review; S.Z.: coding, visualization, and writing—review; X.Y. writing—reviewing and editing. All authors have read and agreed to the published version of the manuscript.

Funding: This research was jointly funded by the National Natural Science Fund of China (No. 41871253), the Open Fund of Key Laboratory of Urban Land Resources Monitoring and Simulation, Ministry of Natural Resources (No. KF-2021-06-081), the Central Guiding Local Science and Technology Development Fund of Shandong—Yellow River Basin Collaborative Science and Technology Innovation Special Project (No. YDZX2023019), the Shandong Natural Science Foundation of China (No. ZR2020QE281, No. ZR2017ZB0422), Central Guiding Local Science and Technology Development Fund of Shandong—Yellow River Basin Collaborative Science and Technology Innovation Special Project (No. YDZX2023019), and the "Taishan Scholar" Project of Shandong Province (No. TSXZ201712).

Data Availability Statement: Not applicable.

Conflicts of Interest: The authors declare no conflict of interest.

References

1. Bendix, J. A Satellite-Based Climatology of Fog and Low-Level Stratus in Germany and Adjacent Areas. *Atmos. Res.* **2002**, *64*, 3–18. [CrossRef]
2. Gultepe, I.; Tardif, R.; Michaelides, S.C.; Cermak, J.; Bott, A.; Bendix, J.; Mueller, M.D.; Pagowski, M.; Hansen, B.; Ellrod, G.; et al. Fog Research: A Review of Past Achievements and Future Perspectives. *Pure Appl. Geophys.* **2007**, *164*, 1121–1159. [CrossRef]
3. Zhang, S.; Li, M.; Meng, X.; Fu, G.; Ren, Z.; Gao, S. A Comparison Study Between Spring and Summer Fogs in the Yellow Sea-Observations and Mechanisms. *Pure Appl. Geophys.* **2012**, *169*, 1001–1017. [CrossRef]
4. Han, J.H.; Kim, K.J.; Joo, H.S.; Han, Y.H.; Kim, Y.T.; Kwon, S.J. Sea Fog Dissipation Prediction in Incheon Port and Haeundae Beach Using Machine Learning and Deep Learning. *Sensors* **2021**, *21*, 5232. [CrossRef] [PubMed]
5. Fu, G.; Guo, J.; Xie, S.-P.; Duane, Y.; Zhang, M. Analysis and High-Resolution Modeling of a Dense Sea Fog Event over the Yellow Sea. *Atmos. Res.* **2006**, *81*, 293–303. [CrossRef]
6. Heo, K.-Y.; Park, S.; Ha, K.-J.; Shim, J.-S. Algorithm for Sea Fog Monitoring with the Use of Information Technologies. *Meteorol. Appl.* **2014**, *21*, 350–359. [CrossRef]
7. Mahdavi, S.; Amani, M.; Bullock, T.; Beale, S. A Probability-Based Daytime Algorithm for Sea Fog Detection Using GOES-16 Imagery. *IEEE J. Sel. Top. Appl. Earth Obs. Remote Sens.* **2021**, *14*, 1363–1373. [CrossRef]

8. Du, P.; Zeng, Z.; Zhang, J.; Liu, L.; Yang, J.; Qu, C.; Jiang, L.; Liu, S. Fog Season Risk Assessment for Maritime Transportation Systems Exploiting Himawari-8 Data: A Case Study in Bohai Sea, China. *Remote Sens.* **2021**, *13*, 3530. [CrossRef]
9. Wang, Y.; Qiu, Z.; Zhao, D.; Ali, M.A.; Hu, C.; Zhang, Y.; Liao, K. Automatic Detection of Daytime Sea Fog Based on Supervised Classification Techniques for FY-3D Satellite. *Remote Sens.* **2023**, *15*, 2283. [CrossRef]
10. Fu, G.; Guo, J.; Pendergrass, A.; Li, P. An Analysis and Modeling Study of a Sea Fog Event over the Yellow and Bohai Seas. *J. Ocean Univ. China* **2008**, *7*, 27–34. [CrossRef]
11. Yang, J.-H.; Yoo, J.-M.; Choi, Y.-S. Advanced Dual-Satellite Method for Detection of Low Stratus and Fog near Japan at Dawn from FY-4A and Himawari-8. *Remote Sens.* **2021**, *13*, 1042. [CrossRef]
12. Ahn, M.; Sohn, E.; Hwang, B. A New Algorithm for Sea Fog/Stratus Detection Using GMS-5 IR Data. *Adv. Atmos. Sci.* **2003**, *20*, 899–913. [CrossRef]
13. Eyre, J.R. Detection of fog at night using Advanced Very High Resolution Radiometer (AVHRR) imagery. *Meteorol. Mag.* **1984**, *113*, 266–271.
14. Ellrod, G.P. Advances in the detection and analysis of fog at night using GOES multispectral infrared imagery. *Weather. Forecast.* **1995**, *10*, 606–619. [CrossRef]
15. Zhang, S.; Yi, L. A Comprehensive Dynamic Threshold Algorithm for Daytime Sea Fog Retrieval over the Chinese Adjacent Seas. *Pure Appl. Geophys.* **2013**, *170*, 1931–1944. [CrossRef]
16. Han, J.-H.; Suh, M.-S.; Yu, H.-Y.; Roh, N.-Y. Development of Fog Detection Algorithm Using GK2A/AMI and Ground Data. *Remote Sens.* **2020**, *12*, 3181. [CrossRef]
17. Ryu, H.-S.; Hong, S. Sea Fog Detection Based on Normalized Difference Snow Index Using Advanced Himawari Imager Observations. *Remote Sens.* **2020**, *12*, 1521. [CrossRef]
18. Wu, X.; Li, S. Automatic Sea Fog Detection over Chinese Adjacent Oceans Using Terra/MODIS Data. *Int. J. Remote Sens.* **2014**, *35*, 7430–7457. [CrossRef]
19. LeCun, Y.; Bengio, Y.; Hinton, G. Deep learning. *Nature* **2015**, *521*, 436. [CrossRef]
20. Zhang, L.; Zhang, L.; Du, B. Deep Learning for Remote Sensing Data A Technical Tutorial on the State of the Art. *IEEE Geosci. Remote Sens. Mag.* **2016**, *4*, 22–40. [CrossRef]
21. Luo, H.; Chen, C.; Fang, L.; Khoshelham, K.; Shen, G. MS-RRFSegNet: Multiscale Regional Relation Feature Segmentation Network for Semantic Segmentation of Urban Scene Point Clouds. *IEEE Trans. Geosci. Remote Sens.* **2020**, *58*, 8301–8315. [CrossRef]
22. Zhao, J.; Zhou, Y.; Shi, B.; Yang, J.; Zhang, D.; Yao, R. Multi-Stage Fusion and Multi-Source Attention Network for Multi-Modal Remote Sensing Image Segmentation. *ACM Trans. Intell. Syst. Technol.* **2021**, *12*, 1–20. [CrossRef]
23. Ding, L.; Zhang, J.; Bruzzone, L. Semantic Segmentation of Large-Size VHR Remote Sensing Images Using a Two-Stage Multiscale Training Architecture. *IEEE Trans. Geosci. Remote Sens.* **2020**, *58*, 5367–5376. [CrossRef]
24. Yu, Y.; Bao, Y.; Wang, J.; Chu, H.; Zhao, N.; He, Y.; Liu, Y. Crop Row Segmentation and Detection in Paddy Fields Based on Treble-Classification Otsu and Double-Dimensional Clustering Method. *Remote Sens.* **2021**, *13*, 901. [CrossRef]
25. Bi, H.; Xu, L.; Cao, X.; Xue, Y.; Xu, Z. Polarimetric SAR Image Semantic Segmentation With 3D Discrete Wavelet Transform and Markov Random Field. *IEEE Trans. Image Process.* **2020**, *29*, 6601–6614. [CrossRef]
26. Zhang, J.; Lu, H.; Xia, Y.; Han, T.; Miao, K.; Yao, Y.; Liu, C.; Zhou, J.P.; Chen, P.; Wang, B. Deep convolutional neural network for fog detection. In *Intelligent Computing Theories and Application, Proceedings of the 14th International Conference on Intelligent Computing, Wuhan, China, 15–18 August 2018*; Springer: Berlin/Heidelberg, Germany, 2018; pp. 1–10.
27. Jeon, H.-K.; Kim, S.; Edwin, J.; Yang, C.-S. Sea Fog Identification from GOCI Images Using CNN Transfer Learning Models. *Electronics* **2020**, *9*, 311. [CrossRef]
28. Zhu, C.; Wan, J.; Liu, S.; Xiao, Y. Sea Fog Detection Using U-Net Deep Learning Model Based on Modis Data. In Proceedings of the 2019 10th Workshop on Hyperspectral Imaging and Signal Processing: Evolution in Remote Sensing (WHISPERS), Amsterdam, The Netherlands, 24–26 September 2019.
29. Olaf, R.; Philipp, F.; Thomas, B. U-Net: Convolutional networks for biomedical image segmentation. In Proceedings of the Medical Image Computing and Computer-Assisted Intervention (MICCAI), Munich, Germany, 5–9 October 2015; pp. 234–241.
30. Zhou, Y.; Chen, K.; Li, X. Dual-Branch Neural Network for Sea Fog Detection in Geostationary Ocean Color Imager. *IEEE Trans. Geosci. Remote Sens.* **2022**, *60*, 4208617. [CrossRef]
31. Huang, Y.; Wu, M.; Guo, J.; Zhang, C.; Xu, M. A Correlation Context-Driven Method for Sea Fog Detection in Meteorological Satellite Imagery. *IEEE Geosci. Remote Sens. Lett.* **2022**, *19*, 1003105. [CrossRef]
32. Goodfellow, I.; Pouget-Abadie, J.; Mirza, M.; Xu, B.; Warde-Farley, D.; Ozair, S.; Courville, A.; Bengio, Y. Generative adversarial networks. *Commun. ACM* **2020**, *63*, 139–144. [CrossRef]
33. Yu, Q.; Zheng, N.; Huang, J.; Zhao, F. CNSNet: A Cleanness-Navigated-Shadow Network for Shadow Removal. *arXiv* **2022**, arXiv:2209.02174.
34. Han, W.; Zhang, Z.; Zhang, Y.; Yu, J.; Chiu, C.-C.; Qin, J.; Gulati, A.; Pang, R.; Wu, Y. ContextNet: Improving convolutional neural networks for automatic speech recognition with global context. In Proceedings of the Annual Conference of the International Speech Communication Association (Interspeech), Shanghai, China, 25–29 October 2020; pp. 3610–3614.
35. Dosovitskiy, A.; Beyer, L.; Kolesnikov, A.; Weissenborn, D.; Zhai, X.; Unterthiner, T.; Dehghani, M.; Minderer, M.; Heigold, G.; Gelly, S.; et al. An image is worth 16 $\times$ 16 words: Transformers for image recognition at scale. *arXiv* **2020**, arXiv:2010.11929.

36. Song, R.; Feng, Y.; Cheng, W.; Mu, Z.; Wang, X. BS2T: Bottleneck Spatial–Spectral Transformer for Hyperspectral Image Classification. *IEEE Trans. Geosci. Remote Sens.* **2022**, *60*, 5532117. [CrossRef]
37. He, X.; Zhou, Y.; Zhao, J.; Zhang, D.; Yao, R.; Xue, Y. Swin Transformer Embedding UNet for Remote Sensing Image Semantic Segmentation. *IEEE Trans. Geosci. Remote Sens.* **2022**, *60*, 4408715. [CrossRef]
38. Zou, J.; He, W.; Zhang, H. LESSFormer: Local-Enhanced Spectral-Spatial Transformer for Hyperspectral Image Classification. *IEEE Trans. Geosci. Remote Sens.* **2022**, *60*, 5535416. [CrossRef]
39. Yang, X.; Cao, W.; Lu, Y.; Zhou, Y. Hyperspectral Image Transformer Classification Networks. *IEEE Trans. Geosci. Remote Sens.* **2022**, *60*, 5528715. [CrossRef]
40. Chen, J.; Lu, Y.; Yu, Q.; Luo, X.; Adeli, E.; Wang, Y.; Lu, L.; Yuille, A.L.; Zhou, Y. Transunet: Transformers make strong encoders for medical image segmentation. *arXiv* **2021**, arXiv:2102.04306.
41. Wu, D.; Lu, B.; Zhang, T.; Yan, F. A Method of Detecting Sea Fogs Using CALIOP Data and Its Application to Improve MODIS-Based Sea Fog Detection. *J. Quant. Spectrosc. Radiat. Transf.* **2015**, *153*, 88–94. [CrossRef]
42. Kim, D.; Park, M.-S.; Park, Y.-J.; Kim, W. Geostationary Ocean Color Imager (GOCI) Marine Fog Detection in Combination with Himawari-8 Based on the Decision Tree. *Remote Sens.* **2020**, *12*, 149. [CrossRef]
43. Shin, D.; Kim, J.-H. A New Application of Unsupervised Learning to Nighttime Sea Fog Detection. *ASIA Pac. J. Atmos. Sci.* **2018**, *54*, 527–544. [CrossRef]
44. Wan, J.; Su, J.; Sheng, H.; Liu, S.; Li, J.J. Spatial and Temporal Characteristics of Sea Fog in Yellow Sea and Bohai Sea Based on Active and Passive Remote Sensing. In Proceedings of the IGARSS 2020—2020 IEEE International Geoscience and Remote Sensing Symposium, Waikoloa, HI, USA, 26 September–2 October 2020.
45. Zhang, S.-P.; Xie, S.-P.; Liu, Q.-Y.; Yang, Y.-Q.; Wang, X.-G.; Ren, Z.-P. Seasonal Variations of Yellow Sea Fog: Observations and Mechanisms. *J. Clim.* **2009**, *22*, 6758–6772. [CrossRef]
46. Holz, R.E.; Ackerman, S.A.; Nagle, F.W.; Frey, R.; Dutcher, S.; Kuehn, R.E.; Vaughan, M.A.; Baum, B. Global Moderate Resolution Imaging Spectroradiometer (MODIS) Cloud Detection and Height Evaluation Using CALIOP. *J. Geophys. Res. Atmos.* **2008**, *113*, D00A19. [CrossRef]
47. Vaswani, A.; Shazeer, N.; Parmar, N.; Uszkoreit, J.; Jones, L.; Gomez, A.N.; Kaiser, Ł.; Polosukhin, I. Attention is all you need. In *Advances in Neural Information Processing Systems*; MIT Press: Cambridge, MA, USA, 2017; pp. 5998–6008.
48. Hu, J.; Shen, L.; Albanie, S.; Sun, G.; Wu, E. Squeeze-and-Excitation Networks. *IEEE Trans. Pattern Anal. Mach. Intell.* **2020**, *42*, 2011–2023.
49. Wang, Q.; Wu, B.; Zhu, P.; Li, P.; Zuo, W.; Hu, Q. ECA-Net: Efficient Channel Attention for Deep Convolutional Neural Networks. *arXiv* **2019**, arXiv:1910.03151.
50. Powers, D.M.W. Evaluation: From Precision, Recall and F-Measure to ROC, Informedness, Markedness and Correlation. *arXiv* **2020**, arXiv:2010.16061.
51. Chen, L.; Niu, S.; Zhong, L. Detection and Analysis of Fog Based on MODIS Data. *J. Nanjing Inst. Meteorol.* **2006**, *29*, 448–454.
52. Shorten, C.; Khoshgoftaar, T.M. A Survey on Image Data Augmentation for Deep Learning. *J. Big Data* **2019**, *6*, 60. [CrossRef]
53. Lin, T.; Goyal, P.; Girshick, R.; He, K.; Dollár, P. Focal Loss for Dense Object Detection. *IEEE Trans. Pattern Anal. Mach. Intell.* **2020**, *42*, 318–327. [PubMed]
54. Chen, L.C.; Zhu, Y.; Papandreou, G.; Schroff, F.; Adam, H. Encoder-decoder with atrous separable convolution for semantic image segmentation. In Proceedings of the European Conference on Computer Vision (ECCV), Munich, Germany, 8–14 September 2018; pp. 801–818.
55. Long, J.; Shelhamer, E.; Darrell, T. Fully convolutional networks for semantic segmentation. In Proceedings of the IEEE Conference on Computer Vision and Pattern Recognition, Boston, MA, USA, 7–12 June 2015; pp. 3431–3440.
56. Cermak, J.; Bendix, J. A novel approach to fog/low stratus detection using Meteosat 8 data. *Atmos. Res.* **2008**, *87*, 279–292. [CrossRef]
57. Gao, S.H.; Wu, W.; Zhu, L.; Fu, G. Detection of nighttime sea fog/stratus over the Huang-hai Sea using MTSAT-1R IR data. *Acta Oceanol. Sin.* **2009**, *28*, 23–35.

remote sensing

MDPI

Article

MCPT: Mixed Convolutional Parallel Transformer for Polarimetric SAR Image Classification

Wenke Wang [1], Jianlong Wang [1,*], Bibo Lu [1], Boyuan Liu [1], Yake Zhang [2] and Chunyang Wang [1]

1 School of Computer Science and Technology, Henan Polytechnic University, Jiaozuo 454003, China; pomeloru3126@home.hpu.edu.cn (W.W.); lubibo@hpu.edu.cn (B.L.); liuboyuan20040320@home.hpu.edu.cn (B.L.); wcy@hpu.edu.cn (C.W.)
2 School of Computer and Information Engineering, Henan Normal University, Xinxiang 453007, China; 2023094@htu.edu.cn
* Correspondence: wangjianlong24@hpu.edu.cn

Abstract: Vision transformers (ViT) have the characteristics of massive training data and complex model, which cannot be directly applied to polarimetric synthetic aperture radar (PolSAR) image classification tasks. Therefore, a mixed convolutional parallel transformer (MCPT) model based on ViT is proposed for fast PolSAR image classification. First of all, a mixed depthwise convolution tokenization is introduced. It replaces the learnable linear projection in the original ViT to obtain patch embeddings. The process of tokenization can reduce computational and parameter complexity and extract features of different receptive fields as input to the encoder. Furthermore, combining the idea of shallow networks with lower latency and easier optimization, a parallel encoder is implemented by pairing the same modules and recombining to form parallel blocks, which can decrease the network depth and computing power requirement. In addition, the original class embedding and position embedding are removed during tokenization, and a global average pooling layer is added after the encoder for category feature extraction. Finally, the experimental results on AIRSAR Flevoland and RADARSAT-2 San Francisco datasets show that the proposed method achieves a significant improvement in training and prediction speed. Meanwhile, the overall accuracy achieved was 97.9% and 96.77%, respectively.

Keywords: polarimetric SAR; convolutional neural network; vision transformer; mixed depthwise convolution tokenization; parallel encoder; global average pooling

Citation: Wang, W.; Wang, J.; Lu, B.; Liu, B.; Zhang, Y.; Wang, C. MCPT: Mixed Convolutional Parallel Transformer for Polarimetric SAR Image Classification. *Remote Sens.* **2023**, *15*, 2936. https://doi.org/10.3390/rs15112936

Academic Editors: Moulay A. Akhloufi and Mozhdeh Shahbazi

Received: 5 May 2023
Revised: 31 May 2023
Accepted: 2 June 2023
Published: 5 June 2023

1. Introduction

Synthetic Aperture Radar (SAR) [1] is a remote sensing sensor belonging to the active imaging technology that operates in the microwave band. It has many extraordinary characteristics, such as full-time work, being less affected by weather and certain penetration ability to ground objects [2–5]. Therefore, SAR-related applications affect various aspects of human production and life, such as land use, urban planning and crop yield assessment [6]. Polarimetric SAR (PolSAR) is a type of SAR that works in different forms of polarization combinations. It captures information on object material composition, geometric features and azimuth, and provides more comprehensive object descriptions for different deployment scenarios. Thus, it has become a research hotspot in the field of SAR remote sensing [7]. PolSAR image classification [8] refers to the process of dividing all pixels in an image into a specific category according to specific rules. It is a fundamental research direction in PolSAR image understanding and interpretation techniques. Automatically analyzing radar image features to classify and recognize images is very beneficial in improving the efficiency of human image understanding, while also greatly enhancing the ability of people to obtain image information.

In traditional PolSAR image feature extraction methods, such as polarization target decomposition [9], the scattering mechanisms of targets are interpreted based on practical physical constraints. The polarimetric data are decomposed into several physically meaningful parameters to facilitate the analysis of complex scattering processes. So far, several decomposition methods have been proposed, including Cameron decomposition [10], Cloude–Pottier decomposition [11,12], Krogager decomposition [13], and so on. However, these methods can only determine target scattering types and not specifically classify them into land cover types. Moreover, large amounts of PolSAR data are being obtained with the maturing development of various earth observation platforms. Traditional methods based on polarization decomposition and data characteristics cannot meet the expanding interpretation requirements. At the same time, deep learning technology has continued to evolve, demonstrating powerful capabilities for processing data and extracting features. As a result, more and more researchers are applying it to PolSAR image classification research [14]. Compared to traditional methods, deep learning methods use labeled data to train neural networks to obtain features, establish relationships between data feature information and categories, and predict the category of unknown pixels. Such data-driven feature extraction frameworks are also known as representation learning [15].

At present, convolutional neural networks (CNNs) [16] are the most widely used representation learning method in PolSAR image classification tasks. Zhou et al. [17] first studied the application of deep CNNs in PolSAR image classification and designed a four-layer CNN customized for PolSAR image classification that automatically learns different levels of polarized spatial features from data through two cascaded convolutional layers. By taking the spatial characteristics of PolSAR images into account, it is able to successfully distinguish between sloping built-up areas with vegetation. These are often mixed in the polarization feature space. Since then, various CNN variants for PolSAR image classification have been proposed. For example, Chen et al. [18] attempted to exploit a limited number of labeled samples to train deep CNNs for PolSAR image classification and proposed a polarimetric-feature-driven deep CNN classification scheme. Classic roll invariant polarimetric features and hidden polarimetric features in the rotation domain were used to support the training of the deep CNN classifier and to enhance the final classification performance. The model has better performance compared with traditional CNN. However, the polarization features used for training are derived from field knowledge [19–21], which makes the acquisition of labeled data for the method a time-consuming and difficult task. Yang et al. [22] pointed out that selecting an appropriate number of high-quality features obtained by target decomposition of PolSAR images is crucial for the tasks. Thus, a CNN-based PolSAR image classification feature selection method was proposed to select feature subsets. The best-performing one is selected as the final result. Taking the performance of feature combinations into account, the selected subset of features performs better in both traditional and deep learning classification methods. In order to improve the performance of CNNs on limited training data, Shang et al. [23] proposed a new densely connected and depthwise separable convolutional neural network. Depthwise separable convolution extracts features independently from each channel of the PolSAR image and dense connection is introduced to directly connect non-adjacent layers. In this way, it can avoid extracting redundant features, reuse different levels of feature maps from PolSAR images and reduce the number of training parameters. Nonetheless, as research on CNN-based PolSAR image classification continues to deepen, the locality caused by inductive biases has become a bottleneck restricting its performance, making it difficult for CNN-based PolSAR image classification methods to make further improvements.

The transformer [24] model was originally proposed in the field of natural language processing (NLP), mainly by pre-training on a large text corpus, followed by fine-tuning on a smaller specific task dataset. Inspired by the success of the transformer in NLP, Dosovitskiy et al. [25] proposed a vision transformer model that can be easily applied to the image domain with minimal modifications to the standard transformer model. ViT utilizes a self-attention mechanism (SA) [24] to replace tokens (words) in languages with image patches, which can achieve remote interaction between pixels and capture global correlations in the image domain [26]. Dong et al. [27] explored the ViT model for PolSAR image classification in detail and proposed a representation learning framework based on ViT, which included both supervised and unsupervised learning. In the supervised learning framework, image patches are used as input, and SA is utilized to extract global features. In addition, an improved contrastive-based strategy is introduced to achieve simple unsupervised representation learning. Compared to CNNs and their variants, ViT improves classification performance by building more global representations, while also demonstrating robustness to the initial input form. These studies may prompt people to rethink the dominant position of CNNs in PolSAR image classification. As research on ViT in PolSAR image classification continues to deepen, many ViT-based classification methods have emerged. For example, Wang et al. [28] pointed out that the scarcity of PolSAR-labeled samples and the small receptive field of models limit the performance of deep learning methods for PolSAR image classification and then proposed a ViT-based classification method. It can receive PolSAR images of different resolutions and is pre-trained with masked autoencoders with unlabeled data to address the problem of scarce labeled data. Correspondingly, Jamali et al. [29] demonstrated that the main problem with applying ViT to PolSAR image classification is the scarcity of labeled data and proposed a ViT-based framework. Both 3-D and 2-D CNNs are adopted as feature extractors and a local window attention is implemented to enhance local feature representation power for effective classification of PolSAR data. Therefore, it can greatly reduce annotation costs and hardware requirements.

As can be seen from the existing work, most of the ViT-based classification methods proposed so far are basic applications of ViT to PolSAR image classification tasks. There is a lack of deeper integration and exploration between ViT and PolSAR image classification. However, convolutional structures still dominate in PolSAR image representation learning and classification. The local nature of convolution is not always advantageous as it extracts local neighborhood features and does not take the global information into account. Consequently, when extracting features from PolSAR patches centered around the input pixels, there is a large amount of local information in the extracted features that leads to low efficiency or accuracy. Although the application of ViT partially addresses these issues, research on ViT in PolSAR image classification is still limited and not deep enough. In addition, the pre-training of ViT requires sufficient labeled data but annotating every pixel in a PolSAR dataset is time-consuming and costly. Based on the above analysis and inspired by previous works, a mixed convolutional parallel transformer model is proposed for PolSAR image classification. It improves the original ViT model with better targeting for PolSAR image classification. Furthermore, the model achieves higher training and prediction speed while maintaining high accuracy with fewer parameters and lower computational cost. There are three main improvements to the proposed model:

(1) Mixed depthwise convolution tokenization. A mixed depthwise convolution (MixConv) is introduced in the data pre-processing part of the model for the tokenization of the input data, which replaces the linear projection used in the original ViT. MixConv naturally mixes multiple kernel sizes in a single convolution and can extract feature maps with different receptive fields at the same time. This tokenization process makes the proposed method more flexible than the original ViT. It is no longer limited by the input resolution, which must be strictly divisible by a pre-defined patch size. It also facilitates the removal of position embedding and enriches training data information. Additionally, the class and position embeddings in ViT are removed, further reducing the number of parameters and computation cost.

(2) Parallel encoder. The idea of parallel structure is introduced to implement an parallel encoder. It can still have a relatively low depth when superimposed with multiple encoders during training, thus achieving a lower latency and making it easier to optimize. Consequently, the speed of training and prediction is accelerated and good training results can be achieved even on less powerful personal platforms.

(3) Global average pooling. The global average pooling (GAP) method is used as a substitute for class embedding. A GAP layer is added after the encoder and the

average of all the pixels in the feature maps of each channel is taken to give an output. This operation is simple and effective, does not increase additional parameters, and avoids over-fitting. It summarizes spatial information and is more stable to spatial transformations of the input.

The remaining sections of this article are organized as follows. Section 2 provides a brief introduction to the relevant technical background and describes in detail the specific improvements proposed by our method. In Section 3, the experimental results are presented and analyzed comprehensively. Finally, Section 4 concludes the study and offers prospects for future works.

2. The Proposed Method

As input data for the method proposed in this paper, the initial form of the PolSAR data is a 9-D real vector [30]: $[T_{11}, T_{22}, T_{33}, \mathrm{Re}[T_{12}], \mathrm{Im}[T_{12}], \mathrm{Re}[T_{13}], \mathrm{Im}[T_{13}], \mathrm{Re}[T_{23}], \mathrm{Im}[T_{23}]]$, where $\mathrm{Re}[\bullet]$ and $\mathrm{Im}[\bullet]$ denote real and imaginary parts, respectively. Its rows represent the feature dimensions, while columns indicate the total number of pixels in the image. The overall structure of the proposed method is illustrated in Figure 1. First of all, we extract pixel-centric neighborhoods of the PolSAR data as image patches of the same size, which serve as the initial input to the model. These image patches are then transformed into 1-D token vectors through mixed depthwise convolution tokenization, without additional class and position embeddings. Moreover, a multi-layer parallel encoder with two branches extracts global information from the input data and obtains the importance weights of each token vector relative to others. Finally, a global average pooling layer is applied to output the features of each class. After incorporating a softmax classifier, pixels can be classified into specific categories and the final output results are obtained. These improvements are described in more detail below.

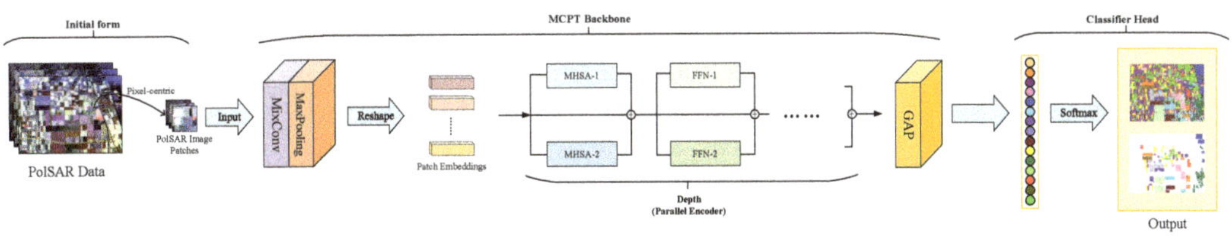

Figure 1. The scheme of the proposed PolSAR image classification method.

2.1. Mixed Depthwise Convolution Tokenization

CNNs have been widely used in image classification [31–33], segmentation [34–36], detection [37–40] and many other applications [41–44]. The latest design trend is to improve accuracy or efficiency. Following this trend, depthwise convolutions have become more prevalent in CNNs, such as MobileNet [45], ShuffleNet [46], NASNet [47], etc. Different from conventional convolutions, depthwise convolutions apply the kernel separately to each channel, thereby reducing the computational cost by a factor of c, where c is the number of channels. When designing convolutional networks with depthwise convolutional kernels, the kernel size is an important factor but is often overlooked. Traditionally, a simple 3×3 kernel is used, but recent research indicates that larger kernel sizes such as 5×5 and 7×7 kernels can probably improve the accuracy and efficiency of the model. Tan et al. [48] systematically studied the impact of kernel size and found that a single kernel size is limited. On this basis, the MixConv was proposed, which mixes different kernel sizes in a single convolutional operation, making it easy to capture different patterns at different resolutions. Figure 2 shows the structure of MixConv. It divides the channels into several groups and applies different kernel sizes to each group of channels.

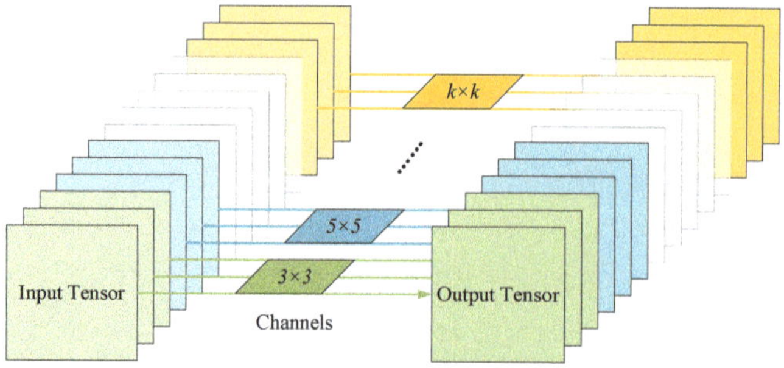

Figure 2. Mixed depthwise convolution (MixConv).

ViT splits the original image into equally sized patches and treats them as words in the text. Since the standard transformer receives token embeddings of 1-D sequences as input, a learnable linear projection is utilized to flatten the patches and map to D dimensions. The processed patches are referred to as patch embeddings, where D is a constant latent vector size that is used throughout all of the ViT layers. To simplify the image representation output by the encoder, ViT introduces a class token by adding a learnable class embedding to the previous patch embeddings. In addition, ViT adds a trainable 1-D position embedding to the patch embeddings to preserve position information. The resulting sequence of patch embedding vectors containing both class and position embeddings is applied as the input to the encoder. During training, the class embedding interacts with each patch embedding to obtain the image representation and serves as the output of the encoder. A classification head is then connected to obtain the final output categories of the model. However, the additional class embedding increases both computational and parameter requirements, while the position embedding fixes the sequence length and constrains the feature information to a immovable scale. To address these issues, we implement tokenization with a modified MixConv. During tokenization, the receptive field of the model is effectively expanded to capture data information at different levels by employing convolution kernels of various sizes. Moreover, tokenization serves as a feature extraction process, where the resulting multi-scale feature maps produced by different receptive fields are presented as input to the encoder. Such tokenization enriches the training data and allows the input resolution to no longer be strictly limited by the preset patch size. Furthermore, the convolution blocks can also be stacked and further down-sampled, even facilitating the removal of position embedding in the model [49]. The implementation details are described below.

As shown in Figure 3, PolSAR patches of size $m \times m \times 9$ are first extracted around the center pixels, and these patches are then used as the input data for processing. Since ViT only accepts a sequence of token embeddings, tokenization is essential to transform the 3-D image patches and map them to 2-D patch embeddings. Firstly, the input image patches are grouped according to the number of output channels, with an average division method selected such that d channels are equally divided into 3 groups, denoted as G_1, G_2, and G_3, respectively, where $G_1 + G_2 + G_3 = d$ and d refers to the dimension of the 2-D patch embeddings. For each group, convolution operations are performed with kernels of sizes $[3 \times 3, 5 \times 5, 7 \times 7]$ and a stride of 3, producing feature maps of the shape $\left(\frac{m}{3}, \frac{m}{3}, d\right)$. After passing through a max-pooling layer with a size of 3×3 and stride of 1, the original PolSAR patch is split into N^2 feature patches of shape $\left(\frac{m}{N}, \frac{m}{N}, 9\right)$. Eventually, each feature patch is flattened along the spatial dimension to form $\left(\frac{m}{N} \cdot \frac{m}{N} \cdot 9\right)$ -D vectors. These vectors are stacked to reconstruct the input from $\mathbb{R}^{m \times m \times 9}$ to $\mathbb{R}^{N^2 \times \left(\frac{m}{N} \cdot \frac{m}{N} \cdot 9\right)}$, which completes the tokenization process. These vectors are called patch embeddings and serve as input to the subsequent parallel encoder. Unlike the mainstream ViT model [26], class embedding and position embedding are not employed in this work. The class embedding can be replaced by a GAP operation. Furthermore, the classification task does not focus on the location information in the image and there is no strict positional relationship between the pixels in the input PolSAR image patches [50]. Therefore, these two learnable parameters are removed.

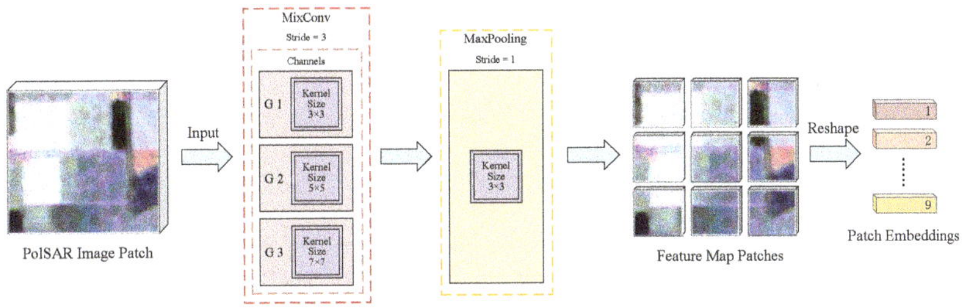

Figure 3. Mixed depthwise convolution tokenization.

2.2. Parallel Encoder

In ViT, the encoder architecture follows the standard transformer structure, consisting of multi-head self-attention (MHSA) and multi-layer perceptron (MLP) layers [25]. Layer normalization (LN) is applied before each layer and residual connection is applied after each layer. The MLP layer contains two layers with Gaussian error linear unit (GELU) [51] activation functions. Compared with CNNs, ViT has fewer image-specific biases as only the MLP layers are local and translation-invariant. Whereas, the MHSA operates globally and utilizes 2-D neighborhoods rarely. In addition, the position embedding is initialized without 2-D position information about the image patches, which requires the model to learn the spatial relationships for all of the image patches from scratch. ViT treats the image as a sequence of image patches and processes them with the transformer architecture that is used in NLP. The strategy is simple and scalable and has shown significant performance gains when combined with pre-training on large data sets. Consequently, ViT achieves state-of-the-art results on many image classification datasets [25]. The overall structure of the ViT model is shown in Figure 4 below.

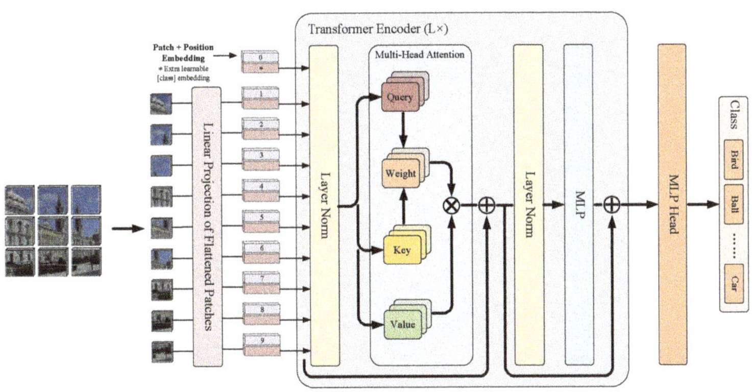

Figure 4. Vision Transformer model overview.

In recent years, transformer architecture has made a significant impact on the design of the computer vision field. Many architectures [52–56] have directly inherited some designs from transformers or have been inspired by their latest discoveries. Despite the significant progress made by transformers, there is still considerable work to be done in optimizing how they design and train. In neural network architecture design, there is often a debate about how to balance width and depth. Since the introduction of residual networks [57], the obstacles of optimization for deep networks have been significantly reduced. As a result, neural networks have evolved in a deeper direction. However, some studies have shown that shallower networks have lower latency and are easier to optimize [58,59].

The original ViT encoder consists of a series of layers, including normalization layers, multi-head self-attention layers, feed-forward network (FFN) layers, residual structures, etc. All of these layers connected in a fixed order, which results in an ever-growing network depth as the number of encoders continues to increase. Such a design makes the model more powerful but also increases the computational load. Therefore, Touvron et al. [60] proposed a parallel ViT approach by duplicating matching blocks to parallelize the architecture, which can be achieved for any number of parallel blocks. Then, a wider and shallower architecture was created with the same number of parameters and calculations. Depending on the implementation, the design allows for more parallel processing, simplifies the optimization process, and can reduce latency. The idea of shallow networks is adopted by designing the encoder with a parallelized structure, which maintains a relatively low depth even when multiple encoders are stacked for training. Therefore, the proposed method can achieve lower latency and better optimization effects. When trained on platforms with lower computational power, the model produces a good result which demonstrates its applicability.

To capture the long-range interactions between pixels in PolSAR image patches, it is necessary to obtain importance weights for each pixel relative to other pixels. The self-attention mechanism is the main way to accomplish that goal. For each element in the input sequence $X \in \mathbb{R}^{N\times D}$, it is mapped into three variables with the specific dimensions query q, key k, and value v by a learnable linear projection matrix W_{qkv}. The importance weight of this element relative to other elements is obtained by calculating the weighted sum of all values v in the sequence. The attention weights matrix A_{ij} is based on the pairwise similarity between two elements in the sequence and their respective representations of query q^i and key k^j.

$$[q,k,v] = XW_{qkv} \qquad W_{qkv} \in \mathbb{R}^{D\times 3D_h}, \tag{1}$$

$$A = \mathrm{softmax}\left(\frac{qk^T}{\sqrt{D_h}}\right) \qquad A \in \mathbb{R}^{N\times N}, \tag{2}$$

$$\mathrm{SA}(X) = Av. \tag{3}$$

MHSA is an extension of SA. When the input patches are embedded, they are linearly projected into queries and keys of dimensions d_k, and values of dimensions d_v. These queries, keys, and values are then transformed by h groups of different learnable linear projections. Next, these h sets of transformed queries, keys, and values are processed in parallel through attention pooling. Finally, the outputs of these h attention pooling operations are concatenated together and transformed by another learnable linear projection to produce the final output. This design is referred to as multi-head self-attention, where each of the h outputs of the attention pooling is referred to as a head. If these queries, keys, and values are packed into matrices Q, K, and V, respectively, the output matrix can be represented as:

$$\mathrm{MHSA}(Q,K,V) = \mathrm{Concat}(\mathrm{h}_1,\ldots,\mathrm{h}_\mathrm{h})W^O \tag{4}$$

$$\mathrm{h_i} = \mathrm{SA}_i(QW_i^Q, KW_i^K, VW_i^V) \tag{5}$$

where $W_i^Q \in \mathbb{R}^{d_{model}\times d_k}, W_i^K \in \mathbb{R}^{d_{model}\times d_k}, W_i^V \in \mathbb{R}^{d_{model}\times d_v}$, and $W^O \in \mathbb{R}^{hd_v\times d_{model}}$ are parameter matrices for the linear projections. The FFN consists of two fully connected layers (FC) and a GELU activation function, which can be represented as:

$$\mathrm{FFN}(x) = \max(0, xW_1 + b_1)W_2 + b_2 \tag{6}$$

Therefore, the original ViT encoder can be represented as follows:

$$x'_{l+1} = x_l + \mathrm{MHSA}_l(x_l), \qquad x_{l+1} = x'_{l+1} + \mathrm{FFN}_l(x'_{l+1}), \tag{7}$$

$$x'_{l+2} = x_{l+1} + \mathrm{MHSA}_{l+1}(x_{l+1}), \qquad x_{l+2} = x'_{l+2} + \mathrm{FFN}_{l+1}(x'_{l+2}), \tag{8}$$

where $MHSA_l(\bullet)$ and $MHSA_{l+1}(\bullet)$ represent MHSA residual blocks, and $FFN_l(\bullet)$ and $FFN_{l+1}(\bullet)$ represent FFN residual blocks. This series of operations can be replaced by a set of parallel structures:

$$x_{l+1} = x_l + MHSA_{l,1}(x_l) + MHSA_{l,2}(x_l) \quad (9)$$

$$x_{l+2} = x_{l+1} + FFN_{l,1}(x_{l+1}) + FFN_{l,2}(x_{l+1}) \quad (10)$$

For a given number of MHSA and FFN blocks, this will reduce the number of layers by two. As shown in Figure 5, the architecture can be parallelized by pairing identical blocks to achieve any number of parallel blocks. The tokenized patch embeddings are fed into the encoder via three parallel paths to extract global features, which are processed by a number of parallel blocks to obtain the final feature maps.

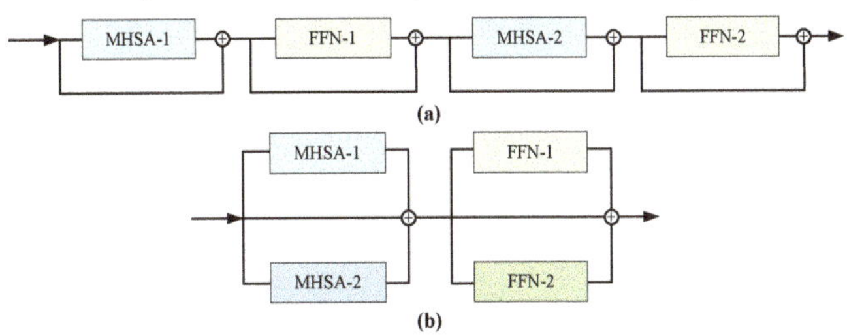

Figure 5. The structure of different encoders. (**a**) Original serial structure encoder block. (**b**) Proposed parallel structure encoder block.

2.3. Global Average Pooling

When a convolutional neural network is applied to a classification task, the feature map of the last convolutional layer in the network is usually vectorized and mapped to the sample label space by a fully connected (FC) layer, and finally, the final output class is obtained by a softmax classifier [61,62]. Such a structure treats the convolutional layers as feature extractors that can be flexibly applied to different tasks. However, the FC layer has a large number of parameters, which leads to a significant increase in the computational power of the network. It also tends to cause overfitting and affects the generalization ability of the network. Therefore, global average pooling [63] is proposed to replace the FC layer to solve the above problem. As shown in Figure 6, unlike traditional CNNs that add FC layers at the end of the network, GAP takes the average of each feature map and feeds the resulting vector directly into the softmax classifier to obtain the category output. By emphasizing the correspondence between feature maps and categories, GAP is more suitable for convolutional structures and easily interprets feature maps as category confidence maps. In addition, GAP is simple and effective. It does not add any additional parameters, which can prevent overfitting. Furthermore, the spatial information is summed to improve the stability of the model to the input spatial transformation.

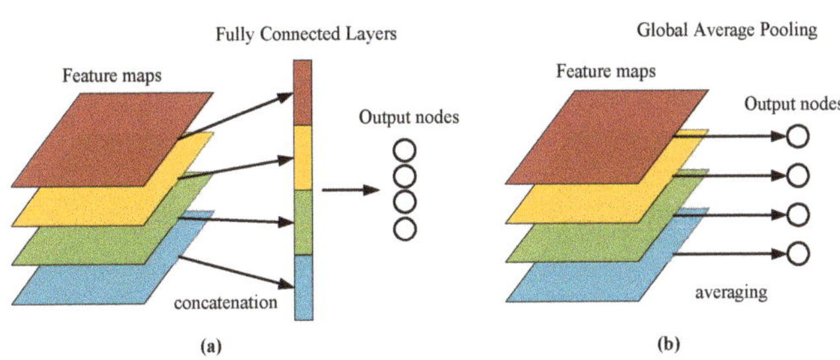

Figure 6. Comparison diagram using the fully connected layer and the global average pooling layer. (**a**) Obtain output using fully connected layers. (**b**) Obtain output using global average pooling layer.

In the original ViT model, the class embedding was trained together with other features as part of the global interaction characteristics of ViT. As a result, the trained class embedding can be used as the final output of the model. As the learnable class and positional embeddings are removed in the tokenization process, it is impossible to obtain feature representations of the original data through class embedding. Therefore, a GAP layer is added after the parallel encoder to replace the class embedding. By adding a GAP layer after the parallel encoder, the mean value of all pixels in each channel of the feature map is calculated to obtain one output per feature map. A softmax classification head is connected at the end of the model, which consists of an MLP layer, and the final classification task is achieved by using this classification head to obtain the corresponding category.

3. Experimental Analysis and Results

3.1. Datasets Description

(1) AIRSAR Flevoland: The Flevoland image is a 750 × 1024 subimage of the L-band multi-view PolSAR dataset acquired by the AIRSAR platform on 16 August 1989. The ground resolution of the image is 6.6 m × 12.1 m, and it includes 15 kinds of ground objects, each represented by a unique color. Figure 7a illustrates the Pauli map and Figure 7b shows the ground truth map of the dataset. Figure 7c shows the corresponding ground truth map and legend of the dataset, which consists of 167,712 labeled pixels [64].

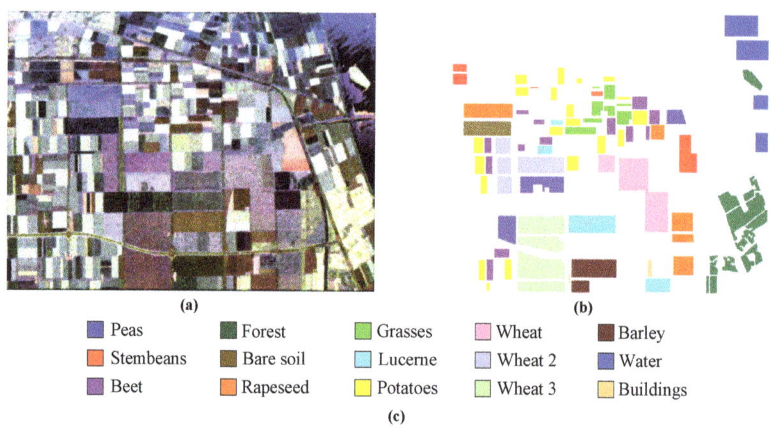

Figure 7. AIRSAR Flevoland dataset and its color code. (**a**) Pauli RGB map. (**b**) Ground truth map. (**c**) Legend of the dataset.

(2) RADARSAT-2 San Francisco: The second dataset is a San Francisco Bay Area image with C-band acquired by the RADARSAT-2 satellite. Figure 8a displays the PauliRGB image of the selected scene with a size of 1380 × 1800, which primarily contains five land cover types: high-density urban areas, water, vegetation, developed urban areas, and low-density urban areas. Figure 8b shows the ground truth map consisting of 1804087 pixels with known label information. Figure 8c respectively show the corresponding ground truth map and the legend explaining the land cover types [65].

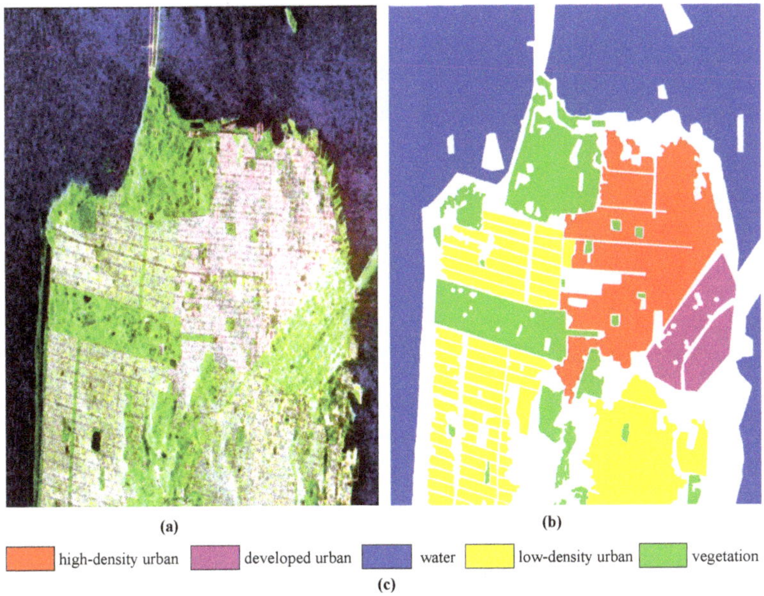

Figure 8. RADARSAT-2 San Francisco dataset and its color code. (**a**) Pauli RGB map. (**b**) Ground truth map. (**c**) Legend of the dataset.

3.2. Experimental Setup

The experiments were conducted on a personal laptop computer with the model Legion Y9000P IAH7H, equipped with an NVIDIA GeForce RTX3060 Laptop GPU with 6 GB of graphics memory. The proposed method was implemented using the PyTorch-GPU deep learning framework with PyTorch version 1.11.0 and CUDA version 12.0. Specific details of the experimental setup will be discussed in the following sections.

Since the input for ViT is in image format, a neighborhood with a size of 14×14 is extracted centered on pixel. Therefore, the value of the input space size m is set to 15. The MixConv kernel sizes are set to $(3, 5, 7)$, and the number of output channels d is set to 225. The computation times of the SA, i.e., the number of heads, are set to 4, with the dimension of each head d_h set to 76. The number of parallel branches is set to 2, and the depth of the parallel encoder, i.e., the number of parallel blocks, is set to 3. The output of each MHSA and FFN layer in parallel branches is connected to its input through a residual connection. The hidden node numbers in the FC layer of the FFN are set to 900, and the residual structure is also used. Layer normalization is applied before each layer of the parallel encoder to accelerate convergence.

The Flevoland and San Francisco datasets have 167,712 and 1,804,087 labeled data, respectively. For different dataset experiments, 375 labeled data from each category (3.35% and 0.1% of the labeled data in the Flevoland and San Francisco datasets, respectively) are randomly selected for training and validation, while the remaining labeled data are used for testing. To better measure the model's performance, 5-fold cross-validation is used for training. During training, each model in the 5 folds is trained for 150 epochs with a batch size of 256. The Adam optimizer is used and the initial learning rate is set to 0.001. The loss function is the original cross-entropy loss function in ViT. Moreover, class balance is ensured in each subset of data for each fold. By testing on labeled data and selecting the fold with the highest overall accuracy (OA) among the 5-fold models, the best-performing fold is chosen as the final training model for subsequent result prediction.

To further demonstrate the effectiveness of the proposed method, several PolSAR image classification methods based on CNN or ViT are selected and tested in the following experiments. The CNN-based methods included CV-FCN [66], CV-MLPs [67], and CV-

3D-CNN [68]. The ViT-based methods included SViT [27], and PolSARFormer [29], for a total of 5 methods used in the comparative experiments. To ensure the experimental effectiveness of the comparison methods, each comparison method is set up and trained in detail according to the parameters specified in its paper.

3.3. Classification Results

The experimental results on the two datasets are shown in Tables 1 and 2 and the predicted classification results for all data are shown in Figures 9 and 10. The bolded numbers in both tables indicate the best indicator results in the comparison methods. The colors in both figures are consistent with the meaning indicated by the legend of the corresponding dataset. The experimental results demonstrate that the proposed method can achieve good test accuracy and prediction performance with a relatively small amount of training data. Furthermore, both the training and prediction time are significantly reduced compared with other methods. The following is a detailed discussion of the classification results for each of the datasets, where the training time for the proposed method includes a 5-fold cross-validation training.

Table 1. Objective evaluation indicators of six methods on the AIRSAR Flevoland dataset.

Method	CV-FCN	CV-MLPs	CV-3D-CNN	SViT	PolSARFormer	MCPT
Water	96.49	49.96	99.83	**100.00**	99.87	99.16
Forest	99.59	60.98	**99.87**	97.76	96.33	99.05
Lucerne	99.77	73.37	97.90	99.87	81.21	**99.92**
Grass	99.10	0.00	**99.69**	98.24	62.75	96.36
Peas	99.79	90.59	99.95	99.36	92.14	**99.80**
Barley	97.79	0.00	97.03	**99.88**	97.57	98.55
Bare Soil	99.53	0.00	98.90	99.61	93.80	**100.00**
Beet	98.80	81.47	**98.84**	97.74	91.20	98.65
Wheat 2	**99.34**	56.37	95.54	97.28	69.65	95.95
Wheat 3	99.70	36.66	**99.80**	99.95	97.88	98.91
Steambeans	95.93	89.70	99.10	**99.81**	95.98	97.35
Rapeseed	**99.86**	80.46	98.74	98.88	75.98	92.85
Wheat	**99.94**	66.19	98.92	95.47	91.28	97.37
Buildings	99.42	0.00	**100.00**	93.74	83.35	98.23
Potatoes	99.72	87.63	**99.99**	98.89	89.04	97.47
AA	**98.98**	51.56	98.94	98.49	87.87	97.97
Kappa	**99.33**	51.67	98.92	98.43	88.22	97.71
OA	**99.39**	56.62	99.01	98.62	89.19	97.90
Training time(s)	9034.43	**65.79**	2180.03	4294.07	93,703.70	482.52
Predicting time(s)	**15.85**	38.46	1856.65	85.97	1697.50	21.93

Table 2. Objective evaluation indicators of six methods on the RADARSAT-2 San Francisco dataset.

Method	CV-FCN	CV-MLPs	CV-3D-CNN	SViT	PolSARFormer	MCPT
Water	99.86	99.60	99.90	99.97	98.12	**99.99**
Vagetation	**97.48**	87.89	95.42	94.79	77.31	91.00
High-Density Urban	**99.70**	88.97	95.14	95.57	87.24	96.17
Developed	**97.49**	93.39	93.94	95.68	84.72	92.79
Low-Density Urban	**99.49**	84.82	92.57	97.83	83.14	94.29
AA	**98.80**	90.93	95.39	96.76	86.11	94.84
Kappa	**99.02**	90.16	95.47	97.10	85.83	95.35
OA	**99.28**	93.17	96.85	97.98	90.16	96.77
Training time(s)	8227.01	**153.35**	4865.09	1506.58	77,593.12	160.33
Predicting time(s)	103.31	114.66	12,210.48	338.28	6038.70	**85.79**

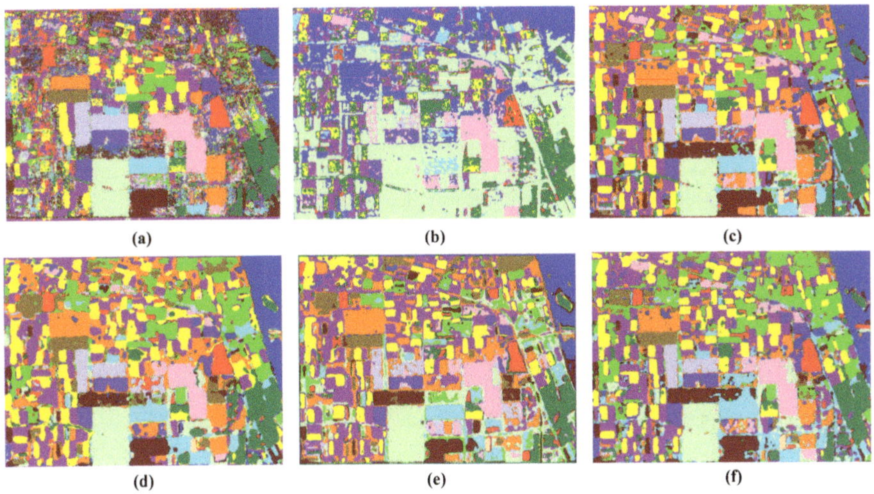

Figure 9. Prediction results of the whole map on AIRSAR Flevoland dataset. (**a**) Result of CV-FCN. (**b**) Result of CV-MLPs. (**c**) Result of CV-3D-CNN. (**d**) Result of SViT. (**e**) Result of PolSARFormer. (**f**) Result of MCPT.

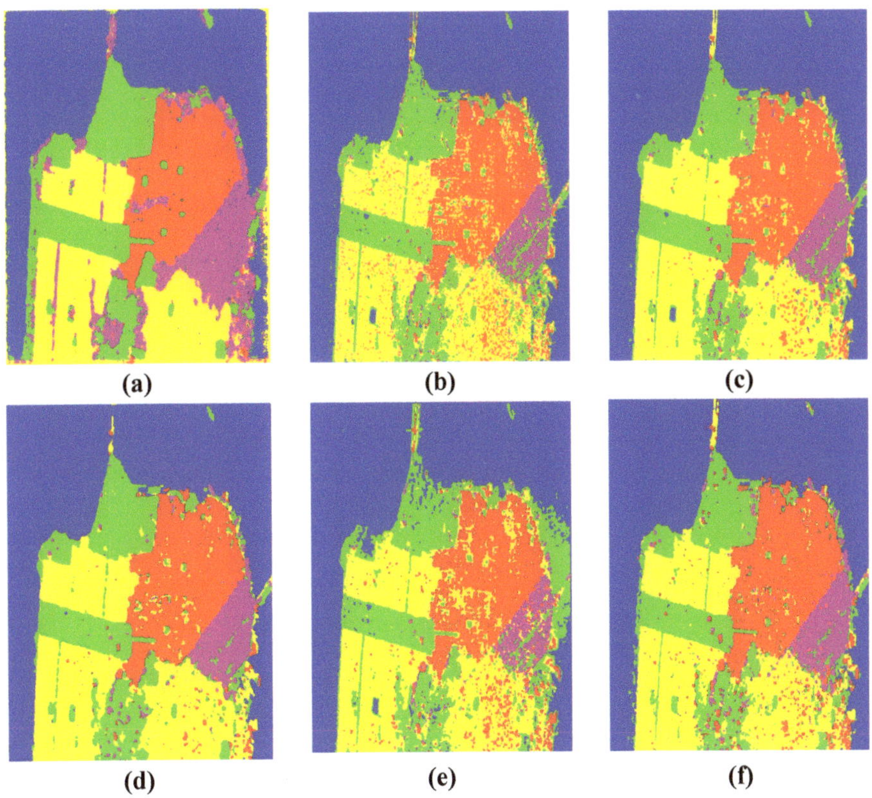

Figure 10. Prediction results of the whole map on RADARSAT-2 San Francisco dataset. (**a**) Result of CV-FCN. (**b**) Result of CV-MLPs. (**c**) Result of CV-3D-CNN. (**d**) Result of SViT. (**e**) Result of PolSARFormer. (**f**) Result of MCPT.

3.3.1. Experimental Results on The AIRSAR Flevoland Dataset

It can be observed from Table 1 that among the CNN-based methods, the CV-FCN method has the highest OA (99.39%), AA (98.98%) and Kappa (99.33%). However, its training time is excessively long at 9034.43 s. The experimental results of the CV-MLPs method are not acceptable compared with other methods. Among the 15 land types, the testing accuracy of four classes, grass, barley, bare doil, and buildings is 0. This indicates that the representation learning ability of this method on the AIRSAR Flevoland dataset is poor and it is not competitive. The CV-3D-CNN method has shown good performance with an OA of 99.01%, AA of 98.94% and Kappa of 98.92%. However, its prediction time (1856.65 s) is the longest of all methods. Among the ViT-based methods, the SViT method also achieves good performance with an OA of 98.62%, AA of 98.49% and Kappa of 98.43%. However, its training time (4294.07 s) and prediction time (85.97 s) are not advantageous. The PolSARFormer method does not have significant advantages in terms of accuracy, and its training and prediction times are too long compared with other methods. The proposed method demonstrates high-level performance in OA (97.90%), AA (97.97%), and Kappa (97.71%). Moreover, it shows strong advantages in terms of both training and prediction time. The training time is only 482.52 s and the prediction time is only 21.93 s.

Figure 9 displays the predicted results of different classification methods on the Flevoland image. It can be seen from Figure 9a that the CV-FCN method can make good predictions on the labeled data but there are irregular and noisy results on the unlabeled part. It has significant overfitting and does not achieve the desired effect. In Figure 9b, it can be observed that the classes of grass, barley, bare soil, and buildings cannot be clearly perceived in the prediction map, while water and wheat 3 occupy most of the image space. It indicates that the CV-MLPs method performs poorly on this dataset, with most regions showing prediction errors. The results of each class in Figure 9d are relatively pure with fewer instances of other categories. However, its classification boundaries are irregular with obvious stickiness. Figure 9e shows slightly worse results. The classification boundaries are unclear and many cases of marginal parts are predicted to be other categories. The prediction results in Figure 9c,f are relatively good and can also provide positive category judgments for the unlabeled part. From some details, it can be seen that the proposed method performs marginally worse than the CV-3D-CNN method in some local classifications. However, the proposed method achieves 100% accuracy in the classification of the bare soil class. Meanwhile, it has an advantage in the prediction results of the building class compared with all other comparison methods. In conclusion, the subjective evaluation of the predicted images based on the corresponding image ground truth and the PauliRGB image leads to the same conclusion as the objective evaluation.

3.3.2. Experiment Results on The RADARSAT-2 San Francisco Dataset

The experiment results in the Table 2 show that the CV-FCN method still achieves the highest OA (99.28%), AA (98.80%), and Kappa (99.02%) among the CNN-based methods. However, it requires a training time of up to 8227.01 s. The CV-MLPs method exhibits a good performance on this dataset. However, its accuracy is low for the vegetation, high-density urban, and low-density urban classes, resulting in a relatively low overall accuracy. The performance of the CV-3D-CNN method on this dataset is worse than on the AIRSAR Flevoland dataset. Not only all three indicators OA (96.85%), AA (95.39%), and Kappa (95.47%) decrease, but also both training time (4865.09 s) and predicting time (12,210.48 s) increase dramatically. This indicates that the CV-3D-CNN method performs less favorably when dealing with large amounts of data. Among the ViT-based methods, the SViT method achieves high accuracy and has advantages in terms of training time (1506.58 s) and prediction time (338.28 s) compared with CNN-based methods. The PolSARFormer method has the lowest accuracy of all the methods used in the experiment. It is not competitive compared with other methods with a training time of 77593.12 s and prediction time of 6038.70 s. The proposed method achieves impressive results in terms of OA (96.77%),

AA(94.84%), and Kappa (95.35%). Furthermore, both the training and prediction time are significantly reduced.

Figure 10 illustrates the prediction results of different classification methods on the San Francisco image. According to the predicted image in Figure 10a, the CV-FCN method is the only one with prediction errors at the outer boundary. It is difficult to clearly distinguish the specific shape of the boundary between water and land. In addition, its prediction of the developed urban category is too aggressive, resulting in a wider distribution of this category. The prediction results in Figure 10b are not pure for each category. Figure 10e has relatively poor prediction results. It has more prediction errors at the boundary of water and land, where vegetation at the edge is misclassified as water or the opposite. There is too much of the phenomenon of being predicted as another category in each category area. The predicted results in Figure 10c,d,f are relatively good. Among them, the image in Figure 10d is purer with fewer classification errors in each category. Figure 10c,f are slightly inferior to Figure 10d. However, the difference is not significant. In summary, subjective visual evaluation of the predicted image leads to the same conclusion as the objective evaluation, which again confirms the feasibility of the proposed method.

On the AIRSAR Flevoland dataset, CV-3D-CNN outperforms other comparative methods when all factors are considered. The proposed method achieves a remarkable increase of approximately 5 times in training speed and 85 times in prediction speed compared with the CV-3D-CNN. Meanwhile, it exhibits some minor decrease of 1.11%, 0.97%, and 1.21% in terms of OA, AA, and Kappa, respectively. Similarly, on the RADARSAT-2 San Francisco dataset, the SViT performs best among other comparative methods. The proposed method shows a respective decrease of 1.21%, 1.92%, and 1.75% in OA, AA, and Kappa compared to the SViT. However, there is an acceleration of about 9 times in training speed and 4 times in prediction speed.

The experimental results on both datasets demonstrate that the proposed method achieves good test accuracy and prediction performance with relatively small amounts of training data. There is a significant reduction in training and prediction time in comparison to other methods. Sacrificing a slightly lower accuracy for considerably improved training and prediction speed leads to excellent time efficiency in practical applications. However, when other methods use the same amount of data as the proposed methods, some methods cannot be trained well and fail to learn useful features that cause a decrease in classification accuracy. Some other methods are slower and have longer training time and prediction time.

4. Discussion

4.1. Ablation Experiments

To verify the effectiveness of each mechanism introduced in the proposed method, ablation experiments are performed on the AIRSAR Flevoland dataset. Experimenting with the three mechanisms will generate many control experiments. Due to time and space constraints, only two mechanisms, mixed depthwise convolution tokenization and parallel encoder, will be discussed here. In the ablation experiments, five metrics are chosen as the judging standard, namely, overall accuracy (OA), training time, prediction time, number of Floating Point Operations (FLOPs), and number of parameters (Params). The first three of the metrics are the same as used in the above comparison experiments to visualize the actual effect of each control group in the ablation experiments. FLOPs is used to measure the model computational complexity and can indirectly measure the model speed. Params is the total number of parameters to be trained in the network model, which is used to measure the size of the model. Table 3 shows the results of the ablation experiments.

To facilitate the description of the experimental results, each experiment is numbered in the table. Experiment (1) is first performed on the original ViT, and the results of this experiment are used as the baseline for the overall ablation experiments. As can be seen from Table 3, the original ViT can achieve a high level of OA on the PolSAR image classification. However, its training time and prediction time are longer, which can also be seen from its FLOPs and Params corresponding to the correlation. The FLOPs of the original ViT is as high as 1703.363 M and the Params also reaches 85.241 M, indicating that its computational and parametric quantities are large. Thus, its direct application to PolSAR image classification has problems such as model complexity and parameter redundancy. Experiment (2) shows a slight decrease in OA after replacing the ViT tokenization process with the mixed depthwise convolution tokenization mechanism introduced in this paper. However, the training time and prediction time are dramatically reduced, and the number of computations and parameters are significantly reduced. The results of this experiment demonstrate the effectiveness of the mixed depthwise convolution tokenization mechanism, which is consistent with the theory described in the previous section. Experiment (3) changes the encoder to the parallel encoder used in the proposed method on the basis of ViT. From the experimental results, it can be seen that adding only the parallel encoder has a large impact on the OA of the model, reducing the accuracy by 2.51%. It is due to the fact that the width network shows a weaker learning ability at a shallower network level compared to the depth network. However, as the number of network levels increases, the width network will have the same model performance as the depth network, and it is easier to optimize than the depth network. The addition of parallel encoder is also able to significantly reduce the training time and prediction time, and the reduction is slightly larger compared to experiment (2). Moreover, the number of computations and parameters are significantly reduced. Experiment (3) shows that the introduction of parallel encoder can effectively reduce the training time and prediction time, as well as the computational and parametric quantities, in accordance with the above theory. Experiment (4) introduces both mixed depthwise convolution tokenization and parallel encoder mechanisms. From the results, it can be seen that the accuracy has improved compared to experiment (3) where only parallel encoder is introduced, and basically reaches the same level as ViT. The training time and prediction time are between experiment (2) and experiment (3), and the computation and the number of parameters are the same as those in experiment (2), which are in accordance with the theoretical requirements. For experiment (5), i.e., the test results of the proposed method MCPT are basically the same as experiment (4) with slight optimization. It shows that the overall impact of GAP on the proposed method is smaller than the other two mechanisms. Therefore, it is not discussed in detail here.

Table 3. Results of the proposed method for ablation experiments on AIRSAR Flevoland dataset.

Method	OA	Training Time (s)	Prediction Time (s)	FLOPs (M)	Params (M)
(1) ViT	97.96	4615.73	284.09	1703.363	85.241
(2) ViT + Mixed Depthwise Convolution Tokenization	97.48	465.93	24.42	74.919	4.103
(3) ViT + Parallel Encoder	95.45	436.99	22.08	74.014	4.118
(4) ViT + Mixed Depthwise Convolution Tokenization + Parallel Encoder	97.84	458.14	23.24	74.919	4.103
(5) MCPT	97.82	457.53	23.09	74.919	4.103

In conclusion, the results of the ablation experiments demonstrate the effectiveness of the three mechanisms introduced in this paper. It is further verified that the proposed method is able to maintain a high level of accuracy while significantly improving the training speed and prediction speed, reducing the model complexity, and enabling ViT to be better applied to PolSAR image classification tasks.

4.2. Impact of Training Data Amount

In deep learning methods, the size of the training data plays an important role in the final effect of the model. Figure 11a shows the amount of training data used by all comparison methods and the proposed method, where the numbers represent the weight of the amount of training data used by the different methods on the two datasets relative to all labeled data. As can be seen from Figure 11a, the training data used by the comparison methods are more than those of the proposed method. Among them, the CV-FCN method even uses 80% of the labeled data for training, which enables it to achieve good test results, but also raises the problem of poor prediction of such unlabeled regions as the ones shown in Figure 9a.

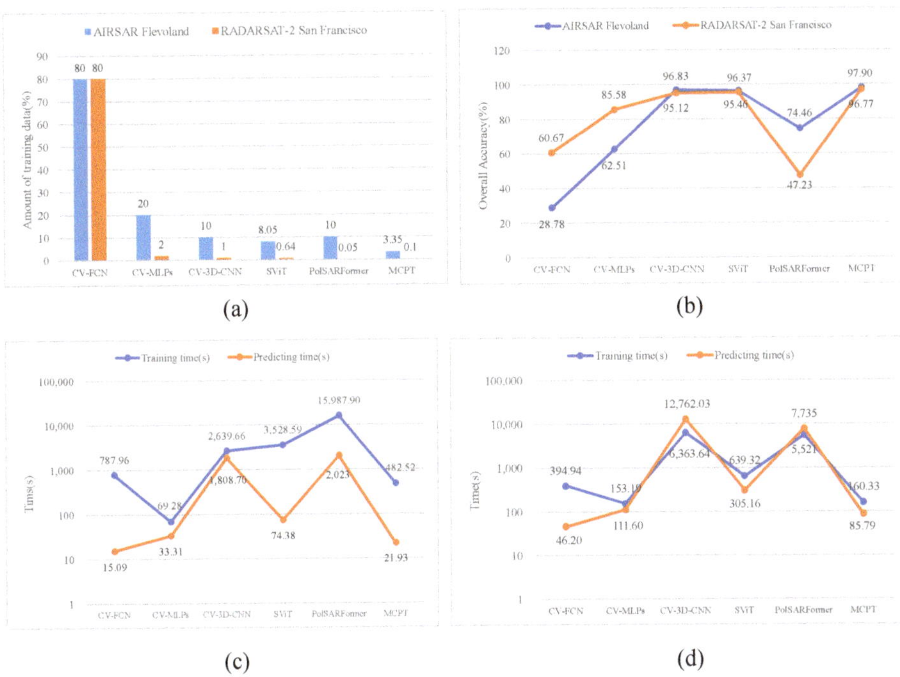

Figure 11. Impact of data amount. (**a**) The amount of training data for the comparison methods. (**b**) Overall accuracy of comparison methods on two datasets using the same data amount of the proposed method. (**c**) Training and predicting time of comparison methods on AIRSAR Flevoland dataset using the same data amount of the proposed method. (**d**) Training and predicting time of comparison methods on RADARSAT-2 San Francisco dataset using the same data amount as in the proposed method.

When all comparison methods are trained according to the amount of data used in this paper, the results are unsatisfactory. Figure 11b–d present the results of different comparison methods trained on the two datasets according to the amount of data used in this paper. Where Figure 11b shows the OA of the test on the two datasets. It can be seen that after reducing the amount of training data, the OA of the comparison methods for testing on both datasets is lower than that of the proposed method. The results show that the reduction in the amount of training data affects the accuracy performance of the comparison methods. Meanwhile, it is demonstrated that the proposed method can achieve a high level of accuracy performance when using a small amount of training data. Figure 11c,d shows the training time and prediction time performance on the AIRSAR Flevoland and RADARSAT-2 San Francisco datasets, respectively. From the two figures, it is observed that the proposed method in this paper still shows a strong advantage in training time and prediction time when the same amount of data is used.

The impact of the amount of training data on the different methods can be illustrated in the figures above. Although the degree of impact varies, more or less all bring about a loss of accuracy. The effectiveness of the proposed method can be verified by the fact that the proposed method can achieve a high level of overall accuracy with fewer training data and less training time and prediction time. While using the same amount of training data as the proposed method, the results of the comparison methods all become worse to different degrees, which further verifies the effectiveness of the proposed method.

4.3. About PolSAR Image Classification Metrics

The commonly used evaluation metrics in PolSAR image classification are overall accuracy (OA), average accuracy per class (AA), and kappa coefficient (Kappa). OA indicates the ratio of the number of correctly classified samples to the number of all samples, and it is used to evaluate the overall performance of the model. AA represents the ratio between the number of correct predictions in each category and the overall number in each category, and ultimately then averages the accuracy of each category for measuring the performance of the model on a given land cover type. The Kappa coefficient is a measure of classification accuracy for consistency testing and is calculated based on the confusion matrix. Its value usually ranges from 0 to 1. Larger values indicate higher consistency and better model classification performance. OA, AA, and Kappa are calculated as follows, respectively:

$$OA = \frac{TP + TN}{TP + FN + FP + TN} \tag{11}$$

$$AA = \frac{\sum Recall_i}{N_i} \tag{12}$$

$$Recall = \frac{TP}{TP + FN} \tag{13}$$

$$Kappa = \frac{p_o - p_e}{1 - p_e} \tag{14}$$

where TP denotes true positive, FN denotes false negative, FP denotes false positive, and TN denotes true negative; i denotes the category and N_i denotes the number of categories; *Recall* denotes the recall rate, i.e., the ratio of the number of correctly classified positive samples to the number of positive samples; p_o denotes the overall classification accuracy, $a_1, a_2, \ldots, a_c$ denotes the number of true samples per class, $b_1, b_2, \ldots, b_c$ denotes the number of predicted samples per class, and n denotes the total number of samples, then p_e can be expressed as:

$$p_e = \frac{a_1 \times b_1 + a_2 \times b_2 + \cdots + a_c \times b_c}{n^2} \tag{15}$$

From the above three metrics, it can be seen that AA contains the meaning represented by recall, and recall is mostly used for binary classification. However, this paper mainly focuses on multi-classification tasks, so recall is not a good measure of the effect of the model used in this paper. As for Kappa, it is calculated from the confusion matrix and its result already contains the information about the confusion matrix. Therefore, the performance of the PolSAR image classification effect can be basically evaluated completely using these three metrics.

5. Conclusions

By investigating the limitations of CNN application on PolSAR image classification and exploring in depth the different implementations of ViT for the tasks, this paper proposes an MCPT model based on ViT for PolSAR image classification. The proposed model employs a mixed depthwise convolution for tokenization and parallel encoders to learn representations of PolSAR images. In addition, the class enbedding is replaced by a GAP operation and the position embedding is removed. All of these improvements reduce the need for extensive training data and computational complexity. Moreover, the model significantly enhances training and prediction speed while maintaining a high level of accuracy. The experimental results on both datasets demonstrate that the proposed method achieves good test accuracy and prediction performance with relatively small amounts of training data. This results in sacrificing a slightly lower accuracy for considerably improved training and prediction speed. Future research will focus on proposing a more reasonable ViT-based classification method that improves classification accuracy while maintaining the existing training and prediction speed. In addition, it is also a good research direction to better combine CNN with ViT to extract both local and global features in PolSAR images.

Author Contributions: Conceptualization, W.W.; Data curation, J.W. and B.L. (Boyuan Liu); Formal analysis, W.W.; Funding acquisition, J.W. and C.W.; Methodology, W.W.; Project administration, B.L. (Bibo Lu); Resources, B.L. (Bibo Lu); Software, W.W.; Supervision, J.W. and Y.Z.; Validation, C.W.; Visualization, B.L. (Boyuan Liu); Writing—original draft, W.W.; Writing—review and editing, J.W. and B.L. (Bibo Lu). All authors have read and agreed to the published version of the manuscript.

Funding: This research was funded partly by the National Natural Science Foundation of China under Grant 62201201; the Doctoral Foundation of Henan Polytechnic University under Grant B2022-15; the Henan Provincial Science and Technology Research Project under Grant 232102211019; and the Key Research Project Fund of Institution of Higher Education in Henan Province under Grant 23A520029.

Data Availability Statement: The Flevoland dataset acquired by NASA/JPL AIRSAR is openly available in the official website of AIRSAR at https://airsar.jpl.nasa.gov/ (accessed on 16 August 1989). The San Francisco dataset acquired by CSA RADARSAT-2 is openly available in the website at https://ietr-lab.univ-rennes1.fr/polsarpro-bio/san-francisco/ (accessed on 9 April 2008).

Acknowledgments: The authors would like to thank the anonymous reviewers for their constructive comments and suggestions which strengthened a lot this paper.

Conflicts of Interest: The authors declare no conflict of interest.

Abbreviations

The following abbreviations are used in this manuscript:

ViT	Vision transformer
MCPT	Mixed convolutional parallel transformer
SAR	Synthetic aperture radar
PolSAR	Polarimetric synthetic aperture image
CNN	Convolutional neural network
NLP	Natural language processing
SA	Self-attention
MixConv	Mixed depthwise convolution
GAP	Global average pooling
MHSA	Multi-head self-attention
MLP	Multi-layer perceptron
LN	Layer normalization
GELU	Gaussian error linear unit
FFN	Feed-forward network
FC	Fully connected
OA	Overall accuracy
AA	Average accuracy

References

1. Chan, Y.K.; Koo, V.C. An introduction to synthetic aperture radar (SAR). *Prog. Electromagn. Res. B* **2008**, *2*, 27–60. [CrossRef]
2. Bamler, R. Principles of Synthetic Aperture Radar. *Surv. Geophys.* **2000**, *21*, 147–157. [CrossRef]
3. Pasmurov, A.; Zinoviev, J. Radar Imaging Application. In *Radar Imaging and Holography*; IET Digital Library: London, UK, 2005; pp. 191–230. [CrossRef]
4. Ulander, L.; Barmettler, A.; Flood, B.; Frölind, P.O.; Gustavsson, A.; Jonsson, T.; Meier, E.; Rasmusson, J.; Stenström, G. Signal-to-Clutter Ratio Enhancement in Bistatic Very High Frequency (VHF)-Band SAR Images of Truck Vehicles in Forested and Urban Terrain. *IET Radar Sonar Navig.* **2010**, *4*, 438. [CrossRef]
5. Zhang, X.; Jiao, L.; Liu, F.; Bo, L.; Gong, M. Spectral Clustering Ensemble Applied to SAR Image Segmentation. *IEEE Trans. Geosci. Remote Sens.* **2008**, *46*, 2126–2136. [CrossRef]

6. Chai, H.; Yan, C.; Zou, Y.; Chen, Z. Land Cover Classification of Remote Sensing Image of Hubei Province by Using PSP Net. *Geomat. Inf. Sci. Wuhan Univ.* **2021**, *46*, 1224–1232. [CrossRef]
7. Zhang, L.; Duan, B.; Zou, B. Research Development on Target Decomposition Method of Polarimetric SAR Image. *J. Electron. Inf. Technol.* **2016**, *38*, 3289–3297. [CrossRef]
8. West, R.D.; Riley, R.M. Polarimetric Interferometric SAR Change Detection Discrimination. *IEEE Trans. Geosci. Remote Sens.* **2019**, *57*, 3091–3104. [CrossRef]
9. Holm, W.; Barnes, R. On Radar Polarization Mixed Target State Decomposition Techniques. In Proceedings of the 1988 IEEE National Radar Conference, Ann Arbor, MI, USA, 20–21 April 1988; pp. 249–254. [CrossRef]
10. Cameron, W.; Leung, L. Feature Motivated Polarization Scattering Matrix Decomposition. In Proceedings of the IEEE International Conference on Radar, Arlington, VA, USA, 7–10 May 1990; pp. 549–557. [CrossRef]
11. Cloude, S. Target Decomposition Theorems in Radar Scattering. *Electron. Lett.* **1985**, *21*, 22–24. .:19850018. [CrossRef]
12. Cloude, S.; Pottier, E. An Entropy Based Classification Scheme for Land Applications of Polarimetric SAR. *IEEE Trans. Geosci. Remote Sens.* **1997**, *35*, 68–78. [CrossRef]
13. Krogager, E. New Decomposition of the Radar Target Scattering Matrix. *Electron. Lett.* **1990**, *26*, 1525. .:19900979. [CrossRef]
14. Parikh, H.; Patel, S.; Patel, V. Classification of SAR and PolSAR Images Using Deep Learning: A Review. *Int. J. Image Data Fusion* **2020**, *11*, 1–32. [CrossRef]
15. Wang, H.; Xu, F.; Jin, Y.Q. A Review of Polsar Image Classification: From Polarimetry to Deep Learning. In Proceedings of the IGARSS 2019–2019 IEEE International Geoscience and Remote Sensing Symposium, Yokohama, Japan, 28 July–2 August 2019; pp. 3189–3192. [CrossRef]
16. Chua, L.; Roska, T. The CNN Paradigm. *IEEE Trans. Circuits Syst. I* **1993**, *40*, 147–156. [CrossRef]
17. Zhou, Y.; Wang, H.; Xu, F.; Jin, Y. Polarimetric SAR Image Classification Using Deep Convolutional Neural Networks. *IEEE Geosci. Remote Sens. Lett.* **2016**, *13*, 1935–1939. [CrossRef]
18. Chen, S.; Tao, C. PolSAR Image Classification Using Polarimetric-Feature-Driven Deep Convolutional Neural Network. *IEEE Geosci. Remote Sens. Lett.* **2018**, *15*, 627–631. [CrossRef]
19. Lee, H.; Kwon, H. Going Deeper With Contextual CNN for Hyperspectral Image Classification. *IEEE Trans. Image Process.* **2017**, *26*, 4843–4855. [CrossRef] [PubMed]
20. Chen, S.; Li, Y.; Wang, X.; Xiao, S.; Sato, M. Modeling and Interpretation of Scattering Mechanisms in Polarimetric Synthetic Aperture Radar: Advances and Perspectives. *IEEE Signal Process. Mag.* **2014**, *31*, 79–89. [CrossRef]
21. Chen, S.; Wang, X.; Sato, M. Uniform Polarimetric Matrix Rotation Theory and Its Applications. *IEEE Trans. Geosci. Remote Sens.* **2014**, *52*, 4756–4770. [CrossRef]
22. Yang, C.; Hou, B.; Ren, B.; Hu, Y.; Jiao, L. CNN-Based Polarimetric Decomposition Feature Selection for PolSAR Image Classification. *IEEE Trans. Geosci. Remote Sens.* **2019**, *57*, 8796–8812. [CrossRef]
23. Shang, R.; He, J.; Wang, J.; Xu, K.; Jiao, L.; Stolkin, R. Dense Connection and Depthwise Separable Convolution Based CNN for Polarimetric SAR Image Classification. *Knowl.-Based Syst.* **2020**, *194*, 105542. [CrossRef]
24. Vaswani, A.; Shazeer, N.; Parmar, N.; Uszkoreit, J.; Jones, L.; Gomez, A.N.; Kaiser, L.; Polosukhin, I. Attention Is All You Need. *arXiv* **2017**, arXiv:1706.03762.
25. Dosovitskiy, A.; Beyer, L.; Kolesnikov, A.; Weissenborn, D.; Zhai, X.; Unterthiner, T.; Dehghani, M.; Minderer, M.; Heigold, G.; Gelly, S.; et al. An Image Is Worth 16 × 16 Words: Transformers for Image Recognition at Scale. In Proceedings of the International Conference on Learning Representations, Kigali, Rwanda, 1–5 May 2023.
26. Han, K.; Wang, Y.; Chen, H.; Chen, X.; Guo, J.; Liu, Z.; Tang, Y.; Xiao, A.; Xu, C.; Xu, Y.; et al. A Survey on Vision Transformer. *IEEE Trans. Pattern Anal. Mach. Intell.* **2023**, *45*, 87–110. [CrossRef]
27. Dong, H.; Zhang, L.; Zou, B. Exploring Vision Transformers for Polarimetric SAR Image Classification. *IEEE Trans. Geosci. Remote Sens.* **2022**, *60*, 1–15. [CrossRef]
28. Wang, H.; Xing, C.; Yin, J.; Yang, J. Land Cover Classification for Polarimetric SAR Images Based on Vision Transformer. *Remote Sens.* **2022**, *14*, 4656. [CrossRef]
29. Jamali, A.; Roy, S.K.; Bhattacharya, A.; Ghamisi, P. Local Window Attention Transformer for Polarimetric SAR Image Classification. *IEEE Geosci. Remote Sens. Lett.* **2023**, *20*, 1–5. [CrossRef]
30. Zhang, Z.; Wang, H.; Xu, F.; Jin, Y.Q. Complex-Valued Convolutional Neural Network and Its Application in Polarimetric SAR Image Classification. *IEEE Trans. Geosci. Remote Sens.* **2017**, *55*, 7177–7188. [CrossRef]
31. Li, Q.; Cai, W.; Wang, X.; Zhou, Y.; Feng, D.D.; Chen, M. Medical Image Classification with Convolutional Neural Network. In Proceedings of the 2014 13th International Conference on Control Automation Robotics & Vision (ICARCV), Singapore, 10–12 December 2014; pp. 844–848. [CrossRef]
32. Qin, J.; Pan, W.; Xiang, X.; Tan, Y.; Hou, G. A Biological Image Classification Method Based on Improved CNN. *Ecol. Inform.* **2020**, *58*, 101093. [CrossRef]
33. Sultana, F.; Sufian, A.; Dutta, P. Advancements in Image Classification Using Convolutional Neural Network. In Proceedings of the 2018 Fourth International Conference on Research in Computational Intelligence and Communication Networks (ICRCICN), Kolkata, India, 22–23 November 2018; pp. 122–129. [CrossRef]
34. Dolz, J.; Gopinath, K.; Yuan, J.; Lombaert, H.; Desrosiers, C.; Ben Ayed, I. HyperDense-Net: A Hyper-Densely Connected CNN for Multi-Modal Image Segmentation. *IEEE Trans. Med. Imaging* **2019**, *38*, 1116–1126. [CrossRef]

35. Liu, F.; Lin, G.; Shen, C. CRF Learning with CNN Features for Image Segmentation. *Pattern Recognit.* **2015**, *48*, 2983–2992. [CrossRef]
36. Mortazi, A.; Bagci, U. Automatically Designing CNN Architectures for Medical Image Segmentation. In Proceedings of the Machine Learning in Medical Imaging, Granada, Spain, 16 September 2018; Shi, Y., Suk, H.I., Liu, M., Eds.; Lecture Notes in Computer Science; Springer: Cham, Switzerland, 2018; pp. 98–106. ._12. [CrossRef]
37. Chandrasegaran, K.; Tran, N.T.; Cheung, N.M. A Closer Look at Fourier Spectrum Discrepancies for CNN-generated Images Detection. In Proceedings of the 2021 IEEE/CVF Conference on Computer Vision and Pattern Recognition (CVPR), Nashville, TN, USA, 20–25 June 2021; pp. 7196–7205. [CrossRef]
38. Chattopadhyay, A.; Maitra, M. MRI-based Brain Tumour Image Detection Using CNN Based Deep Learning Method. *Neurosci. Inform.* **2022**, *2*, 100060. [CrossRef]
39. Chauhan, R.; Ghanshala, K.K.; Joshi, R. Convolutional Neural Network (CNN) for Image Detection and Recognition. In Proceedings of the 2018 First International Conference on Secure Cyber Computing and Communication (ICSCCC), Jalandhar, India, 15–17 December 2018; pp. 278–282. [CrossRef]
40. Zhou, Z.; Wu, Q.M.J.; Wan, S.; Sun, W.; Sun, X. Integrating SIFT and CNN Feature Matching for Partial-Duplicate Image Detection. *IEEE Trans. Emerg. Top. Comput. Intell.* **2020**, *4*, 593–604. [CrossRef]
41. Bhatt, D.; Patel, C.; Talsania, H.; Patel, J.; Vaghela, R.; Pandya, S.; Modi, K.; Ghayvat, H. CNN Variants for Computer Vision: History, Architecture, Application, Challenges and Future Scope. *Electronics* **2021**, *10*, 2470. [CrossRef]
42. Jia, W.; Tian, Y.; Luo, R.; Zhang, Z.; Lian, J.; Zheng, Y. Detection and Segmentation of Overlapped Fruits Based on Optimized Mask R-CNN Application in Apple Harvesting Robot. *Comput. Electron. Agric.* **2020**, *172*, 105380. [CrossRef]
43. Ravanbakhsh, M.; Nabi, M.; Mousavi, H.; Sangineto, E.; Sebe, N. Plug-and-Play CNN for Crowd Motion Analysis: An Application in Abnormal Event Detection. In Proceedings of the 2018 IEEE Winter Conference on Applications of Computer Vision (WACV), Lake Tahoe, NV, USA, 12–15 March 2018; pp. 1689–1698. [CrossRef]
44. Xie, W.; Zhang, C.; Zhang, Y.; Hu, C.; Jiang, H.; Wang, Z. An Energy-Efficient FPGA-Based Embedded System for CNN Application. In Proceedings of the 2018 IEEE International Conference on Electron Devices and Solid State Circuits (EDSSC), Shenzhen, China, 6–8 June 2018; pp. 1–2. [CrossRef]
45. Sandler, M.; Howard, A.; Zhu, M.; Zhmoginov, A.; Chen, L.C. MobileNetV2: Inverted Residuals and Linear Bottlenecks. In Proceedings of the 2018 IEEE/CVF Conference on Computer Vision and Pattern Recognition, Salt Lake City, UT, USA, 18–23 June 2018; pp. 4510–4520. [CrossRef]
46. Ma, N.; Zhang, X.; Zheng, H.T.; Sun, J. ShuffleNet V2: Practical Guidelines for Efficient CNN Architecture Design. In Proceedings of the Computer Vision—ECCV, Munich, Germany, 8–14 September 2018; Springer: Cham, Switzerland, 2018; pp. 122–138. [CrossRef]
47. Zoph, B.; Vasudevan, V.; Shlens, J.; Le, Q.V. Learning Transferable Architectures for Scalable Image Recognition. In Proceedings of the 2018 IEEE/CVF Conference on Computer Vision and Pattern Recognition, Salt Lake City, UT, USA, 18–23 June 2018; pp. 8697–8710. [CrossRef]
48. Tan, M.; Le, Q.V. MixConv: Mixed Depthwise Convolutional Kernels. *arXiv* **2019**, arXiv:1907.09595.
49. Hassani, A.; Walton, S.; Shah, N.; Abuduweili, A.; Li, J.; Shi, H. Escaping the Big Data Paradigm with Compact Transformers. *arXiv* **2022**, arXiv:2104.05704.
50. Chen, X.; Xie, S.; He, K. An Empirical Study of Training Self-Supervised Vision Transformers. In Proceedings of the 2021 IEEE/CVF International Conference on Computer Vision (ICCV), Montreal, BC, Canada, 11–17 October 2021; pp. 9620–9629. [CrossRef]
51. Hendrycks, D.; Gimpel, K. Gaussian Error Linear Units (GELUs). *arXiv* **2020**, arXiv:1606.08415.
52. Chen, C.F.R.; Fan, Q.; Panda, R. CrossViT: Cross-Attention Multi-Scale Vision Transformer for Image Classification. In Proceedings of the 2021 IEEE/CVF International Conference on Computer Vision (ICCV), Montreal, BC, Canada, 11–17 October 2021; pp. 347–356. [CrossRef]
53. Chu, X.; Tian, Z.; Wang, Y.; Zhang, B.; Ren, H.; Wei, X.; Xia, H.; Shen, C. Twins: Revisiting the Design of Spatial Attention in Vision Transformers. In Proceedings of the Advances in Neural Information Processing Systems 34, Online, 7 December 2021; Curran Associates, Inc.: New York City, NY, USA, 2021; pp. 9355–9366.
54. Heo, B.; Yun, S.; Han, D.; Chun, S.; Choe, J.; Oh, S.J. Rethinking Spatial Dimensions of Vision Transformers. In Proceedings of the 2021 IEEE/CVF International Conference on Computer Vision (ICCV), Montreal, BC, Canada, 11–17 October 2021; pp. 11916–11925. [CrossRef]
55. Tu, Z.; Talebi, H.; Zhang, H.; Yang, F.; Milanfar, P.; Bovik, A.; Li, Y. MaxViT: Multi-axis Vision Transformer. In *Computer Vision—ECCV 2022. ECCV 2022*; Avidan, S., Brostow, G., Cissé, M., Farinella, G.M., Hassner, T., Eds.; Lecture Notes in Computer Science; Springer: Cham, Switzerland, 2022; pp. 459–479. [CrossRef]
56. Yang, R.; Ma, H.; Wu, J.; Tang, Y.; Xiao, X.; Zheng, M.; Li, X. ScalableViT: Rethinking the Context-Oriented Generalization of Vision Transformer. In *Computer Vision—ECCV 2022. ECCV 2022*; Avidan, S., Brostow, G., Cissé, M., Farinella, G.M., Hassner, T., Eds.; Lecture Notes in Computer Science; Springer: Cham, Switzerland, 2022; pp. 480–496. [CrossRef]
57. He, K.; Zhang, X.; Ren, S.; Sun, J. Deep Residual Learning for Image Recognition. In Proceedings of the 2016 IEEE Conference on Computer Vision and Pattern Recognition (CVPR), Las Vegas, Nevada, USA, 27–30 June 2016; pp. 770–778. [CrossRef]
58. Goyal, A.; Bochkovskiy, A.; Deng, J.; Koltun, V. Non-Deep Networks. In Proceedings of the Advances in Neural Information Processing Systems 35, New Orleans, LA, USA, 28 November–9 December 2022.

59. Zhou, J.; Wei, C.; Wang, H.; Shen, W.; Xie, C.; Yuille, A.; Kong, T. Image BERT Pre-training with Online Tokenizer. In Proceedings of the International Conference on Learning Representations, Virtual, 25–29 April 2022.
60. Touvron, H.; Cord, M.; El-Nouby, A.; Verbeek, J.; Jégou, H. Three Things Everyone Should Know About Vision Transformers. In Proceedings of the 17th European Conference, Tel Aviv, Israel, 23–27 October 2022; Avidan, S., Brostow, G., Cissé, M., Farinella, G.M., Hassner, T., Eds.; Lecture Notes in Computer Science; Springer: Cham, Switzerland, 2022; pp. 497–515. ._29. [CrossRef]
61. Goodfellow, I.; Warde-Farley, D.; Mirza, M.; Courville, A.; Bengio, Y. Maxout Networks. In Proceedings of the 30th International Conference on Machine Learning (PMLR), Atlanta, GA, USA, 17–19 June 2013; pp. 1319–1327.
62. Krizhevsky, A.; Sutskever, I.; Hinton, G.E. ImageNet Classification with Deep Convolutional Neural Networks. *Commun. ACM* **2017**, *60*, 84–90. [CrossRef]
63. Lin, M.; Chen, Q.; Yan, S. Network in Network. *arXiv* **2013**, arXiv:1312.4400.
64. Liu, F. PolSAR Image Classification and Change Detection Based on Deep Learning. Ph.D. Thesis, Xidian University, Xi'an, China, 2017.
65. Liu, X.; Jiao, L.; Liu, F.; Zhang, D.; Tang, X. PolSF: PolSAR Image Datasets on San Francisco. In Proceedings of the IFIP Advances in Information and Communication Technology, Xi'an, China, 28–31 October 2022; Shi, Z., Jin, Y., Zhang, X., Eds.; Springer: Cham, Switzerland, 2022; pp. 214–219. ._23. [CrossRef]
66. Cao, Y.; Wu, Y.; Zhang, P.; Liang, W.; Li, M. Pixel-Wise PolSAR Image Classification via a Novel Complex-Valued Deep Fully Convolutional Network. *Remote Sens.* **2019**, *11*, 2653. [CrossRef]
67. Ronny, H. Complex-Valued Multi-Layer Perceptrons—An Application to Polarimetric SAR Data. *Photogramm. Eng. Remote Sens.* **2010**, *76*, 1081–1088. [CrossRef]
68. Tan, X.; Li, M.; Zhang, P.; Wu, Y.; Song, W. Complex-Valued 3-D Convolutional Neural Network for PolSAR Image Classification. *IEEE Geosci. Remote Sens. Lett.* **2020**, *17*, 1022–1026. [CrossRef]

remote sensing MDPI

Technical Note

A Seabed Terrain Feature Extraction Transformer for the Super-Resolution of the Digital Bathymetric Model

Wuxu Cai [1], Yanxiong Liu [1,2,*], Yilan Chen [1,2], Zhipeng Dong [1], Hanxiao Yuan [1] and Ningning Li [3]

1 The First Institute of Oceanography, Ministry of Natural Resources, Qingdao 266061, China; caiwuxu@fio.org.cn (W.C.); chenyilan@fio.org.cn (Y.C.); zhipengdong@fio.org.cn (Z.D.); yuanhanxiao@fio.org.cn (H.Y.)
2 The Key Laboratory of Ocean Geomatics, Ministry of Natural Resources, Qingdao 266590, China
3 State Key Laboratory of Information Engineering in Surveying, Mapping and Remote Sensing, Wuhan University, Wuhan 430079, China; liningning@whu.edu.cn
* Correspondence: yxliu@fio.org.cn

Abstract: The acquisition of high-resolution (HR) digital bathymetric models (DBMs) is crucial for oceanic research activities. However, obtaining HR DBM data is challenging, which has led to the use of super-resolution (SR) methods to improve the DBM's resolution, as, unfortunately, existing interpolation methods for DBMs suffer from low precision, which limits their practicality. To address this issue, we propose a seabed terrain feature extraction transform model that combines the seabed terrain feature extraction module with the efficient transform module, focusing on the terrain characteristics of DBMs. By taking advantage of these two modules, we improved the efficient extraction of seabed terrain features both locally and globally, and as a result, we obtained a highly accurate SR reconstruction of DBM data within the study area, including the Mariana Trench in the Pacific Ocean and the adjacent sea. A comparative analysis with bicubic interpolation, SRCNN, SRGAN, and SRResNet shows that the proposed method decreases the root mean square error (RMSE) by 16%, 10%, 13%, and 12%, respectively. These experimental results confirm the high accuracy of the proposed method in terms of reconstructing HR DBMs.

Keywords: digital bathymetry model; seabed terrain feature; deformable convolutional layers; transformer; super-resolution

Citation: Cai, W.; Liu, Y.; Chen, Y.; Dong, Z.; Yuan, H.; Li, N. A Seabed Terrain Feature Extraction Transformer for the Super-Resolution of the Digital Bathymetric Model. *Remote Sens.* **2023**, *15*, 4906. https://doi.org/10.3390/rs15204906

Academic Editors: Moulay A. Akhloufi and Mozhdeh Shahbazi

Received: 27 July 2023
Revised: 25 September 2023
Accepted: 26 September 2023
Published: 11 October 2023

1. Introduction

Bathymetry provides fundamental information for all marine activities, and numerous marine scientific research endeavors rely on digital bathymetry model (DBM) data. High-resolution (HR) DBMs are commonly used for charting purposes [1]; at the same time, they play a crucial role in establishing hydrodynamic fluid models [2], constructing marine biological habitats and ecosystems [3], and other significant studies. In addition, HR DBMs are valuable for maritime search and rescue operations. For example, the search and rescue operations for Malaysia Airlines flight MH370 were severely hampered by the lack of accurate bathymetric data in the corresponding area [4]. Therefore, the acquisition of HR DBMs is of immense importance not only for scientific research but also for various social activities. However, the scarcity of HR DBMs severely limits their application and development.

HR DBMs are typically derived from ship-based acoustic surveys. On the one hand, these surveys face challenges and inefficiencies, resulting in the limited coverage of HR bathymetric data. According to statistics [5], ship-borne sonar-based HR data ($\leq$800 m) cover only approximately 6.2% of the global seafloor. Consequently, the availability of HR bathymetric data to construct DBMs is severely limited for most of the global ocean. On the other hand, many institutions have obtained global low-resolution (LR) DBM data, including EPOTO and GEBCO. To address the shortage of HR DBMs, researchers have employed various super-resolution (SR) methods to improve the resolution of DBMs; DBM SR aims to recover detailed topographic information from LR DBMs to generate HR DBMs. Currently, SR methods for DBMs mainly rely on traditional interpolation techniques, such as inverse distance-weighted interpolation [6], bilinear interpolation [7], bicubic spline interpolation [8], continuous curvature tensor spline interpolation [9], natural neighborhood interpolation [10], and kriging interpolation [11]. Although these interpolation methods are widely used due to their simplicity and speed, they can present certain problems, particularly when generating large-scale DBMs. One of their main drawbacks is the potential to be over-smooth, as these methods do not adequately exploit local terrain features during the interpolation process [12]. In addition to traditional interpolation methods, many researchers have employed deep learning-based methods to enhance the resolution of terrestrial digital elevation models (DEMs). These methods have the advantage of not relying on additional multisource data and instead generate HR DEMs solely through the reconstruction of LR DEMs [13]. Similarly, in situations in which data are limited, the resolution of DBMs can theoretically be improved using deep learning-based SR methods. This approach not only addresses the issue of insufficient data but also serves as a valuable source of HR data for DBM data fusion.

Current deep learning-based SR models are mainly based on convolutional neural networks (CNNs) and other networks based on CNNs, such as residual networks (ResNets), generative adversarial networks (GANs), and transformers. The pioneering CNN model for image SR is SRCNN, which consists of three convolutional layers that perform feature extraction, non-linear mapping, and image reconstruction to enhance image resolution [14]. The advancement of neural networks has highlighted their crucial role in feature extraction; however, the stacking of network layers can reduce the effectiveness of this network and lead to degradation problems [15]. To address this problem, He et al. [16] employed ResNet to extract shallow features and transmit mapping to the deep network. This identification mapping does not require any additional parameters or calculations, ensuring that the characteristics of this network do not deteriorate as the number of layers increases. In the context of DEM SR, Chen et al. [17] were the first to apply SRCNN to DEM data for SR reconstruction. Subsequently, other researchers have demonstrated the strong feature extraction capabilities and applicability of ResNet to DEM SR [18,19]. Peak signal-to-noise ratio (PSNR) evaluations of SR reconstruction accuracy are frequently performed in the CV domain. However, studies have revealed a mixed relationship between PSNR and visual quality, suggesting that a greater PSNR does not always translate into superior visual perception [20]. In order to enhance perception, Ledig et al. [20] integrated a perceptual loss function with a generative adversarial network for SR. Given the distinct characteristics of DEM data compared to natural images, Zhang and Yu [21] applied SRGAN, ESRGAN, and CEDGAN to DEM SR, and the experimental results showed that SRGAN was the most effective in terms of extracting terrain features, outperforming other GANs. This result suggests that neural networks with superior performance in the CV domain might not be equally effective when extracting topographic features. The current focus of DEM SR research is centered on fully and effectively extracting terrain features. Zhou et al. [22] introduced dual filters into ResNet to accomplish this goal, while Chen et al. [12] incorporated a spatial attention mechanism module into CNN to improve its feature extraction efficiency. Zhang et al. [23] proposed the addition of slope loss to the loss function, which significantly improved feature extraction effectiveness. Zhou et al. [24] integrated vector terrain features before outputting HR DEM, which significantly improved the accuracy of the model. To solve practical problems, researchers are now placing their research objects in hard-to-access topographic data. Jiang et al. [25] used high mountain data for the study data and constructed a new loss function in ResNet, which combined the terrain parameters of slope and curvature.

However, the development of global high-precision HR DBMs has been slow due to the challenges experienced when obtaining bathymetric data, particularly the lack of HR data. The resolution of HR DBMs is significantly lower than that of terrestrial DEMs, and the application of SR research based on deep learning in DBMs is limited. Yutani et al. [26] utilized sparse coding and dictionary learning techniques for SR in DBMs, resulting in improved results compared to bicubic interpolation, and Hidaka et al. [27] employed deep learning for SR in DBMs for the first time and compared the SR reconstruction effects of various deep learning-based methods, including SRCNN, FSRCNN, ESPCN, SRGAN, and ESRGAN. The results were superior to bicubic interpolation but did not contribute to building a deep learning-based model for DBM terrain characteristics. Additionally, Zhang et al. [28] enhanced the resolution of the global GEBCO_2021 dataset from 15 arc seconds to 3 arc seconds using ResNet and migration learning. Although many deep learning-based SR methods utilize ResNet for DEMs, there are some general issues with this approach. First, the use of regular convolution kernels in ResNet might not be optimal for recovering different topographic features due to the irregular terrain. Second, the convolution operation does not adequately consider the influence of distant pixels. The transformer, with its self-attention mechanism, possesses strong capabilities in terms of capturing global information and long-range interactions among similar features, making it successful in various visual tasks. Some researchers [29] have employed this transformer in the computer vision domain to enhance SR accuracy. However, this transformer is currently underutilized in SR for terrain models. Zheng et al. [30] designed a transformer model for SR in DEMs via leveraging terrain self-similarity. The reconstruction effect was superior to SRCNN and SRGAN, but improvements specific to local features were not made. Additionally, SR is a low-level vision task, and using a complete transformer structure can reduce computational efficiency. Therefore, modifications to the transformer structure are necessary for its application in DBM SR.

In this paper, considering the complex and extensive nature of DBM seabed terrain features, a seabed terrain feature extraction transformer (STFET) is proposed to achieve DBM SR, combining ResNet with deformable convolutional layers and an efficient transformer. The following are this paper's key contributions:

1. This study is an early attempt to restore HR DBMs from LR DBMs using a transformer. In addition, we utilize the proposed transformer-based model, combined with ResNet and deformable convolutional layers, which ensures that STFET can capture both local and global seafloor topographic features.
2. Given the characteristics of large changes and rapid fluctuations in the DBM terrain, the traditional convolutional layers in ResNet are replaced by deformable convolutional layers, which have the ability to flexibly modify the sampling position of the convolution kernel to align with irregular features, improving the extraction of local seabed terrain features.
3. The transformer's self-attention mechanism allows long-term dependencies to be established between similar regions in the DBM, improving global terrain feature reconstruction. In addition, the matrix is broken down into smaller matrices for parameter manipulation, increasing the speed of model operations.

2. Methodology

The StfeT method includes the following: (1) data-processing; (2) training in the proposed model; and (3) testing and evaluation. The workflow is summarized in Figure 1. More detail is provided in other sections.

2.1. Network Architecture

In view of the complex terrain characteristics of DBMs, in this paper, we propose a seabed terrain feature extraction transformer model for DBM SR. This model mainly consists of three parts: shallow terrain feature extraction, deep feature extraction, including the seabed terrain feature extraction module and the efficient transformer, and sub-pixel convolution reconstruction, as shown in Figure 2. The utilization of convolutional layers offers a straightforward and efficient approach to incorporating feature maps, demonstrating a consistent performance across various computer vision tasks.

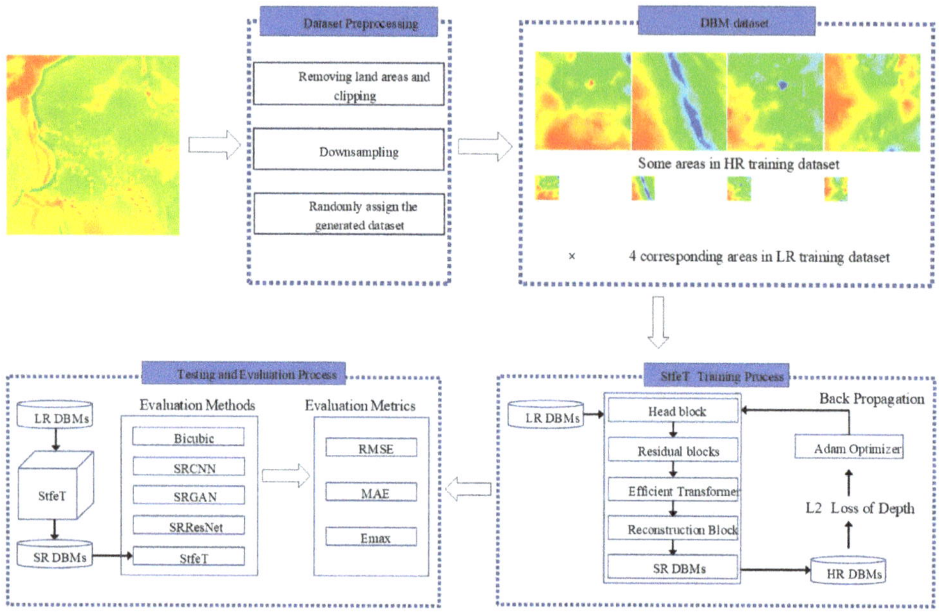

Figure 1. The StfeT workflow.

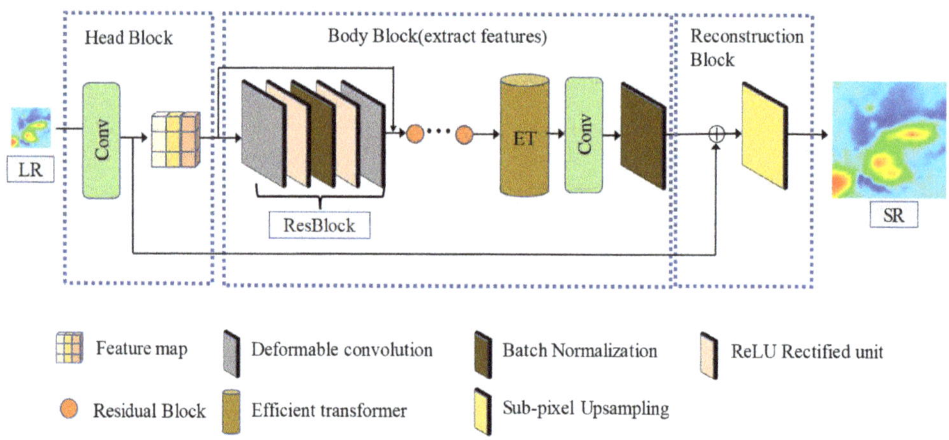

Figure 2. Framework of the proposed STFET.

To extract the shallow features, we employed a regular convolutional layer f_N (channel = 64, kernel size = 9, stride = 1) as the first STFET module to extract the DBM feature $F_{shallow}$:

$$F_{shallow} = f_N(I_{LR}) \tag{1}$$

where I_{LR} represents LR DBM. After extracting the shallow features to extract deep features, we employed the seabed terrain feature extraction module and the effective transformer module. Seabed terrain features were extracted adaptively using ResNet with deformable convolutional layers, and the intermediate features are represented by A_1, A_2,. . ., A_n:

$$A_i = Res(A_{i-1}) + A_{i-1}, i = 1,2,\ldots,n \tag{2}$$

$$A_1 = Res(F_{shallow}) + F_{shallow} \tag{3}$$

$$F_{deep} = ET(A_n) \tag{4}$$

where *Res* represents a residual block with deformable convolutional layers. Through residual learning, seabed terrain features could be effectively extracted, and the deformable convolutional layer could focus on high-frequency information, thereby enhancing the extraction of local features. *ET* represents the efficient transformer, and the feature maps after residual learning were learned using *ET*. Through the self-attention mechanism, the correlation between similar seabed terrain features was modeled, allowing the model to focus on other positions with similar seabed terrain features to extract global seabed terrain features. The deep feature F_{deep} was obtained through these two modules.

After obtaining the deep seabed terrain features, the conventional convolutional layers were used to extract and transform the high-frequency features in the residual information. *BN* is batch normalization, which is then used to hasten model convergence and improve the model's robustness. Furthermore, a skip connection is used to connect the shallow topographic features with deep topographic features to obtain all topographic features F_{all}.

$$F_{all} = BN\left(f_N\left(F_{deep}\right) + F_{shallow}\right) \tag{5}$$

Finally, the LR feature maps were transformed into HR feature maps using sub-pixel convolution which were reconstructed to generate I_{SR}, which is SR DBM. During sub-pixel convolution, the HR feature maps were obtained using convolution and multichannel recombination, avoiding a large number of zero-filled areas in general deconvolution upsampling. Moreover, we used the dropout to prevent model overfitting before reconstruction.

$$I_{SR} = f_P\left(F_{deep}\right) \tag{6}$$

where f_P is a sub-pixel convolutional layer. STFET uses a large number of convolutional layers to extract and transform features. The local seabed terrain features and global seabed terrain features were extracted mainly through ResNet with deformable convolutional layers and the efficient transformer. Finally, a sub-pixel convolutional layer was used to reconstruct the SR DBM.

2.2. Seabed Terrain Feature Extraction Module

DBM often has clear topographic undulations in local areas, generating different topographies, and the depth distribution of DBM is closely related to the topographic features of local areas. The effective extraction of seafloor topographic features can provide more accurate feature maps, thereby improving the quality of SR DBM. Very deep convolutional networks (VDSRs) [31] were the first to use ResNet for the SR domain of images, as they can utilize both deeper neural networks and mitigate the problems of disappearing or exploding gradients and network degradation. Seabed terrain features have different shapes, and the convolution kernel used in the traditional convolution layer is usually in the shape of an n × n regular grid, which makes it difficult to extract complicated seabed terrain features. To extract complex-shaped objects or terrain features, researchers have replaced conventional convolutional layers with deformable convolutional layers [32,33]. Deformable convolution adds an offset to the regular convolutional layer, which is implemented using an additional convolutional layer, as shown in Figure 3. In order to always be able to cover complicated terrain features, the deformable convolution mechanism makes sure that offsets can be learned during the learning process. Therefore, the deformable convolution layer can improve its ability to extract terrain features. In this paper, we replaced regular convolutional layers in the residual network with deformable convolutional layers, and the feature extraction process of the conventional convolutional layer and the deformable convolutional layer in the DBM is shown in Figure 4. When the conventional convolutional layer extracts terrain features because the convolution kernel is regular, the edge terrain features are easily overlooked, whereas the deformable convolutional layer can adaptively extract seabed terrain features by learning during training.

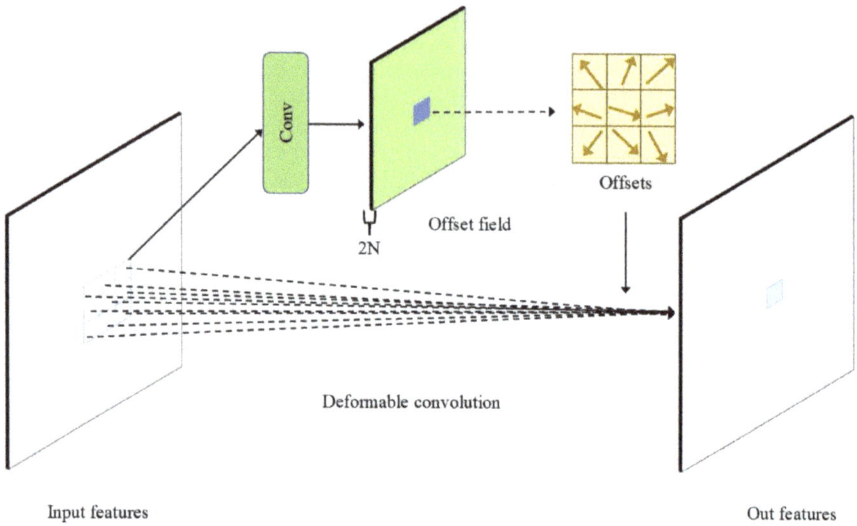

Figure 3. Schematic of the deformable convolution principle.

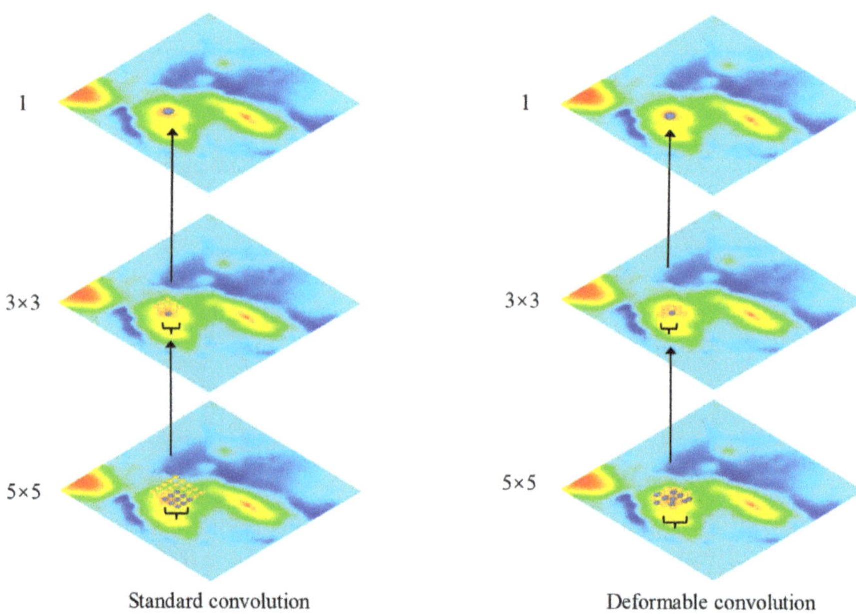

Figure 4. Deformable convolution and regular convolution for feature extraction on DBM.

The seabed terrain feature extraction module consists of N residual blocks. In previous studies, the higher the number of residual blocks, the higher the accuracy, but at the same time, it increases the computation time. Therefore, it is necessary to balance the reconstruction accuracy and computation time to choose an appropriate number of residual blocks. Each residual block has five layers: a ReLU activation function, two deformable convolutional layers (channel = 64, kernel size = 3, stride = 1), and two batch normalization layers. The deformable convolutional layer is used to extract high-frequency details of the DBM to adaptively extract seabed terrain features. DBMs are different from natural images because the difference between the maximum value and the minimum value is often on the kilometer scale, and batch normalization can reduce the impact of extreme water depth. Furthermore, the ReLU activation function can enhance the non-linear relationship between layers of the neural network, which can improve the model's fitting ability. Therefore, the seabed terrain feature extraction module can effectively extract topographic features from the seafloor.

2.3. Efficient Transformer Module

According to the first law of geography, the closer things are, the more related they are. Therefore, in close regions, DBMs generally have a similar topography globally. The texture details of the current image block can be recovered by referring to other image blocks in single-image super-resolution (SISR), which allows similar image blocks in an image to be used as references for one another. The transformer self-attention mechanism can realize this function very well, so the transformer has a theoretical basis to recover the current area by referring to other areas to extract similar terrain features.

The structure of the efficient transformer used in this paper is shown in Figure 5. Similar to the Vision Transformer (ViT) [34], only the encoding part in the standard transformer is used, and it is composed of Multi-Layer Perceptron (MLP) and efficient multi-head self-attention (EMSA). At the same time, we used layer normalization (LN) before each module to accelerate the convergence speed of the model. Then, after each module, the residual connection was achieved on an element-by-element basis. This process can be represented using the following formula:

$$E = EMSA(LN(Un(A_n))) + Un(A_n) \tag{7}$$

$$F_{deep} = Fold(MLP(LN(E)) + E) \tag{8}$$

where Un denotes the unfold function, whose purpose is to generate a local sliding block from the feature map extracted by the seabed terrain feature extraction module. E represents the feature extracted using $EMSA$, and $Fold$ denotes the inverse operation of Un, which converts the sliding block into a feature map.

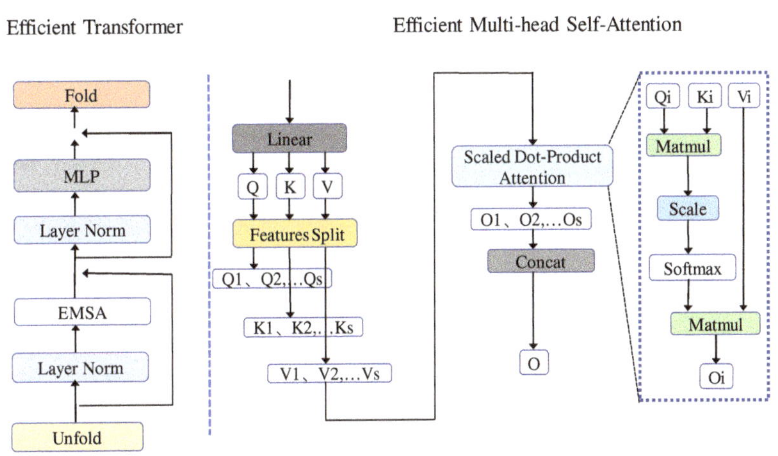

Figure 5. Structure of the efficient transformer and EMSA.

The EMSA structure is shown in Figure 5. First, features were embedded in query Q, key K, and value V using a linear layer. As in the transformer, we projected Q, K, and V m times, where m is the multi-head number, and the multi-head self-attention (MSA) directly computes Q, K, and V. The computations typically use a large matrix, which consumes a large amount of GPU memory and computation. Unlike MSA, the DBM matrix we used for training to extract terrain features is relatively large due to the LR of DBM and the large scale of the terrain features. Therefore, we used feature splitting to divide Q, K, and V into s parts to obtain Q_s, K_s, and V_s, where s is the splitting factor. Then, we use Scaled Dot Product Attention (SDPA) to perform attention to generate multiple outputs $O_1 ... O_s$. This process is shown on the right side of Figure 5. Finally, all outputs were concatenated to produce the overall output feature O.

3. Experimental Setup

3.1. Network Hyperparameters

In the proposed STFET, to ensure accuracy and computational efficiency, the number of residual blocks N was set to 16. The multi-head number in the efficient transformer module was set to 8, and the hidden layer size of the MLP was $mlpdim$ = 85, and was obtained using the following formula:

$$mlpdim = (inchannel + outchannl) \times 2/3 \quad (9)$$

where $inchannel$ and $outchannel$ are the numbers of MLP input channels and output channels, respectively, which are both 64. To train the proposed STFET, we used adaptive moment estimation (Adam) with a 0.0002 learning rate, a random seed of 42, and a dropout rate of 0.2. The loss in STFET was the L2 loss MSE. For CNN-based algorithms, all the models were coded using the PyTorch framework. In addition, we used 8 GB of RAM and trained the models on a deep-learning server with an Nvidia RTX 2070S GPU. Moreover, the hyperparameters were adjusted to retrain the comparison models in the optimal way possible.

3.2. Study Area and Data

Unlike natural images, DBM is a grayscale image with only one channel. DBMs tend to have topographic features. In this study, the Mariana Trench in the Pacific Ocean and the surrounding waters to the east were selected as the experimental area, constituting a range of 0°–45°N, 135°–180°E. This area is large and rich in seafloor topography and essentially includes most of the typical seafloor topographic features, such as ridges, trenches, seamounts, sea basins, and other representative seafloor topographic features. The values of natural images are pixel values, ranging from 0 to 255. However, the elevation of our study area was between −10,923 and 3659 m. The study area is shown in Figure 6.

The DBM data used in this paper were gebco_2022_sub_ice, sourced from https://www.gebco.net/data_and_products/gridded_bathymetry_data/gebco_2022 (accessed on 10 September 2023), which has a resolution of 15 arc seconds. Since original DBM data are large, it was necessary to preprocess them to train the data. We cropped the original DBM data into 128 × 128 patches and removed most of the land area, and a total of 6386 DBM images were obtained as HR DBMs. To obtain paired HR DBMs and LR DBMs as datasets for training, many previous DEM SR studies were scaled 4 times for verification experiments [12,22,24]. In this study, we also used bicubic interpolation to reduce the data size 4-fold and obtain 128 × 128 patch DBM images with resolutions of 60 arc seconds as LR DBMs. The ratio of the training set data, verification set data, and test set data was 8:1:1 through random allocation. Before training, due to the large difference in the DBM water depth values DBMs needed to be standardized, and therefore, each DBM water depth value was standardized to [−1,1]. The normalization formula is expressed as follows:

$$DBM_i = 2 \times (DBM_i - H_{min})/(H_{max} - H_{min}) - 1 \quad (10)$$

where H_{min} is the minimum value of DBM and H_{max} is the maximum value of DBM.

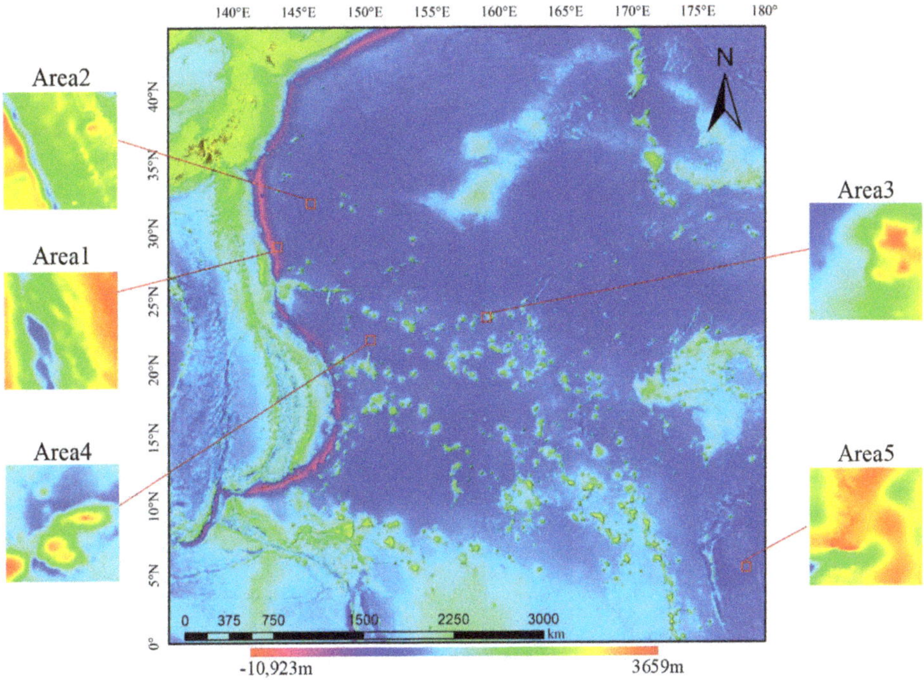

Figure 6. Study area.

Since the format of DBMs is usually .tif, there was only one channel, while natural images had three channels. Therefore, before inputting data into the model, we needed to set the number of input channels in the convolutional layer to 1 for training. Similarly, it was necessary to set the number of output channels of the convolutional layer to 1 before reconstruction.

3.3. Evaluation Methods

To test the effectiveness of STFET, we selected bicubic interpolation [35], SRCNN, SRGAN, and SRResNet (SRGAN generator) as comparison methods. Bicubic interpolation is one of the simplest and most effective interpolation methods and is often used as a baseline method for SR [21]. The first deep learning-based SR method was SRCNN. This network is simple and effective, and some scholars have proven its practicality in DEM SR. SRGAN is the first model to employ the generation confrontation network for the SR task, and it emphasizes that high accuracy does not always equate to superior visual quality using the perceptual loss function to enhance the picture. This finding has been supported by later research [36,37], and the focus of several studies has shifted from improving image accuracy to improving image data quality. Therefore, in this paper, we chose SRGAN to verify whether the perceptual loss function could acquire better visual quality in DBMs. SRResNet is the generator component of SRGAN, and it improves SRGAN's accuracy while using only the mean squared error (MSE) loss function to avoid imbalance during the training process.

DBM water depth values generally vary widely, reaching several thousand meters underwater or even tens of thousands of meters underwater. Therefore, to quantitatively compare the performance of various DBM SR methods, we used common DBM evaluation indices, the root mean squared error (RMSE) and mean absolute error (MAE), as evaluation indices and used the maximum bathymetric error (Emax) as an evaluation index to reflect the stability of the SR methods. These evaluation indices can be calculated as follows:

$$RMSE = \sqrt{\frac{1}{N}\sum_{i=1}^{N}(h_i - \hat{h}_i)^2} \quad (11)$$

$$MAE = \frac{1}{N}\sum_{i}^{N}|h_i - \hat{h}_i| \quad (12)$$

$$E_{max} = max(abs(h_i - \hat{h}_i)) \quad (13)$$

where N is the total number of pixels in a DBM image. h_i and $\hat{h}_i$ represent the elevation values of the HR DBM and the corresponding SR DBM, respectively.

4. Results and Discussions

4.1. Results on DBMs

Table 1 shows the quantitative results of 638 DBMs in the test set in terms of the RMSE, MAE, and Emax indices. Among these results, those in bold highlight the best performance of each indicator, and the underlined values are the second-best results. From Table 1, we can see that, compared to other methods, STFET had the best effect on all evaluation indices. Moreover, the accuracy of all deep learning-based methods was better than bicubic interpolation, which shows that deep learning methods have an application value in DBM SR. Similar to the results for natural images [20], SRGAN achieves a lower accuracy than SRResNet based on the RMSE accuracy index.

Table 1. Quantitative results on the test set. Results in bold are best, and those underlined are second best.

Method	RMSE (m)	MAE (m)	Emax (m)
Bicubic	15.85	8.08	295.80
SRCNN	<u>14.78</u>	<u>7.26</u>	<u>293.48</u>
SRGAN	15.27	7.87	306.98
SRResNet	15.13	7.89	300.07
STFET	**13.30**	**6.88**	**262.05**

From Table 1, it can also be seen that SRCNN performs better than SRGAN and SRResNet; these results are different from those obtained in the CV domain. STFET uses the seabed terrain feature extraction module and effective transformer module, and SR's accuracy is significantly improved. Compared with bicubic interpolation, SRCNN, SRGAN, and SRResNet, the performance results in terms of RMSE decreased by 16%, 10%, 13%, and 12%, respectively. The test set includes complex-area and flat-area DBMs, and the STFET results are optimal, proving that this method is universal and reliable.

4.2. Quantitative Evaluation Results on Complex Regions

To evaluate the performance of DBM in different terrain regions, we selected five regions with more significant topographic features from the test set to conduct an accuracy comparison. These five regions contained highly undulating and complex terrain, including trenches and seamounts. The locations are shown in Figure 6. Among them, the extremum between area 2 and area 5 was small, and the terrain undulation was relatively moderate, while the topographic features of areas 1, 3, and 4 were more obvious. The original DBM for the five regions and the SR DBM reconstructed using different SR methods were combined for comparison, and the quantitative results are shown in Table 2.

Table 2. Quantitative results for complex regions. Results in bold are the best, and those underlined are the second best.

Area	Method	RMSE (m)	MAE (m)	Emax (m)
1	Bicubic	32.26	24.90	<u>157.34</u>
	SRCNN	31.69	24.72	201.30
	SRGAN	<u>15.11</u>	<u>11.03</u>	175.39
	SRResNet	21.40	16.25	**140.28**
	STFET	**11.98**	**8.41**	198.65
2	Bicubic	17.58	9.78	494.95
	SRCNN	17.26	9.73	**483.92**
	SRGAN	<u>14.49</u>	<u>7.44</u>	<u>484.46</u>
	SRResNet	14.69	7.87	492.02
	STFET	**12.57**	**6.12**	491.90
3	Bicubic	30.08	17.66	189.57
	SRCNN	29.77	18.04	196.98
	SRGAN	<u>12.44</u>	<u>9.27</u>	136.85
	SRResNet	15.22	10.97	<u>126.56</u>
	STFET	**10.92**	**8.18**	**125.52**
4	Bicubic	27.10	14.30	314.91
	SRCNN	26.42	14.63	**295.98**
	SRGAN	<u>13.99</u>	<u>7.94</u>	357.04
	SRResNet	16.69	9.50	<u>310.11</u>
	STFET	**12.84**	**6.99**	360.42
5	Bicubic	16.63	6.42	268.70
	SRCNN	15.63	5.05	279.80
	SRGAN	14.16	5.45	259.50
	SRResNet	<u>12.93</u>	<u>5.27</u>	<u>207.15</u>
	STFET	**11.29**	**4.58**	**194.97**

From Table 2, it can be seen that the STFET results were still optimal for complex terrain. In area 2 and area 5, all methods obtained quite accurate results, and STFET exhibited the greatest improvement. However, in area 1, area 3, and area 4, the terrain features were complex. For example, area 1 comprised a large area containing trenches with topographic features. The reconstruction effect of bicubic interpolation and SRCNN was poor, while the reconstruction effect of SRGAN and SRResNet was much better in area 1. This observation illustrates the ability of residual networks to extract features significantly. STFET has a seabed terrain feature extraction module and an efficient transformer module and, therefore, it can enhance the extraction of local terrain features and similar global terrain features so that the reconstruction results are significantly improved. Compared with bicubic interpolation, SRCNN, SRGAN and SRResNet, the accuracy in terms of the RMSE decreased by 63%, 62%, 21%, and 44%, respectively, in area 1. These results demonstrate the effectiveness of STFET in terms of extracting seafloor topographic features.

4.3. Visual Evaluation of Different Methods

To compare the visual quality of reconstructions using different methods, we combined HR DBMs with SR DBMs generated using different methods for the corresponding regions, and these results are shown in Figure 7. We can see that, on the whole, the SR DBM reconstructed by all methods was relatively close to the HR DBM. However, in some areas with large terrain fluctuations and obvious topographic features, subtle differences in terrain could still be observed. As shown in Figure 8, we zoomed in on the area with obvious seabed terrain features in Figure 7; the reconstruction effect of bicubic interpolation and SRCNN in areas with large terrain fluctuations was too smooth, and the visual quality was poor, which is consistent with the quantitative results in Table 2. At the same time, the proposed method had a better reconstruction effect compared to other methods, and the result is closer to an HR DBM. Therefore, the results confirm the effectiveness of STFET in terms of both visual quality and reconstruction accuracy.

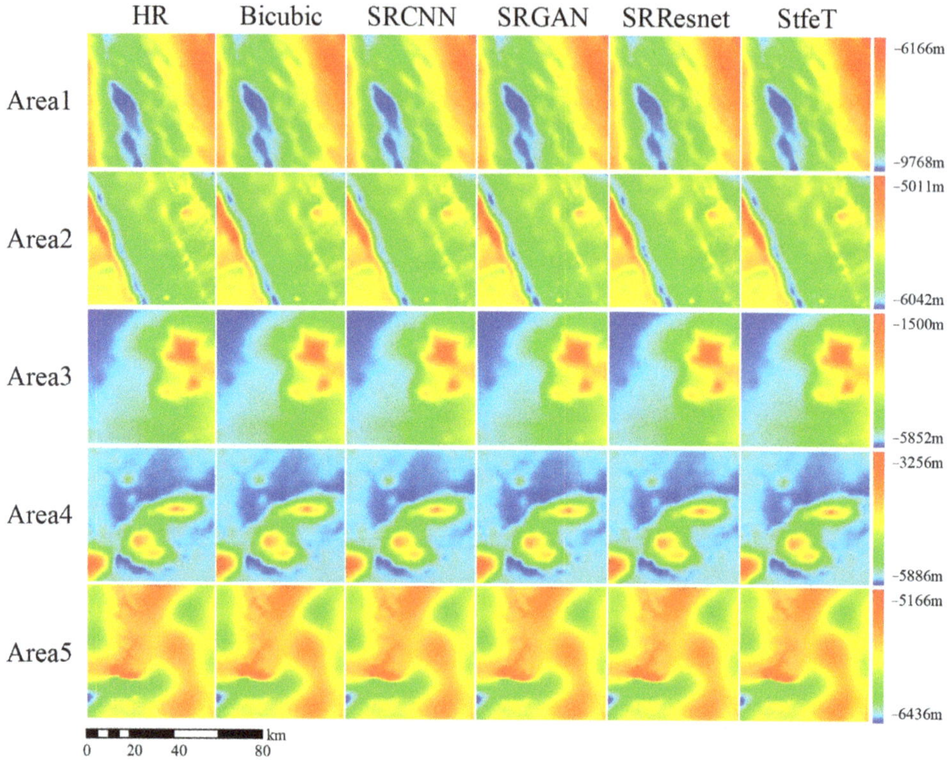

Figure 7. Comparison of results using different SR methods in various topographic areas.

4.4. Residual Evaluation of Different Methods

To visually compare the differences between the DBM and HR DBM generated using each SR method, we plotted the water depth difference between the SR DBM and HR DBM in the five districts as an error map, and the results are shown in Figure 9. The main difference in the reconstructed DBM lies in special terrain areas, such as the trench in area 1 and the seamount in area 4 in the figure, where the error is relatively large. From the error map results, we can clearly see that the error area generated using STFET was much smaller overall. The error in the local area was also greatly reduced closer to the HR DBM, which indicates that the effectiveness of the seabed terrain extraction module and the efficient transformer module was further validated. Meanwhile, similar to natural images, SRGAN outperformed SRResNet visually. However, the performance of bicubic interpolation and SRCNN was poor, and their error range was large; therefore, they were not suitable for DBM SR in complex terrain areas. As a result, the main challenge of DBM SR was to restore these areas. At the same time, it was necessary to ensure the reconstruction effect of the overall terrain.

To quantitatively analyze the error accuracy of various methods, we took 0–25 m, 25–100 m, 100–200 m, and more than 200 m as the statistical conditions and counted the number of error points generated using different methods of DBMs in Figure 9. The results are shown in Table 3. From the results, we can clearly see that the DBM reconstruction errors generated by StfeT were more concentrated at 0–25 m, and this reconstruction accuracy was better than other methods.

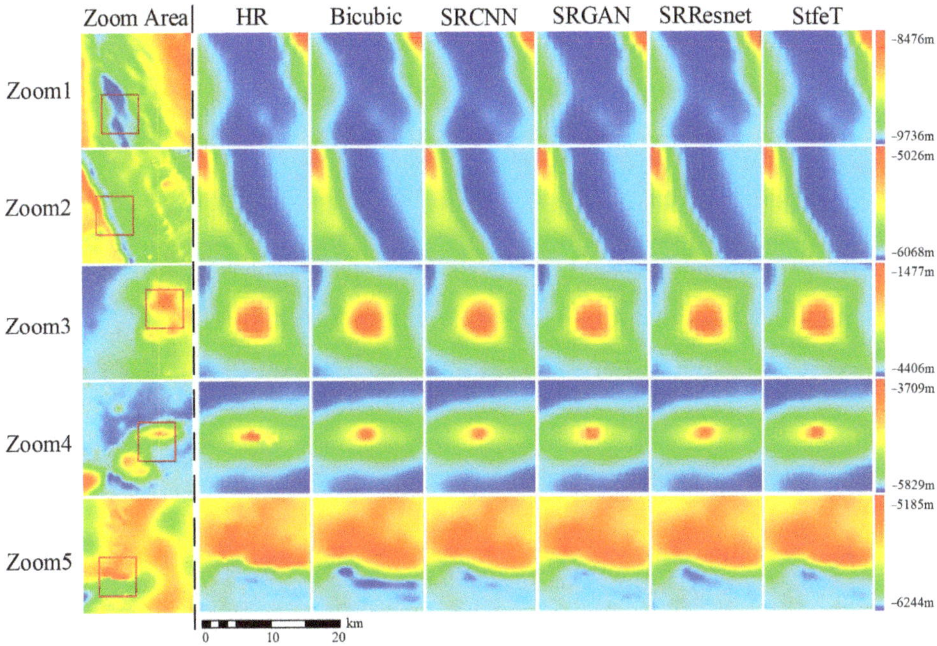

Figure 8. Comparison of the results after zooming in to Figure 7.

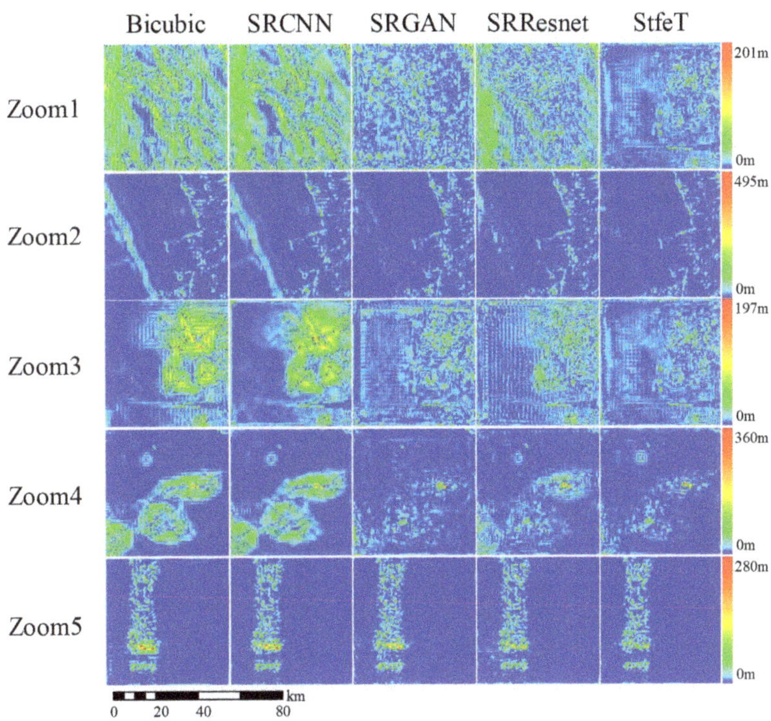

Figure 9. Comparison of SR DBM and HR DBM error maps for different methods in various regions.

Table 3. The number of errors generated using different methods.

Method	Number (Error < 25 m)	Number (25 m < Error < 100 m)	Number (100 m < Error < 200 m	Number (Error > 200 m
Bicubic	61,711	19,390	793	26
SRCNN	61,378	19,876	642	24
SRGAN	77,494	4344	60	22
SRResNet	71,029	10,704	163	24
StfeT	78,893	2949	56	22

4.5. Ablation Study

In this section, we verify the effectiveness of the seabed terrain feature extraction module and the efficient transformer module in DBM SR. We added the seabed terrain feature extraction module and the efficient transformer module into other models, with SRResNet as the baseline. SRResNet-dconv represents SRResNet with the seabed terrain feature extraction module, and SRResNet-ET represents SRResNet with the efficient transformer module. The comparison results are shown in Table 4.

Table 4. Ablation study results for the proposed STFET. Results in bold are best.

Method	RMSE (m)	MAE (m)	Emax (m)
SRResNet	15.13	7.89	300.07
SRResNet-dconv	15.10	7.70	322.98
SRResNet-ET	15.01	7.70	319.30
STFET	**13.30**	**6.88**	**262.05**

Common convolutional layers were used in conventional residual networks, and their convolution kernels were usually regular; however, the DBM seabed terrain features were irregular. By using deformable convolutional layers, features that more closely resembled the actual terrain could be obtained to some extent, thereby improving the accuracy of super-resolution. In addition, the efficient transformer could use a self-attention mechanism that allowed each pixel to interact with all pixels in the entire image, and this nonlocal operation could better capture long-range correlations in images, helping to recover missing details more accurately. The results in Table 3 show that the seabed terrain feature extraction module and the efficient transformer module have the potential to improve the accuracy of different deep learning-based SR methods. The effectiveness of the seabed terrain extraction module and the efficient transformer is demonstrated. However, when the seabed terrain feature extraction module and the efficient transformer module were used alone, this improvement effect was not very satisfactory. This observation shows that when using deep learning-based SR methods to extract features from DBMs, it is necessary to both extract local features and strengthen the extraction of similar global features. Only in this way can high-accuracy HR DBM be obtained more effectively.

5. Conclusions

In most previous studies, scholars have often used interpolation methods to improve DBM resolution, but the accuracy of generating HR DBM was poor. In this paper, we propose a seabed terrain feature extraction transformer for DBM SR, which mainly consists of the seabed terrain feature extraction module and the efficient transformer module. First, the seabed terrain feature extraction module has an important function, and it is able to adaptively extract DBM seabed terrain features. This adaptive ability enables the model to perceive DBM seabed terrain features at different scales and strengthen the extraction of local DBM seabed terrain features. Second, the efficient transformer module finds similar seabed terrain features through the self-attention mechanism, which enhances the extraction effect of similar global seabed terrain features. The effectiveness of these two modules was verified through ablation experiments, which demonstrated that, compared with bicubic interpolation and SRCNN, SRGAN, and SRResNet, STFET decreased the RMSE by 16%, 10%, 13%, and 12%, respectively. Furthermore, especially in areas with complex seabed terrain features, the accuracy of SR DBM generated using STFET reconstruction was significantly improved.

The method proposed in this paper considers global and local seabed terrain features for DBM SR but is not limited to DBM SR. While making up for the lack of interpolation accuracy, it can also be combined with the interpolation method for data fusion. Our next step considers that the proposed STFET can be used to transfer the trained model to the sea domain with only LR DBM, and we plan to rebuild and generate HR DBMs to compensate for the deficiency of LR DBMs and meet the needs of practical applications.

Author Contributions: W.C. collected and processed the data, performed analysis, and wrote the Paper; Y.L. proposed the main idea and made suggestions on the experiments; Y.C. and Z.D. helped to write and edit the article; H.Y. and N.L. contributed to the validation. All authors have read and agreed to the published version of the manuscript.

Funding: This work was supported by the Shandong Natural Science Foundation (grant number ZR2023QD113), the Shandong Postdoctoral Innovation Project (grant number SDCX-ZG-202202041), and the Qingdao Natural Science Foundation (grant number 23-2-1-73-zyyd-jch).

Data Availability Statement: The digital bathymetric model data used in this paper can be downloaded at: https://www.gebco.net/data_and_products/gridded_bathymetry_data/gebco_2022/ (accessed on 10 September 2023).

Conflicts of Interest: The authors declare that they have no known competing financial interest or personal relationships that could have appeared to influence the work reported in this paper.

References

1. Wu, Z.; Li, J.; Jin, X.; Shang, J.; Li, S.; Jin, X. Distribution, features, and influence factors of the submarine topographic boundaries of the Okinawa Trough. *Sci. China Earth Sci.* **2014**, *57*, 1885–1896. [CrossRef]
2. Sandwell, D.T.; Smith, W.H. Marine gravity anomaly from Geosat and ERS 1 satellite altimetry. *J. Geophys. Res. Solid Earth* **1997**, *102*, 10039–10054. [CrossRef]
3. Wang, M.; Wu, Z.; Yang, F.; Ma, Y.; Wang, X.H.; Zhao, D. Multifeature extraction and seafloor classification combining LiDAR and MBES data around Yuanzhi Island in the South China Sea. *Sensors* **2018**, *18*, 3828. [CrossRef]
4. Picard, K.; Brooke, B.P.; Harris, P.T.; Siwabessy, P.J.; Coffin, M.F.; Tran, M.; Spinoccia, M.; Weales, J.; Macmillan-Lawler, M.; Sullivan, J. Malaysia Airlines flight MH370 search data reveal geomorphology and seafloor processes in the remote southeast Indian Ocean. *Mar. Geol.* **2018**, *395*, 301–319. [CrossRef]
5. Mayer, L.; Jakobsson, M.; Allen, G.; Dorschel, B.; Falconer, R.; Ferrini, V.; Lamarche, G.; Snaith, H.; Weatherall, P. The Nippon Foundation—GEBCO seabed 2030 project: The quest to see the world's oceans completely mapped by 2030. *Geosciences* **2018**, *8*, 63. [CrossRef]
6. Schaffer, J.; Timmermann, R.; Arndt, J.E.; Kristensen, S.S.; Mayer, C.; Morlighem, M.; Steinhage, D. A global, high-resolution data set of ice sheet topography, cavity geometry, and ocean bathymetry. *Earth Syst. Sci. Data* **2016**, *8*, 543–557. [CrossRef]
7. Ramillien, G.; Cazenave, A. Global bathymetry derived from altimeter data of the ERS-1 geodetic mission. *J. Geodyn.* **1997**, *23*, 129–149. [CrossRef]
8. Briggs, I.C. Machine contouring using minimum curvature. *Geophysics* **1974**, *39*, 39–48. [CrossRef]
9. Smith, W.H.; Wessel, P. Gridding with continuous curvature splines in tension. *Geophysics* **1990**, *55*, 293–305. [CrossRef]
10. Glenn, J.; Tonina, D.; Morehead, M.D.; Fiedler, F.; Benjankar, R. Effect of transect location, transect spacing and interpolation methods on river bathymetry accuracy. *Earth Surf. Process. Landforms* **2016**, *41*, 1185–1198. [CrossRef]
11. Merwade, V. Effect of spatial trends on interpolation of river bathymetry. *J. Hydrol.* **2009**, *371*, 169–181. [CrossRef]
12. Chen, G.; Chen, Y.; Wilson, J.P.; Zhou, A.; Chen, Y.; Su, H. An Enhanced Residual Feature Fusion Network Integrated with a Terrain Weight Module for Digital Elevation Model Super-Resolution. *Remote Sens.* **2023**, *15*, 1038. [CrossRef]
13. Habib, M. Evaluation of DEM interpolation techniques for characterizing terrain roughness. *Catena* **2021**, *198*, 105072. [CrossRef]
14. Dong, C.; Loy, C.C.; He, K.; Tang, X. Learning a Deep Convolutional Network for Image Super-Resolution. In Proceedings of the ECCV: 13th European Conference, Zurich, Switzerland, 6–12 September 2014.
15. Yang, W.; Zhang, X.; Tian, Y.; Wang, W.; Xue, J.H.; Liao, Q. Deep learning for single image super-resolution: A brief review. *IEEE Trans. Multimed.* **2019**, *21*, 3106–3121. [CrossRef]
16. He, K.; Zhang, X.; Ren, S.; Sun, J. Deep residual learning for image recognition. In Proceedings of the 2016 IEEE Conference on Computer Vision and Pattern Recognition (CVPR), Las Vegas, NV, USA, 27–30 June 2016; pp. 770–778. [CrossRef]
17. Chen, Z.; Wang, X.; Xu, Z.; Hou, W. Convolutional neural network based dem super resolution. *Int. Arch. Photogramm. Remote Sens. Spat. Inf. Sci. ISPRS Arch.* **2016**, *41*, 247–250. [CrossRef]

18. Wu, Z.; Shen, C.; Van Den Hengel, A. Wider or deeper: Revisiting the resnet model for visual recognition. *Pattern Recognit.* **2019**, *90*, 119–133. [CrossRef]
19. Xu, Z.; Chen, Z.; Yi, W.; Gui, Q.; Hou, W.; Ding, M. Deep gradient prior network for DEM super-resolution: Transfer learning from image to DEM. *ISPRS J. Photogramm. Remote Sens.* **2019**, *150*, 80–90. [CrossRef]
20. Ledig, C.; Theis, L.; Husz'ar, F.; Caballero, J.; Cunningham, A.; Acosta, A.; Aitken, A.; Tejani, A.; Totz, J.; Wang, Z.; et al. Photo-Realistic Single Image Super-Resolution Using a Generative Adversarial Network. In Proceedings of the 2017 IEEE Conference on Computer Vision and Pattern Recognition (CVPR), Honolulu, HI, USA, 21–26 July 2017; pp. 105–114. [CrossRef]
21. Zhang, Y.; Yu, W. Comparison of DEM Super-Resolution Methods Based on Interpolation and Neural Networks. *Sensors* **2022**, *22*, 745. [CrossRef]
22. Zhou, A.; Chen, Y.; Wilson, J.P.; Su, H.; Xiong, Z.; Cheng, Q. An Enhanced double-filter deep residual neural network for generating super resolution DEMs. *Remote Sens.* **2021**, *13*, 3089. [CrossRef]
23. Zhang, B.; Xiong, W.; Ma, M.; Wang, M.; Wang, D.; Huang, X.; Yu, L.; Zhang, Q.; Lu, H.; Hong, D. Super-resolution reconstruction of a 3 arc-second global DEM dataset. *Sci. Bull.* **2022**, *67*, 2526–2530. [CrossRef]
24. Zhou, A.; Chen, Y.; Wilson, J.P.; Chen, G.; Min, W.; Xu, R. A multi-terrain feature-based deep convolutional neural network for constructing super-resolution DEMs. *Int. J. Appl. Earth Obs. Geoinf.* **2023**, *120*, 103338. [CrossRef]
25. Jiang, Y.; Xiong, L.; Huang, X.; Li, S.; Shen, W. Super-resolution for terrain modeling using deep learning in high mountain Asia. *Int. J. Appl. Earth Obs. Geoinf.* **2023**, *118*, 103296. [CrossRef]
26. Yutani, T.; Yono, O.; Kuwatani, T.; Matsuoka, D.; Kaneko, J.; Hidaka, M.; Kasaya, T.; Kido, Y.; Ishikawa, Y.; Ueki, T. Super-Resolution and Feature Extraction for Ocean Bathymetric Maps Using Sparse Coding. *Sensors* **2022**, *22*, 3198. [CrossRef] [PubMed]
27. Hidaka, M.; Matsuoka, D.; Kuwatani, T.; Kaneko, J.; Kasaya, T.; Kido, Y.; Ishikawa, Y.; Kikawa, E. Super-resolution for Ocean Bathymetric Maps Using Deep Learning Approaches: A Comparison and Validation. *Geoinformatics* **2021**, *32*, 3–13. (In Japanese) [CrossRef]
28. Zhang, Y.; Yu, W.; Zhu, D. Terrain feature-aware deep learning network for digital elevation model superresolution. *ISPRS J. Photogramm. Remote Sens.* **2022**, *189*, 143–162. [CrossRef]
29. Lu, Z.; Li, J.; Liu, H.; Huang, C.; Zhang, L.; Zeng, T. Transformer for single image super-resolution. In Proceedings of the 2022 IEEE/CVF Conference on Computer Vision and Pattern Recognition (CVPR), New Orleans, LA, USA, 19–20 June 2022; pp. 457–466. [CrossRef]
30. Zheng, X.; Bao, Z.; Yin, Q. Terrain Self-Similarity-Based Transformer for Generating Super Resolution DEMs. *Remote Sens.* **2023**, *15*, 1954. [CrossRef]
31. Kim, J.; Lee, J.K.; Lee, K.M. Accurate Image Super-Resolution Using Very DeeConvolutional Networks. In Proceedings of the 2016 IEEE Conference on Computer Vision and PatternRecognition (CVPR), Las Vegas, NV, USA, 27–30 June 2016; pp. 1646–1654. [CrossRef]
32. Dai, J.; Qi, H.; Xiong, Y.; Li, Y.; Zhang, G.; Hu, H.; Wei, Y. Deformable convolutional networks. In Proceedings of the 2017 IEEE International Conference on Computer Vision (ICCV), Venice, Italy, 22–29 October 2017; pp. 764–773. [CrossRef]
33. Zhu, X.; Hu, H.; Lin, S.; Dai, J. Deformable convnets v2: More deformable, better results. In Proceedings of the 2019 IEEE/CVF Conference on Computer Vision and Pattern Recognition (CVPR), Long Beach, CA, USA, 15–20 June 2019; pp. 9308–9316. [CrossRef]
34. Dosovitskiy, A.; Beyer, L.; Kolesnikov, A.; Weissenborn, D.; Zhai, X.; Unterthiner, T.; Dehghani, M.; Minderer, M.; Heigold, G.; Gelly, S. An image is worth 16 × 16 words: Transformers for image recognition at scale. *arXiv* **2020**, arXiv:2010.11929.
35. Han, D. Comparison of Commonly Used Image Interpolation Methods. In Proceedings of the 2nd International Conference on Computer Science and Electronics Engineering (ICCSEE 2013), Hangzhou, China, 22–23 March 2013; pp. 1556–1559.
36. Wang, X.; Yu, K.; Wu, S.; Gu, J.; Liu, Y.; Dong, C.; Qiao, Y.; Change Loy, C. ESRGAN: Enhanced Super-Resolution Generative Adversarial Networks. In Proceedings of the European Conference on Computer Vision (ECCV) Workshops, Munich, Germany, 8–14 September 2018; pp. 63–79.
37. Zhang, W.; Liu, Y.; Dong, C.; Qiao, Y. Ranksrgan: Generative adversarial networks with ranker for image super-resolution. In Proceedings of the 2019 IEEE/CVF International Conference on Computer Vision (ICCV), Seoul, Republic of Korea, 27 October–2 November 2019; pp. 3096–3105. [CrossRef]

remote sensing

MDPI

Article

Super Resolution of Satellite-Derived Sea Surface Temperature Using a Transformer-Based Model

Runtai Zou [1], Li Wei [1,2,*] and Lei Guan [1,2,3]

[1] College of Marine Technology, Faculty of Information Science and Engineering, Ocean University of China, Qingdao 266100, China; zouruntai@stu.ouc.edu.cn (R.Z.); leiguan@ouc.edu.cn (L.G.)
[2] Key Laboratory of Ocean Observation and Information of Hainan Province, Sanya Oceanographic Institution, Ocean University of China, Sanya 572024, China
[3] Laboratory for Regional Oceanography and Numerical Modeling, Qingdao National Laboratory for Marine Science and Technology, Qingdao 266237, China
* Correspondence: weili2068@ouc.edu.cn

Abstract: Sea surface temperature (SST) is one of the most important factors related to the ocean and the climate. In studying the domains of eddies, fronts, and current systems, high-resolution SST data are required. However, the passive microwave radiometer achieves a higher spatial coverage but lower resolution, while the thermal infrared radiometer has a lower spatial coverage but higher resolution. In this paper, in order to improve the performance of the super-resolution SST images derived from microwave SST data, we propose a transformer-based SST reconstruction model comprising the transformer block and the residual block, rather than purely convolutional approaches. The outputs of the transformer model are then compared with those of the other three deep learning super-resolution models, and the transformer model obtains lower root-mean-squared error (RMSE), mean bias (Bias), and robust standard deviation (RSD) values than the other three models, as well as higher entropy and definition, making it the better performing model of all those compared.

Keywords: super-resolution; transformer-based model; sea surface temperature

Citation: Zou, R.; Wei, L.; Guan, L. Super Resolution of Satellite-Derived Sea Surface Temperature Using a Transformer-Based Model. *Remote Sens.* **2023**, *15*, 5376. https://doi.org/10.3390/rs15225376

Academic Editors: Moulay A. Akhloufi and Mozhdeh Shahbazi

Received: 17 September 2023
Revised: 6 November 2023
Accepted: 14 November 2023
Published: 16 November 2023

1. Introduction

Sea surface temperature (SST), an essential variable related to the ocean and the climate, plays a crucial role in the heat, freshwater, and momentum flux exchange at the ocean–atmosphere interface [1]. Changes in sea surface temperature reveal fronts and eddies caused by subsurface heat changes, and eddies are associated with weak SST gradients [2]. Satellite remote sensing is an effective approach to deriving SST with dense spatial resolution on regional and global scales, making eddies, fronts, and current systems visible in SST imagery [3].

Satellite infrared observations of SST have shown capabilities for finer sampling, with a spatial resolution of about 1–4 km; examples include the Visible Infrared Imaging Radiometer Suite (VIIRS) onboard the Suomi National Polar-orbiting Partnership (Suomi NPP) satellite, the advanced geosynchronous radiation imager (AGRI) onboard the Fengyun-4A (FY-4A), and the Moderate Resolution Imaging Spectroradiometer (MODIS) onboard the Terra and Aqua satellites [4]. However, the SST data derived from satellite infrared radiometers are susceptible to interference from weather factors such as clouds, resulting in low spatial coverage. The measurements of SST from passive microwave radiometers are able to penetrate thick cloud layers, enabling all-weather observations, but they have a coarser spatial resolution of about 50–75 km [5]. The identification and tracking of small SST features requires high spatial resolution and frequent revisits [6]. Reconstructing the high-resolution sea surface temperature fields from passive microwave remote sensing SST observations would greatly facilitate our development of a comprehensive understanding of intricate SST features.

Deep learning-based super-resolution image reconstruction is a novel image reconstruction technique that could improve low-resolution images by downscaling with high fidelity [7]. Dong et al. [8] first used the super-resolution convolutional neural network (SRCNN) method with three convolution layers to solve the super-resolution (SR) problem, and achieved great results. In order to enhance the speed and resolution of the model, the fast super-resolution convolutional neural network (FSRCNN) [9] was proposed by Dong et al., which achieved a better super-resolution quality while being tens of times faster than previous models. Kim et al. [10] proposed the very deep SR network (VDSR) method, with 20 layers, to enhance SR performance. They found that deeper work can benefit from the capacity of deep learning-based SR, and the residual learning framework can be useful in solving the gradient explosion problem. A method called the deep compendium model (DCM) was proposed by Haut et al. [11]. It is a novel SR method based on a deep efficient model that integrates improvements on previous network designs, including residual units, skip connections, and network-in-network (NIN), to efficiently acquire high-quality super-resolution remote sensing data while avoiding undesirable visual artifacts.

Deep learning-based super-resolution image reconstruction has been explored in the field of SST and has shown outstanding achievements. In some approaches, the Level 4 SST data are downscaled by interpolating them into low-resolution (LR) SST images, and subsequently upscaling the LR SST images to SR SST images via the deep learning method, which shows that deep learning methods can be used to deal with SST SR problems. For example, Aurelien et al. [12] downscaled the Operational Sea Surface Temperature and Ice Analysis (OSTIA) SST data into LR SST images via bicubic interpolation, and they then chose the SRCNN model to upscale the LR SST images into SR SST images. Khoo et al. [13] proposed the Spectral Normalization–Enhanced Super Resolution Generative Adversarial Network (SN-ESRGAN) to upscale LR SST images into SR SST images. The LR SST images were downscaled from the OSTIA SST data by nearest interpolation. By validating the low-resolution data from the South China Sea, this method can achieve a higher and more realistic resolution than other methods. In other approaches, infrared SST data are selected as high-resolution (HR) data and microwave SST data are selected as LR data to train deep learning methods. Lloyd et al. [14] adopted the VDSR method to upscale the 1 km resolution brightness temperature data from the Sentinel 3 satellite into 200 m resolution SST data, and the target SST data came from the Landsat 8 satellite. Tomoki et al. [15] used the 125 km-resolution ERA20C SST data as the LR SST data and the 25 km resolution OISST SST data as the HR SST data to train the residual in residual dense block (RRDB) net, and the results exhibited a high quality. Bo et al. [5] proposed the oceanic data reconstruction (ODRE) network and trained it with 3-day-averaged $0.25^\circ \times 0.25^\circ$ grid advanced microwave scanning radiometer 2 (AMSR2) SST data and daily 4 km Lever-3 mapped MODIS Terra SST data. The article demonstrates the excellent performance of the ODRE model compared to the FDSR, DRRN, SRCNN, and VDSR models.

The studies mentioned above primarily used CNN as the main network. It is worth noting, however, that the transformer-based model, which incorporates the attention mechanism, has gained significant attention in computer vision applications and has achieved remarkable outcomes. The transformer model has been sought to apply to computer vision tasks, rather than convolutional approaches alone [16,17]. The focus of this paper is on improving the performance of super-resolution SST images derived from microwave SST data. We propose a transformer-based SST reconstruction model, which can upscale the microwave SST data 12.5 times.

The summary is organized as follows: The experimental data are introduced in Section 2.1. The overall framework of the proposed transformer model is presented in Section 2.2.1 and the other three SR models are introduced in Section 2.2.2. Besides, a brief introduction is given to the loss function and implementation details in Section 2.2.3. The results are represented in Section 3 and are discussed in Section 4. The conclusion is given in Section 5.

2. Materials and Methods

2.1. Training Data

The VIIRS_L3S_LEO_PM SST dataset comprises the 0.02° gridded L3S sub-skin SST data produced by the National Oceanic and Atmospheric Administration, Center for Satellite Applications and Research (NOAA STAR) [18]. The data were sourced from the VIIRS onboard the Joint Polar Satellite System (JPSS) satellites; specifically, the Suomi NPP and NOAA20 (N20) satellites. The dataset covers the period from February 2012 to the present, and is reported in two files per 24-hour interval, daytime and nighttime, approximately at the local equator crossing times around 01:30/13:30. The data were provided in NetCDF4 format. The data were downloaded from https://search.earthdata.nasa.gov (accessed on 18 September 2023). The SST image yielded by the VIIRS_L3S_LEO_PM SST dataset on 1 January 2022 is shown in Figure 1.

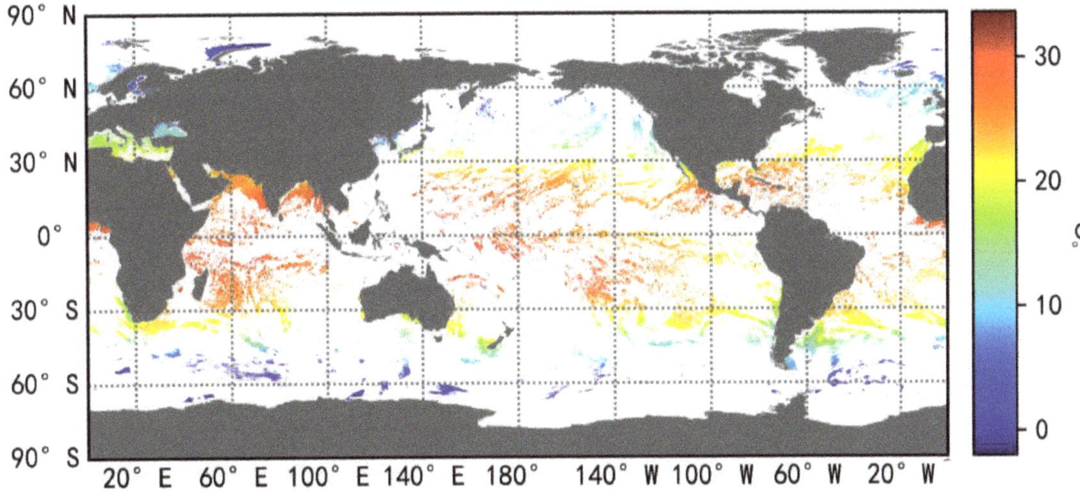

Figure 1. The SST image of VIIRS on 1 January 2022.

The AMSR2_L3U SST dataset comprises the 0.25° gridded L3U sub-skin SST data produced by Remote Sensing Systems (REMSS). The data were sourced from the advanced microwave scanning radiometer 2 (AMSR2) sensor onboard the Global Change Observation Mission–Water (GCOM-W) satellite developed by the Japan Aerospace Exploration Agency (JAXA). The dataset covers the period from 2 July 2012 to the present day. The local equator crossing times are at approximately 01:30 and 13:30. The data are provided in NetCDF4 format. The data were downloaded from the website https://data.remss.com/amsr2/ocean/L3/ (accessed on 18 September 2023). The SST image yielded by the AMSR2_L3U SST dataset on 1 January 2022 is shown in Figure 2.

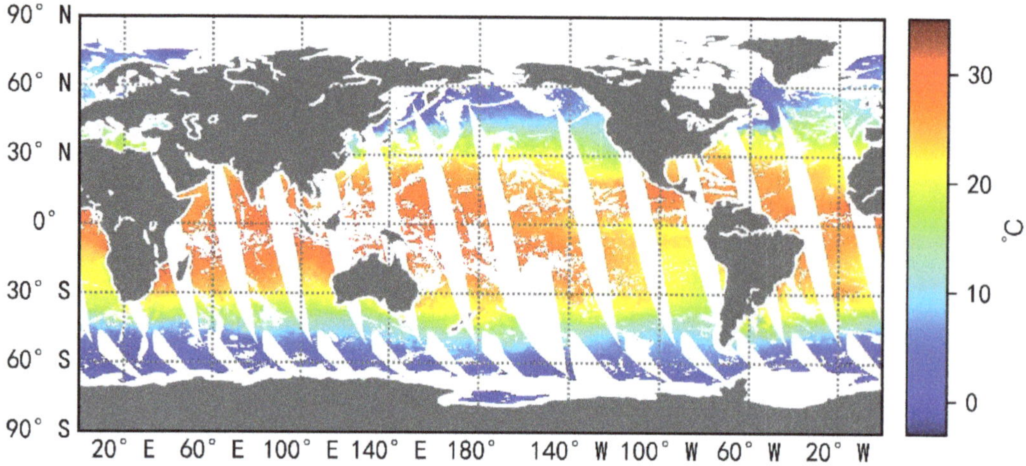

Figure 2. The SST image of AMSR2 on 1 January 2022.

The left and right patches in Figure 3 are truncated from the SST images at the same locations as the AMSR2 and VIIRS, respectively. Figures 1 and 2 clearly show that the passive microwave radiometer represented by the AMSR2 achieves a higher spatial coverage, while the thermal infrared radiometer represented by the VIIRS has a lower spatial coverage. Nevertheless, as shown in Figure 3, the SST image retrieved from the AMSR2 has a lower resolution, while the SST image retrieved from the VIIRS has a higher resolution. Therefore, by adopting the super-resolution method, the resolution of the AMSR2 SST data can be improved, yielding a higher spatial coverage.

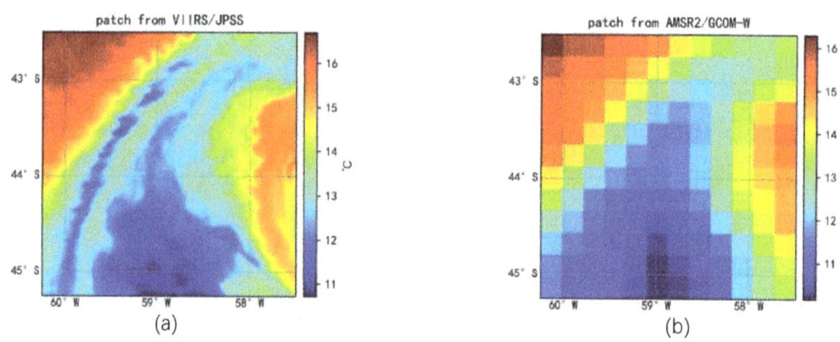

Figure 3. The patches of (**a**) VIIRS and (**b**) AMSR2 on 1 January 2022.

Considering the SST for the VIIRS and AMSR2 differ in depth, the value of the VIIRS SST and the AMSR2 SST at daytime is different from that at nighttime because of the strong diurnal warming effect. Therefore, in order to improve the accuracy of results, the nighttime data and daytime data should be taken to train the models, respectively. In this paper, we only discuss the model trained by the nighttime data. There are two years of global nighttime data, 2021 and 2022 SST, used for the training dataset in this study. The global nighttime SST data from January 2023 were used for validation. The AMSR2 and VIIRS data taken on the same day can be interpreted as a low-resolution and high-resolution image pair. The training image pairs were split into 12-by-12 patches in AMSR2 and 150-by-150 patches in VIIRS, and the patches with missing data were excluded. After filtering out missing data, there are 258,738 full patches, which were completely employed as training data in this experiment. Likewise, after filtering out missing data, there are 9469 full patches, which were completely employed as testing data in this experiment. All the training patches and testing images were normalized to the interval [0, 1] before training.

2.2. Methods

Transformer-based models have been widely adopted in the field of computer vision. However, transformer-based deep learning SR models are rarely applied to the SST field at present. Therefore, a transformer-based model is proposed to reconstruct the high-resolution SST fields from passive microwave remote sensing SST observations. In order to verify the accuracy and quality of the outputs of the transformer-based model, three CNN-based SR models are used to compare the statistics of SST difference between the outputs and the VIIRS SST data. The detailed framework of the proposed model is shown as follows.

2.2.1. The Proposed Model

The overall framework of the transformer model is shown in Figure 4. The reason for choosing two groups of ResNet blocks, linear layers, and transformer encoders in parallel is that the DCM has achieved good results using jump connections [11], so we also used jump connections in the design of the proposed model.

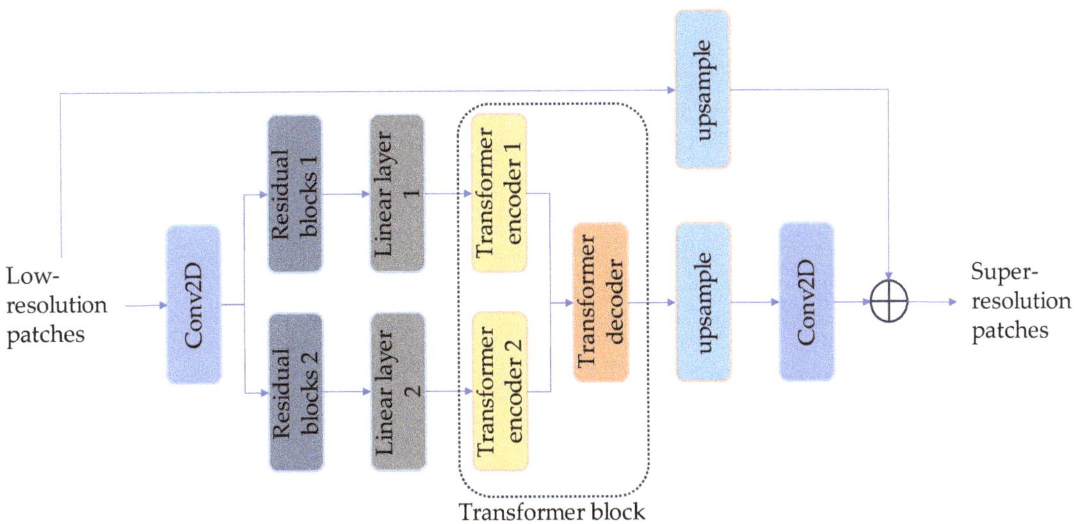

Figure 4. Flowchart of the proposed model.

Given an LR patch, one convolution is utilized to transform the pixel space input from the AMSR2_L3U SST dataset to feature space.

$$f_{in} = \mathrm{conv}(x_{LR}) \tag{1}$$

here, conv denotes a convolutional operation and x_{LR} represents the LR patch, while f_{in} represents the output tensor of the convolution.

Two residual blocks are borrowed from Res-Net [19] and play a crucial role in image deep learning by aiding in gradient vanishing and exploding. With the introduction of skip connections, these blocks allow for the seamless flow of gradients, mitigating the issues of gradient instability. Furthermore, the residual blocks enhance the fitting capacity of the network by effectively learning residual information, thereby capturing the difference between the input and output features. Therefore, residual blocks not only improve the ability of the model to fit complex patterns, but also accelerate convergence during training. As shown in Figure 5, residual blocks 1 and residual blocks 2 can be constructed by stacking multiple residual blocks, enabling the expression of intricate features and enhancing overall model performance. The convolutions and rectified linear units (ReLUs) are included in the residual blocks, which are defined as

$$x_1 = Res_1(f_{in}) \tag{2}$$

$$x_2 = Res_2(x_1) \tag{3}$$

where Res_1 and Res_2 represent the first and second residual blocks, and x_1 and x_2 denote the outputs of the first residual block and the second residual block, respectively.

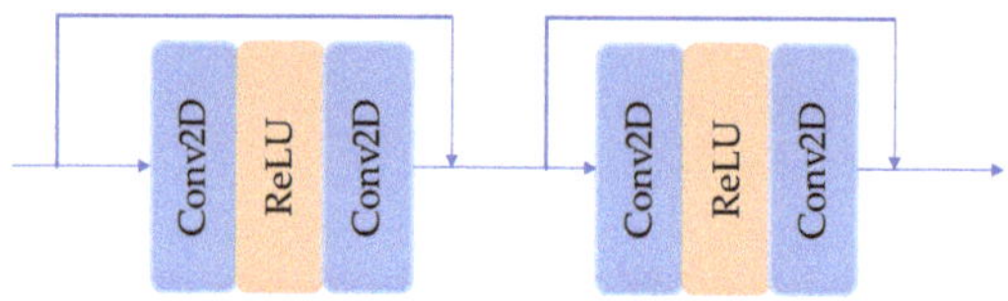

Figure 5. Illustration of the residual blocks.

After feature extraction in the feature space, two linear layers can be used to project the input tensor into a higher-dimensional space, thereby enhancing the expressive power of the model. The weights and biases in the linear layer are model parameters that need to be learned through the backpropagation algorithm. By utilizing training data and a loss function, the model can automatically adjust the parameters in the linear layer, enabling the model to fit the training data better and achieve greater generalization. The linear layers are defined as

$$l_1 = Linear_1(x_1) \tag{4}$$

$$l_2 = Linear_2(x_2) \tag{5}$$

where $Linear_1$ and $Linear_2$ represent the first and second linear layers, and l_1 and l_2 denote the outputs of the first linear layer and the second linear layer, respectively.

Following the idea of Alexey et al. [16], the standard transformer here receives a one-dimensional sequence of token embeddings as the input. To handle the three-dimensional (3-D) features of patches, we should reshape the patches $l \in R^{H\times W\times C}$ into sequences of flattened two-dimensional (2-D) vectors $l_{p_i} \in R^{P_H\times P_W\times C}$, $i = \{1,\ldots,N\}$, where (H, W, C) represents the height, the width, and the number of channels of the features, $(P_H \times P_W \times C)$ represents the height, the width, and the number of channels of the vectors, and $N = (HW/P_H P_W)$ is the number of vectors, and also denotes the length of the input sequence. The input vector size is typically fixed to D dimensions, and we need to use a trainable linear projection to map l_{p_i} to D dimensions. The formula for reshaping the patches is defined as

$$l_k = \left[l_{p_0}^{(k)}, l_{p_1}^{(k)}, \ldots, l_{p_N}^{(k)}\right], k = 1,2 \tag{6}$$

$$y_k = \left[l_{p_0}^{(k)}S, l_{p_1}^{(k)}S, \ldots, l_{p_N}^{(k)}S\right], k = 1,2 \tag{7}$$

where S is the matrix of the linear projection.

Following the original design of Vaswani et al. [20], the multi-headed self-attention (MSA) module enhances the modeling capacity of the transformer encoder by learning different attention focuses from distinct subspaces through the application of the attention mechanism on different linear projection spaces. This enables the model to capture diverse aspects of the input sequence, thereby improving its representational power and generalization performance. The combination process that takes place in the multi-input MSA module can be formulated as

$$\text{Atten} = \text{softmax}\left(\frac{\text{QK}^T}{\sqrt{\text{d}_\text{k}}}\right) \tag{8}$$

$$\begin{gathered}\text{MultiHead}(\text{Q}, \text{K}, \text{V}) = \text{Concat}(\text{head}_1, \ldots, \text{head}_\text{h})\text{W}^\text{O} \\ \text{where head}_\text{i} = \text{Atten}\left(\text{QW}_\text{i}^\text{Q}, \text{KW}_\text{i}^\text{K}\right)\text{VW}_\text{i}^\text{V}\end{gathered} \tag{9}$$

where d_k is the dimensions of the features and head denotes the heads of the MSA module. The W^O, $\text{W}_\text{i}^\text{Q}$, $\text{W}_\text{i}^\text{K}$ and $\text{W}_\text{i}^\text{V}$ are parameter matrices and Q, K, and V are the variables decided by the index of related components.

The transformer encoder 1 and decoder are set as examples to provide a clear description about the process of the transformer block, which are carefully shown in Figure 6. And the transformer encoder 2 in the proposed model has the same structure as the encoder 1. The transformer encoder contains a MSA module and a multilayer perceptron (MLP) network. The layer normalization (LN) [21], which helps mitigate the issue of internal covariate shift during the training process and stabilizes the learning procedure of the model, is set before the MSA module and the MLP module. The MLP module has two linear layers as well as a gaussian error linear unit (GELU) [22], which enables the model to learn and represent complex features, thus improving its ability to generalize and make accurate predictions. The overall formula of the encoder can be represented as

$$\text{y}'_\text{k} = \text{MSA}(\text{LN}(\text{y}_\text{k})) + \text{y}_\text{k}, \text{k} = 1, 2 \tag{10}$$

$$\text{z}_\text{k} = \text{MLP}(\text{LN}(\text{y}'_\text{k})) + \text{y}'_\text{k}, \text{k} = 1, 2 \tag{11}$$

where z_k is the output of the transformer encoder, with the same dimension as y_k.

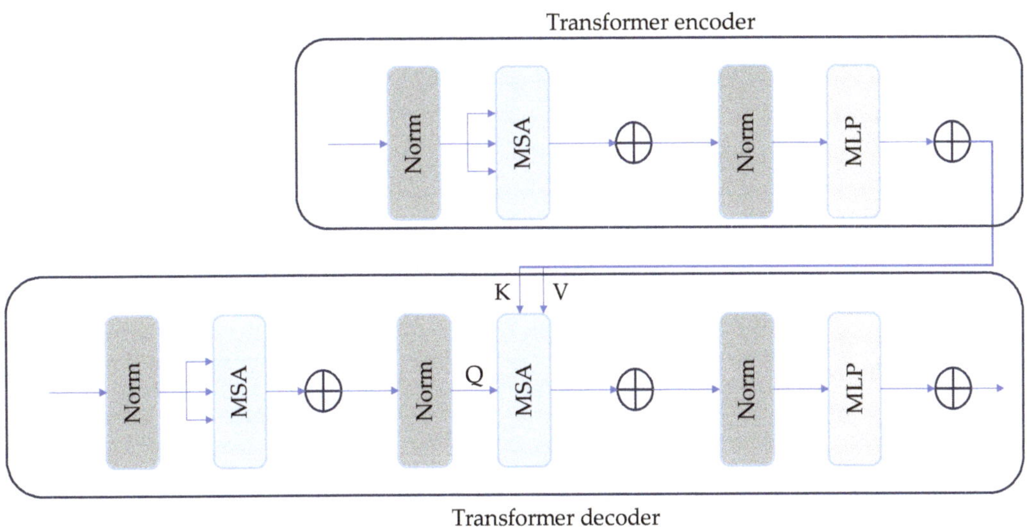

Figure 6. Illustration of the transformer encoder 1 and decoder.

In addition to the MSA module and MLP network, the transformer decoder also includes a specific MSA module with cross-attention. By calculating attention weights, this module can simultaneously handle the features input by the decoder and connect them with the output derived from encoder 1. This mechanism allows the decoder to fully utilize the contextual information, resulting in the generation of an output sequence corresponding more accurately to the input sequence. It serves as the core component of the decoder. The output of the decoder can be obtained using the following equation

$$z_2' = \mathrm{MSA}(\mathrm{LN}(z_2)) + z_2 \tag{12}$$

$$z_2'' = \mathrm{MSA}(\mathrm{LN}(z_2'), \mathrm{LN}(z_1)) + z_2' \tag{13}$$

$$z_0 = \mathrm{MLP}(\mathrm{LN}(z_2'')) + z_2'' \tag{14}$$

where z_0 is the output of the decoder. Then, we reshape the sequences of the flattened 2-D vectors z_0 back into patches and use the linear layer to project the higher-dimensional space back to the tensor. After that, we can receive the output q_0.

In the up-sample layer, a function of the PyTorch [23] deep learning framework is employed to perform the up-sample operation. Its main purpose is to increase the size of the input tensor to the target size, thereby altering the spatial resolution of the data to meet the requirements. In this paper, the bicubic method is utilized for the up-sampling operation. The up-sample layer can be represented as

$$u_0 = \mathrm{upsample}(q_0) \tag{15}$$

$$h = \mathrm{upsample}(x_{LR}) \tag{16}$$

where u_0 is the output of the up-sample layer derived from the front layers, and the h is directly up-sampled from the low-resolution patches.

Lastly, one convolution is utilized to transform the feature space into the SST pixel space. It can be represented as

$$f_{out} = \mathrm{conv}(u_0) \tag{17}$$

where f_{out} represents the output tensor of convolution.

Finally, after one convolutional layer is applied, the super-resolution HR patch x_{SR} is obtained by adding the residual and f_{out}. The final result can be expressed as follows:

$$x_{SR} = h + f_{out} \tag{18}$$

2.2.2. Other SR Models

In order to compare the quality of the outputs of the transformer model, three famous deep learning-based SR models, including the FSRCNN, DCM, and VDSR models, are used to generate high-quality super-resolution outputs, which can be compared with the outputs of the transformer model in different comparisons.

As shown in Figure 7, the DCM model [11] consists of two parts, including the feature extractor part and the reconstruction part. In the feature extractor part, there are 12 convolution layers, which can be used to extract the corresponding feature maps. The parametric rectified linear unit (PReLU) functions [24] are set after every convolution layer, which enables us to deal with the decaying ReLU effect and the vanishing gradient problem. In the reconstruction part, an up-sample layer is used to increase the size of the input tensor to the target size. Then, the two branches with the convolution layer and the PReLU function are set to reduce the depth of the input volume. Lastly, the SR patch is the sum of the outputs of the previous layers and of the up-sample layer derived directly from the low-resolution patch.

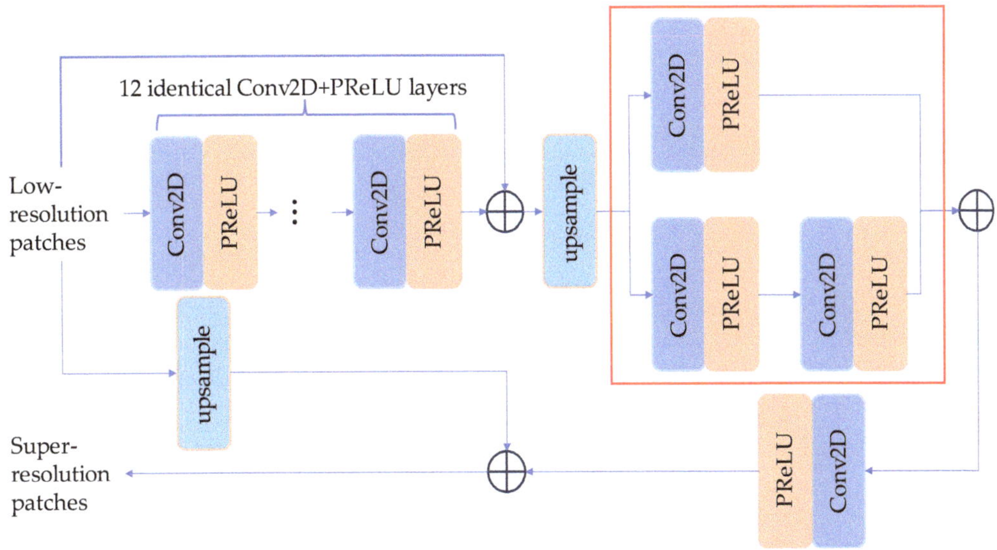

Figure 7. Flowchart of the DCM model.

As shown in Figure 8, the FSRCNN model [9] is typically used in the area of super-resolution imagery. It consists of patch extraction, representation, non-linear mapping, an up-sample layer, and reconstruction. Firstly, a convolution and PReLU function is set in the patch extraction and representation regions. There are five convolution layers and PReLU functions in the non-linear mapping stage. In the up-sample layer, the up-sample function of PyTorch [23] is also used in this model to increase the input size to the target size. At last, a convolution layer without a PReLU function is used for the reconstruction.

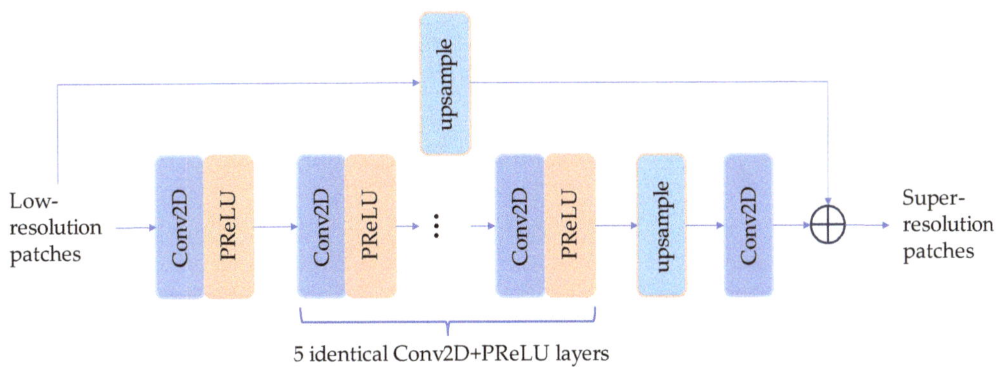

Figure 8. Flowchart of the FSRCNN model.

As shown in Figure 9, the architecture of the VDSR model primarily comprises patch extraction, non-linear mapping, and reconstruction [10]. Twenty convolution layers are used in this model, except for in the last part, which is followed by batch normalization (BN) and the ReLU. An up-sample layer is set before the last convolution. Similar to the other models, the SR patch is the sum of the outputs of the previous layers and of the up-sample layer taken directly from the low-resolution patch.

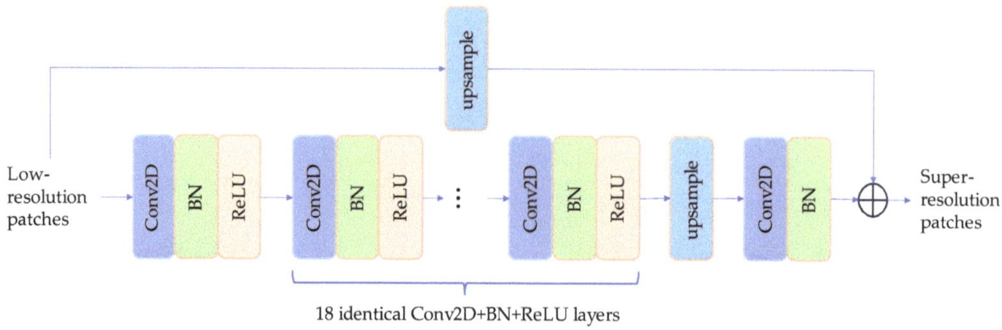

Figure 9. Flowchart of the VDSR model.

2.2.3. Loss Function and Implementation Details

The loss function of all four models is composed of the mean squared error (MSE). Given a low-resolution patch x_{LR} and the corresponding high-resolution reference patch x_{HR}, the loss function can be attained via

$$L(\theta) = \frac{1}{N}\sum_{i=1}^{N}\left(G_{\theta}\left(x_{LR}^{(i)}\right) - x_{HR}^{(i)}\right)^2 \tag{19}$$

where G_{θ} refers to the models with parameters θ, $G_{\theta}\left(x_{LR}^{(i)}\right)$ is the aforementioned $x_{SR}^{(i)}$, and N is the number of training patches.

In the optimization, we used the Adaptive Moment Estimation (Adam) optimizer [25] to train four models, where $\beta_1 = 0.9$, $\beta_2 = 0.99$, and $\epsilon = 10^{-8}$. The initial learning rate was set to 3×10^{-4}, and the mini-batch size was set to 5. When the loss of a patch is less than 10^{-5}, the training is stopped. These methods were implemented in PyTorch [23], and all experiments were run on an NVIDIA GeForce RTX 3090 graphics card.

3. Results

In this section, numerous measures are used to verify the quality of the transformer model. At first, the scatterplots are presented to show the minimax errors of all four models directly, and the proportion of errors are calculated to clarify the distribution of errors. Then, the statistics of SST difference between the outputs of the VDSR, DCM, FSRCNN and transformer models and the VIIRS SST data are computed to illustrate the accuracy of all four models. At last, the SST images outputted by all four models are displayed to compare the quality of all four models.

To intuitively compare the stability of the results of all four models, we plotted the scatterplots of all the outputs from January 2023 and calculated the proportion of the error distribution. Figure 10 shows that the scatterplots of the output results of the transformer and DCM models are good, and most of the points are concentrated on the 1:1 straight line. However, compared with the outputs of the transformer and DCM models, the outputs of the VDSR and FSRCNN models show certain points with large differences, reaching as high as 5 °C. According to the distributions of the errors, the transformer model yielded a higher percentage of SST difference within the range of 0.1, 0.2, 0.5, and 1 °C compared with the other three models. Conclusively, the transformer model demonstrates distinct characteristics of better overall stability.

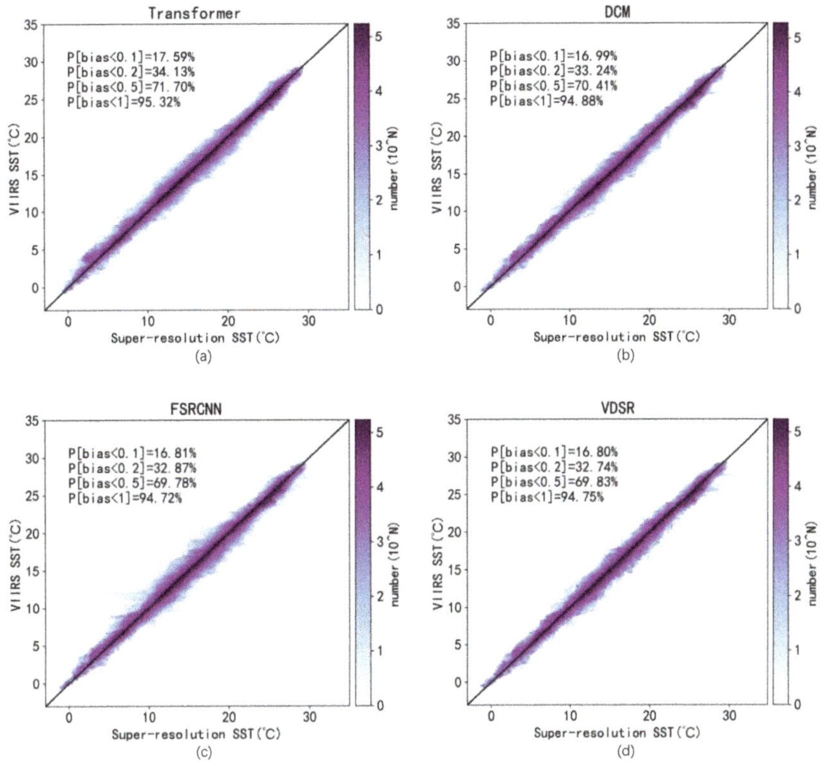

Figure 10. Density scatterplots of the VIIRS SST and the super-resolution SST values yielded using the (**a**) transformer, (**b**) DCM, (**c**) FSCNN, and (**d**) VDSR models, where the line is the 1:1 line. The number is shown in logarithmic form.

To evaluate the accuracy of the super-resolution SST obtained from the four models, the root mean square error (RMSE), mean bias (Bias), and robust standard deviation (RSD) between the outputs and VIIRS SST were calculated. As shown in Table 1, the bilinear and cubic interpolation showed a higher bias, RSD, and RMS compared to the deep learning models. Overall, the accuracy of the super-resolution SST values obtained from the deep learning model is slightly higher than that of the bilinear and cubic interpolation. In the comparison of four deep learning models, the transformer model showed the smallest bias and RMS, with values of 0.1 °C and 0.48 °C, respectively. The bias of the VDSR model is relatively high compared to the other three models, with the value of 0.15 °C. Comparing the RSD values of the models helps to exclude some extreme values from interfering with the error analysis, and so it is conducive to deriving a better understanding of the overall error of the data; here, the VDSR model and the transformer model showed the smallest values and thus the better results. Overall, the accuracy of the super-resolution SST values obtained from the transformer model is slightly higher than that of the other three models.

Table 1. Statistics of SST difference between the outputs of the VDSR, DCM, FSRCNN, and transformer models and VIIRS SST data. The number in boldface indicates the best performance.

Model	Mean Bias (°C)	RSD (°C)	RMS (°C)	N
Transformer	**0.10**	**0.35**	**0.48**	213,052,500
DCM	0.12	0.37	0.50	213,052,500
FSRCNN	0.13	0.36	0.49	213,052,500
VDSR	0.15	**0.35**	0.49	213,052,500
Bilinear	0.21	0.43	0.55	213,052,500
Cubic	0.18	0.41	0.53	213,052,500

Furthermore, the entropy and definition were employed to assess the quality of the SST images. The definition was measured by the gradient method, and a higher gradient magnitude indicates more pronounced edges or contour changes in the image [26]. Entropy is an important metric for measuring the richness of information [27]. As shown in Table 2, all models showed similar entropy values, while those of the transformer model and FSRCNN were higher compared to the other two deep learning models. And the entropy of the cubic is the highest. The definition of results of the transformer model are much better than those of the other three models, with a larger value of 75.46. The DCM and FSRCNN showed similar gradients, around 57. Compared with the traditional interpolation method, of bilinear and cubic, the deep learning model performs better in improving the image resolution. Moreover, the definition of the AMSR2 and VIIRS SST images are 3.99 and 128.08. All models can significantly improve the definition of the AMSR2 SST images, but there is still a gap with the definition of those of VIIRS. Overall, the SST image quality was improved through the deep learning models.

Table 2. Entropy and definition of the SST outputs of the VDSR, DCM, FSRCNN, transformer models and AMSR2 SST and VIIRS SST.

	Transformer	**DCM**	**FSRCNN**	**VDSR**	**AMSR2**	**VIIRS**	**Bilinear**	**Cubic**
Entropy	4.20	4.19	4.19	4.18	4.12	4.24	4.18	4.22
Definition	75.46	57.69	57.40	50.10	3.99	128.08	48.85	52.80

In order to demonstrate the spatial pattern of the results, we randomly selected the output from three regions. The super-resolution SST patches from the four models, as well as the AMSR2 SST and VIIRS SST patches, are displayed in Figure 11a. The reconstructions of the transformer and FSRCNN models in the northeastern area of the low-SST region are better than those of the other two models. The SST image generated by the transformer model is better than those of the other three models in the southern low-SST zone. In Figure 11b, the SST image output from the transformer model retains the details of the fronts a little more effectively than the other three models. The fronts of the other three models are either broken or notably weakened, resulting in deviations from actual conditions. In Figure 11c, we can see that the VDSR and DCM models have been over-smoothed, resulting in losses of detail. The SST images outputted by the transformer and FSRCNN models reflect the low-temperature region in the middle of the image, indicating that the quality of their reconstruction is high in this patch.

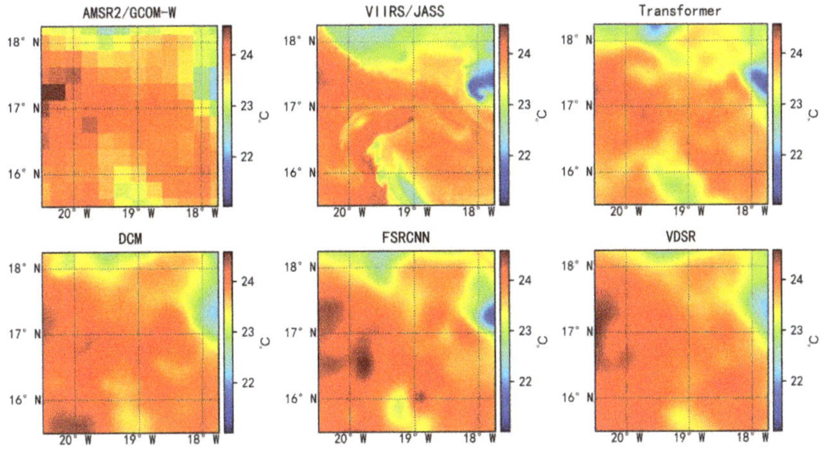

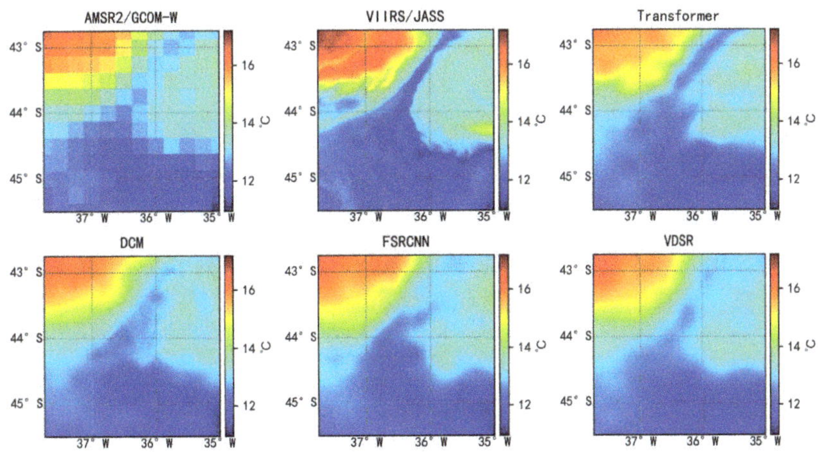

Figure 11. *Cont.*

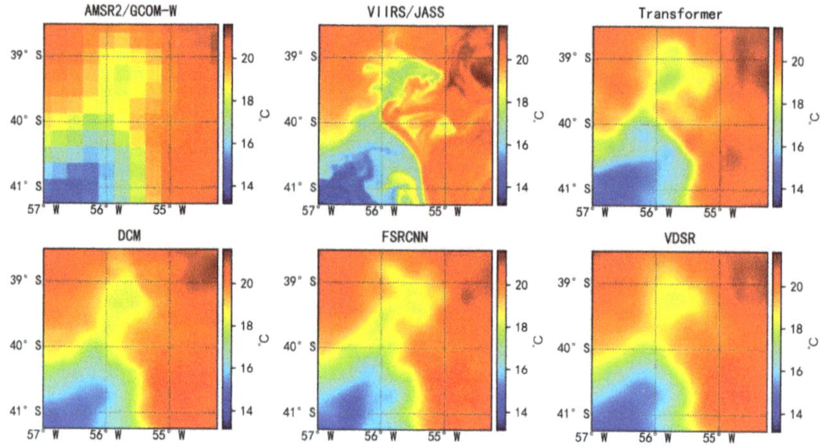

(c)

Figure 11. Exemplary SST patches for AMSR2 SST, VIIRS SST, and the super-resolution SSTs of FSRCNN, transformer, VDSR, and DCM on (**a**) 12 January 2023, (**b**) 7 January 2023, and (**c**) 28 January 2023.

4. Discussion

The transformer model achieves the lower bias, RSD, and RMS values, while obtaining higher values of entropy and definition, suggesting that the transformer block and the residual block benefit in SST reconstruction. Additionally, it effectively generates the distributions of absent temperature values in low-resolution data. These findings suggest that the transformer block and residual block included in the transformer-based model are useful for reducing pixel-by-pixel inaccuracies in the super-resolution sea surface temperature data. The accuracy of the DCM model is lower than that of the transformer model and higher than that of the VDSR model. This is because the DCM model features skip connections and NIN, allowing the model to acquire a better understanding of the relation between low-resolution SST data and high-resolution SST data. As a result, the DCM model is capable of generating more accurate outputs than the FSRCNN and VDSR models.

To further investigate the accuracy of the transformer model, we have drawn patches from the outputs of the transformer model, the AMSR2 SST, and the VIIRS SST, and we have also plotted the corresponding SST histograms of the outputs of the transformer models, and of the VIIRS SST and SSIM images. The patches of three days are arbitrarily chosen as examples. The patches of 10 January 2023 are shown in Figure 12a–e. The comparison of the AMSR2 SST, VIIRS SST, and SR SST images shows that the transformer model removes the artificial low-temperature region in the southeastern part of the AMSR2 SST image, and at the same time enhances the real high-temperature region in the southwest part of the SST image, making the result more consistent with that of the VIIRS. The histogram of the output SST also shows that its distribution is basically consistent with the distribution shown in the VIIRS SST. At the same time, the SSIM images also show a high degree of similarity, with the SSIM values of most regions being above 0.98, and only a small number of regions having lower SSIM values; however, these latter are still above 0.9, which reflects the excellent performance of the transformer model. The patches of 20 January 2023 are shown in Figure 12f–l. The reconstruction of the low-temperature region in the northeastern area of the output of the transformer model is perfect, and the low-temperature region in the middle of the image also matches closely to the trend of VIIRS, which indicates the good performance of the transformer model. Meanwhile, we can see that the distribution of the output SST is similar to that of the VIIRS SST, and we can also see from the SSIM image that the overall SSIM value is still above 0.95, indicating similar results to those of VIIRS. The patches of 9 January 2023 are shown in Figure 12m–o. We can see that the transformer model is corrected in the high-temperature area in the southeastern region compared with the AMSR2 SST, yielding a result that is close to that of the VIIRS SST image. Meanwhile, in the low-temperature region that can be seen in the northwestern region, the trend of the output SST is closer to that of the VIIRS SST, which indicates that the reconstruction is more effective. Additionally, in the SST distribution histogram, there are slight differences amongst the highest values, but on the whole, the SST distribution of the output results is consistent with that of the VIIRS SST. In the SSIM image, all the distributions are basically above 0.94, which indicates the high quality of the output images. The output of the transformer model gained fine features in comparison to that of the AMSR2 SST image, while it remains close to the VIIRS SST image. The histogram of the distribution of SST closely matches the VIIRS SST histogram, and the difference in the SSIM value between the VIIRS SST and the output SST is large, indicating that the transformer model achieves better results in SST super-resolution.

As revealed by Figure 12, the output of the transformer model yields a similar temperature distribution to that seen in the high-resolution SST data and yields a higher accuracy and perceptual quality. In order to assess the contribution of the individual structures of the model, we designed three ablation experiments. Whether two groups of ResNet blocks, linear layers, and transformer encoders or not are included in the transformer model in parallel is the first thing we think about. The result of this experiment indicates that the accuracy of the super-resolution SST values obtained from the transformer model with two groups is higher than that obtained from the transformer model with one group in ablation studies. Therefore, it makes sense that there are two groups of ResNet blocks, linear layers, and transformer encoders in parallel in the structure of the transformer. Secondly, we designed an ablation study regarding the transformer blocks, the result of which showed that when comparing results before removing the transformer blocks, the model without transformer blocks achieves higher bias, RSD, and RMS values, while obtaining lower values of entropy and definition, suggesting that the transformer block learns various attentional focuses from different subspaces, enhancing the modeling capability of the transformer encoder. Lastly, we added an ablation study regarding the residual blocks. The result of the experiment is that when comparing results before removing the residual blocks, the model without residual blocks achieves higher bias, RSD, and RMS values, while obtaining lower values of entropy and definition, suggesting that the residual blocks address the issue of vanishing and exploding gradients by implementing jump connections, which facilitate a smooth gradient flow. Residual blocks are also capable of efficiently learning residual information and capturing distinctions between input and output properties, resulting in an improved network fitting capability and faster convergence rates during training. Incorporating multiple residual blocks to the learn complex features of SST data enhances the overall performance of the model. Nevertheless, in certain regions, a contrast arises between the output of the proposed model and the high-resolution SST data. This discrepancy is primarily found in the near-shore area, and is caused by the error of microwave SST data. As the models used in this study were trained using global SST data, it is possible that a single model may not provide better results for every sea. Furthermore, as in prior research, we can infer that models designed for a specific region tend to perform better than globally trained models, which underscores the importance of regional specialization.

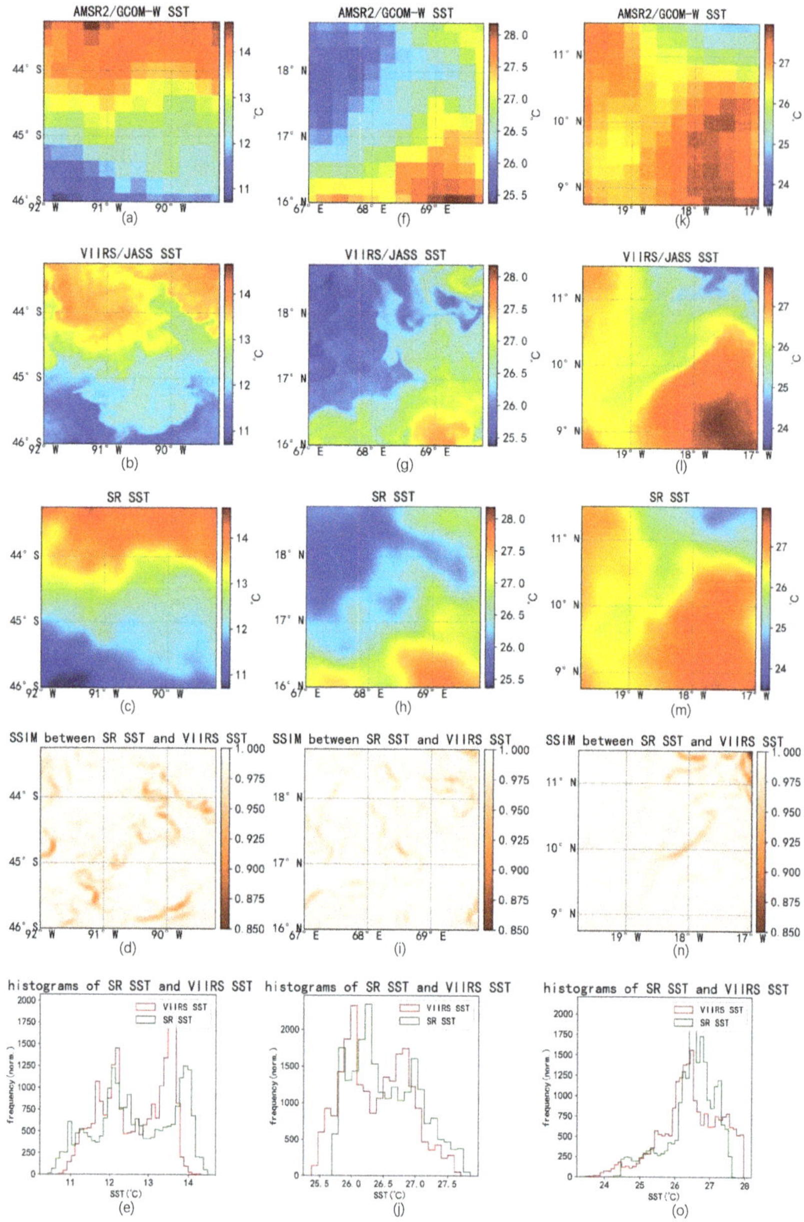

Figure 12. Exemplary SST patches for AMSR2 SST, VIIRS SST, and the super-resolution SSTs of transformer model, histograms of SST between super-resolution SST and VIIRS SST, and SSIM between super-resolution SST and VIIRS SST on (**a–e**) 10 January 2023, (**f–j**) 20 January 2023, (**k–o**) 9 January 2023. In the histograms, the pink bars represent the frequency of VIIRS SST values, while the green bars represent the frequency of SR SST values. In the SSIM image, the lighter the color, the greater the similarity.

5. Conclusions

It is widely acknowledged that high-resolution SST has a crucial role in the research of the domains of eddies, fronts, and current systems. The SST data from satellite infrared radiometers has a finer sampling but acquires a lower spatial coverage, resulting from the clouds. Nevertheless, satellite measurements of SST from passive microwave radiometers have demonstrated the capability of allowing all-weather observations but have a coarser spatial resolution. Therefore, in order to understand the intricate SST features, the transformer-based SR model proposed in this paper offers a flexible and efficient way to extract the high-resolution sea surface temperature fields from passive microwave remote sensing SST observations. The proposed model comprises the transformer block and the residual block, rather than purely convolutional approaches. The transformer block learns various attentional focuses from different subspaces, enhancing the modeling capability of the transformer encoder. And the residual blocks are capable of efficiently learning residual information and capturing distinctions between input and output properties, resulting in an improved network fitting capability and faster convergence rates during training. Because of the transformer block and the residual block, this transformer-based model obtains lower RMSE, Bias, and RSD values than the other three models, as well as a higher entropy and definition, making it the better-performing model of all compared. The results of this study demonstrate that this transformer-based model represents a viable means to produce SR SST data from LR microwave SST data, the outcomes of which align relatively with the SST values derived from the target dataset, indicating that advanced deep learning methods are suitable not only for the image field, but also for the SST field.

Author Contributions: Conceptualization, L.G., L.W. and R.Z.; methodology, R.Z.; software, R.Z.; validation, R.Z.; formal analysis, R.Z.; data curation, L.W. and R.Z.; writing—original draft preparation, R.Z.; writing—review and editing, L.G. and L.W.; visualization, R.Z.; supervision, L.G. and L.W.; project administration, L.G.; funding acquisition, L.G. and L.W. All authors have read and agreed to the published version of the manuscript.

Funding: This research was funded by the Hainan Provincial Joint Project of Sanya Yazhou Bay Science and Technology City, grant number 2021JJLH0081, the Hainan Provincial Natural Science Foundation of China, grant number 122CXTD519, and the National Natural Science Foundation of China, grant number 42206176.

Data Availability Statement: VIIRS_L3S_LEO_PM SST data were downloaded from https://search.earthdata.nasa.gov (accessed on 18 September 2023). AMSR2_L3U SST data were downloaded from the website https://data.remss.com/amsr2/ocean/L3/ (accessed on 18 September 2023).

Acknowledgments: VIIRS_L3S_LEO_PM SST data were provided by the National Oceanic and Atmospheric Administration, Center for Satellite Applications and Research. AMSR2_L3U SST data were provided by Remote Sensing Systems.

Conflicts of Interest: The authors declare no conflict of interest.

References

1. Yang, C.; Leonelli, F.E.; Marullo, S.; Artale, V.; Beggs, H.; Nardelli, B.B.; Chin, T.M.; De Toma, V.; Good, S.; Huang, B.; et al. Sea Surface Temperature Intercomparison in the Framework of the Copernicus Climate Change Service (C3S). *J. Clim.* **2021**, *34*, 5257–5283. [CrossRef]
2. Tandeo, P.; Chapron, B.; Ba, S.; Autret, E.; Fablet, R. Segmentation of Mesoscale Ocean Surface Dynamics Using Satellite SST and SSH Observations. *IEEE Trans. Geosci. Remote Sens.* **2014**, *52*, 4227–4235. [CrossRef]
3. Carroll, A.G.; Armstrong, E.M.; Beggs, H.M.; Bouali, M.; Casey, K.S.; Corlett, G.K.; Dash, P.; Donlon, C.J.; Gentemann, C.L.; Høyer, J.L.; et al. Observational Needs of Sea Surface Temperature. *Front. Mar. Sci.* **2019**, *6*, 420. [CrossRef]
4. Chin, T.M.; Vazquez-Cuervo, J.; Armstrong, E.M. A multi-scale high-resolution analysis of global sea surface temperature. *Remote Sens. Env.* **2017**, *200*, 154–169. [CrossRef]
5. Ping, B.; Su, F.; Han, X.; Meng, Y. Applications of Deep Learning-Based Super-Resolution for Sea Surface Temperature Reconstruction. *IEEE J. Stars* **2021**, *14*, 887–896. [CrossRef]
6. Martin, S. *An Introduction to Ocean Remote Sensing*; Cambridge University Press: Cambridge, UK, 2014; pp. 86–89. ISBN 978-1-107-01938-6.

7. Sato, H.; Fujimoto, S.; Tomizawa, N.; Inage, H.; Yokota, T.; Kudo, H.; Fan, R.; Kawamoto, K.; Honda, Y.; Kobayashi, T.; et al. Impact of a Deep Learning-based Super-resolution Image Reconstruction Technique on High-contrast Computed Tomography: A Phantom Study. *Acad. Radiol.* **2023**, *30*, 2657–2665. [CrossRef] [PubMed]
8. Dong, C.; Loy, C.C.; He, K.; Tang, X. Image Super-Resolution Using Deep Convolutional Networks. *IEEE Trans. Pattern Anal. Mach. Intell.* **2016**, *38*, 295–307. [CrossRef] [PubMed]
9. Dong, C.; Loy, C.C.; Tang, X. Accelerating the Super-Resolution Convolutional Neural Network. In Proceedings of the Computer Vision–ECCV 2016: 14th European Conference, Amsterdam, The Netherlands, 11–14 October 2016; Leibe, B., Matas, J., Sebe, N., Welling, M., Eds.; Springer International Publishing: Cham, Switzerland, 2016; pp. 391–407. [CrossRef]
10. Kim, J.; Lee, J.K.; Lee, K.M. Accurate Image Super-Resolution Using Very Deep Convolutional Networks. In Proceedings of the 2016 IEEE Conference on Computer Vision and Pattern Recognition (CVPR), Las Vegas, NV, USA, 27–30 June 2016; pp. 1646–1654.
11. Haut, J.M.; Paoletti, M.E.; Fernandez-Beltran, R.; Plaza, J.; Plaza, A.; Li, J. Remote Sensing Single-Image Superreso-lution Based on a Deep Compendium Model. *IEEE Geosci. Remote Sens. Lett.* **2019**, *16*, 1432–1436. [CrossRef]
12. Ducournau, A.; Fablet, R. Deep Learning for Ocean Remote Sensing: An Application of Convolutional Neural Networks for Super-Resolution on Satellite-Derived SST Data. In Proceedings of the 2016 9th Iapr Workshop On Pattern Recognition in Remote Sensing (Prrs), Cancun, Mexico, 4 December 2016. [CrossRef]
13. Khoo, J.J.D.; Lim, K.H.; Pang, P.K. Deep Learning Super Resolution of Sea Surface Temperature on South China Sea. In Proceedings of the 2022 International Conference on Green Energy, Computing and Sustainable Technology (GECOST), Miri Sarawak, Malaysia, 26–28 October 2022; pp. 176–180. [CrossRef]
14. Lloyd, D.T.; Abela, A.; Farrugia, R.A.; Galea, A.; Valentino, G. Optically Enhanced Super-Resolution of Sea Surface Temperature Using Deep Learning. *IEEE Trans. Geosci. Remote Sens.* **2022**, *60*, 1–14. [CrossRef]
15. Izumi, T.; Amagasaki, M.; Ishida, K.; Kiyama, M. Super-resolution of sea surface temperature with convolutional neural network and generative adversarial network-based methods. *J. Water Clim. Chang.* **2022**, *13*, 1673–1683. [CrossRef]
16. Dosovitskiy, A.; Beyer, L.; Kolesnikov, A.; Weissenborn, D.; Zhai, X.; Unterthiner, T.; Dehghani, M.; Minderer, M.; Heigold, G.; Gelly, S.; et al. An Image is Worth 16 × 16 Words: Transformers for Image Recognition at Scale. *arXiv* **2020**, arXiv:2010.11929. [CrossRef]
17. Lei, S.; Shi, Z.; Mo, W. Transformer-Based Multistage Enhancement for Remote Sensing Image Super-Resolution. *IEEE Trans. Geosci. Remote Sens.* **2022**, *60*, 1–11. [CrossRef]
18. Jonasson, O.; Gladkova, I.; Ignatov, A.; Kihai, Y. Algorithmic Improvements and Consistency Checks of the NOAA Global Gridded Super-Collated SSTs from Low Earth Orbiting Satellites (L3S-LEO). In Proceedings of the Ocean Sensing and Monitoring XIII, Online. 12–16 April 2021; Volume 11752. [CrossRef]
19. He, K.; Zhang, X.; Ren, S.; Sun, J. Deep Residual Learning for Image Recognition. In Proceedings of the 2016 IEEE Conference on Computer Vision and Pattern Recognition (CVPR), Las Vegas, NV, USA, 27–30 June 2016; pp. 770–778. [CrossRef]
20. Vaswani, A.; Shazeer, N.; Parmar, N.; Uszkoreit, J.; Jones, L.; Gomez, A.N.; Kaiser, A.; Polosukhin, I. Attention is all you need. *arXiv* **2017**, arXiv:1706.03762. [CrossRef]
21. Ba, J.L.; Kiros, J.R.; Hinton, G.E. Layer normalization. *arXiv* **2016**, arXiv:1607.06450. [CrossRef]
22. Hendrycks, D.; Gimpel, K. Gaussian error linear units (gelus). *arXiv* **2016**, arXiv:1606.08415. [CrossRef]
23. Paszke, A.; Gross, S.; Massa, F.; Lerer, A.; Bradbury, J.; Chanan, G.; Killeen, T.; Lin, Z.; Gimelshein, N.; Antiga, L.; et al. PyTorch: An Imperative Style, High-Performance Deep Learning Library. *arXiv* **2019**, arXiv:1912.01703. [CrossRef]
24. Xu, B.; Wang, N.; Chen, T.; Li, M. Empirical evaluation of rectified activations in convolutional network. *arXiv* **2015**, arXiv:1505.00853. [CrossRef]
25. Kingma, D.P.; Ba, J. Adam. A method for stochastic optimization. *arXiv* **2014**, arXiv:1412.6980. [CrossRef]
26. Huang, J.; Mumford, D. Statistics of natural images and models. In Proceedings of the 1999 IEEE Computer Society Conference on Computer Vision and Pattern Recognition (Cat. No PR00149), Fort Collins, CO, USA, 23–25 June 1999; pp. 541–547. [CrossRef]
27. Jost, L. Entropy and diversity. *Oikos* **2006**, *113*, 363–365. [CrossRef]

remote sensing

MDPI

Article

Multi-Attention Multi-Image Super-Resolution Transformer (MAST) for Remote Sensing

Jiaao Li [1,2,3,†], Qunbo Lv [1,2,3,†], Wenjian Zhang [1,2,3], Baoyu Zhu [1,2,3], Guiyu Zhang [1,2,3] and Zheng Tan [1,2,3,*]

1 Aerospace Information Research Institute, Chinese Academy of Sciences, No.9 Dengzhuang South Road, Haidian District, Beijing 100094, China; lijiaao21@mails.ucas.ac.cn (J.L.); lvqunbo@aoe.ac.cn (Q.L.); zhangwenjian21@mails.ucas.ac.cn (W.Z.); zhubaoyu20@mails.ucas.ac.cn (B.Z.); zhangguiyu21@mails.ucas.ac.cn (G.Z.)

2 School of Optoelectronics, University of Chinese Academy of Sciences, No.19(A) Yuquan Road, Shijingshan District, Beijing 100049, China

3 Department of Key Laboratory of Computational Optical Imagine Technology, CAS, No.9 Dengzhuang South Road, Haidian District, Beijing 100094, China

* Correspondence: tanzheng@aircas.ac.cn; Tel.: +86-1591-063-9123

† These authors contributed equally to this work.

Abstract: Deep-learning-driven multi-image super-resolution (MISR) reconstruction techniques have significant application value in the field of aerospace remote sensing. In particular, Transformer-based models have shown outstanding performance in super-resolution tasks. However, current MISR models have some deficiencies in the application of multi-scale information and the modeling of the attention mechanism, leading to an insufficient utilization of complementary information in multiple images. In this context, we innovatively propose a Multi-Attention Multi-Image Super-Resolution Transformer (MAST), which involves improvements in two main aspects. Firstly, we present a Multi-Scale and Mixed Attention Block (MMAB). With its multi-scale structure, the network is able to extract image features from different scales to obtain more contextual information. Additionally, the introduction of mixed attention allows the network to fully explore high-frequency features of the images in both channel and spatial dimensions. Secondly, we propose a Collaborative Attention Fusion Block (CAFB). By incorporating channel attention into the self-attention layer of the Transformer, we aim to better establish global correlations between multiple images. To improve the network's perception ability of local detailed features, we introduce a Residual Local Attention Block (RLAB). With the aforementioned improvements, our model can better extract and utilize non-redundant information, achieving a superior restoration effect that balances the global structure and local details of the image. The results from the comparative experiments reveal that our approach demonstrated a notable enhancement in cPSNR, with improvements of 0.91 dB and 0.81 dB observed in the NIR and RED bands of the PROBA-V dataset, respectively, in comparison to the existing state-of-the-art methods. Extensive experiments demonstrate that the method proposed in this paper can provide a valuable reference for solving multi-image super-resolution tasks for remote sensing.

Keywords: deep learning; remote sensing; multi-attention; Transformer; multi-scale; multi-image super-resolution

Citation: Li, J.; Lv, Q.; Zhang, W.; Zhu, B.; Zhang, G.; Tan, Z. Multi-Attention Multi-Image Super-Resolution Transformer (MAST) for Remote Sensing. *Remote Sens.* **2023**, *15*, 4183. https://doi.org/10.3390/rs15174183

Academic Editor: Salah Bourennane

Received: 16 July 2023
Revised: 17 August 2023
Accepted: 24 August 2023
Published: 25 August 2023

1. Introduction

Space optical remote sensing utilizes optical cameras in space to observe the Earth or space targets, capturing remote sensing images that reflect their reflection or the radiation information of the targets. It enables applications such as classification, identification, and comprehensive detection, serving as an important means of obtaining information for global change monitoring, regional surveillance, resource exploration, urban planning, and other fields [1]. The higher the geometric resolution of a space optical camera, the greater the precision in differentiating, identifying, and classifying targets.

Although increasing the optical aperture and focal length can directly improve the resolution, it significantly increases the cost, weight, volume, and development complexity of the camera. Additionally, regardless of the size, aperture, or focal length of the imaging system, the imaging resolution is inevitably limited by physical factors in the imaging chain. This leads to a decrease in the resolution index and is primarily caused by atmospheric blurring, optical aberrations and diffraction-induced blurring, aliasing due to an undersampling of CCD/CMOS detectors, and electronic noise [2].

Therefore, overcoming the physical limitations mentioned above and improving the resolution without changing the optical hardware is an important issue that needs to be addressed in space-based optical remote sensing. Image super-resolution reconstruction technology can overcome the limitations of traditional optical imaging and is an effective approach to solve the associated problems [3].

Image super-resolution reconstruction is the reconstruction of a high-resolution (HR) image from one or multiple low-resolution (LR) images. It possesses ill-posedness, falls within the realm of inverse problems, and represents a challenging research direction [4]. Since the problem was proposed by Harris [5] in 1964, scholars have proposed various super-resolution methods around the inverse convolution algorithm [6]. With the rapid development of deep learning technology, the use of neural networks to accomplish super-resolution reconstruction tasks has become a frontier approach in academic research, and a series of excellent super-resolution algorithms have emerged, often achieving state-of-the-art performance in various benchmarks of super-resolution [7]. In terms of network models, they can be divided into convolutional neural network algorithms [8–10], generative adversarial neural network algorithms [11–13], Transformer network algorithms [14–17], etc.

Single-image super-resolution aims to reconstruct an HR image from a noisy and low-quality LR image [18]. Multi-image super-resolution utilizes complementary information between multiple LR images of the same scene [19] to reconstruct an HR image through sub-pixel level information fusion [20,21]. Compared to single-image super-resolution algorithms, multi-image super-resolution (MISR) algorithms leverage complementary sub-pixel information [22], which can yield better reconstruction results [23].

The principle of the MISR algorithm is to extract and fuse sub-pixel features from multiple images to reconstruct the high-frequency information of the image and improve the resolution. It primarily focuses on three key aspects: (1) establishing a mechanism for extracting and fusing complementary sub-pixel features; (2) reducing the impact of registration errors between multiple images on the reconstruction; and (3) suppressing the influence of blurring and noise contamination on the reconstruction. In recent years, researchers in the field of remote sensing have conducted innovative research in high-frequency information feature extraction [24,25], registration error suppression [26–28], noise error suppression [29,30], and other aspects related to these three key aspects. However, there are still some issues that need to be addressed.

In terms of feature extraction, the existing algorithms are still inadequate in extracting the depth features that contribute to the super-resolution task in multiple images; the complementarity of features between each image is not sufficient, and features that do not contribute to the reconstruction task are also passed to the feature fusion step, which affects the reconstruction effect. In terms of feature fusion, the existing algorithms are still inadequate in terms of the local detail fusion mechanism, which leads to the fusion of sub-pixel features of multiple images while also fusing noise and blurred features, affecting the reconstruction effect. Finally, no optimization algorithm has been proposed that can comprehensively address the three abovementioned key aspects.

Based on the analysis of the limitations of the existing MISR methods for remote sensing, this paper innovatively proposes a synthetically improved Multi-Attention Multi-Image Super-Resolution Transformer (MAST). This network mitigates the issue of effective information loss that is present in existing methods by employing a multi-scale and multi-attention approach. Comparative experiments showed that the proposed algorithm achieved state-of-the-art super-resolution results in the NIR and RED bands of the PROBA-V MISR dataset, which can provide valuable references and insights for future studies in this field. In brief, the main contributions of this work are as follows:

1. We innovatively proposed a Multi-Scale and Mixed Attention Block (MMAB) that enabled a model to fully extract complementary information from multiple images. Specifically, we introduced a Multi-Scale Extraction Block (MSEB) to obtain more reference information and effectively suppress registration errors. We also presented a Channel Attention Guided Block (CAGB) and a Spatial Attention Guided Block (SAGB) to enhance the model's perception of high-frequency features with high contributions to the super-resolution task in each image.
2. We also presented a Collaborative Attention Fusion Block (CAFB) based on the Transformer, which constructed a more comprehensive feature fusion mechanism incorporating self-attention, channel attention, and local attention. The embedding of channel attention enhanced the advantage of the Transformer in establishing global correlations between multiple frames. The introduction of the Residual Local Attention Block (RLAB) largely broke through the limitations of the Transformer in capturing local high-frequency information, enabling the model to simultaneously restore the overall structure and local details of the image.

2. Related Work

2.1. Multi-Image Super-Resolution

The multi-image degradation model for the same scene imaged k times can be expressed as follows:

$$y_k = [A \otimes H_k \otimes C_k] \otimes x + n_k \tag{1}$$

where y_k represents the k-th LR image, x represents the HR image, the symbol $\otimes$ denotes convolution, n_k represents additive system noise, A represents the detector sampling coefficients, H_k denotes the blur coefficients, and C_k represents the inter-frame displacement relationship function. Multi-image super-resolution reconstruction involves solving for the HR image from this equation.

Tsai and Huang [31] first proposed the concept of multi-frame image super-resolution in 1984. They presented a frequency domain method that improves the spatial resolution of an image by utilizing the complementary information brought by the relative displacement between multiple LR images. However, the limitation of frequency domain methods is that they are restricted to global translational motion and linear space-invariant degradation models. Spatial domain methods, on the other hand, possess strong capabilities of incorporating prior constraints. These methods include iterative back-projection methods, convex set projection methods, maximum a posteriori methods, and more [32].

Although the deconvolution algorithm provides a solution for MISR, using it to address the multi-image super-resolution reconstruction problem involves solving an ill-posed problem that is inherently unstable. Registration errors and modeling errors are gradually amplified during the iterative process, making it difficult to effectively suppress them. With the continuous development of artificial intelligence technology, deep learning has provided a new approach to solve the multi-image super-resolution problem using deep neural networks. This approach avoids the instability caused by the ill-posed nature of deconvolution and provides a more effective solution.

2.2. Deep Learning for Multi-Image Super-Resolution

Deep learning methods have shown great potential in addressing multi-frame super-resolution problems [33–37]. Deudon et al. [22] proposed HighRes-net, which first utilizes deep learning to solve the multi-frame super-resolution task. They achieved implicit registration by setting a reference frame channel and recursively fused LR image pairs using a global fusion operator. Molini et al. [38] introduced the registration filter RegNet with feature learning capabilities in DeepSUM, integrating the image registration task into a CNN. They further incorporated non-local characteristics into the network using graph convolution methods in DeepSUM++. Arefin et al. [39] presented MISR-GRU, which employs the ConvGRU module to compute the hidden state of the feature representation generated by the encoder and utilizes global average pooling to obtain fusion features. Dorr et al. [40] introduced 3DWDSRNet, a 3D wide activation residual network that replaces 3D convolutions with wide activation blocks to enhance the spatio-temporal correlation of multiple images. They also employed frame permutation techniques for data augmentation.

By introducing attention mechanisms, the model's perception and utilization of high-frequency features can be enhanced, leading to a better reconstruction of image details and further improving the performance of super-resolution tasks [41–44]. Salvetti et al. [45] introduced a 3D convolutional feature attention mechanism with embedded residual connections in the RAMS network, enabling the network to focus on extracting high-frequency information in both the temporal and spatial dimensions. Valsesia et al. [46] utilized self-attention mechanisms in the temporal dimension and studied the relationship between the uncertainty of super-resolution images and temporal variations. They constructed an MISR model, PIUnet, which possesses permutation invariance for temporal images. In the burst super-resolution task, Bhat et al. [47] aligned multiple frames using optical flow and employed attention-based adaptive weighting and fusion methods to generate denoised super-resolution RGB images.

2.3. *Transformers for Multi-Image Super-Resolution*

The Transformer model is a deep learning model based on the encoder–decoder architecture with a self-attention mechanism [48]. Compared to RNNs and CNNs, the Transformer model has two key advantages. Firstly, it enables fully parallel computation, which improves the efficiency of training and inference. Secondly, the Transformer has a global receptive field, which allows it to capture global information more effectively. Recently, significant progress has been made in image super-resolution research using the Transformer model [14,15,17,36,49].

An et al. [50] were the first to apply a Transformer to the remote sensing image MISR task, proposing TR-MISR, and through extensive experiments, they demonstrated the effectiveness of the Transformer model in multi-frame fusion and in improving the model's robustness to noise. Luo et al. [36] reduced alignment errors by combining optical flow estimation and deformable convolution and proposed BSRT, utilizing blocks and groups based on the Swin Transformer as the backbone structure to more effectively extract global contextual information. Dudhane et al. [37] introduced Burstormer, a Transformer-based hierarchical model, for burst image restoration and enhancement tasks. This model improved feature alignment and fusion by introducing enhanced deformable alignment modules and reference-free feature enrichment modules, utilizing multi-scale local and non-local features.

3. Methodology

3.1. *Network Architecture*

The overall network of the Multi-Attention Multi-Image Super-Resolution Transformer (MAST) is illustrated in Figure 1, which consists of four parts: shallow feature extraction, deep feature extraction, collaborative attention feature fusion, and HR image reconstruction.

In particular, the MAST network takes I_{LR} as its input, which signifies a compilation of LR images captured from the identical scene. The shape of I_{LR} is $H \times W \times k \times C_0$, with H, W, k, and C_0 denoting the image's height, width, the number of LR images, and the channels per image, respectively. It is noteworthy that the value of C_0 is 1 in this context. To streamline our notation for simplicity, we abbreviate the shape of I_{LR} as $H \times W \times k$.

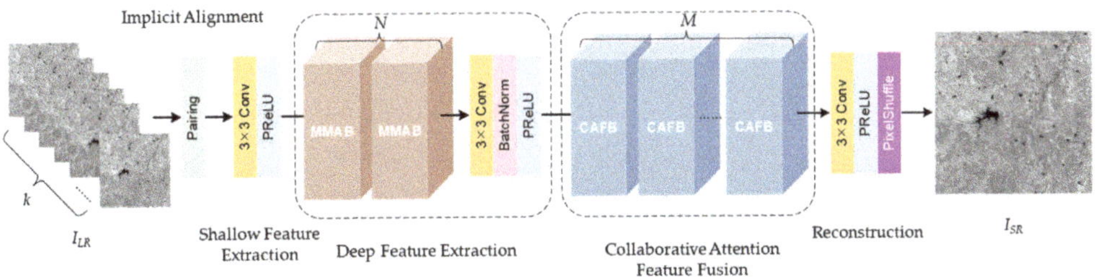

Figure 1. The overall architecture of MAST. In this context, k indicates the number of LR images. N and M represent the number of proposed modules of MMAB and CAFB, respectively.

Initially, we determined the median image of the input image collection using the median operation. This step allowed us to establish a reference image, denoted as Ref, for each individual LR image. The process is illustrated below. In this context, x and y correspond to the pixel positions within the image, while i represents the index of the i-th among the initial 9 LR images.

$$\text{Ref}[x,y] = \text{Median}(I_{LR}[x,y,i]),\ i = 1,2,\ldots,9 \tag{2}$$

Following this, we proceeded to concatenate each LR image with its corresponding reference image along the channel dimension, leading to the generation of LR-Ref image pairs referred to as I_{PR}. These pairs possess a shape of $H \times W \times 2k$ and are represented below. Here, j indicates the j-th LR image within the image set I_{LR}, while z signifies the channel of I_{PR}.

$$I_{PR}[x,y,z] = \begin{cases} \text{LR}_j[x,y,z], & z[0,k-1] \\ \text{Ref}[x,y,z-k], & z[k,2k-1] \end{cases},\ j = 1,2,\ldots,k \tag{3}$$

This form of image pairing can be seen as an implicit initial image registration condition, and it exhibits a degree of resilience against challenges such as noise, distortion, or localized occlusion that are present in the images. It does not depend on explicit feature point matching or correspondence, thereby effectively mitigating some issues related to image misalignment.

The shallow feature extraction layer comprises a 3×3 convolutional layer, followed by a PReLU activation function layer. Through the process of feature extraction applied to the input LR images, we obtained the shallow feature F_S, which has the shape of $H \times W \times C$, where C ($C > k$) represents the channels of F_S.

$$F_S = \text{PReLU}(H_{3\times3\ Conv}(I_{PR})) \tag{4}$$

Introducing convolutional layers into the Transformer model not only enhances the network's training process, but also provides a more comprehensive feature representation for subsequent deep feature extraction and multi-frame information fusion [17]. Convolutional layers excel at capturing local patterns and structures in images, which are crucial for many visual tasks. Incorporating the convolutional layer into the Transformer model enhances the model's ability to perceive these local features, resulting in a richer feature representation [51]. The important characteristics of convolutional layers are parameter sharing and local connectivity, which, to some extent, reduce the model's parameter count, making it more amenable to training and optimization. Furthermore, convolutional layers can provide more stable gradient information during the early stages of training, aiding in accelerating the network's convergence process [52].

Next, we proceeded to conduct deep feature extraction on the input feature, facilitating the network's capacity to draw out additional complementary information through the integration of multi-scale and mixed attention guidance. This procedure resulted in a more intricate feature representation F_D, depicted as follows:

$$F_i = \text{MMAB}_i(F_{i-1}),\ i = 1,2,\ldots,N, \\ F_D = \text{BN}(H_{3\times3\ Conv}(F_N)) \tag{5}$$

where F_{i-1} and F_i indicate the input feature and output feature of i-th MMAB, respectively. BN represents the batch normalization operation.

By incorporating collaborative attention, the network in the fusion stage gains the ability to simultaneously grasp both global and local information within the image. This fosters a more comprehensive feature fusion, achieving a harmonious restoration of both the overall image structure and its intricate local details. F_M represents the fusion feature derived from the deep features, as demonstrated below. In this context, F_{j-1} and F_j indicate the input and output features of the CAFB module, respectively.

$$F_j = \text{CAFB}_j(F_{j-1}),\ j = 1,2,\ldots,M, \tag{6}$$

The HR image reconstruction segment comprises a 3 × 3 convolutional layer, a PReLU layer (the two are denoted together as $H_{3\times3\ Conv}(\cdot)$ below for brevity), and a pixel shuffle layer. Through this reconstruction process, we derived the super-resolution image I_{SR}, which can be represented as

$$I_{SR} = \text{PixelShuffle}(H_{3\times3\ Conv}(F_M)) \tag{7}$$

3.2. Deep Feature Extraction Module

The Multi-Scale and Mixed Attention Block (MMAB) serves as the pivotal module within the network's deep feature extraction segment, with its structure illustrated in Figure 2. In our pursuit of effectively acquiring complementary high-frequency feature information, we harnessed three distinct mechanism-driven feature extraction modules: MESB, CAGB, and SAGB. Among these, MESB operates on the foundation of multi-scale feature extraction principles, CAGB amalgamates channel dimension compression and expansion techniques with channel attention mechanisms, and SAGB prioritizes the utilization of spatial attention for feature extraction.

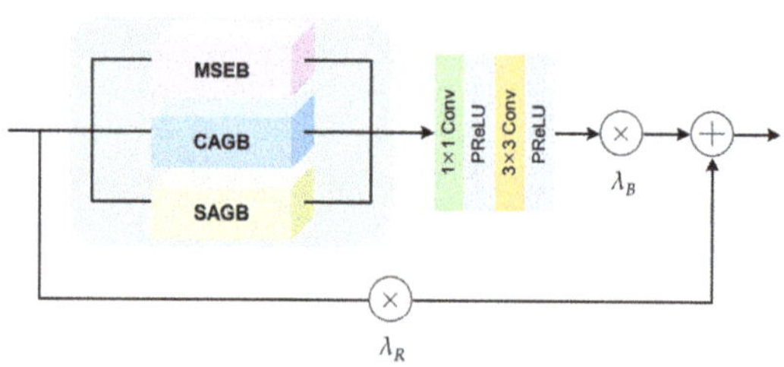

Figure 2. The structure of an MMAB.

By harmonizing the outputs of these three modules in a parallel fashion, our model can apprehend an expansive spectrum of complementary information across various frequencies during the feature extraction phase. This, in turn, enriches the contextual information available for image reconstruction. Additionally, existing research indicates that parallel fusion modules contribute to alleviating the adverse effects of motion blur on the outcomes, thereby enhancing the super-resolution capabilities of the model [53].

Moreover, we incorporated two learnable adaptive weight scaling parameters to dynamically fine-tune the weights of both the residual path and the backbone path [54]. λ_B signifies the weight assigned to the backbone path, whereas λ_R represents the weight attributed to the residual path. Throughout the entire training phase of the neural network, as the adaptive weights undergo continuous updates, they not only enhance the network's focus on high-frequency features, but also improve the propagation of gradient information. This leads to an acceleration in the convergence of the network's training process. The complete computational process of an MMAB can be succinctly described as follows:

$$G_1 = \text{MSEB}(Y),\ G_2 = \text{CAGB}(Y),\ G_3 = \text{SAGB}(Y)$$
$$F_C[x,y,z] = \begin{cases} G_1[x,y,z], & z \in [0, C-1] \\ G_2[x,y,z-C], & z \in [C, 2C-1] \\ G_3[x,y,z-2C], & z \in [2C, 3C-1] \end{cases} \tag{8}$$

$$Z = \lambda_R \cdot Y + \lambda_B \cdot H_{3\times 3Conv}(H_{1\times 1Conv}(F_C)) \tag{9}$$

where MSEB(·), CAGB(·), and SAGB(·) correspond to the three sub-blocks present within the MMAB module. G_1, G_2, and G_3 denote the output features of the three modules. Y symbolizes the input feature. F_C indicates the concatenated output feature derived from the three branches, whereas Z signifies the resulting feature following adaptive weighting. The shapes of G_1, G_2, G_3, Y, and Z all maintain dimensions of $H \times W \times C$, while the shape of F_C adopts $H \times W \times 3C$. Additionally, $H_{3\times 3\,Conv}(\cdot)$ and $H_{1\times 1\,Conv}(\cdot)$ refer to the 3×3 and 1×1 convolution layers, respectively.

3.2.1. Multi-Scale Extraction Block (MSEB)

Objects exhibit different detailed characteristics at various scales. Multi-scale structures can capture contextual information at different scales through receptive fields of different sizes, thereby enhancing feature representation. Moreover, it was shown that the use of multi-scale structures in networks can help to suppress alignment errors [26–28]. We borrowed the idea of a Pyramid Pooling Module (PPM) [55] in our design and introduced adaptive weight scaling parameters to create a residual-structured multiscale feature extraction block (MSEB), the structure of which is shown in Figure 3. By incorporating the adaptive weight scaling parameter [54], the MSEB can dynamically fine-tune the feature contents within both the residual and backbone paths, enabling the model to extract complementary information more efficiently and optimize the network's training process.

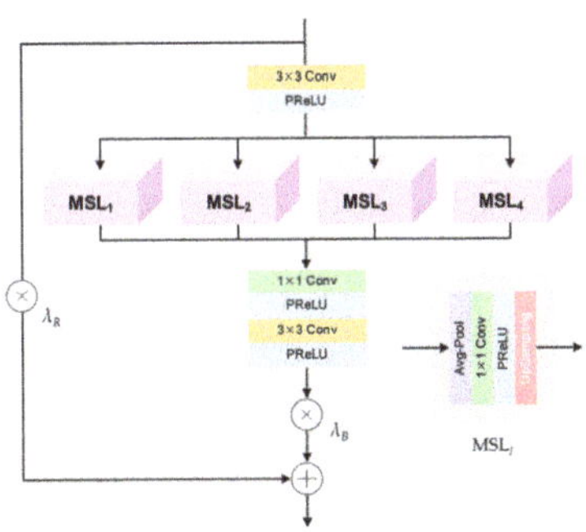

Figure 3. The structure of the MSEB.

To elaborate, F_0 represents the input feature, and F_H signifies the features extracted via $H_{3\times 3\,Conv}(\cdot)$. $\text{MSL}_i(\cdot)$ denotes the feature extraction layer with varying pooling kernels. The specific procedure entails initiating average pooling to generate feature maps with sizes of 1×1, 2×2, 3×3, and 6×6. Among these, the 1×1 feature map encapsulates global information, while the remaining three encompass local details. Local features are employed to capture subtle local variations, while global features are utilized to maintain overall coherence. Subsequently, these feature maps are restored to their original sizes using bilinear interpolation. Concat(·) indicates the concatenation of feature maps of identical shape, following the same process as outlined in Equation (8). By fusing multi-scale contextual features, the adverse impacts of local noise and inaccuracies on the overall registration outcome can be effectively alleviated. F_A represents the fused multi-scale feature, while F_B signifies the output feature after the addition of residual connection. The overall calculation process is described as follows:

$$F_A = H_{1\times1\ Conv}(\text{Concat}(\text{MSL}_1(F_H), \ldots, \text{MSL}_4(F_H))) \tag{10}$$

$$F_B = \lambda_B \cdot H_{3\times3\ Conv}(F_A) + \lambda_R \cdot F_0 \tag{11}$$

3.2.2. Channel Attention Guided Block (CAGB)

The Channel Attention Guided Block (CAGB) is a sub-module of our proposed MMAB, the structure of which is shown in Figure 4, designed to direct the network to thoroughly harness high-frequency feature information within the channel dimension. By increasing the dimensionality, we can introduce a greater variety of distinct feature channels, enabling the model to comprehensively capture richer feature information [56]. Additionally, by reducing the dimensionality of the feature map in the channel dimension, the network can focus on more meaningful high-frequency features while reducing data redundancy [57].

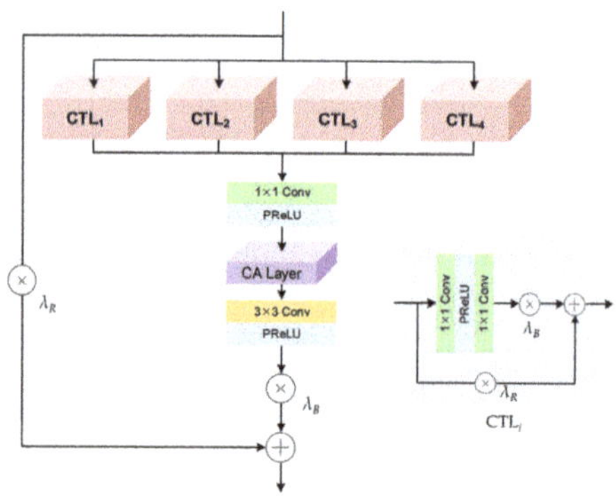

Figure 4. The structure of the CAGB.

Through the application of various dimension transformations, we obtained multi-level contextual information and performed information fusion to extract effective complementary information from the fused features. Furthermore, we introduced the channel attention mechanism, enabling the network to adaptively learn relationships between different feature channels, thus enhancing the perception of crucial high-frequency features.

In the depicted diagram, CTL_i denotes the channel attention layer. Among these, $\text{CTL}_1(\cdot)$ and $\text{CTL}_2(\cdot)$ indicate the channel expansion of input feature F_0, with expansion factors of 2 and 4, respectively, followed by a reduction to match the input channels. On the other hand, $\text{CTL}_3(\cdot)$ and $\text{CTL}_4(\cdot)$ signify the dimension reduction in F_0, with reduction factors of $1/2$ and $1/4$, respectively, followed by an expansion. F_K and F_Q represent the context-fused feature and the output feature after the addition of residual connection, respectively. $H_{CA}(\cdot)$ stands for the channel attention layer. The entire process unfolds as follows:

$$F_K = H_{1\times1\ Conv}(\text{Concat}(\text{CTL}_1(F_0), \ldots, \text{CTL}_4(F_0))) \tag{12}$$

$$F_Q = \lambda_B \cdot (H_{3\times3\ Conv}(H_{CA}(F_K))) + \lambda_R \cdot F_0 \tag{13}$$

3.2.3. Spatial Attention Guided Block (SAGB)

The Spatial Attention Guided Block (SAGB) is another constituent of our proposed MMAB, the structure of which is shown in Figure 5, dedicated to delving more deeply into high-frequency feature information from the spatial dimension. By subtracting smoothed filtered feature maps from the original ones, we derived high-frequency filtered features. This process of enhancing high-frequency information can accentuate image details and edges. Subsequently, we applied computed spatial attention weights to the feature map. These weights prioritize different locations, thereby enhancing feature representation in significant image regions and subduing less important areas. This strategy effectively extracts high-frequency information. The computation process of the SAGB can be presented as follows:

$$F_L = H_{UP}(H_{Avg}(F_G)) \tag{14}$$

$$F_E = H_{3\times3\ Conv}(H_{SA}(F_G - F_L)) \tag{15}$$

where F_G represents the feature obtained after $H_{3\times3\ Conv}(\cdot)$ with a size of $H \times W \times C$. F_L signifies the smoothed feature with a size of $H/2 \times W/2 \times C$. F_E denotes the output features. $H_{Avg}(\cdot)$ indicates the average pooling operation, while $H_{UP}(\cdot)$ represents bilinear interpolation. $H_{SA}(\cdot)$ represents the spatial attention layer, which adjusts the attention distribution in the spatial dimension, enabling the network to focus on high-frequency information regions.

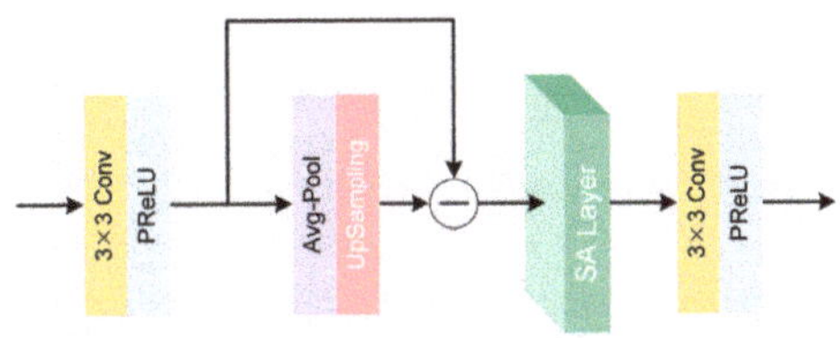

Figure 5. The structure of the SAGB.

3.3. Collaborative Attention Feature Fusion Module

The Collaborative Attention Fusion Block (CAFB) is the feature fusion module proposed by us based on the Transformer architecture, and its structure is illustrated in Figure 6. The reshaping process depicted in the figure below involves transforming the input feature map into a suitable format for the Transformer to process tokens. The Transformer model is suitable for multi-frame image processing as it is range-independent in computing attention weights and can capture distant dependencies throughout the input sequence. It demonstrates significant advantages in capturing correlation information between multiple images and enhancing robustness to noise. Our enhancement stems from the incorporation of both channel attention and residual local attention within the Transformer model. Compared to traditional attention mechanisms like CBAM, we adopted a parallel combination approach to simultaneously incorporate both global and local information.

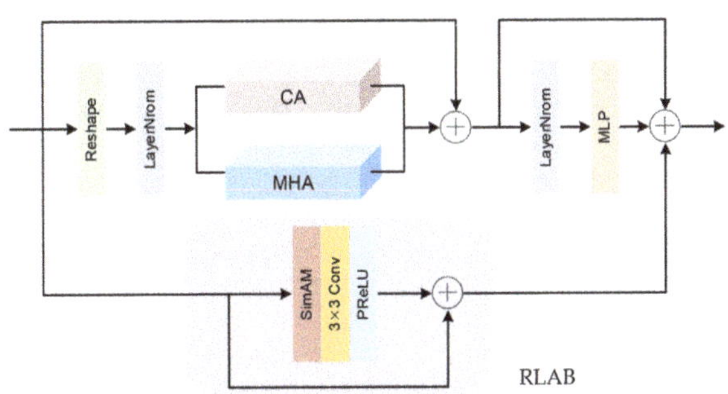

Figure 6. The structure of the CAFB.

By incorporating the channel attention mechanism to extract importance and correlation information from diverse channels and embedding this information into the Transformer, we further amplified the Transformer's ability to capture comprehensive correlations among multi-frame information [58–60].

Meanwhile, the Residual Local Attention Block (RLAB) enhances the network's ability to capture local high-frequency features. It does so by employing SimAM attention to extract the importance of neurons through the analytical solution of the energy function, directly inferring the 3D attention weights of the feature map [61]. Ultimately, through the application of weighted processing to the outputs of the two branches, an effective fusion of global and local information is accomplished. This empowers the super-resolution network to effectively restore both the overarching image structure and intricate details with proficiency during the reconstruction process. The complete computation process of CAFB unfolds as follows:

$$X_M = \mathrm{MHA}(\mathrm{LN}(X)) + \alpha \cdot \mathrm{CA}(\mathrm{LN}(X)) + X \quad (16)$$

$$X_N = \mathrm{MLP}(\mathrm{LN}(X_M)) + X_M \quad (17)$$

$$X_U = X_N + \alpha \cdot (H_{3\times3\ Conv}(\mathrm{RLAB}(X)) + X) \quad (18)$$

where X signifies the input of the CAFB module, constituting the feature acquired through deep extraction. X_U represents the resulting feature, incorporating insights from both the Transformer and RLAB. X_M and X_N correspond to the interim features derived during the processing stages. MHA corresponds to the standard multi-head attention, LN stands for the LayerNorm layer, and MLP represents the multi-layer perceptron.

To attain optimal fusion performance, we introduced the hyperparameter α. Through judicious weight allocation, we can harmonize the contributions of global structural information and local detail information. This allows for the extraction of comprehensive global structural insights while accentuating crucial local details, culminating in an enhanced feature representation. In the experimental context, we set α to a value of 0.01.

4. Experiments

In this section, we introduce the PROBA-V Kelvin Dataset, which is the MISR dataset used in the satellite imagery experiments. Furthermore, we delve into the exploration of evaluation metrics employed to gauge the quality of super-resolution images. Additionally, we offer a comprehensive introduction to both the data preprocessing process and the configurations of the model parameter settings.

4.1. PROBA-V Kelvin Dataset

The PROBA-V dataset is a real remote sensing image dataset provided by the PROBA-V satellite and serves as a benchmark dataset for multi-frame super-resolution tasks [62]. The dataset consists of satellite data collected from 74 manually selected regions worldwide. It includes imagery captured in two bands: NIR and RED. These images were acquired at different times. The dataset provides images with two different spatial resolutions. The LR images have a resolution of 300 m and a size of 128 × 128 pixels. The HR images, on the other hand, have a resolution of 100 m and a size of 384 × 384 pixels.

The dataset consists of a total of 1450 scenes, with 1160 scenes used for training and 290 scenes used for testing. Each scene includes an HR image and multiple LR images. The number of LR images in each scene ranges from 9 to 35. Since these images were acquired at different times and under different weather conditions, the occlusion of variable features such as clouds and snow may affect the results of the image alignment. The PROBA-V dataset provides a manually annotated quality map for each image in the scene. The quality map is encoded in binary format, using 0 to represent occluded pixels and using 1 to represent non-obscured pixels. Only non-obscured pixels are considered clear pixels, and we performed super-resolution reconstruction only for the portion of the image that is clear.

The PROBA-V dataset is a significant publicly available benchmark dataset that is currently used for remote sensing image tasks and multi-image super-resolution tasks. Several MISR algorithms have been trained on this dataset, providing reliable solutions to the tasks [22,45,46,50]. This dataset serves as compelling evidence to validate the superiority of our algorithms. In the future, we will actively strive to acquire more datasets to verify the generalizability of our model across various data distributions.

4.2. Evaluation Metric

In super-resolution tasks, PSNR and SSIM are commonly used to evaluate objective image quality metrics, but they may not be effective enough for evaluating the quality of super-resolution remote sensing images, especially when there are problems such as pixel shift and brightness bias in the images, and they have less reference value for assessing the quality of reconstructed images. To assess the reconstruction quality of remote sensing images more accurately, we introduced cPSNR and cSSIM, which are metrics that are used in the PROBA-V challenge. These metrics have been improved to address pixel offset, brightness deviation, and cloud-related factors that are present in remote sensing images.

The size of the LR in the validation set is 128 × 128 pixels, and the size of the super-resolution image SR is 384 × 384. As the acquisition of LR and HR images involves separate processes, there might be a slight misalignment at the pixel level between them. It was assumed that the maximum potential offset between the two images is 3 pixels. To account for the potential errors due to misalignment, a 3-pixel-wide strip along the edges of the SR was removed. This resulted in a cropped image denoted as $I_{3,3}^{SR}$, which has a size of 378 × 378 pixels. Subsequently, we proceeded to crop the HR image from the upper-left coordinate (u, v), resulting in a cropped image denoted as $I_{u,v}^{HR}$ that is the same size as $I_{3,3}^{SR}$. The procedure for computing the brightness deviation between $I_{3,3}^{SR}$ and $I_{u,v}^{HR}$ is outlined in Equation (19). In this equation, $M_{u,v}$ represents a binary mask image. $\|\cdot\|$denotes the L_1 norm of the matrix. When calculating various metrics, we use the mask to restrict the range of statistics and only consider the results from clear pixels.

$$b_{u,v} = \frac{1}{\|M_{u,v}\|} \sum (I_{u,v}^{HR} - I_{3,3}^{SR}) \cdot M_{u,v} \tag{19}$$

In the experiments of this paper, we adopted negative cPSNR as the loss function for network training. The cPSNR and cSSIM are calculated as follows:

$$\text{cMSE} = \max_{u,v[0,6]} \frac{1}{\|M_{u,v}\|} \sum \left[(I_{u,v}^{HR} - I_{3,3}^{SR}) \cdot M_{u,v} + b_{u,v}\right]^2 \tag{20}$$

$$\text{cPSNR} = -10\,\lg(\text{cMSE}) \tag{21}$$

$$\text{cSSIM} = \max_{u,v \in [0,6]} \text{SSIM}(I_{u,v}^{HR} \cdot M_{u,v},\ I_{u,v}^{SR} \cdot M_{u,v}) \tag{22}$$

4.3. Experimental Settings

4.3.1. Data Pre-Processing

We used 70% of the training set for model training, while the remaining 30% was used to validate the performance of the model. To obtain more image scenes, we used image cropping for data enhancement. By cropping, we could obtain more LR and HR image patches with sizes of 64 × 64 and 192 × 192 pixels, respectively. To ensure the quality of the reconstructed images and to maintain a balance of computational effort during the training process, we chose to use only the scenes with an average reliable pixel count higher than 85% to participate in the model training.

We sorted the images in the dataset from high to low based on the number of clear pixels in the LR images. For the experimental setup, the network took k LR images as the input. If there were more images in the scene than k, we selected the top k LR images with the highest number of clear pixels as the input for the network. If the number of LR images in the scene was less than k, we filled the missing images with zero matrices to meet the input requirements.

4.3.2. Parameter Settings

In the experiment, the input number of LR images k accepted by our network was 32. The LR images are grayscale images with a channel number C_0 of 1. The channels C for shallow feature F_S, deep feature F_D, and fusion feature F_M were all set to 64. In the MAST network, we set the number of the MMAB and CAFB to 2 and 6, respectively. The number of heads in the multi-head attention of the Transformer model was set to 8, and the dimension of the hidden layer in the MLP was set to 128.

We conducted experiments on the Ubuntu 18.04 system. The hardware setup included four Nvidia RTX A5000 GPUs (with a total memory of 96 GB) and an Intel Xeon (R) Gold 5320 CPU @ 2.20 GHz. The entire code framework was implemented using Python 3.7 and PyTorch 1.7.1. In the experiments, we set the batch size to 8 based on the computational capacity of the devices. For the optimizer, we used Adam with an initial learning rate of 5×10^{-4}. The learning rate decay strategy employed cosine decay, with a minimum learning rate of 5×10^{-6}. During the training process, we trained the model for a total of 400 epochs, which took approximately 76 h. Subsequently, we fine-tuned the model with a learning rate of 1×10^{-5} and performed 80 epochs of fine-tuning training. The model had a parameter count of 3.3 million.

5. Discussion

5.1. Comparison with State-of-the-Art Methods

We compared MAST with other existing MISR methods of remote sensing. Since the PROBA-V dataset did not provide ground-truth for the test set, we evaluated the reconstruction performance of different methods on the validation set using cPSNR and cSSIM as performance metrics to assess the quality of the reconstructed images. Table 1 presents the quantitative results obtained using various methods on the NIR and RED bands.

Table 1. Quantitative comparison with different methods on the validation set.

Band	NIR		RED	
Metric	**cPSNR**	**cSSIM**	**cPSNR**	**cSSIM**
Bicubic	45.44	0.9770	47.33	0.9840
BTV [63]	45.93	0.9794	48.12	0.9861
IBP [64]	45.96	0.9796	48.21	0.9965
RCAN [9]	45.66	0.9798	49.22	0.9870
VSR-DUF [65]	47.20	0.9850	49.59	0.9902
HighRes-net [22]	47.55	0.9855	49.75	0.9904
3DWDSRNet [40]	47.58	0.9856	49.90	0.9908
DeepSUM [38]	47.84	0.9858	50.00	0.9908
MISR-GRU [39]	47.88	0.9861	50.11	0.9910
DeepSUM++ [66]	47.93	0.9862	50.08	0.9912
RAMS [45]	48.23	0.9875	50.17	0.9913
$RAMS_{+20}$ [45]	48.51	0.9880	50.44	0.9917
PIUnet [46]	48.72	0.9883	50.62	0.9921
TR-MISR [50]	48.54	0.9882	50.67	0.9921
MAST (this study)	**49.63**	**0.9884**	**51.48**	**0.9924**

In Figure 7, we present a visual comparison of the test results between MAST and the baseline method using Bicubic interpolation in two bands: NIR and RED. The results show that MAST exhibited excellent reconstruction performance in almost all scenarios. Specifically, in 98.23% of the NIR band, the reconstruction performance of the MAST method surpassed that of the baseline method. Similarly, in the RED band, our MAST method outperformed the baseline method in as many as 98.86% of the scenarios in terms of reconstruction performance.

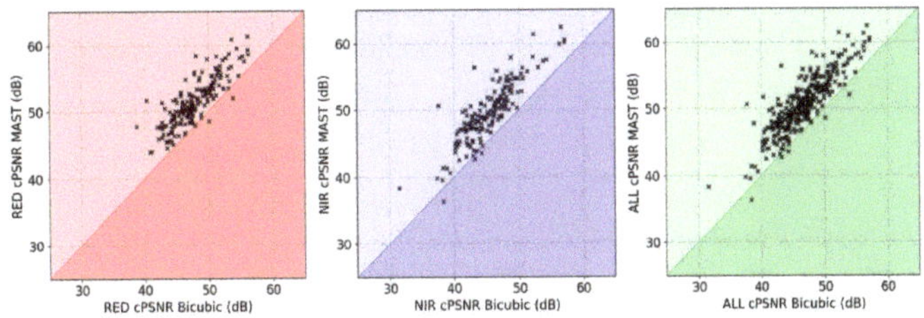

Figure 7. Comparison of cPSNR for MAST and Bicubic on the validation set. Each point in the figure represents a scene. The point above the line $y = x$ represents the scene in which the MAST reconstruction outperforms the baseline method.

From the image regions shown in Figures 8 and 9, it is visually apparent that our method significantly improved the preservation of textures and fine details, resulting in better visual results. These results further validate the excellent performance of the MAST method in the super-resolution task of multiple remote sensing images.

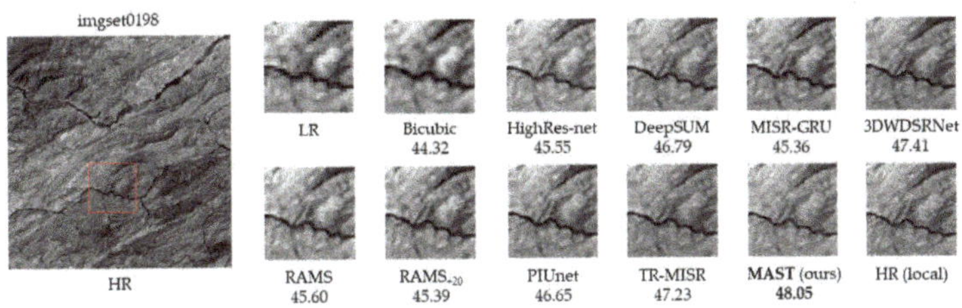

Figure 8. Comparison between different MISR methods on the imgset0198 scene of the RED band.

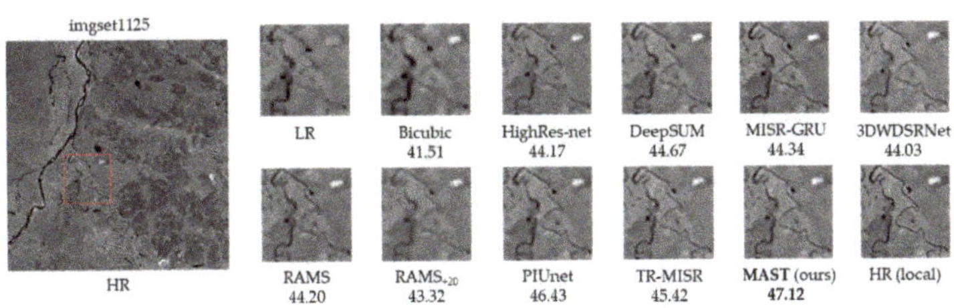

Figure 9. Comparison between different MISR methods on the imgset1125 scene of the NIR band.

We used Bicubic interpolation as the baseline method for super-resolution reconstruction. For each scene, we applied Bicubic interpolation to the clearest image to obtain an HR image. The red boxes in the figures indicate the reconstructed regions we are comparing.

1. Classical MISR Methods:

IBP obtains an initial image through Bicubic interpolation and performs registration using phase correlation. It gradually improves the estimation results by leveraging the image's own information and reconstruction errors. BTV combines the characteristics of total variation regularization and bilateral filtering to solve the bilateral total variation minimization problem. This method can remove noise while preserving the edge information in the image.

2. SISR Method:

RCAN learns the high-frequency details of images through residual learning and enhances the high-frequency features using the channel attention mechanism. The RCAN framework consists of five residual groups, each containing five residual channel attention blocks.

3. Video Super-Resolution Method:

VSR-DUF utilizes dynamic upsampling filters to reconstruct HR images and enhances the image details by calculating residuals. We chose a base network consisting of 16 layers and used nine LR images as the input for the network.

4. Deep Learning Super-Resolution Methods:

HighRes-net utilizes the reference frame for implicit alignment and recursively fuses multiple images. The experiments were conducted using the default framework, and training and testing were conducted with 16 LR images.

The 3DWDSRNet method utilizes 3D WDSR blocks instead of 3D convolutions to fuse multiple images, which helps to maintain lower GPU memory requirements. The experiments were conducted using seven LR images as the input.

DeepSUM introduces a dynamic registration subnetwork and utilizes 3D convolutions to fuse multiple images. The experiments were conducted using the clearest nine LR images to train the network. DeepSUM++ improves the non-local characteristics of the network by introducing graph convolutions.

MISR-GRU utilizes 24 LR images for implicit co-registration. It employs a fusion strategy based on ConvGRU and performs registration on super-resolution images using the ShiftNet network.

RAMS introduces an attention mechanism with 3D convolutions to better extract high-frequency features. The input consists of the clearest nine LR images. $RAMS_{+20}$ improves upon the inference process by adopting a temporal self-ensemble strategy. It randomly permutes the input images 20 times and takes the average of the super-resolution results as the output.

PIUnet exhibits permutation invariance in the temporal dimension. However, it only performs well when the number of test images is close to the number of input images. Based on the parameter settings in the paper, we set the input to be nine LR images.

TR-MISR is currently the top-ranked algorithm in the leaderboard, and its network's core is a fusion module based on a Transformer. In the experiments, we set the number of LR images to 32.

To address the issue of the insufficient utilization of complementary information from multiple frames in current MISR models, caused by a single scale and a single attention mechanism, this paper proposes an improved super-resolution network that combines multi-scale and multi-attention approaches. In terms of feature extraction, we introduce the Multi-Scale and Mixed Attention Block (MMAB) to enhance the extraction and utilization of complementary information from multiple images. Additionally, in the aspect of feature fusion, we propose the Collaborative Attention Fusion Block (CAFB) that simultaneously utilizes global and local information to achieve a balance between the restoration of the overall structure and the preservation of local details in the image.

Both the qualitative and quantitative results from the comparative experiments demonstrate that our method achieved state-of-the-art performance in the super-resolution reconstruction task on the PROBA-V dataset, further validating the effectiveness of our approach.

5.2. Analysis of Ablation Experiments

5.2.1. Effectiveness of Each Module

To validate the effectiveness of the two key modules, MMAB and CAFB, in the MAST network structure, we conducted a series of ablation experiments. We evaluated the contribution of each submodule to the overall performance, and the experimental results are recorded in Tables 2 and 3. The results of the ablation experiments intuitively demonstrate the importance of each submodule and provide guidance for further improvement and optimization of the network.

Table 2. Ablation results of individual components of MMAB on the validation set.

MMAB			NIR		RED	
MSEB	**SAGB**	**CAGB**	**cPSNR**	**cSSIM**	**cPSNR**	**cSSIM**
√	×	×	48.93	0.9867	50.96	0.9916
√	√	×	48.98	0.9869	50.99	0.9919
√	√	√	50.99	0.9879	51.22	0.9921

Table 3. Ablation results of individual components of CAFB on the validation set.

CAFB			NIR		RED	
Transformer	**CA**	**RLAB**	**cPSNR**	**cSSIM**	**cPSNR**	**cSSIM**
√	√	×	48.89	0.9874	50.84	0.9914
√	√	√	49.23	0.9881	51.19	0.9920

1. Importance of MMAB

To evaluate the effectiveness of the MMAB in feature extraction, we conducted a series of ablation experiments with TR-MISR as the baseline. The feature extraction part of TR-MISR consists of two residual blocks. Apart from the differences in the feature extraction part, we maintained the same network structure as TR-MISR. In Table 2, we compare the cPSNR values of the models using three different versions of the feature extraction block on the validation set: (1) MSEB, (2) MSEB + SAGB, and (3) MSEB + SAGB + CAGB.

The multi-scale structure has a stronger ability to adapt to different scale feature information, allowing for more comprehensive feature extraction. The results in Table 2 show that when using only the MSEB for feature extraction, there was an average improvement of 0.2 dB on the validation set for both NIR and RED. The inclusion of attention mechanisms can guide the network to explore high-frequency information in the images and extract more complementary information from multiple images. The complete MMAB for feature extraction achieved cPSNR improvements of 0.52 dB and 0.55 dB on NIR and RED, respectively.

2. Importance of CAFB

To evaluate the effectiveness of the CAFB, we conducted ablation experiments with TR-MISR as the baseline. Apart from the difference in the fusion module, we maintained the same network structure as TR-MISR. In Table 3, we compare the cPSNR values of the models using two different versions of the fusion module on the validation set: one with the Transformer + CA network and the other with the complete network (Transformer + CA + RLAB). Here, CA represents channel attention. RLAB represents the Residual Local Attention Block that we designed.

By incorporating channel attention, the ability of the Transformer model to capture more contextual features is effectively enhanced. Compared to the TR-MISR model that solely relies on the Transformer-based fusion block, our network achieved a surpassing improvement of 0.17 dB on both the NIR and RED validation sets when channel attention was added. Additionally, the inclusion of the local attention mechanism enables the model to focus more on high-frequency information. The complete fusion network showed a significant improvement in the model's performance, with an average improvement of over 0.5 dB.

5.2.2. Analysis of the Effect of the Number of Input Images

In theory, as the number of input images increases, the MISR model can utilize more complementary information. However, it also introduces more noise and redundant information. The increase in noise can potentially degrade the quality of the reconstructed images. MISR utilizes complementary information between multiple images. If there is excessive redundant information in the network, it will increase the computational burden of the model and reduce the convergence speed of the network. The robustness of the model to noise has a significant impact on the quality of image reconstruction.

Table 4 presents the cPSNR comparison results of our proposed method on the validation set using different numbers of input images (without fine-tuning). When the number of input images k was 16, our method achieved higher cPSNR in both the NIR and RED bands compared to the other comparative methods. As the number of input images increased to 32, the reconstruction performance of the model reached its best, significantly outperforming the current state-of-the-art MISR method. Specifically, in the NIR band, our method improved cPSNR by 0.71 dB, while in the RED band, it showed a relative improvement of 0.69 dB in cPSNR.

Table 4. Comparison of evaluation metrics on the validation dataset for different numbers of input LR images.

Band	**NIR**		**RED**	
k	**cPSNR**	**cSSIM**	**cPSNR**	**cSSIM**
32	49.43	0.9883	51.36	0.9922
24	49.26	0.9884	51.27	0.9922
16	48.87	0.9879	51.02	0.9920
8	48.41	0.9865	50.64	0.9908

Figure 10 illustrate specific regions of the super-resolution images under different input conditions. It is visually evident that as the number of input images increased, the reconstructed images exhibited richer details and texture information. The experimental results strongly demonstrate that our proposed method can better utilize the complementary information from multiple images and exhibit improved robustness in handling noise. Our method achieved superior reconstruction performance on the PROBA-V dataset.

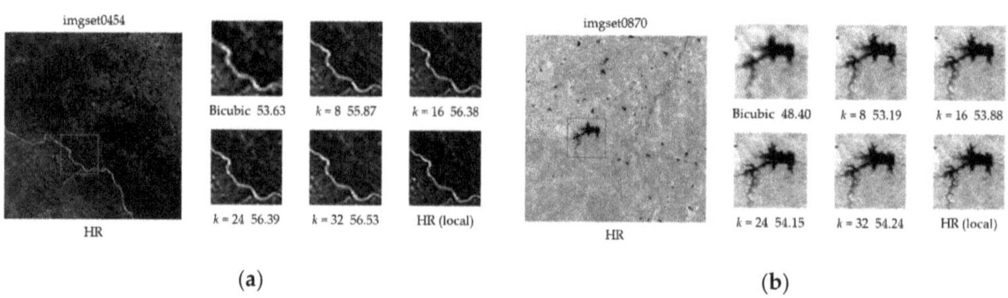

Figure 10. (**a**) The effect of different input images k on the imgset0454 scene of the RED band. The reconstruction outcome achieves the best result when $k = 32$. (**b**) The effect of different input images k on the imgset0870 scene of the NIR band. The reconstruction outcome achieves the best result when $k = 32$.

5.2.3. Impact of the Number of Blocks

To investigate the impact of the number of the proposed two main blocks on the model performance, we conducted a series of comparative experiments by varying the number of MMABs and CAFBs (denoted by N and M, respectively) in the network structure based on our current hardware configuration. The quantitative results obtained on the validation set are recorded in Table 5.

Table 5. Comparison of evaluation metrics on the validation dataset for different numbers of modules.

Band		NIR		RED	
N_{MMAB}	N_{CAFB}	cPSNR	cSSIM	cPSNR	cSSIM
2	6	49.43	0.9883	51.36	0.9922
2	4	49.21	0.9873	51.15	0.9918
2	2	48.96	0.9868	50.82	0.9912
3	4	49.30	0.9877	51.25	0.9923
1	6	48.50	0.9862	50.35	0.9908

The experimental results indicate that within a certain range, there was an improvement in the reconstruction performance as the number of blocks increased. It is noteworthy that the model achieved optimal reconstruction performance when N and M were set to 2 and 6, respectively. This implies that, in a fixed dataset size, appropriately increasing the number of modules can enhance the model's capacity, leading to a better fitting of the input features and an improved performance in super-resolution reconstruction. The results further validate the effectiveness of the MMAB and CAFB modules.

5.2.4. Impact of the Size of the Training Images

The size of the input images often has an impact on the overall performance and speed of the model. To investigate the influence of image size in the training data on the super-resolution results, we conducted three sets of comparative experiments. Within the constraints of the computational conditions, we cropped the LR images in the training set to 64×64, 68×68, and 72×72, respectively. Correspondingly, the paired HR images were also cropped to three times the size of the corresponding LR images. The quantitative results of the validation set are reported in Table 6.

Table 6. Comparison of evaluation metrics on the validation dataset for different input image sizes.

Band	NIR		RED	
Size	cPSNR	cSSIM	cPSNR	cSSIM
64	49.63	0.9884	51.48	0.9924
68	49.68	0.9892	51.52	0.9925
72	49.77	0.9889	51.54	0.9925

Larger images usually contain more details and textures, which can provide more realistic information for the super-resolution algorithm and help to generate more accurate high-resolution details. The computational complexity of the Vision Transformer is proportional to the square of the image size, and it requires more computational resources and time to deal with larger images. Therefore, it is particularly important to select the appropriate image size in the model training.

6. Conclusions

In this article, we proposed a Multi-Attention Multi-Image Super-Resolution Transformer (MAST), a novel multi-image super-resolution network of remote sensing, which is innovative in two aspects. Firstly, we introduced the MMAB for deep feature extraction. Through the design of a multi-scale architecture, we enhanced the network's ability to capture features at various scales. Additionally, we employed a method of mixed attention to guide the network in exploring high-frequency features across both channel and spatial dimensions. Through these improvements, the network's capability to extract complementary high-frequency information from multiple images was elevated. Secondly, we presented the CAFB for feature fusion. By incorporating channel attention within the Transformer architecture, the MISR model gained the ability to effectively harness correlation information among multiple images. Furthermore, the introduction of RLAB enhanced the network's capacity to discern local intricate features. These combined enhancements empowered the network to achieve a harmonious equilibrium between the utilization of local and global information during fusion. On the PROBA-V dataset, MAST achieved cPSNR values of 49.63 dB and 51.48 dB for the NIR and RED bands, respectively, along with cSSIM values of 0.9884 and 0.9924, demonstrating state-of-the-art super-resolution performance. The extensive experiments validated the superiority of MAST compared to other advanced methods. In the future, our work will focus on establishing more accurate registration mechanisms to further improve the super-resolution results of multiple images by reducing registration errors.

Author Contributions: Conceptualization, J.L. and Z.T.; methodology, J.L.; software, J.L.; investigation, J.L., W.Z., B.Z., and G.Z.; writing—original draft preparation, J.L.; writing—review and editing, J.L. and Z.T.; project administration, Q.L.; funding acquisition, Z.T. All authors have read and agreed to the published version of the manuscript.

Funding: This research was funded by the Key Program Project of Science and Technology Innovation of the Chinese Academy of Sciences (no. KGFZD-135-20-03-02) and by the Innovation Foundation of the Key Laboratory of Computational Optical Imaging Technology, CAS (no. CXJJ-23S016).

Data Availability Statement: The data presented in this study are available upon request from the corresponding authors.

Conflicts of Interest: The authors declare no conflict of interest.

References

1. Hussain, S.; Lu, L.; Mubeen, M.; Nasim, W.; Karuppannan, S.; Fahad, S.; Tariq, A.; Mousa, B.; Mumtaz, F.; Aslam, M. Spatiotemporal variation in land use land cover in the response to local climate change using multispectral remote sensing data. *Land* **2022**, *11*, 595. [CrossRef]
2. Ngo, T.D.; Bui, T.T.; Pham, T.M.; Thai, H.T.; Nguyen, G.L.; Nguyen, T.N. Image deconvolution for optical small satellite with deep learning and real-time GPU acceleration. *J. Real-Time Image Process.* **2021**, *18*, 1697–1710. [CrossRef]
3. Wang, X.; Yi, J.; Guo, J.; Song, Y.; Lyu, J.; Xu, J.; Yan, W.; Zhao, J.; Cai, Q.; Min, H. A review of image super-resolution approaches based on deep learning and applications in remote sensing. *Remote Sens.* **2022**, *14*, 5423. [CrossRef]
4. Jo, Y.; Oh, S.W.; Vajda, P.; Kim, S.J. Tackling the ill-posedness of super-resolution through adaptive target generation. In Proceedings of the IEEE/CVF Conference on Computer Vision and Pattern Recognition, Nashville, TN, USA, 20–25 June 2021; pp. 16236–16245.
5. Harris, J. Diffraction and Resolving Power. *J. Opt. Soc. Am.* **1964**, *54*, 931. [CrossRef]
6. Milanfar, P. *Super-Resolution Imaging*; CRC Press: Boca Raton, FL, USA, 2011.
7. Wang, Z.; Chen, J.; Hoi, S.C. Deep learning for image super-resolution: A survey. *IEEE Trans. Pattern Anal. Mach. Intell.* **2020**, *43*, 3365–3387. [CrossRef] [PubMed]
8. Dong, C.; Loy, C.C.; He, K.; Tang, X. Image super-resolution using deep convolutional networks. *IEEE Trans. Pattern Anal. Mach. Intell.* **2015**, *38*, 295–307. [CrossRef] [PubMed]
9. Zhang, Y.; Li, K.; Li, K.; Wang, L.; Zhong, B.; Fu, Y. Image super-resolution using very deep residual channel attention networks. In Proceedings of the European Conference on Computer Vision (ECCV), Munich, Germany, 8–14 September 2018; pp. 286–301.
10. Mei, Y.; Fan, Y.; Zhou, Y. Image super-resolution with non-local sparse attention. In Proceedings of the IEEE/CVF Conference on Computer Vision and Pattern Recognition, Nashville, TN, USA, 20–25 June 2021; pp. 3517–3526.
11. Chan, K.C.; Wang, X.; Xu, X.; Gu, J.; Loy, C.C. Glean: Generative latent bank for large-factor image super-resolution. In Proceedings of the IEEE/CVF Conference on Computer Vision and Pattern Recognition, Nashville, TN, USA, 20–25 June 2021; pp. 14245–14254.
12. Wang, X.; Xie, L.; Dong, C.; Shan, Y. Real-esrgan: Training real-world blind super-resolution with pure synthetic data. In Proceedings of the IEEE/CVF International Conference on Computer Vision, Virtual, 11–17 October 2021; pp. 1905–1914.
13. Shi, Y.; Han, L.; Han, L.; Chang, S.; Hu, T.; Dancey, D. A latent encoder coupled generative adversarial network (le-gan) for efficient hyperspectral image super-resolution. *IEEE Trans. Geosci. Remote Sens.* **2022**, *60*, 1–19. [CrossRef]

14. Yang, F.; Yang, H.; Fu, J.; Lu, H.; Guo, B. Learning texture transformer network for image super-resolution. In Proceedings of the IEEE/CVF Conference on Computer Vision and Pattern Recognition, Seattle, WA, USA, 13–19 June 2020; pp. 5791–5800.
15. Chen, H.; Wang, Y.; Guo, T.; Xu, C.; Deng, Y.; Liu, Z.; Ma, S.; Xu, C.; Xu, C.; Gao, W. Pre-trained image processing transformer. In Proceedings of the IEEE/CVF Conference on Computer Vision and Pattern Recognition, Nashville, TN, USA, 20–25 June 2021; pp. 12299–12310.
16. Liang, J.; Cao, J.; Sun, G.; Zhang, K.; Van Gool, L.; Timofte, R. Swinir: Image restoration using swin transformer. In Proceedings of the IEEE/CVF International Conference on Computer Vision, Virtual, 11–17 October 2021; pp. 1833–1844.
17. Chen, X.; Wang, X.; Zhou, J.; Qiao, Y.; Dong, C. Activating more pixels in image super-resolution transformer. In Proceedings of the IEEE/CVF Conference on Computer Vision and Pattern Recognition, Vancouver, BC, Canada, 18–22 June 2023; pp. 22367–22377.
18. Zhang, J.; Xu, T.; Li, J.; Jiang, S.; Zhang, Y. Single-Image Super Resolution of Remote Sensing Images with Real-World Degradation Modeling. *Remote Sens.* **2022**, *14*, 2895. [CrossRef]
19. Yue, L.; Shen, H.; Li, J.; Yuan, Q.; Zhang, H.; Zhang, L. Image super-resolution: The techniques, applications, and future. *Signal Process.* **2016**, *128*, 389–408. [CrossRef]
20. Wronski, B.; Garcia-Dorado, I.; Ernst, M.; Kelly, D.; Krainin, M.; Liang, C.-K.; Levoy, M.; Milanfar, P. Handheld multi-frame super-resolution. *ACM Trans. Graph. (ToG)* **2019**, *38*, 1–18. [CrossRef]
21. Tarasiewicz, T.; Nalepa, J.; Kawulok, M. A graph neural network for multiple-image super-resolution. In Proceedings of the 2021 IEEE International Conference on Image Processing (ICIP), Virtual, 19–22 September 2021; pp. 1824–1828.
22. Deudon, M.; Kalaitzis, A.; Goytom, I.; Arefin, M.R.; Lin, Z.; Sankaran, K.; Michalski, V.; Kahou, S.E.; Cornebise, J.; Bengio, Y. Highres-net: Recursive fusion for multi-frame super-resolution of satellite imagery. *arXiv* **2020**, arXiv:2002.06460.
23. Bhat, G.; Danelljan, M.; Timofte, R.; Cao, Y.; Cao, Y.; Chen, M.; Chen, X.; Cheng, S.; Dudhane, A.; Fan, H. NTIRE 2022 burst super-resolution challenge. In Proceedings of the IEEE/CVF Conference on Computer Vision and Pattern Recognition, New Orleans, LA, USA, 19–24 June 2022; pp. 1041–1061.
24. Huang, B.; He, B.; Wu, L.; Guo, Z. Deep residual dual-attention network for super-resolution reconstruction of remote sensing images. *Remote Sens.* **2021**, *13*, 2784. [CrossRef]
25. Jia, S.; Wang, Z.; Li, Q.; Jia, X.; Xu, M. Multiattention generative adversarial network for remote sensing image super-resolution. *IEEE Trans. Geosci. Remote Sens.* **2022**, *60*, 1–15. [CrossRef]
26. Yang, Z.; Dan, T.; Yang, Y. Multi-temporal remote sensing image registration using deep convolutional features. *IEEE Access* **2018**, *6*, 38544–38555. [CrossRef]
27. Qin, P.; Huang, H.; Tang, H.; Wang, J.; Liu, C. MUSTFN: A spatiotemporal fusion method for multi-scale and multi-sensor remote sensing images based on a convolutional neural network. *Int. J. Appl. Earth Obs. Geoinf.* **2022**, *115*, 103113. [CrossRef]
28. Ye, Y.; Tang, T.; Zhu, B.; Yang, C.; Li, B.; Hao, S. A multiscale framework with unsupervised learning for remote sensing image registration. *IEEE Trans. Geosci. Remote Sens.* **2022**, *60*, 1–15. [CrossRef]
29. Qiao, B.; Xu, B.; Xie, Y.; Lin, Y.; Liu, Y.; Zuo, X. HMFT: Hyperspectral and Multispectral Image Fusion Super-Resolution Method Based on Efficient Transformer and Spatial-Spectral Attention Mechanism. *Comput. Intell. Neurosci.* **2023**, *2023*, 4725986. [CrossRef]
30. Qiu, Z.; Shen, H.; Yue, L.; Zheng, G. Cross-sensor remote sensing imagery super-resolution via an edge-guided attention-based network. *ISPRS J. Photogramm. Remote Sens.* **2023**, *199*, 226–241. [CrossRef]
31. TSAI, R. Multiframe Image Restoraition and Registration. *Adv. Comput. Vis. Image Process.* **1984**, *1*, 317–339.
32. Guo, M.; Zhang, Z.; Liu, H.; Huang, Y. Ndsrgan: A novel dense generative adversarial network for real aerial imagery super-resolution reconstruction. *Remote Sens.* **2022**, *14*, 1574. [CrossRef]
33. Bhat, G.; Danelljan, M.; Yu, F.; Van Gool, L.; Timofte, R. Deep reparametrization of multi-frame super-resolution and denoising. In Proceedings of the IEEE/CVF International Conference on Computer Vision, Virtual, 11–17 October 2021; pp. 2460–2470.
34. Luo, Z.; Yu, L.; Mo, X.; Li, Y.; Jia, L.; Fan, H.; Sun, J.; Liu, S. Ebsr: Feature enhanced burst super-resolution with deformable alignment. In Proceedings of the IEEE/CVF Conference on Computer Vision and Pattern Recognition, Nashville, TN, USA, 20–25 June 2021; pp. 471–478.
35. Dudhane, A.; Zamir, S.W.; Khan, S.; Khan, F.S.; Yang, M.-H. Burst image restoration and enhancement. In Proceedings of the IEEE/CVF Conference on Computer Vision and Pattern Recognition, New Orleans, LA, USA, 19–24 June 2022; pp. 5759–5768.
36. Luo, Z.; Li, Y.; Cheng, S.; Yu, L.; Wu, Q.; Wen, Z.; Fan, H.; Sun, J.; Liu, S. BSRT: Improving burst super-resolution with swin transformer and flow-guided deformable alignment. In Proceedings of the IEEE/CVF Conference on Computer Vision and Pattern Recognition, New Orleans, LA, USA, 19–24 June 2022; pp. 998–1008.
37. Dudhane, A.; Zamir, S.W.; Khan, S.; Khan, F.S.; Yang, M.-H. Burstormer: Burst Image Restoration and Enhancement Transformer. *arXiv* **2023**, arXiv:2304.01194.
38. Molini, A.B.; Valsesia, D.; Fracastoro, G.; Magli, E. Deepsum: Deep neural network for super-resolution of unregistered multitemporal images. *IEEE Trans. Geosci. Remote Sens.* **2019**, *58*, 3644–3656. [CrossRef]
39. Arefin, M.R.; Michalski, V.; St-Charles, P.-L.; Kalaitzis, A.; Kim, S.; Kahou, S.E.; Bengio, Y. Multi-image super-resolution for remote sensing using deep recurrent networks. In Proceedings of the IEEE/CVF Conference on Computer Vision and Pattern Recognition Workshops, Seattle, WA, USA, 14–19 June 2020; pp. 206–207.
40. Dorr, F. Satellite image multi-frame super resolution using 3D wide-activation neural networks. *Remote Sens.* **2020**, *12*, 3812. [CrossRef]

41. Ghaffarian, S.; Valente, J.; Van Der Voort, M.; Tekinerdogan, B. Effect of attention mechanism in deep learning-based remote sensing image processing: A systematic literature review. *Remote Sens.* **2021**, *13*, 2965. [CrossRef]
42. Lu, E.; Hu, X. Image super-resolution via channel attention and spatial attention. *Appl. Intell.* **2022**, *52*, 2260–2268. [CrossRef]
43. Xia, B.; Hang, Y.; Tian, Y.; Yang, W.; Liao, Q.; Zhou, J. Efficient non-local contrastive attention for image super-resolution. In Proceedings of the AAAI Conference on Artificial Intelligence, Virtual, 22 February–1 March 2022; pp. 2759–2767.
44. Zhang, X.; Zeng, H.; Guo, S.; Zhang, L. Efficient long-range attention network for image super-resolution. In Proceedings of the Computer Vision–ECCV 2022: 17th European Conference, Tel Aviv, Israel, 23–27 October 2022; Part XVII. pp. 649–667.
45. Salvetti, F.; Mazzia, V.; Khaliq, A.; Chiaberge, M. Multi-image super resolution of remotely sensed images using residual attention deep neural networks. *Remote Sens.* **2020**, *12*, 2207. [CrossRef]
46. Valsesia, D.; Magli, E. Permutation invariance and uncertainty in multitemporal image super-resolution. *IEEE Trans. Geosci. Remote Sens.* **2021**, *60*, 1–12. [CrossRef]
47. Bhat, G.; Danelljan, M.; Van Gool, L.; Timofte, R. Deep burst super-resolution. In Proceedings of the IEEE/CVF Conference on Computer Vision and Pattern Recognition, Nashville, TN, USA, 20–25 June 2021; pp. 9209–9218.
48. Vaswani, A.; Shazeer, N.; Parmar, N.; Uszkoreit, J.; Jones, L.; Gomez, A.N.; Kaiser, Ł.; Polosukhin, I. Attention is all you need. *Adv. Neural Inf. Process. Syst.* **2017**, *30*, 5998–6008.
49. Zhang, D.; Huang, F.; Liu, S.; Wang, X.; Jin, Z. SwinFIR: Revisiting the SWINIR with fast Fourier convolution and improved training for image super-resolution. *arXiv* **2022**, arXiv:2208.11247.
50. An, T.; Zhang, X.; Huo, C.; Xue, B.; Wang, L.; Pan, C. TR-MISR: Multiimage super-resolution based on feature fusion with transformers. *IEEE J. Sel. Top. Appl. Earth Obs. Remote Sens.* **2022**, *15*, 1373–1388. [CrossRef]
51. Li, K.; Wang, Y.; Zhang, J.; Gao, P.; Song, G.; Liu, Y.; Li, H.; Qiao, Y. Uniformer: Unifying convolution and self-attention for visual recognition. *IEEE Trans. Pattern Anal. Mach. Intell.* **2023**, 1–18. [CrossRef] [PubMed]
52. Xiao, T.; Singh, M.; Mintun, E.; Darrell, T.; DollÃ¡r, P.; Girshick, R. Early convolutions help transformers see better. *Adv. Neural Inf. Process. Syst.* **2021**, *34*, 30392–30400.
53. Fang, N.; Zhan, Z. High-resolution optical flow and frame-recurrent network for video super-resolution and deblurring. *Neurocomputing* **2022**, *489*, 128–138. [CrossRef]
54. Lu, Z.; Li, J.; Liu, H.; Huang, C.; Zhang, L.; Zeng, T. Transformer for single image super-resolution. In Proceedings of the IEEE/CVF Conference on Computer Vision and Pattern Recognition, New Orleans, LA, USA, 19–24 June 2022; pp. 457–466.
55. Zhao, H.; Shi, J.; Qi, X.; Wang, X.; Jia, J. Pyramid scene parsing network. In Proceedings of the IEEE Conference on Computer Vision and Pattern Recognition, Honolulu, HI, USA, 21–26 July 2017; pp. 2881–2890.
56. Recanatesi, S.; Farrell, M.; Advani, M.; Moore, T.; Lajoie, G.; Shea-Brown, E. Dimensionality compression and expansion in deep neural networks. *arXiv* **2019**, arXiv:1906.00443.
57. Zebari, R.; Abdulazeez, A.; Zeebaree, D.; Zebari, D.; Saeed, J. A comprehensive review of dimensionality reduction techniques for feature selection and feature extraction. *J. Appl. Sci. Technol. Trends* **2020**, *1*, 56–70. [CrossRef]
58. d'Ascoli, S.; Touvron, H.; Leavitt, M.L.; Morcos, A.S.; Biroli, G.; Sagun, L. Convit: Improving vision transformers with soft convolutional inductive biases. In Proceedings of the International Conference on Machine Learning, Virtual, 18–24 July 2021; pp. 2286–2296.
59. Han, K.; Xiao, A.; Wu, E.; Guo, J.; Xu, C.; Wang, Y. Transformer in transformer. *Adv. Neural Inf. Process. Syst.* **2021**, *34*, 15908–15919.
60. Patel, K.; Bur, A.M.; Li, F.; Wang, G. Aggregating global features into local vision transformer. In Proceedings of the 2022 26th International Conference on Pattern Recognition (ICPR), Montreal, QC, Canada, 21–25 August 2022; pp. 1141–1147.
61. Yang, L.; Zhang, R.-Y.; Li, L.; Xie, X. Simam: A simple, parameter-free attention module for convolutional neural networks. In Proceedings of the International Conference on Machine Learning, Virtual, 18–24 July 2021; pp. 11863–11874.
62. Märtens, M.; Izzo, D.; Krzic, A.; Cox, D. Super-resolution of PROBA-V images using convolutional neural networks. *Astrodynamics* **2019**, *3*, 387–402. [CrossRef]
63. Farsiu, S.; Robinson, M.D.; Elad, M.; Milanfar, P. Fast and robust multiframe super resolution. *IEEE Trans. Image Process.* **2004**, *13*, 1327–1344. [CrossRef]
64. Irani, M.; Peleg, S. Improving resolution by image registration. *CVGIP Graph. Models Image Process.* **1991**, *53*, 231–239. [CrossRef]
65. Jo, Y.; Oh, S.W.; Kang, J.; Kim, S.J. Deep video super-resolution network using dynamic upsampling filters without explicit motion compensation. In Proceedings of the IEEE Conference on Computer Vision and Pattern Recognition, Salt Lake City, UT, USA, 18–22 June 2018; pp. 3224–3232.
66. Molini, A.B.; Valsesia, D.; Fracastoro, G.; Magli, E. Deepsum++: Non-local deep neural network for super-resolution of unregistered multitemporal images. In Proceedings of the IGARSS 2020—2020 IEEE International Geoscience and Remote Sensing Symposium, Waikoloa, HI, USA, 26 September–2 October 2020; pp. 609–612.

remote sensing MDPI

Article

FERA-Net: A Building Change Detection Method for High-Resolution Remote Sensing Imagery Based on Residual Attention and High-Frequency Features

Xuwei Xu [1,2], Yuan Zhou [3], Xiechun Lu [4] and Zhanlong Chen [1,2,3,5,*]

1 School of Geography and Information Engineering, China University of Geosciences, Wuhan 430074, China
2 National Engineering Research Center of Geographic Information System, Wuhan 430074, China
3 State Key Laboratory of Information Engineering in Surveying, Mapping and Remote Sensing, Wuhan University, Wuhan 430072, China
4 Key Laboratory of Geological Survey and Evaluation of Ministry of Education, China University of Geosciences, Wuhan 430074, China
5 School of Computer Science, China University of Geosciences, Wuhan 430074, China
* Correspondence: chenzl@cug.edu.cn

Abstract: Buildings can represent the process of urban development, and building change detection can support land use management and urban planning. However, existing building change detection models are unable to extract multi-scale building features effectively or fully utilize the local and global information of the feature maps, such as building edges. These defections affect the detection accuracy and may restrict further applications of the models. In this paper, we propose the feature-enhanced residual attention network (FERA-Net) to improve the performance of the ultrahigh-resolution remote sensing image change detection task. The FERA-Net is an end-to-end network with a U-shaped encoder–decoder structure. The Siamese network is used as the encoder with an attention-guided high-frequency feature extraction module (AGFM) extracting building features and enriching detail information, and the decoder applies a feature-enhanced skip connection module (FESCM) to aggregate the enhanced multi-level differential feature maps and gradually recover the change feature maps in this structure. The FERA-Net can generate predicted building change maps by the joint supervision of building change information and building edge information. The performance of the proposed model is tested on the WHU-CD dataset and the LEVIR-CD dataset. The experimental results show that our model outperforms the state-of-the-art models, with 93.51% precision and a 92.48% F1 score on the WHU-CD dataset, and 91.57% precision and an 89.58% F1 score on the LEVIR-CD dataset.

Keywords: building change detection; residual attention mechanism; high-frequency feature; feature fusion

Citation: Xu, X.; Zhou, Y.; Lu, X.; Chen, Z. FERA-Net: A Building Change Detection Method for High-Resolution Remote Sensing Imagery Based on Residual Attention and High-Frequency Features. *Remote Sens.* **2023**, *15*, 395. https://doi.org/10.3390/rs15020395

Academic Editors: Moulay A. Akhloufi and Mozhdeh Shahbazi

Received: 8 November 2022
Revised: 29 December 2022
Accepted: 6 January 2023
Published: 9 January 2023

1. Introduction

Buildings are a significant component of urban development, and building change detection is commonly utilized in land use management [1], illegal building management [2], and urban development [3]. Change detection (CD) aims to analyze multitemporal images within the same region and detect the process of object or phenomenon differences within the region [4]. Remote sensing images can be used for identifying and evaluating surface changes utilizing obtained multitemporal images covering the same surface region and its accompanying data [5]; this is one of the most active areas of research in the remote sensing field. With the development of remote sensing technology, a substantial chunk of remote sensing data are now accessible, providing substantial information about land cover [2]. The development of computing and AI technology also provides a great deal of theoretical guidance for CD methods. The CD methods can be divided into two categories: pixel-based and object-based methods [6].

Due to their resolution limitations [7], pixel-based methods for change detection in remote sensing images became mainstream. These methods compare pixels between multiple temporal images [8,9] to produce change maps using clustering techniques [10,11] or threshold segmentation [12,13]. The main classifiers used include support vector machines [14], random forests [15], and k-nearest neighbor classification [16], etc. However, they ignore contextual information and are prone to noise [17]. Object-based methods were introduced to take advantage of the spatial contextual relationships in these images. As a result, these techniques are frequently utilized for jobs involving the CD of low- and medium-resolution remote sensing images [18]. These methods use local clusters of labeled pixels called "objects" to reduce false positives and missing values in change maps [19]. They effectively remove noise and misaligned images by segmenting images into homogeneous regions using spectral and spatial adjacency similarities and then generating change result maps based on the texture, shape, and spatial relationships of neighboring objects [20–23]. However, the precision of object-based CD models is impacted by the segmentation procedure, and they often require manual intervention and are not robust.

Due to breakthroughs in the computer vision field, deep learning networks have been applied to remote sensing image analysis [24,25]. Deep learning networks are applied to the CD task to automatically extract the relevant semantic features and obtain an abstract feature representation of the image information. Compared with traditional methods, deep-learning-based approaches can automatically learn features, which significantly reduces the need for specialized domain knowledge. The performance of CD tasks has greatly improved profits owing to the excellent performance of deep learning models in capturing and representing image features. This has resulted in a new era of CD technologies. The CD network framework consists of three main parts: feature extraction, network optimization, and accuracy evaluation. From the perspective of feature extraction, the CD methods can be classified into five major types: autoencoder (AE)-based, recurrent neural network (RNN)-based, generative adversarial network (GAN)-based, convolution neural network (CNN)-based, and transformers-based methods [7]. AE is an artificial neural network used in semi-supervised and unsupervised learning to learn representations of input information by using it as a learning target. Liu et al. [26] propose a stacked autoencoder for synthetic aperture radar change detection that effectively adapts to the multiplicative noise environment present in SAR change detection. RNN generates CD results by learning multitemporal image features and establishing change relationships between multi-period sequence remote sensing images [27,28]. Lyu et al. [29] build on RNN using an improved LSTM approach to design a change rule with transferability to provide reliable change information for multitemporal images. GANs can be trained with a small amount of real data and generate a large amount of pseudo-data through adversarial learning, thus enhancing the network's generalization ability [30,31]. Peng et al. [18] used multitemporal images as input to UNet++ to generate multi-level feature maps of predicted change maps. Multiple side-output fusion (MSOF) was then introduced to generate the final variogram. Fang et al. [32] proposed SUNNet-CD, a tightly connected Siamese network for CD tasks, with deep supervision of the network training by aggregating and refining semantic information at different levels. Transformer-based methods are an effective attention-based approach to improve CD results by capturing global contextual information and establishing remote target dependencies [33]. Wang et al. [25] proposed TransCD by adding the transformer structure to the feature difference network to improve the robustness to noisy changes by establishing global semantic information. The CD methods are roughly divided into two categories, the early fusion (EF) strategy and the late fusion (LF) strategy, from the perspective of the feature fusion strategy [34]. The authors of [35] design a CD network with EF strategies based on UNet networks, use skip connections to improve accuracy, and explore the impact of different fusion strategies on network performance by varying the feature fusion. In [36], the Siamese network is used to extract the dual-time image features, and the boundary integrity of the resultant map is improved based on an LF strategy using multi-level depth features fused with differential features through an attention mechanism. Studies have shown that late fusion strategies give better CD results than early fusion strategies.

There has been a growing interest in researching multi-scale features for building change detection in remote sensing images in recent years. This is due in part to the increasing resolution of remote sensing images, which allows for more detailed and accurate detection of changes in buildings. Previous methods, such as pixel-based and object-based approaches, have been effective in detecting changes in low- to medium-resolution images, but may not fully utilize the multi-scale information available in high-resolution images [37]. Thus, there is a need to develop methods that can effectively extract multi-scale building features from high-resolution images in order to improve the accuracy of building change detection. Research in this area has included the use of deep learning methods, such as convolutional neural networks, to extract and utilize multi-scale features in building change detection [34]. However, there are still challenges to be addressed, such as the information gap between upper and lower layers of the network and the difficulty in fully utilizing the local and global information of the feature maps [38]. Further research in this area is necessary in order to improve the performance and efficiency of building change detection in high-resolution remote sensing images.

Although deep learning methods have achieved great advantages in CD tasks, there are still some problems to overcome. On the one hand, while current building change detection methods consider multi-scale features, they are often unable to fully utilize both global and local information in multi-scale feature maps. Due to successive downsampling operations in the encoder part and successive upsampling operations in the decoder part, detailed information such as building boundary information and multi-scale buildings is gradually diluted. On the other hand, the existing CD networks tend to connect only sibling feature maps when passing contextual information through skip connections. This leads to an information gap between the upper and lower layers when the network is connected, which cannot fully utilize the local and global information of the feature maps, thus affecting the efficiency of feature extraction and limiting the performance of the model.

The following is a summary of this paper's significant contributions.

(1) In order to fully utilize the high-frequency information in remote sensing images to enhance building edges and extract multi-scale features through multi-level feature maps, we propose a new CD network called FERA-Net, using a Siamese network to detect building changes at the pixel level. The network can extract multi-level and multi-scale features from dual-time remote sensing images.
(2) We designed a hierarchical attention-guided high-frequency feature extraction module (AGFM) to effectively capture building information in remote sensing images. The AGFM uses the residual attention mechanism and the high-frequency information enhancement module to effectively capture building features while enhancing building boundary information and mitigating the effect of noise. We used the feature-enhanced skip connection module (FESCM) to fuse the local and global information of the dual-temporal image differential maps, aggregating differential map features of different levels and scales to capture building information at different scales in terms of fusion strategy selection. To better perform change monitoring, we designed a hybrid loss function to focus on building change information and a priori boundary information.
(3) We conducted experiments on two publicly available datasets to evaluate the performance of FERA-Net in remote sensing CD tasks in multiple aspects. Both quantitative and qualitative comparison results show the significant advantages of FERA-Net compared with other network models.

The remainder of this paper is structured as follows. The work related to CD is described in Section 2. The details of the FERA-Net are described in Section 3. The dataset and experimental setup are presented in Section 4. Comparative and ablation experiments are described and discussed in Section 5. Section 6 summarizes the work conducted in this paper and future perspectives.

2. Related Work

2.1. Change Detection Networks with Siamese Network

Siamese networks are connected architectures of two structurally identical neural networks that share weights. These networks accept two separate inputs and produce representations of the high-dimensional spaces they are embedded in. Change detection tasks aim to discover the difference between changed and unaltered pixels in multitemporal images. The features from multitemporal images are extracted using the two branches of the Siamese network. Then, the network creates the change maps based on the similarity between the two attributes. This makes the Siamese network a strong contender for applications of change detection. Caye Daudt et al. [35] designed a CD network based on a single-branch structure and a CD network based on a Siamese structure and experimentally demonstrated the performance improvement of the Siamese network for the CD task. Chen and Shi et al. [39] exploit the spatial–temporal relationships that exist between remote sensing image pixels, design a self-attentive module to model the spatial–temporal relationships, and extract diachronic image spatial–temporal features based on Siamese networks.

2.2. Change Detection Networks with Skip Connection

The encoder–decoder architecture is a common network architecture for change detection. However, when the encoder part extracts the image features through downsampling operations such as convolution, the detailed information of the remote sensing image is easily lost. Therefore, connecting the encoder and decoder with direct information through skip connection can help the decoder to better recover space information and improve the change detection performance. As a result, UNet and its variants, which were the first to use skip connections, are widely used in CD tasks. Ji et al. [40] designed a UNet structure for pixel segmentation incorporating multi-scale aggregation to help recover spatial information. Based on UNet++, Fang et al. [32] combined NestedUNet and Siamese networks to propose SNUNet-CD, which maintains high-resolution features and fine-grained localization information through a skip connection mechanism. D. Peng et al. [41] added an attention mechanism to the UNet++ dense skip connection to reduce the semantic gap between high-level features and the underlying features and to facilitate the extraction and fusion of multi-scale features. The experiment verified that dense skip connections are used in the decoding phase to efficiently fuse different features. Although using skip connections and dense skip connections in UNet and UNet++ can help to recover spatial information, these networks only transfer information between sibling feature maps and do not make full use of full-size feature map information.

2.3. Change Detection Networks with Attention Mechanism

The fundamental concept behind the attention mechanism is to emphasize certain crucial aspects while suppressing unimportant ones based on the association between the original data [42]. By using cross-channel correlations and spatial interactions between features, creating channel attention maps and spatial attention maps, networks can concentrate on where the information is located. This process enables the rapid filtering of high-value information from the mass of information using limited attentional resources. Attentional mechanisms are widely used with change detection networks. Chen et al. [43] designed DASNet to improve the model's performance by capturing long-range dependencies through a dual-attention mechanism to obtain more feature representations. Ma et al. [44] designed a multi-attentive cued feature fusion network for change detection tasks, introducing coordinate attention (CA) via a Feature Enhancement Module (FEM) to obtain more accurate position and channel relationships. J. Chen et al. [45] proposed MSF-Net, which uses an attention mechanism to fuse contextual information to obtain rich global contextual information about a building and effectively suppress irrelevant features.

3. Methods

3.1. Overview

As shown in Figure 1, we designed the network with a typical encoder–decoder architecture called FERA-Net. The AGFM is stacked together in the encoder part to extract features from bitemporal images. In this module, we use the residual spatial attention mechanism for concentrating building change areas and the residual channel attention mechanism to improve the edge features and high-frequency information extracting. The FESCM is utilized to aggregate the change features from the multi-scale feature maps, and then the module transmits the features to the decoding part. It provides spatially detailed information from the low-level feature maps and the semantic information from the high-level feature maps to make up for the information loss caused by downsampling in the encoder part, so that the network can completely mine the multi-scale characteristics at multiple levels. The final-layer feature map of the decoder part is used to generate the final change result map and is also used to extract building edge estimates and construct the hybrid loss function to accomplish accurate segmentation for the edges of the change map.

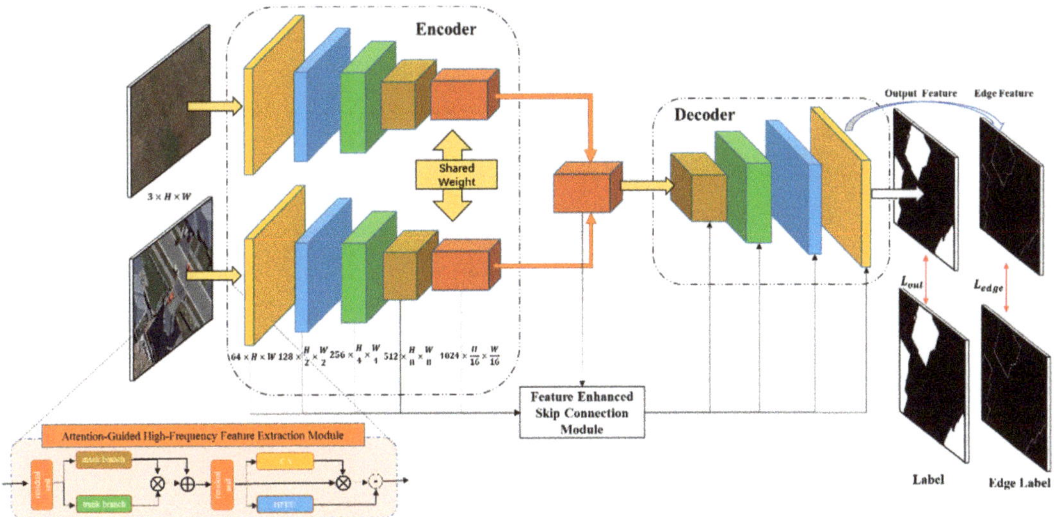

Figure 1. Structure of the FERA-Net. FERA-Net comprises the encoder part and decoder part. The encoding part consists of five attention-guided high-frequency feature extraction modules with dual branches. The decoding part consists of a feature-enhanced skip connection module.

3.2. Encoder Part

Buildings in bitemporal images often possess different features, the changing characteristics of which cannot be fully described by only using one level of feature maps. Thus, we use a multi-layer approach to mine different types of feature information. Considering that we are using a pair of dual-time images as input data, we introduce the Siamese network to implement the feature extraction task in the coding part. Specifically, the Siamese network uses two different inputs to extract deep-level features through two similar subnetworks with the same architecture, parameters, and weights, and can also be used to describe the dual-time image features. As shown in Figure 1, when a pair of dual-time images of size $3 \times H \times W$ is used as input, the network we designed obtains the corresponding five-layer feature maps through five AGFMs, denoted as F_1, F_2, F_3, F_4, and F_5, which have dimensions of $64 \times H \times W$, $128 \times \frac{H}{2} \times \frac{W}{2}$, $256 \times \frac{H}{4} \times \frac{W}{4}$, $512 \times \frac{H}{8} \times \frac{W}{8}$, and $1024 \times \frac{H}{16} \times \frac{W}{16}$, respectively, which are used to generate coarse-grained feature maps with high resolution and fine-grained features with low resolution.

3.2.1. Attention-Guided High-Frequency Feature Extraction Module

Buildings are manufactured objects with obvious edges. In the end-to-end network, the building boundary information is always weakened, and it is difficult to recover the building boundary due to its network characteristics, which often lead to the blurring of the edges of buildings during building change detection [46].

However, the identification of building edges is essential in building change detection and greatly impacts the final detection accuracy. In remote sensing images, the high-frequency signal corresponds to the part of the image that changes drastically, which indicates the edges (contours) or noise in the image. Therefore, if the high-frequency features of remote sensing images are enhanced, we can recover the edge information of buildings to some extent, but this will also enhance the other high-frequency features, such as noise. To address this issue, we integrate the attention method into the building change detection network. The AGFM is designed to identify areas of building change and retain building boundary information. The detailed AGFM structure diagram is shown in Figure 2.

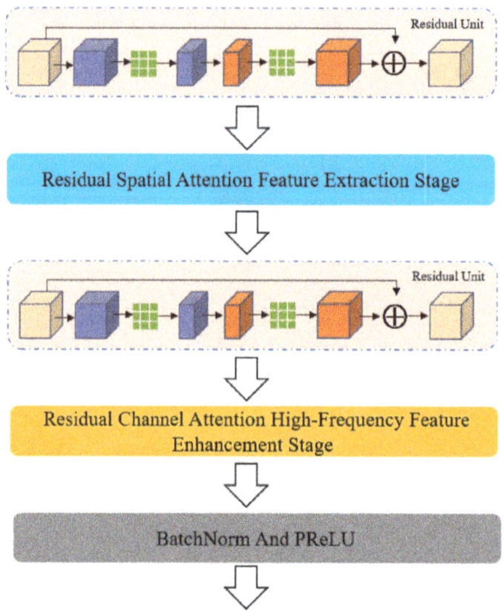

Figure 2. Illustration of Attention-Guided High-Frequency Feature Extraction Module.

3.2.2. Residual Spatial Attention Feature Extraction Stage

Not every region in the image contributes equally to the building change detection task; we focus only on the regions of building change in our study [47]. Therefore, it is necessary to apply spatial attention mechanisms to the task of change detection. The spatial attention mechanism reduces the effect of noise by generating a spatial attention map that focuses on the changing building regions in the remotely sensed image. In the RSAS module, we divide the attention module into the mask branch and the trunk branch. The trunk branch is used for feature extraction, and the mask branch uses a top-down structure to learn to generate spatial attention masks. The trunk branch output $F(x)$ with input x is given to the mask branch. The output of the attention mask is

$$H_{i,c}(x) = M_{i,c}(x) \times F_{i,c}(x) \tag{1}$$

where i ranges over all spatial positions and $c \in \{1, \ldots, C\}$ is the index of the channel.

The attention calculated by the above equation is differentiable, and the attention that can be differentiated can be learned by the neural network by calculating the gradient and by forward propagation and backward feedback to obtain the weight of the attention. Thus, the attention mask can be used as a filter for gradient updates during backpropagation. In the mask branch, its gradient is:

$$\frac{\partial M(x,\theta)F(x,\varnothing)}{\partial\varnothing} = M(x,\theta)\frac{\partial F(x,\varnothing)}{\partial\varnothing} \tag{2}$$

where θ represents the mask branch parameters and $\varnothing$ represents the trunk branch parameters. By generating spatial attention $M(x)$ through the mask branch structure, the effect of noise in the image on the identification of changing regions can be effectively reduced.

However, merely stacking attention modules will lead to performance degradation, so we use residual learning in the spatial attention module. As shown in Figure 3, the output of the modified attention module is

$$H_{i,c}(x) = (1 + M_{i,c}(x)) \times F_{i,c}(x) \tag{3}$$

where $M(x)$ ranges from $[0,1]$, with $M(x)$ approximating 0 and $H(x)$ approximating the original features, $F(x)$. $M(x)$ work as a feature selector, which enhances good features and suppresses noise from trunk features. This process is referred to as attention residual learning.

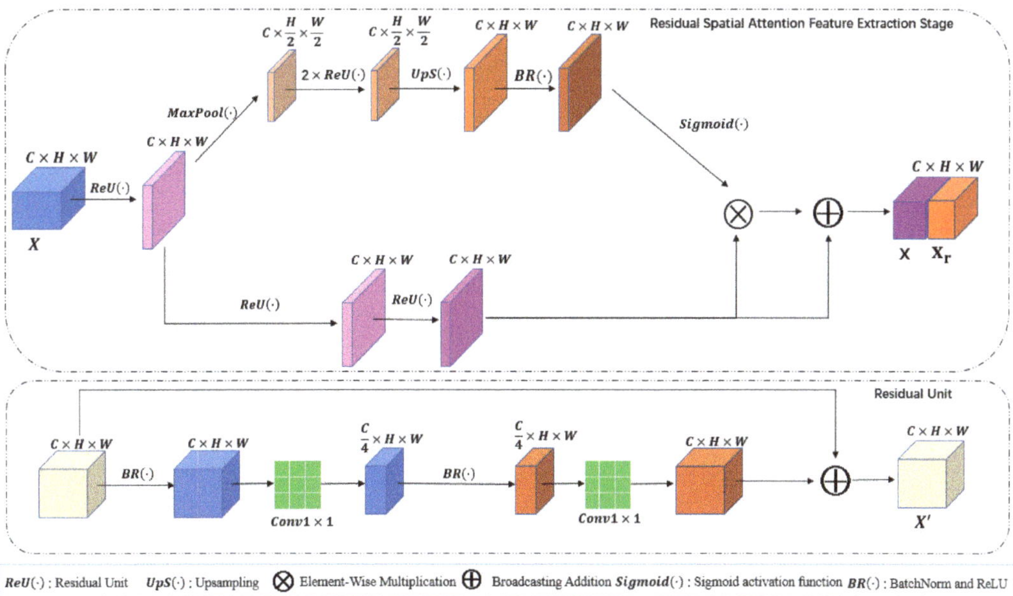

Figure 3. Illustration of Residual Spatial Attention Feature Extraction Stage.

3.2.3. Residual Channel Attention High-Frequency Feature Enhancement Stage

In building CD tasks, the edges of buildings are often areas of dramatic change, often consisting of high-frequency features in the image, and identifying these is important. Due to the variety of building shapes in remote sensing images, the isotropic Sobel operator is more suitable for extracting high-frequency features [46]. As shown in Figure 4, after acquiring the high-frequency features representing the edges of buildings, we use the channel attention to add the extracted high-frequency information to the network. The channel attention typically consists of a direct, global average pooling of information within a channel, and it ignores local information within each channel. The feature compression along the spatial dimension turns each two-dimensional feature channel into a real number with a global perceptual field, in which the output dimension matches the number of input feature channels using global average pooling. Formally, for the output of the modified attention module $X \in \mathbb{R}^{C\times H\times W}$, the global average pooling $G(X)$ can be obtained as follows:

$$G(X) = \frac{1}{H\times W}\sum_{i=1}^{H}\sum_{j=1}^{W} X(i,j) \tag{4}$$

We chose the simple gate control mechanism with a sigmoid activation function to aggregate information for capturing the correlation between channels, as shown in the following formula:

$$F_{CA}(X) = \xi(FC(\delta(FC(G(X))))) \tag{5}$$

where δ denotes the ReLU function, ξ denotes a sigmoid activation function, and FC(•) denotes fully connected layers.

In addition, we use the isotropic Sobel operator to extract high-frequency features, as shown in the following formula:

$$F_{HF}(X) = Mean\left(\sum_{i=1}^{8} S_i(X)\right) \tag{6}$$

Finally, the extracted high-frequency features need to be fused with the channel information in the channel attention, as shown in the following formula:

$$F_{RC}(X) = X \oplus F_{CA}(X) \tag{7}$$

$$F_{RCA}(X) = Conv_{1\times1}(Cat(H(X) \oplus ResC(X))) \tag{8}$$

where $Cat(\bullet)$ denotes channel-wise concatenate, $Conv_{1\times1}$ denotes a 1×1 convolution operation, and $\oplus$ denotes element-wise addition.

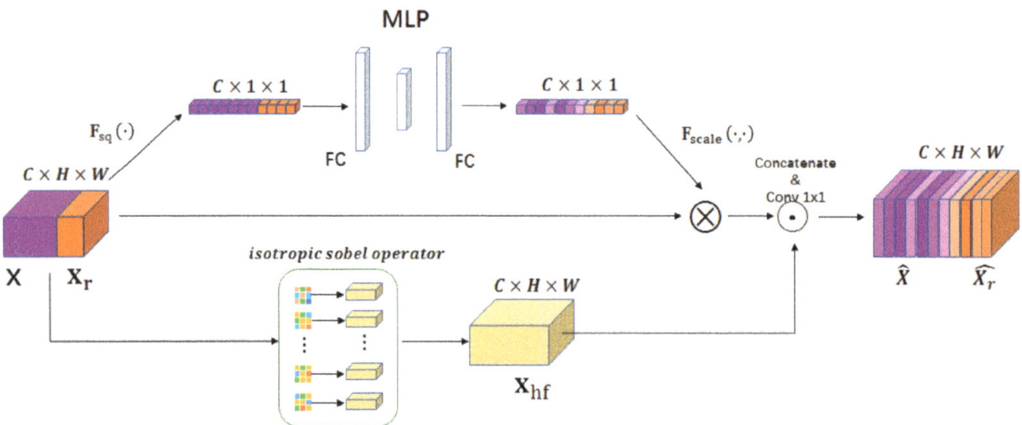

Figure 4. Illustration of Residual Channel Attention High-Frequency Feature Enhancement Stage.

3.3. Decoder Part

In the coding section, we extracted five levels of diachronic remote sensing image feature maps. In the decoding part, we enhanced the predictive power of the model by aggregating full-scale features through the FESCM to mine the building change features between the dual-temporal image feature maps.

3.3.1. Feature-Enhanced Skip Connection Module

Skip connections are frequently employed in change detection tasks to link low-level feature maps and high-level feature maps for improving a network's capacity for prediction. However, the UNet-based model connects only sibling feature maps, which makes it difficult to examine enough data from the full scale and renders the placement and borders of buildings unclear. In order to reduce the differential information of buildings, aggregate local and global feature contexts, and learn feature representations from the multi-scale aggregated feature maps, we propose the FESCM to aggregate the full-scale feature maps produced by F_{t1} and F_{t2}. Furthermore, since different depths of perceptual fields are not equally sensitive to targets of different sizes (shallow layers are more sensitive to small targets, and deep layers are more sensitive to large targets), we stitch them together by feature concatenation to integrate the advantages of both. In addition, as the number of network layers increases, one layer of downsampling will keep losing information (large object edge information and small objects themselves), so the features with small receptive fields are supplemented to make the location information more accurate.

As shown in Figure 5, we take the third layer feature map X_{De}^3 in the network as an example, and the steps are as follows:

1. The low-level feature maps $F_{t1}^1, F_{t2}^1, F_{t1}^2,$ and F_{t2}^2 preserve the detailed information, fusing the fine-grained semantics. $F_{t1}^1, F_{t2}^1, F_{t1}^2,$ and F_{t2}^2 are to be downsampled to the same scale as F_{t1}^3 to unify the size. This is followed by a 3×3 convolution with 64 channels. The downsampling in this part is performed using overlap-free maximum pooling.
2. The feature map of the same scale encoder, the F^3 layer, is directly operated by a 3×3 convolution with 64 channels.
3. The high-level feature maps X_{De}^4 and X_{De}^5 are upsampled by $2\times$ and $4\times$, respectively, to ensure the uniformity of the feature map size. The upsampling of this part is achieved by bilinear interpolation.

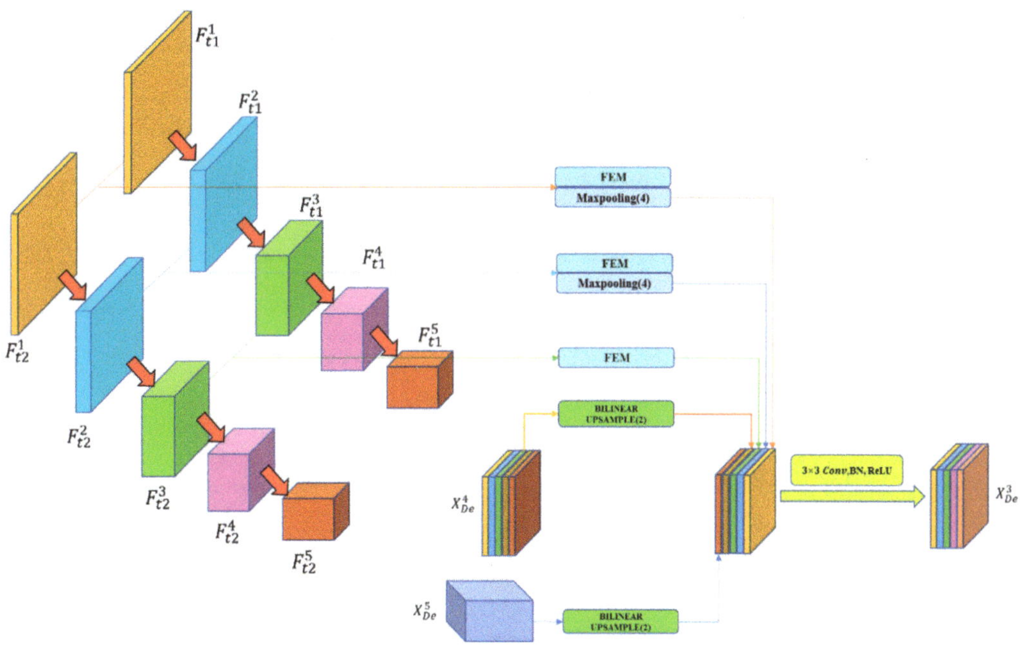

Figure 5. Feature-enhanced skip connection module.

Five layers are stacked (stitched and fused) to form a 320-channel feature map. The feature map generation process of other decoding parts is similar, which is calculated as follows:

$$X_{De}^{i} = \begin{cases} FEM\left(F_{t1}^{i}, F_{t2}^{i}\right), \quad i = N \\ \sum_{k=1}^{i-1} Down\left(FEM\left(F_{t1}^{k}, F_{t2}^{k}\right)\right) + BR\left(FDEM\left(F_{t1}^{i}, F_{t2}^{i}\right)\right) + \sum_{k=i+1}^{N} Up\left(FEM\left(F_{t1}^{k}, F_{t2}^{k}\right)\right), i \in \{1, 2, \ldots, N-1\} \end{cases} \tag{9}$$

where i represents the feature map hierarchy in the encoder, N denotes the number of feature maps, $FEM(\bullet)$ represents the Feature Enhancement Module, $Down(\bullet)$ represents the downsampling operation, $\mathcal{BR}(\bullet)$ denotes a convolution operation, and $Up(\bullet)$ denotes the upsampling operation.

The FESCM can improve both the spatial and semantic information from multiple-scale feature maps and the fused information of dual-temporal remote sensing images. This guarantees that the decoder part can fully utilize multi-scale feature maps to produce high-quality change monitoring maps.

3.3.2. Feature Enhancement Module

In the encoder section, we extract multi-scale bitemporal image feature maps through the Siamese network and AGFM. The FESCM allows the aggregation of feature maps at different scales. However, it is difficult to effectively represent semantic information by element addition or the channel tandem aggregation of biaxial feature maps [48]. In order to make full use of the multi-scale feature map, we designed FEM in FESCM to aggregate local and global information to compensate for the inconsistency of attentional feature fusion at different scales.

As shown in Figure 6, in the FEM section, we first add or subtract the previously obtained feature maps F1 and F2 of the two temporal images to obtain the fusion information. The local contextual information is then aggregated through the local attention module (*LAM*). In the local attention module, we use point-wise convolution (*PWConv*) to aggregate local attention information, which can be obtained as follows:

$$LAM(F) = \mathcal{B}(PWConv_2(\delta(\mathcal{BR}(PWConv_1(x))))) \tag{10}$$

where $\mathcal{B}$ denotes batch normalization (BN), $\mathcal{R}$ denotes rectified linear unit (ReLU) activation, and δ denotes the sigmoid function.

In the global attention module (*GAM*), we use global average pooling (*GAPool*) to aggregate global context information, which can be obtained as follows:

$$GAPool(F) = \frac{1}{H \times W} \sum_{i=1}^{H} \sum_{j=1}^{W} F_{[:,i,j]} \tag{11}$$

To further enhance global information, we use *GAM* to capture the feature map to obtain more complete change information. The formula is as follows:

$$GAM(F) = \mathcal{B}(PWConv_2(\delta(\mathcal{BR}(PWConv_1(\text{GAPool}(F)))))) \tag{12}$$

where $\mathcal{B}$ denotes BN, $\mathcal{R}$ denotes ReLU, and δ denotes the sigmoid function.

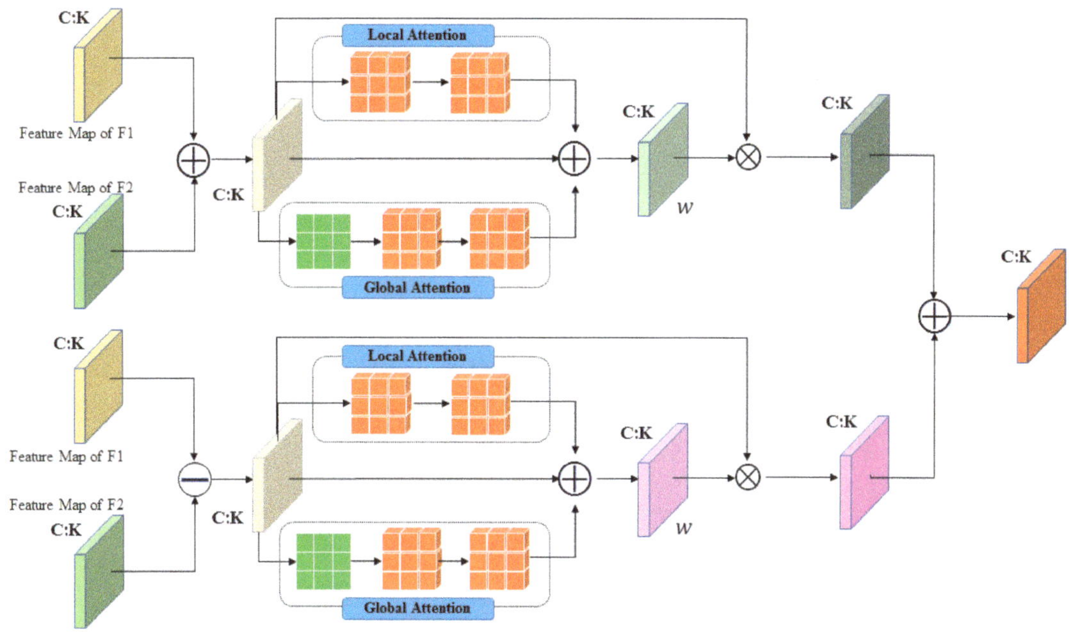

Figure 6. Feature Enhancement Module.

3.4. Loss Function

The loss function is employed to quantify the discrepancy between the output of the model and its true value and to offer recommendations for model optimization. A proper loss function must be chosen in order to make the most of the model's training efficiency. This study uses a hybrid loss function to mitigate the impact of sample imbalance since the amount of changed pixels in the change detection domain is significantly less than the number of background pixels. The subject loss function and the edge loss function are the two fundamental components of the hybrid loss function. The expected building change results are monitored using the subject loss function. The expected edge detection results are monitored using the edge loss function. What follows is a description of the hybrid loss functions for the main body loss function and the edge loss function, which are designated by L_{out} and L_{edge}, respectively.

$$L = \lambda_1 L_{out} + \lambda_2 L_{edge} \tag{13}$$

In the task of building change detection, the number of background pixels is far greater than the number of pixels of changed buildings. During model training, the easily classified negative samples have a great impact on the optimization direction of the model. In order to solve the imbalance problem between positive and negative samples, the focal loss detection method is suggested in an effort to balance out the proportion of challenging and straightforward samples. The focal loss has the following formula:

$$L_{f_1} = \begin{cases} -\alpha(1-y')^{\gamma} log y', & y = 1 \\ -(1-\alpha)(y')^{\gamma} log(1-y')', & y = 0 \end{cases} \tag{14}$$

where y' represents the predicted value and y represents the true ground. α denotes the weights of positive and negative samples. γ denotes the focusing parameter. This paper set $\alpha = 0.25$ and $\gamma = 2.0$.

To minimize the difference between the predicted building edge and the true value, we use the MSE loss function for the prediction of the building edge. The MSE loss function is utilized for the average sum of squares of the discrepancies between the predicted and true values for the edge loss function. The precise formula is given below:

$$MSE = \frac{1}{m}\sum_{i=1}^{m}\left(\hat{y}^i - y^i\right)^2 \quad (15)$$

where $\hat{y}^i$ denotes the model prediction output result, y^i denotes the true label result, and m is the total number of pixels.

4. Datasets and Experimental Implementations

4.1. Datasets

In order to verify the performance and robustness of our proposed model, we conducted comparison experiments on the WHU-CD dataset [49] and the LEVIE-CD dataset [39]. Details of the WHU-CD and LEVIR-CD datasets are shown in Table 1. Figure 7 presents the examples from the datasets. Based on the original dataset, we further cropped the dataset to a size of 256 × 256.

Table 1. Experiment datasets.

Datasets	Data Collection	Bands	Resolution/m	Image Size	Training Set	Validation Set	Test Set
WHU-CD	Aerial Image	3	0.3	32,207 × 15,354	5460	780	1560
LEVIR-CD	Google Earth	3	0.5	1024 × 1024	7120	1024	2048

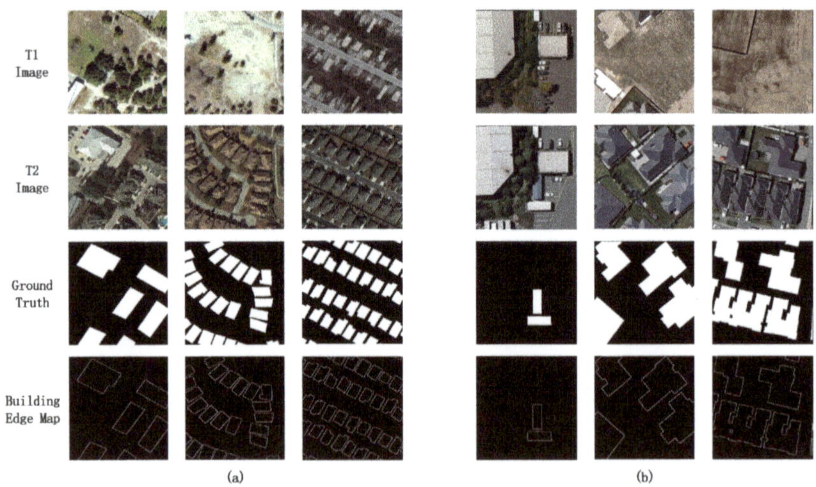

Figure 7. Sample images from the LEVIR-CD dataset (**a**) and the WHU-CD dataset (**b**). The first and second rows represent the remote sensing data for the dual-time images. The third row represents the change in the building ground truth. The fourth row represents the building edges extracted from the third row by the Canny edge detection algorithm.

WHU-CD dataset: This dataset contains two periods of aerial images, with a size of 32,507 × 15,354. The geographic resolution of this dataset, which covers over 22,000 different buildings in Christchurch and was compiled in 2012 and 2016, is 0.2 m per pixel.

LEVIR-CD dataset: The LEVIR-CD dataset consists of 637 groups of ultrahigh-resolution Google Maps pictures with a size of 1024 × 1024 pixels. Each group of images includes the images before and after the building changes and a corresponding label. Different types of structures are covered by LEVIR-CD, including villa homes, high-rise flats, small garages, and enormous warehouses. These diachronic photos, which cover 5 to 14 years, show notable land use changes, particularly the expansion of buildings.

4.2. Comparison Methods

To verify the effectiveness of our proposed model, we selected some classical building change detection models and state-of-the-art building change detection networks for comparison, including the FC-EF [35], FC-SC [35], and FC-SD models [35], the edge-guided cyclic convolution neural network (EGRCNN) [50], and the change detection model based on the transformer (BITNet) [33]. FC-EF, FC-SC, and FC-SD are all UNet-based network models that optimize network performance by adding Siamese structures or changing hopping connections, and are classical CD models. The EGRCNN model focuses on the impact of building boundaries on network performance. The BITNet model uses transformers to model the spatial–temporal contextual information in the image to improve model accuracy.

4.3. Implementation Details

Our proposed model implementation uses the Pytorch 1.10 framework based on the Python 3.8 programming language. This study uses the Adam optimizer to train the model. The initial learning rate is 10-4, and the weight decay is 10-3. The learning rate is adjusted by observing whether the F1 score of the validation set increases within eight epochs. If no increase is observed, the learning rate is reduced by a factor of 0.05. Our proposed method is trained for 100 epochs on the LEVIR-CD dataset and WHU-CD dataset, and the network reaches a state of convergence. All training and testing experiments in the study were performed on a server equipped with an NVIDIA GeForce RTX 3090Ti GPU (with 24 GB of RAM).

4.4. Evaluation Metrics

In order to test the performance of FERA-Net in our experiments, we used the following four evaluation metrics: precision (P), recall (R), F1 score, and intersection over union (IoU). Specifically, the above evaluation metrics can be represented by the following formulas:

$$Precision = \frac{TP}{TP + FP} \quad (16)$$

$$Recall = \frac{TP}{TP + FN} \quad (17)$$

$$F1 = \frac{2 \times Precision \times Recall}{Precision + Recall} \quad (18)$$

$$IoU = \frac{TP}{TP + FP + FN} \quad (19)$$

where TP represents the number of pixels correctly identified as a changed building, FP represents the number of pixels incorrectly identified as a changed building, TN is the number of pixels correctly identified to have changed, and FN is the number of false negatives.

5. Results and Analysis

5.1. Results for WHU-CD

Table 2 shows the quantitative evaluation results of FERA-Net and other CD networks on the WHU-CD dataset. In addition to comparisons with the models presented in Section 4.2, we also present recent experimental data from a number of other state-of-the-art models on the WHU-CD dataset. Among them, the proposed method's p value, F1 score, and IoU are the highest, at 93.51%, 92.52%, and 86.07%, respectively. Despite the fact that it did not achieve the highest R value, FERA-Net achieved the highest p value and F1 score. The results show that our proposed network can successfully balance network performance while concentrating more precisely on the building change area and fully characterizing the change. In particular, FERA-Net's p value of 93.51% is much higher than that of the other change detection networks, indicating that FERA-Net is able to extract individual buildings more completely. In addition, we observed from the data that some other CD networks had large differences in p and R values, suggesting unbalanced network performance, which may have resulted in the network either failing to capture the building in its entirety or failing to identify more areas of change. The p and R values of FERA-Net are relatively close, which indicates that the network is both effective in identifying areas of building change and accurate in characterizing building edges. Regarding the recall metric, the FERA-Net does not achieve the highest score. One of the main reasons is the existence of mislabeled samples, which can be observed visually in the ground truth. As a result, FERA-Net does not classify these mislabeled regions. Thus, a higher FN value will be generated by the FERA-Net, leading to a decrease in recall.

Table 2. Comparison of quantitative evaluation indicators in the WHU-CD dataset, where the best values are in bold.

Method	P (%)	R (%)	F1 (%)	IoU (%)
FC-EF [35]	83.91	79.23	81.50	68.78
FC-SC [35]	71.20	85.10	77.53	63.30
FC-SD [35]	73.36	82.87	77.83	63.70
CDNet [51]	91.75	86.89	89.62	80.60
UNet++MSOF [18]	91.96	89.40	90.66	82.92
STANet [39]	78.73	87.41	82.84	74.57
DASN [52]	88.23	84.62	86.39	76.04
SNUNet [32]	79.25	85.38	82.20	69.78
BITNet [33]	87.32	91.16	89.20	80.50
EGRCNN [50]	84.65	**94.34**	89.23	80.55
FERA-Net	**93.51**	91.47	**92.48**	**86.01**

In the qualitative evaluation results, we visualized eight change detection scenarios in the WHU-CD dataset. Among them, Figure 8 shows the detection effect of FERA-Net for building boundaries. Figure 9 shows the detection of FERA-Net for building change areas in complex scenarios. In order to better represent the detection results, the TP, FP, TN, and FN values are shown in different colors, and the areas where the different networks have significant differences are further highlighted by yellow circles. From the results shown in Figures 8 and 9, the following can be seen:

(1) Accurately describing the shape structures of buildings with ambiguous edges is a great challenge. As shown in Figure 8, FERA-Net can precisely describe the boundary information of the region and segment the border of the changing area more finely. It can also effectively reduce the edge noise of moving buildings.

(2) The interference from various manufactured things, such as roads, containers, and cars, might result in the inaccurate identification of change regions in complicated contexts with building change areas. Figure 9 illustrates that FERA-Net can more precisely capture the building regions of interest.

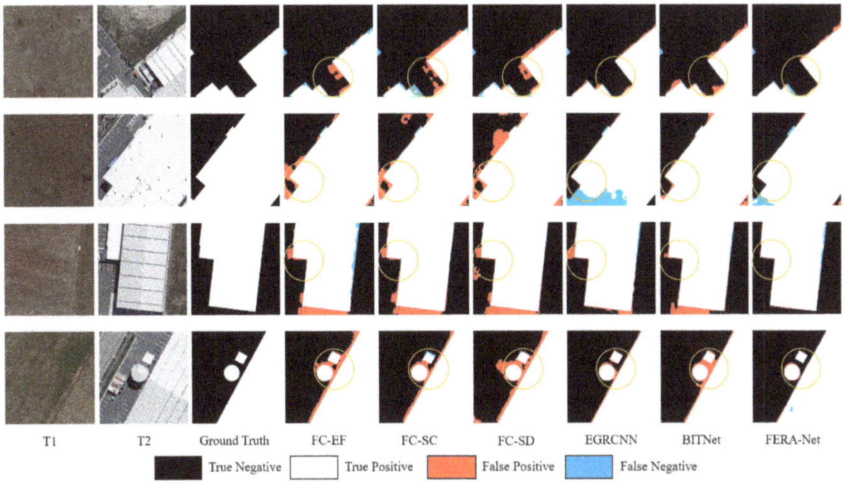

Figure 8. Visual comparison of building edge extraction results in the WHU-CD dataset.

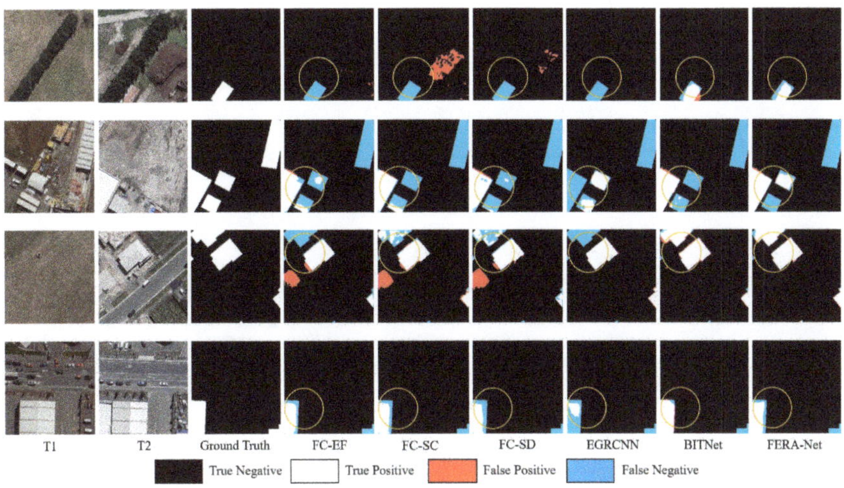

Figure 9. Visualization comparison of results in complex environments in the WHU-CD dataset.

5.2. Results for LEVIR-CD

We also evaluated the performance of FERA-Net on the LEVIR-CD dataset to validate the robustness of the model. In Table 3, the results of the quantitative evaluation of FERA-Net with other change detection networks are presented. From the quantitative evaluation results, we can see that FERA-Net achieved the highest scores on the LEVIR-CD dataset for both evaluation indicators, p value and F1 score, which is consistent with the results on the WHU-CD dataset. It can be seen from the results that FERA-Net achieves the highest p value and F1 score of 91.57% and 89.58%, respectively. Although our proposed model's IoU score for the LEVIR-CD dataset is not as high as on the WHU-CD dataset, it is still the second highest among all models in the experiment. Furthermore, our FERA-Net can improve the p value, F1 score, and IoU, demonstrating our suggested network's ability to balance the p and R values. FERA-Net can precisely identify the building change regions in the remote sensing images and also fully characterize all of the specific properties of the regions.

Table 3. Comparison of quantitative evaluation indicators in the LEVIR-CD dataset, where the best values are in bold.

Method	P (%)	R (%)	F1 (%)	IoU (%)
FC-EF [35]	83.97	84.01	83.99	72.40
FC-SC [35]	86.05	80.19	83.01	70.96
FC-SD [35]	88.57	83.60	86.01	75.46
STANet [39]	85.93	81.57	83.68	74.64
SNUNet [32]	90.32	86.22	88.23	78.94
BITNet [33]	91.41	85.53	88.37	79.17
EGRCNN [50]	88.11	90.39	89.24	**80.57**
CDNet [51]	90.50	84.60	87.50	-
HDFNet [53]	87.54	89.38	88.13	-
HRTNet [54]	85.43	**91.77**	88.48	-
FERA-Net	**91.57**	87.68	**89.58**	80.27

Figures 10 and 11 show the test results of the CD networks. FERA-Net can produce the most visually accurate change detection findings. The detection maps maintain higher internal compactness with more complete building contours. As can be seen from the visualization in Figure 10, FERA-Net is able to identify building edges more accurately for multi-scale buildings. We believe that the high-frequency feature module designed in the network is extremely useful for this result. At the same time, as seen from the results in Figure 11, FERA-Net can reduce interference in remote sensing images and more precisely pinpoint the locations of building changes.

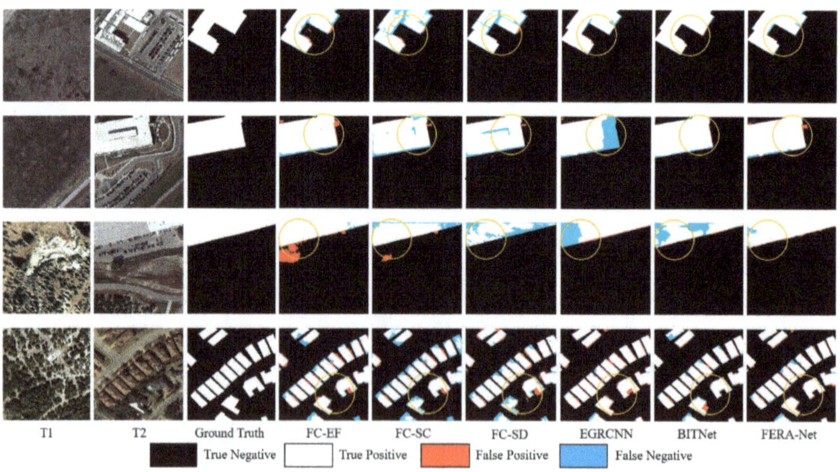

Figure 10. Visual comparison of building edge extraction results in the LEVIR-CD dataset.

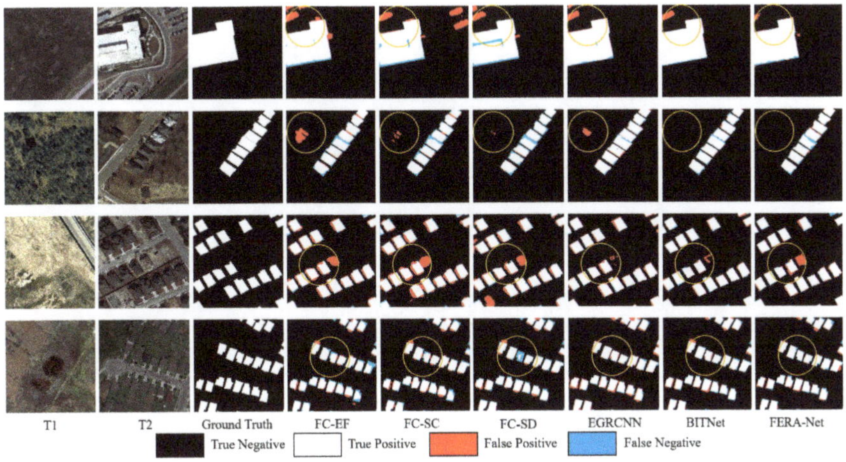

Figure 11. Visualization comparison of results in complex environments in the LEVIR-CD dataset.

5.3. Ablation Study

In order to verify the validity of the key modules, we conducted a series of ablation experiments on WHU-CD datasets in which several FERA-Net network modules were deleted. The effects of various encoders and connection methods on the outcomes of building change detection are specifically covered in this section. The encoder candidates cover CNN and AGFM encoders, and for comparison with other connection methods, full-scale skip connection module (FSCM) and FESCM are chosen. Additionally, our basic model combines the Siam network and UNet3+ models.

The quantitative results are listed in Table 4. Row (a,c) and row (b,d) verify the validity of the AGFM module. Row (a,b) and row (c,d) verify the validity of the FESCM module. The results reveal that utilizing the AGFM as the model's encoder can significantly increase the model's detection accuracy compared with using the CNN module. In comparison to the CNN module, utilizing the AGFM module increases the p value, F1 score, and IoU, which indicates that using the AGFM module is effective for building integrity detection. The experiment with FESCM shows that this module is more effective than the feature-enhanced skip connection module: p value, R value, F1 score, and IoU increase by 0.24%, 4.41%, 1.35%, and 2.28%, respectively. These improvements indicate that the FESCM can capture more information on building change areas and reduces the number of missed detections for changing buildings.

Table 4. Quantitative evaluation metrics for ablation experiments on the WHU-CD dataset, where the best values are in bold.

ID	Encoder Method	Connection Method	P (%)	R (%)	F1 (%)	IoU (%)
(a)	CNN	FSCM	89.12	88.22	88.67	79.64
(b)	CNN	FESCM	88.17	90.50	90.76	83.08
(c)	AGFM	FSCM	92.20	88.90	90.53	82.69
(d)	AGFM	FESCM	**92.44**	**91.31**	**91.88**	**84.97**

Figure 12 shows the ablation experiment result. Dual-time images and the labeled changes are denoted by (a–c). The change detection results from the combinations of the modules in Table 4 are denoted by (d–g). It can be observed that the network can capture the building change area more precisely when it utilizes the AGFM module. It can also minimize boundary noise and characterize the building boundary in more detail. The quantity of missed and false detections gradually decreased in the findings obtained using FESCM.

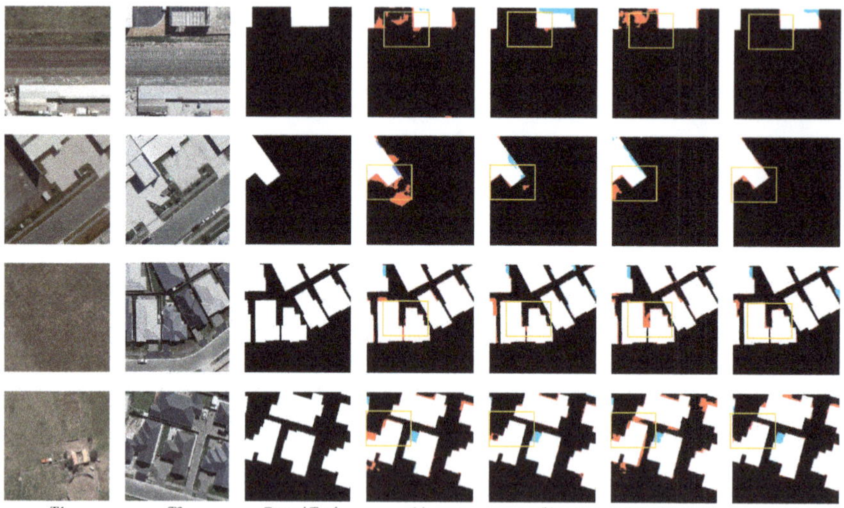

Figure 12. Visualization of ablation experiments on the WHU-CD dataset. (**a**) "CNN + FSCM". (**b**) "CNN + FESCM". (**c**) "AGFM + FSCM". (**d**) "AGFM + FESCM".

6. Discussion

6.1. Effectiveness of the AGFM

In order to analyze the effectiveness of the attention mechanism in the task of building change detection, we examined the performance of a conventional CNN encoder, CBAM encoder (a conventional spatial attention and channel attention module), and AGFM encoder. In this instance, we employed a Siamese network structure, focal loss, and UNet3+ as the baseline. Table 5 displays the quantitative evaluation results: by using the AGFM module, the p value is 3.08% higher than when using the CNN encoder and 6.89% higher than when using the CBAM encoder. This indicates that the AGFM encoder can improve the model in learning the overall building characteristics. Using the AGFM module, the R value is 0.78% higher than the CNN encoder but 2.94% lower than the CBAM encoder. We believe the reason for this is that the AGFM module is more concerned with representing detailed building features but is not the best for extracting changing building areas. Additionally, the AGFM encoder significantly improves the F1 score and IoU assessment metrics by 90.53% and 82.69%, respectively. Overall, the building change detection job benefits more from the AGFM encoder when compared with the CNN encoder and conventional attention encoder.

Table 5. Effectiveness of the AGFM, where the best values are in bold.

ID	Encoder Method	Connection Method	P (%)	R (%)	F1 (%)	IoU (%)
(a)	CNN	FSCM	89.12	88.22	88.67	79.64
(b)	CBAM	FSCM	85.31	**91.84**	88.46	79.30
(c)	AGFM	FSCM	**92.20**	88.90	**90.53**	**82.69**

Figure 13 displays the CD result on the WHU-CD dataset using the CNN, CBAM, and AGFM encoders. According to the figure, the AGFM encoder is superior to the CNN and CBAM encoders in its ability to characterize the main portion of the building and the boundary of the changing building. The other two encoders are susceptible to the building connection part and image classification accuracy in dense building areas, but the AGFM can accurately characterize the building's edge. When describing local information, the

AGFM encoder is not compared with the CBAM encoder, which can explain the R value portion of Table 5.

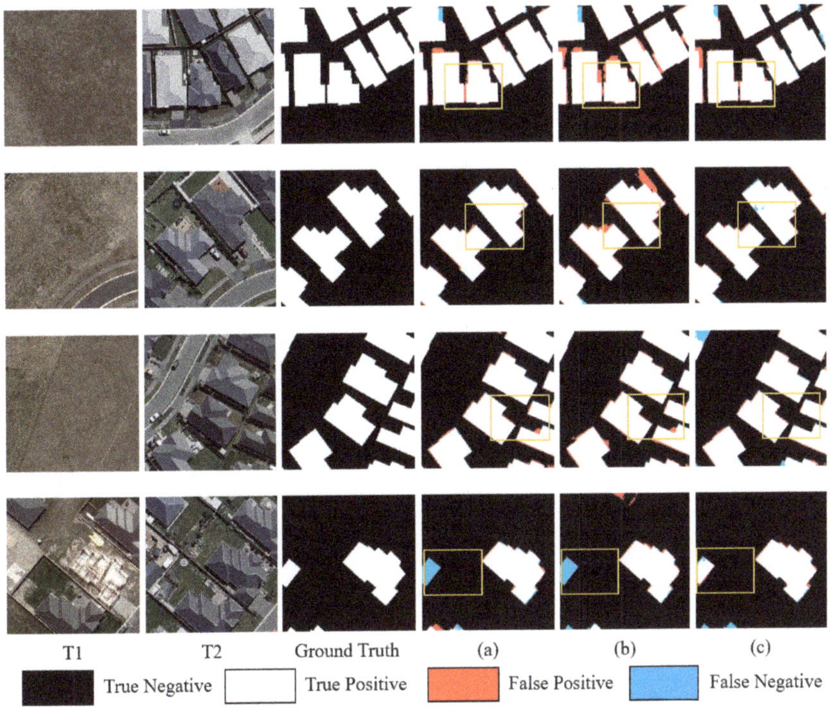

Figure 13. Effectiveness of the AGFM. (**a**) "CNN + FSCM". (**b**) "CBAM + FESCM". (**c**) "AGFM + FSCM".

Considering the good performance of attention mechanisms in the building change detection task in previous studies, we conducted validation experiments on the WHU-CD dataset for the residual spatial attention feature extraction stage (RSAS) and high-frequency residual channel attention enhancement stage (FRCS) in the AGFM module. We aimed to verify the performance of the high-frequency features of the images and attention mechanism for further confirming the mechanism of the AGFM module in FERA-Net in the building CD task.

According to the quantitative analysis in Table 6, adding the FRCS module to the RSAS module can significantly improve the p value of the change detection task, which suggests that we can increase the quality of the building edge information and improve the model detection accuracy by introducing high-frequency features into the transformation detection task.

Table 6. Effectiveness of RSAS and FRCS, where the best values are in bold.

ID	Encoder Method	Connection Method	P (%)	R (%)	F1 (%)	IoU (%)
(a)	RSAS	FESCM	89.53	**90.67**	90.10	81.98
(b)	FRCS	FESCM	91.75	88.64	90.17	82.10
(c)	RSAS + FRCS	FESCM	**92.20**	88.90	**90.53**	**82.69**

6.2. Effectiveness of the FESCM

Since the encoder downsampling process will lose spatial information in the encoder–decoder network structure, it is important to compensate for the lost spatial information in the decoder part by skipping the connection. However, the skip connection used in

UNet only collects data from the feature map of the same level. This skip connection between sibling feature maps allows the network to create a semantic gap between the decoder's higher-level and lower-level feature maps. The lower-level feature maps in UNet mainly acquire fine-grained detailed features (capturing rich spatial information), and the higher-level feature maps mainly acquire coarse-grained semantic features (capturing location information). In contrast, the FERA-Net utilized in this study employs FESCM, which maximizes the utilization of full-scale feature maps and enhances segmentation accuracy by fusing spatial and semantic information from feature maps at different scales.

We designed three additional experiments to verify the efficacy of FESCM described above for the building change detection task. Six comparative experiments were designed to verify the impact of skip connection (SC) in UNet and full-scale skip connection (FSSC) in UNet3+ and the FESCM. As shown in Table 7, the most notable improvements are 5.3% in the p value, 1.56% in the F1 score, and 2.50% in the IoU, but the R value decreased by 2.41%.

Table 7. Effectiveness of the FESCM, where the best values are in bold.

ID	Encoder Method	Connection Method	P (%)	R (%)	F1 (%)	IoU (%)
(a)	CNN	SC	83.82	90.63	87.09	77.14
(b)	CNN	FSSC	89.12	88.22	88.67	79.64
(c)	CNN	FESCM	88.17	90.50	90.76	83.08
(d)	AGFM	SC	88.83	**92.46**	90.61	82.84
(e)	AGFM	FSSC	92.20	88.90	90.53	82.69
(f)	AGFM	FESCM	**92.44**	91.31	**91.88**	**84.97**

Figure 14 shows the visualization results of the experiments: the UNet3+ network with full-scale skip links can describe the change building information more comprehensively than the UNet network, which proves the important role of full-scale skip links in the change detection task. The FESCM obtained the best experimental results, verifying that the FESCM outperforms the FSSC in the fusion of detailed information in low-level feature maps and semantic information in high-level feature maps. This explains the increase in p value and the decrease in R value in Table 6, as it also results in the decoder part of the model incorporating some interference information, which weakens the model's identification of areas of building change. However, in general, adding full-scale skip connections to the building change detection problem helps to improve the model's overall detection performance.

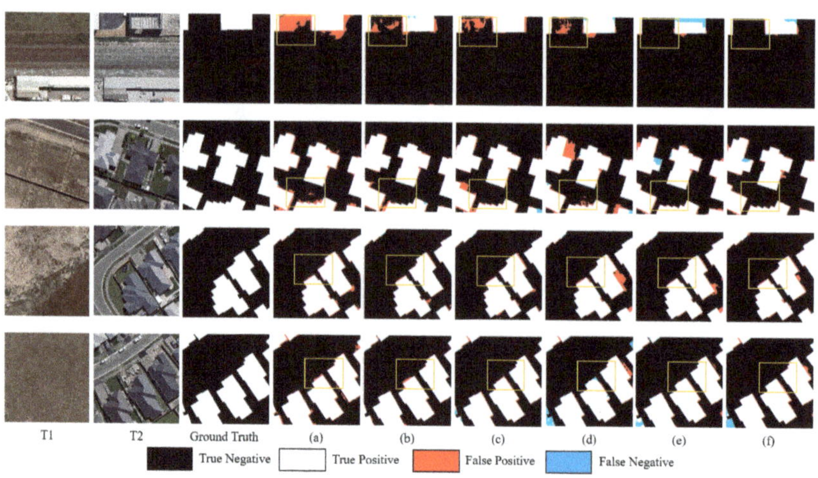

Figure 14. Effectiveness of the FESCM. (**a**) "CNN + SC". (**b**) "CNN + FSSC". (**c**) "CNN + FESCM". (**d**) "AGFM + SC". (**e**) "AGFM + FSSC". (**f**) "AGFM + FESCM".

In addition to the quantitative analysis of the evaluation indicators above, we also visualized the five layers of feature maps in the model, as shown in Figure 15. From the feature maps, we can see that, with the AGFM and FEM, the model focuses on the details of the remote sensing images in the first-, second-, and third-layer feature maps, especially the edges of the buildings, while the last two layers capture the spatial information and focus on the building change areas. The feasibility of the FESCM module was also verified from the visualization results: the different levels of feature maps contain different information, and the FESCM module provides the model's decoder with more comprehensive information about the buildings, which helps to better identify areas of change.

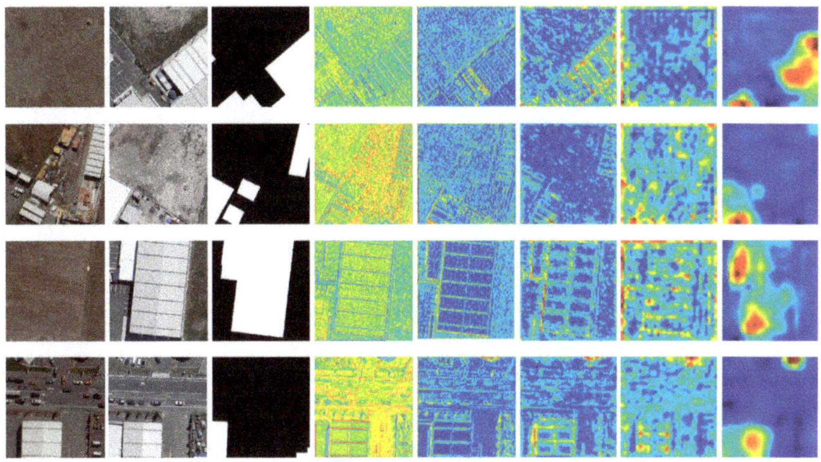

Figure 15. Visualization of the different levels of feature maps. (**a**) The first layer. (**b**) The second layer. (**c**) The third layer. (**d**) The fourth layer. (**e**) The fifth layer.

6.3. *Effectiveness of the Hybrid Loss Function*

At the final layer of the encoder, FERA-Net performs edge extraction to optimize edge supervision and model constraints for improved learning of boundary information. This part focuses on the edge loss's effect on the detection process results. The main body function and the edge loss function constitute the loss function in this study. L_{out} stands for the main body function and L_{edge} stands for the edge loss function. Table 8 shows the impact of our different weighting designs in the hybrid loss function on the performance of the model. As shown in the quantitative evaluation results in Table 8, the model's overall performance was improved by using the hybrid loss function. In particular, the best performance improvement was achieved when the weight of the edge loss function was set to 0.2, where the p value, R value, F1 score, and IoU improved by 1.07%, 0.16%, 0.60%, and 1.04%, respectively. This suggests that the edge loss function L_{edge} can improve the network's ability of extracting boundary information. With the function, the model can more precisely characterize the building border, which enhances the effectiveness of network detection.

Table 8. Effectiveness of hybrid loss function, where the best values are in bold.

ID	Body Loss	Edge Loss	P (%)	R (%)	F1 (%)	IoU (%)
(a)	L_{out}	None	92.44	91.31	91.88	84.97
(b)	$0.9 \times L_{out}$	$0.1 \times L_{edge}$	92.96	91.04	91.99	85.16
(c)	$0.8 \times L_{out}$	$0.2 \times L_{edge}$	**93.51**	**91.47**	**92.48**	**86.01**
(d)	$0.7 \times L_{out}$	$0.3 \times L_{edge}$	90.29	91.16	90.72	83.02

7. Conclusions

This paper proposes a new deep learning network applied to the change detection task named FERA-Net. FERA-Net is a typical encoder–decoder architecture. In the encoder section, we designed the AGFM to extract multi-level feature maps. In the decoder part, we fused the multi-scale feature maps with an FESCM. The AGFM uses a spatial attention mechanism to capture areas of building change and a channel attention mechanism to fuse high-frequency features extracted by the omnidirectional Sobel operator into the feature map. To maximize the use of full-size feature maps and to improve change detection accuracy, we use the FESCM to enhance the global and local information in the building change feature maps and link low-level detailed information with high-level semantic information. In addition, we designed hybrid loss functions to monitor building edges by means of changing the building boundaries extracted at the last layer of the decoder. Compared with other change detection methods, our method can utilize the local and global information of the feature maps and recognize multi-scale feature buildings in highly fractioned remote sensing images. We validated the performance of FERA-Net on the WHU-CD and LEVIR-CD datasets. From the quantitative and qualitative evaluation results, we can see that FERA-Net performs excellently on the two datasets. In the evaluation of quantitative metrics, we compared FERA-Net with classical change detection networks and state-of-the-art change detection networks, achieving the highest p value and F1 score. In the evaluation of qualitative visualization results, FREA-Net can effectively extract building change areas in complex ground environments, while accurately identifying building boundaries and maintaining boundary integrity. However, FERA-Net requires a large number of samples to learn in order to achieve good classification results. The performance of the current supervised building change detection network encounters a bottleneck, and the implementation of semi-supervised change detection based on the FERA-Net network will be the focus of future work.

Author Contributions: X.X. collected and processed the data, performed analysis, and wrote the paper; X.X. and Y.Z. analyzed the results; Y.Z. and X.L. helped to edit the article; X.L. and Z.C. contributed to the validation. All authors have read and agreed to the published version of the manuscript.

Funding: This study was supported by the National Natural Science Foundation of China (41871305) and the Opening Fund of Key Laboratory of Geological Survey and Evaluation of Ministry of Education (GLAB2022ZR06).

Data Availability Statement: Not applicable.

Conflicts of Interest: The authors declare no conflict of interest.

References

1. Zhu, Q.; Guo, X.; Deng, W.; Shi, S.; Guan, Q.; Zhong, Y.; Zhang, L.; Li, D. Land-Use/Land-Cover Change Detection Based on a Siamese Global Learning Framework for High Spatial Resolution Remote Sensing Imagery. *ISPRS J. Photogramm. Remote Sens.* **2022**, *184*, 63–78. [CrossRef]
2. Song, K.; Jiang, J. AGCDetNet:An Attention-Guided Network for Building Change Detection in High-Resolution Remote Sensing Images. *IEEE J. Sel. Top. Appl. Earth Obs. Remote Sens.* **2021**, *14*, 4816–4831. [CrossRef]
3. Jensen, J.R.; Im, J. Remote Sensing Change Detection in Urban Environments. In *Geo-Spatial Technologies in Urban Environments: Policy, Practice, and Pixels*; Jensen, R.R., Gatrell, J.D., McLean, D., Eds.; Springer: Berlin/Heidelberg, Germany, 2007; pp. 7–31. ISBN 978-3-540-69417-5.
4. Singh, A. Review Article Digital Change Detection Techniques Using Remotely-Sensed Data. *Int. J. Remote Sens.* **1989**, *10*, 989–1003. [CrossRef]
5. Bruzzone, L.; Bovolo, F. A Novel Framework for the Design of Change-Detection Systems for Very-High-Resolution Remote Sensing Images. *Proc. IEEE* **2013**, *101*, 609–630. [CrossRef]
6. Hussain, M.; Chen, D.; Cheng, A.; Wei, H.; Stanley, D. Change Detection from Remotely Sensed Images: From Pixel-Based to Object-Based Approaches. *ISPRS J. Photogramm. Remote Sens.* **2013**, *80*, 91–106. [CrossRef]
7. Jiang, H.; Peng, M.; Zhong, Y.; Xie, H.; Hao, Z.; Lin, J.; Ma, X.; Hu, X. A Survey on Deep Learning-Based Change Detection from High-Resolution Remote Sensing Images. *Remote Sens.* **2022**, *31*, 1552. [CrossRef]

8. Quarmby, N.A.; Cushnie, J.L. Monitoring Urban Land Cover Changes at the Urban Fringe from SPOT HRV Imagery in South-East England. *Int. J. Remote Sens.* **1989**, *10*, 953–963. [CrossRef]
9. Zanotta, D.C.; Haertel, V. Gradual Land Cover Change Detection Based on Multitemporal Fraction Images. *Pattern Recognit.* **2012**, *45*, 2927–2937. [CrossRef]
10. Deng, J.S.; Wang, K.; Deng, Y.H.; Qi, G.J. PCA-based Land-use Change Detection and Analysis Using Multitemporal and Multisensor Satellite Data. *Int. J. Remote Sens.* **2008**, *29*, 4823–4838. [CrossRef]
11. Ertürk, A.; Iordache, M.-D.; Plaza, A. Sparse Unmixing-Based Change Detection for Multitemporal Hyperspectral Images. *IEEE J. Sel. Top. Appl. Earth Obs. Remote Sens.* **2016**, *9*, 708–719. [CrossRef]
12. Bruzzone, L.; Cossu, R.; Vernazza, G. Detection of Land-Cover Transitions by Combining Multidate Classifiers. *Pattern Recognit. Lett.* **2004**, *25*, 1491–1500. [CrossRef]
13. Bruzzone, L.; Prieto, D.F. Automatic Analysis of the Difference Image for Unsupervised Change Detection. *IEEE Trans. Geosci. Remote Sens.* **2000**, *38*, 1171–1182. [CrossRef]
14. Habib, T.; Inglada, J.; Mercier, G.; Chanussot, J. Support Vector Reduction in SVM Algorithm for Abrupt Change Detection in Remote Sensing. *IEEE Geosci. Remote Sens. Lett.* **2009**, *6*, 606–610. [CrossRef]
15. Quispe, D.A.J.; Sulla-Torres, J. Automatic Building Change Detection on Aerial Images Using Convolutional Neural Networks and Handcrafted Features. *Int. J. Adv. Comput. Sci. Appl.* **2020**, *11*, 679–684. [CrossRef]
16. Roy, M.; Ghosh, S.; Ghosh, A. A Novel Approach for Change Detection of Remotely Sensed Images Using Semi-Supervised Multiple Classifier System. *Inf. Sci.* **2014**, *269*, 35–47. [CrossRef]
17. Bontemps, S.; Bogaert, P.; Titeux, N.; Defourny, P. An Object-Based Change Detection Method Accounting for Temporal Dependences in Time Series with Medium to Coarse Spatial Resolution. *Remote Sens. Environ.* **2008**, *112*, 3181–3191. [CrossRef]
18. Peng, D.; Zhang, Y.; Guan, H. End-to-End Change Detection for High Resolution Satellite Images Using Improved UNet++. *Remote Sens.* **2019**, *11*, 1382. [CrossRef]
19. Ma, J.; Lu, D.; Li, Y.; Shi, G. CLHF-Net: A Channel-Level Hierarchical Feature Fusion Network for Remote Sensing Image Change Detection. *Symmetry* **2022**, *14*, 1138. [CrossRef]
20. Addink, E.A.; Van Coillie, F.M.B.; De Jong, S.M. Introduction to the GEOBIA 2010 Special Issue: From Pixels to Geographic Objects in Remote Sensing Image Analysis. *Int. J. Appl. Earth Obs. Geoinf.* **2012**, *15*, 1–6. [CrossRef]
21. Huang, X.J.; Xie, Y.W.; Wei, J.J.; Fu, M.; Lv, L.L.; Zhang, L.L. Automatic Recognition of Desertification Information Based on the Pattern of Change Detection-CART Decision Tree. *J. Catastrophol.* **2017**, *32*, 36–42.
22. Javed, A.; Jung, S.; Lee, W.H.; Han, Y. Object-Based Building Change Detection by Fusing Pixel-Level Change Detection Results Generated from Morphological Building Index. *Remote Sens.* **2020**, *12*, 2952. [CrossRef]
23. Lv, P.; Zhong, Y.; Zhao, J.; Zhang, L. Unsupervised Change Detection Based on Hybrid Conditional Random Field Model for High Spatial Resolution Remote Sensing Imagery. *IEEE Trans. Geosci. Remote Sens.* **2018**, *56*, 4002–4015. [CrossRef]
24. Sun, X.; Wang, P.; Wang, C.; Liu, Y.; Fu, K. PBNet: Part-Based Convolutional Neural Network for Complex Composite Object Detection in Remote Sensing Imagery. *ISPRS J. Photogramm. Remote Sens.* **2021**, *173*, 50–65. [CrossRef]
25. Wang, Z.; Zhang, Y.; Luo, L.; Wang, N. TransCD: Scene Change Detection via Transformer-Based Architecture. *Opt. Express* **2021**, *29*, 41409–41427. [CrossRef]
26. Liu, G.; Li, L.; Jiao, L.; Dong, Y.; Li, X. Stacked Fisher Autoencoder for SAR Change Detection. *Pattern Recognit.* **2019**, *96*, 106971. [CrossRef]
27. Chen, H.; Wu, C.; Du, B.; Zhang, L.; Wang, L. Change Detection in Multisource VHR Images via Deep Siamese Convolutional Multiple-Layers Recurrent Neural Network. *IEEE Trans. Geosci. Remote Sens.* **2020**, *58*, 2848–2864. [CrossRef]
28. Liu, R.; Cheng, Z.; Zhang, L.; Li, J. Remote Sensing Image Change Detection Based on Information Transmission and Attention Mechanism. *IEEE Access* **2019**, *7*, 156349–156359. [CrossRef]
29. Lyu, H.; Lu, H.; Mou, L. Learning a Transferable Change Rule from a Recurrent Neural Network for Land Cover Change Detection. *Remote Sens.* **2016**, *8*, 506. [CrossRef]
30. Jiang, F.; Gong, M.; Zhan, T.; Fan, X. A Semisupervised GAN-Based Multiple Change Detection Framework in Multi-Spectral Images. *IEEE Geosci. Remote Sens. Lett.* **2020**, *17*, 1223–1227. [CrossRef]
31. Zhao, W.; Mou, L.; Chen, J.; Bo, Y.; Emery, W.J. Incorporating Metric Learning and Adversarial Network for Seasonal Invariant Change Detection. *IEEE Trans. Geosci. Remote Sens.* **2020**, *58*, 2720–2731. [CrossRef]
32. Fang, S.; Li, K.; Shao, J.; Li, Z. SNUNet-CD: A Densely Connected Siamese Network for Change Detection of VHR Images. *IEEE Geosci. Remote Sens. Lett.* **2022**, *19*, 1–5. [CrossRef]
33. Chen, H.; Qi, Z.; Shi, Z. Remote Sensing Image Change Detection with Transformers. *IEEE Trans. Geosci. Remote Sens.* **2022**, *60*, 1–14. [CrossRef]
34. Li, Z.; Yan, C.; Sun, Y.; Xin, Q. A Densely Attentive Refinement Network for Change Detection Based on Very-High-Resolution Bitemporal Remote Sensing Images. *IEEE Trans. Geosci. Remote Sens.* **2022**, *60*, 1–18. [CrossRef]
35. Caye Daudt, R.; Le Saux, B.; Boulch, A. Fully Convolutional Siamese Networks for Change Detection. In Proceedings of the 2018 25th IEEE International Conference on Image Processing (ICIP), Athens, Greece, 7–10 October 2018; pp. 4063–4067.
36. Zhang, C.; Yue, P.; Tapete, D.; Jiang, L.; Shangguan, B.; Huang, L.; Liu, G. A Deeply Supervised Image Fusion Network for Change Detection in High Resolution Bi-Temporal Remote Sensing Images. *ISPRS J. Photogramm. Remote Sens.* **2020**, *166*, 183–200. [CrossRef]

37. Tan, K.; Zhang, Y.; Wang, X.; Chen, Y. Object-Based Change Detection Using Multiple Classifiers and Multi-Scale Uncertainty Analysis. *Remote Sens.* **2019**, *11*, 359. [CrossRef]
38. Zhang, J.; Pan, B.; Zhang, Y.; Liu, Z.; Zheng, X. Building Change Detection in Remote Sensing Images Based on Dual Multi-Scale Attention. *Remote Sens.* **2022**, *14*, 5405. [CrossRef]
39. Chen, H.; Shi, Z. A Spatial-Temporal Attention-Based Method and a New Dataset for Remote Sensing Image Change Detection. *Remote Sens.* **2020**, *12*, 1662. [CrossRef]
40. Ji, S.; Shen, Y.; Lu, M.; Zhang, Y. Building Instance Change Detection from Large-Scale Aerial Images Using Convolutional Neural Networks and Simulated Samples. *Remote Sens.* **2019**, *11*, 1343. [CrossRef]
41. Peng, D.; Bruzzone, L.; Zhang, Y.; Guan, H.; Ding, H.; Huang, X. SemiCDNet: A Semisupervised Convolutional Neural Network for Change Detection in High Resolution Remote-Sensing Images. *IEEE Trans. Geosci. Remote Sens.* **2021**, *59*, 5891–5906. [CrossRef]
42. Zhang, L.; Hu, X.; Zhang, M.; Shu, Z.; Zhou, H. Object-Level Change Detection with a Dual Correlation Attention-Guided Detector. *ISPRS J. Photogramm. Remote Sens.* **2021**, *177*, 147–160. [CrossRef]
43. Chen, J.; Yuan, Z.; Peng, J.; Chen, L.; Huang, H.; Zhu, J.; Liu, Y.; Li, H. DASNet: Dual Attentive Fully Convolutional Siamese Networks for Change Detection of High Resolution Satellite Images. *IEEE J. Sel. Top. Appl. Earth Obs. Remote Sens.* **2021**, *14*, 1194–1206. [CrossRef]
44. Ma, J.; Shi, G.; Li, Y.; Zhao, Z. MAFF-Net: Multi-Attention Guided Feature Fusion Network for Change Detection in Remote Sensing Images. *Sensors* **2022**, *22*, 888. [CrossRef] [PubMed]
45. Chen, J.; Fan, J.; Zhang, M.; Zhou, Y.; Shen, C. MSF-Net: A Multiscale Supervised Fusion Network for Building Change Detection in High-Resolution Remote Sensing Images. *IEEE Access* **2022**, *10*, 30925–30938. [CrossRef]
46. Zheng, H.; Gong, M.; Liu, T.; Jiang, F.; Zhan, T.; Lu, D.; Zhang, M. HFA-Net: High Frequency Attention Siamese Network for Building Change Detection in VHR Remote Sensing Images. *Pattern Recognit.* **2022**, *129*, 108717. [CrossRef]
47. Wang, F.; Jiang, M.; Qian, C.; Yang, S.; Li, C.; Zhang, H.; Wang, X.; Tang, X. Residual Attention Network for Image Classification. In Proceedings of the 2017 IEEE Conference on Computer Vision and Pattern Recognition (CVPR), Honolulu, HI, USA, 21–26 July 2017; pp. 6450–6458.
48. Chen, Z.; Zhou, Y.; Wang, B.; Xu, X.; He, N.; Jin, S.; Jin, S. EGDE-Net: A Building Change Detection Method for High-Resolution Remote Sensing Imagery Based on Edge Guidance and Differential Enhancement. *ISPRS J. Photogramm. Remote Sens.* **2022**, *191*, 203–222. [CrossRef]
49. Ji, S.; Wei, S.; Lu, M. Fully Convolutional Networks for Multisource Building Extraction From an Open Aerial and Satellite Imagery Data Set. *IEEE Trans. Geosci. Remote Sens.* **2019**, *57*, 574–586. [CrossRef]
50. Bai, B.; Fu, W.; Lu, T.; Li, S. Edge-Guided Recurrent Convolutional Neural Network for Multitemporal Remote Sensing Image Building Change Detection. *IEEE Trans. Geosci. Remote Sens.* **2022**, *60*, 1–13. [CrossRef]
51. Alcantarilla, P.F.; Stent, S.; Ros, G.; Arroyo, R.; Gherardi, R. Street-View Change Detection with Deconvolutional Networks. *Auton Robot* **2018**, *42*, 1301–1322. [CrossRef]
52. Peng, X.; Zhong, R.; Li, Z.; Li, Q. Optical Remote Sensing Image Change Detection Based on Attention Mechanism and Image Difference. *IEEE Trans. Geosci. Remote Sens.* **2021**, *59*, 7296–7307. [CrossRef]
53. Zhang, Y.; Fu, L.; Li, Y.; Zhang, Y. HDFNet: Hierarchical Dynamic Fusion Network for Change Detection in Optical Aerial Images. *Remote Sens.* **2021**, *13*, 1440. [CrossRef]
54. Hou, X.; Bai, Y.; Li, Y.; Shang, C.; Shen, Q. High-Resolution Triplet Network with Dynamic Multiscale Feature for Change Detection on Satellite Images. *ISPRS J. Photogramm. Remote Sens.* **2021**, *177*, 103–115. [CrossRef]

remote sensing

MDPI

Article

Gradient Structure Information-Guided Attention Generative Adversarial Networks for Remote Sensing Image Generation

Baoyu Zhu [1,2,3,†], Qunbo Lv [1,2,3,†], Yuanbo Yang [1,2,3,†], Kai Zhang [1,3], Xuefu Sui [1,2,3], Yinhui Tang [1,2,3] and Zheng Tan [1,3,*]

1 Aerospace Information Research Institute, Chinese Academy of Sciences, No.9 Dengzhuang South Road, Haidian District, Beijing 100094, China
2 School of Optoelectronics, University of Chinese Academy of Sciences, No.19(A) Yuquan Road, Shijingshan District, Beijing 100049, China
3 Department of Key Laboratory of Computational Optical Imagine Technology, CAS, No.9 Dengzhuang South Road, Haidian District, Beijing 100094, China
* Correspondence: tanzheng@aircas.ac.cn; Tel.: +86-1591-063-9123
† These authors contributed equally to this work.

Abstract: A rich and effective dataset is an important foundation for the development of AI algorithms, and the quantity and quality of the dataset determine the upper limit level of the algorithms. For aerospace remote sensing datasets, due to the high cost of data collection and susceptibility to meteorological and airway conditions, the existing datasets have two problems: firstly, the number of datasets is obviously insufficient, and, secondly, there is large unevenness between different categories in datasets. One of the effective solutions is to use neural networks to generate fake data by learning from real data, but existing methods still find difficulty in generating remote sensing sample images with good texture detail and geometric distortion. To address the shortcomings of existing image generation algorithms, this paper proposes a gradient structure information-guided attention generative adversarial network (SGA-GAN) for remote sensing image generation, which contains two innovative initiatives: on the one hand, a learnable gradient structure information extraction branch network can be added to the generator network to obtain complex structural information in the sample image, thus alleviating the distortion of the sample geometric structure in remote sensing image generation; on the other hand, a multidimensional self-attention feature selection module is proposed to further improve the quality of the generated remote sensing images by connecting cross-attentive modules as well as spatial and channel attention modules in series to guide the generator to better utilize global information. The algorithm proposed in this paper outperformed other methods, such as StyleGAN-XL and FastGAN, in both the qualitative and quantitative evaluation, whereby the FID on the DOTA dataset decreased by 23.927 and the IS was improved by 2.351. The comparison experiments show that the method proposed in this paper can generate more realistic sample images, and images generated by this method can improve object detection metrics by increasing the number of single-category datasets and the number of targets in fewer categories in multi-category datasets, which means it can be effectively used in the field of intelligent processing of remote sensing images.

Keywords: remote sensing image generation; generative adversarial networks (GANs); structural information; attention mechanism; object detection; deep learning

Citation: Zhu, B.; Lv, Q.; Yang, Y.; Zhang, K.; Sui, X.; Tang, Y.; Tan, Z. Gradient Structure Information-Guided Attention Generative Adversarial Networks for Remote Sensing Image Generation. *Remote Sens.* **2023**, *15*, 2827. https://doi.org/10.3390/rs15112827

Academic Editors: Moulay A. Akhloufi and Mozhdeh Shahbazi

Received: 19 April 2023
Revised: 17 May 2023
Accepted: 26 May 2023
Published: 29 May 2023

1. Introduction

In recent years, aerospace remote sensing imaging technology has been applied to many fields, such as land and mineral resource management and monitoring, traffic and road network safety monitoring, geological disaster early warning, and national defense system construction [1–3]. Meanwhile, deep learning technology has also greatly promoted the research of remote sensing images in detection and classification [4]. A sufficient amount of data is the cornerstone for achieving high-performance deep learning algorithms, and large high-quality datasets can greatly improve the algorithm performance [5]. However, there are two constraints in the construction of existing remote sensing image datasets: on the one hand, compared with natural image datasets captured by ground-based equipment [6,7], the capture of remote sensing images requires high-cost imaging platforms such as aircraft or satellites, and the acquisition process is limited by aircraft routes and satellite orbits; on the other hand, the influence of factors such as light, rain, fog, and clouds [8] makes it difficult to collect effective high-quality images due to the high proportion of invalid data in each acquisition [9]. The above factors mean that existing datasets cannot meet the demand of artificial intelligence algorithm training in the field of remote sensing [10], which is mainly reflected in two aspects:

Issue 1: Insufficient data size. Take the commonly used DOTA dataset as an example. It includes 15 object classes with nearly 190,000 objects, while the non-remote sensing natural image dataset COCO contains 80 classes with more than 1.5 million objects in total, a size 8 times that of DOTA. The lack of data scale leads to a high risk of overfitting the model.

Issue 2: Large differences in the samples within and between classes. There are more than 20,000 ships in the DOTA dataset but only 6000 planes, and the specific classes and sizes of planes are not uniform. The lack of image diversity and unbalanced numbers between classes in the existing dataset [11] can bias the model toward majority class prediction [12], limiting further improvement in the performance of network models in target detection and classification, for example [13].

In recent years, some mitigation approaches have been proposed at the algorithm level for issue 2 [14,15]. Zhou et al. [16] proposed a dynamic balancing weighting method based on the number of effective samples for remote sensing image segmentation tasks with data imbalance. CBCL [11] dynamically constructs a class-balanced memory queue during the training of object detection models by memorizing training samples to alleviate class imbalance. However, these approaches are unable to mitigate issue 1.

Data enhancement is an effective approach that can alleviate both of these problems at the same time [17], can obtain large amounts of data from a limited dataset, and can effectively alleviate the problems of insufficient data size and class imbalance, especially in aerospace remote sensing applications, which are widely used by researchers [18,19]. However, many traditional data enhancement methods, such as flipping, scaling, cropping, rotating, or adding noise, only increase the number of remotely sensed images and cannot improve the quality of semantic information as well as the diversity of remotely sensed images in applications. In interpretation tasks such as object detection of remote sensing images, geometric transformation or randomly varying pixel values can no longer meet the increasing accuracy requirements. Therefore, it has become an urgent and indispensable task to use artificial neural network methods for data enhancement to generate sample data of remote sensing images.

Generative adversarial networks (GANs) [20] have led to technological breakthroughs in many areas of deep learning and have been rapidly applied in many directions in the field of aerospace remote sensing [21,22]. Remote sensing images usually contain feature information with a large amount of texture and structure information, which is complex [23], and existing natural image generation models rarely consider the structure information in the generation process. Therefore, if they are used in remote sensing image generation, they will lead to geometric structure distortion in the generated sample images, and the generated pseudo-sample images are often poorly realistic and insufficiently diverse to be reliably used as the basis for various analyses and applications in remote sensing [24]. Additionally, almost all existing related studies in the direction of remote sensing are focused on tasks such as image classification and segmentation, and there are few studies on object detection tasks.

To address the above problems, based on StyleGANv2 [25], this paper innovatively proposes a gradient structure information-guided attention generative adversarial network improvement method for remote sensing image generation to alleviate the model performance degradation problem due to insufficient remote sensing data. This paper uses a multidimensional self-attentive feature selection module (MAFM) to guide the generator to make better use of global information, which can help the generator to better control the generation process and generate higher-quality remote sensing images. Meanwhile, the gradient structure information branching network is used to guide the generator body network, so that the generated images have more realistic structure information, thus alleviating the structural distortion phenomenon existing in remote sensing image generation. The mode seeking [26] regularization term is introduced to increase the ratio of the distance between the generated images to the distance between the corresponding latent codes to solve the problem of insufficient diversity.

The main contributions of this paper are as follows:

1. Multidimensional self-attentive module

A multidimensional self-attentive module applicable to remote sensing image generation is proposed to enhance the convolution and improve the generator model performance. Contextual information is captured by tandemly connecting two cross-attentive modules and modeling the importance of feature maps and interdependencies in three dimensions of the spatial and channel domains. The attention model is embedded into the generative adversarial network to guide the generator to utilize global information while adaptively focusing on important regions.

2. Gradient structure information guidance model

Adding a branching network to the remote sensing image generation network and using the gradient structure information guidance method to improve the generation quality of remote sensing images can better preserve the structural information of the samples, allowing the generation network to output remote sensing images with high perceptual quality and less geometric distortion.

2. Related Work

2.1. Generative Adversarial Networks

Since they were proposed by Goodfellow et al., generative adversarial networks have become a popular research direction. A large number of variant structures based on generative adversarial networks have emerged and are widely used in various fields such as image generation, transformation, editing, and super-resolution [27–29]. Despite the great success of GANs in various applications, existing methods suffer from the mode collapse problem, which leads to a lack of diversity in the generated images. ModeGAN [30] alleviates the lack of diversity by introducing additional encoders to enhance the bidirectional mapping between the input noise vector and the generated images. Ghosh et al. [31] used multiple generators and forced different generators to generate samples with different patterns, thus increasing the diversity of the generated images. MSGAN [26] encourages generators to explore more patterns by maximizing the ratio of the distance between generated images to the distance between corresponding latent vectors, thus increasing the diversity of the generated images.

2.2. Image Generation

Image generation is an important research direction in the field of computer vision. The early image generation methods mainly focus on the explicit approximate estimation of the probability density function of the image distribution, but it is difficult to learn a model that can fit the data distribution due to the high-dimensional nature of the probability density distribution of the sample data. In contrast, GANs, as generative models based on implicit density estimation, can provide an effective solution to the problem of sampling and training under high-dimensional probability density distributions by means of adversarial learning, meaning that GANs have been widely used in image generation tasks.

With the continuous development and innovation of deep learning technology, GAN-based image generation methods are constantly being updated and improved. ProGAN [32] starts training from low-resolution image generation and gradually increases the image resolution, which both speeds up the training and stabilizes the training process while still being able to generate high-quality images. SAGAN [33] helps GANs to learn the global dependencies of images and generate higher-quality images by introducing an attention mechanism to GANs. BigGAN [34] substantially improves the quality of the generated images by adding orthogonal regularization to the generator, increasing the number of batch processes, and introducing spectral normalization based on SAGAN. StyleGAN [35] achieves controlled image generation by decoupling the hidden space vectors and separating the different style attributes that control the generated images. StyleGANv2 [25] further improves the quality of the generated images by disassembling AdaIN into Norm and Mod parts to eliminate artifacts.

SiftingGAN [36] introduced the GAN method to remote sensing image generation, proposing a GAN-based method to generate and filter labeled samples to increase the sample dataset. MARTA [37] applies DCGAN to remote sensing images and proposes an unsupervised model of multilayer feature matching to generate adversarial networks. EEGAN [38] proposes an edge-enhanced GAN algorithm that can combine the enhanced edges to generate images with clearer content. GAN-RSIGM [39] introduced the Wasserstein distance to the field of remote sensing image generation, proposing a remote sensing image generation method based on generative adversarial networks for creating high-resolution annotated samples for scene classification. Gu et al. [24] proposed an attention mechanism-based pseudo-annotated sample generation method and applied it to the scene classification of remote sensing images, which can be learned from a single natural image and can effectively generate enough pseudo-annotated samples from a single remote sensing scene image sample. MCGAN [40] uses multi-branch expansion convolution and classification branches to help the generator produce diverse and high-quality remote sensing images.

2.3. Feature Attention

Convolutional operations are processed under local neighborhoods with finite perceptual fields, which are prone to lose global information, and self-attention has become the latest advancement in capturing remote interactions [41]. Attention mechanisms enable deep learning to be more targeted in extracting features from samples [42], leading to an improvement in the accuracy of related tasks.

SENet [43] proposes a channel attention substructure, where the network model automatically obtains the importance of each feature channel by learning and models the interdependencies between feature channels. GSoP [44] proposes a simple and effective GSoP module that captures global second-order statistics. ECANet [45] proposes an ultra-lightweight attention module for local channel interactions. GCT [46] implements a channel-gated attention mechanism by gating weights and biases. FcaNet [47] proposes a multispectral channel attention approach to fully exploit image information by introducing more frequency components.

In the field of aerospace remote sensing, attention mechanisms also play an important role in many directions [48–50]. LMNet [51] proposes the Residual Transformer 3D-spatial Attention Module (RT3DsAM), which can learn feature representations from global information and filter important information.

3. Method

In this paper, we propose a gradient structure information-guided attention generative adversarial network (SGA-GAN). The structure sketch is shown in Figure 1, including the generator and discriminator. The generator network generates fake images from some random numbers extracted from a uniform distribution, and the discriminator network, as the adversary of the generator, tries to distinguish real images from fake images; both are trained iteratively. Finally, the generator network can perfectly generate realistic fake images, and the discriminator network can effectively determine whether the images are real or fake [20].

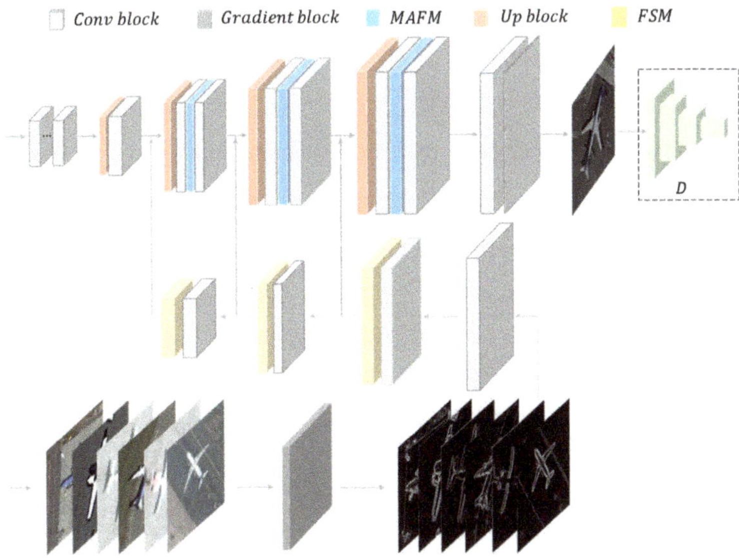

Figure 1. SGA-GAN network structure diagram. D is the discriminator.

In order to improve the performance of the model in remote sensing images, this paper introduces the multidimensional self-attentive feature selection module (MAFM in Figure 1) into the generator network. Additionally, this paper uses the gradient structure information to guide the model to generate more realistic sample images.

3.1. Multidimensional Self-Attentive Module

The generator is the core part of the whole generative adversarial network, and the performance of this part determines the final quality of the generated images. The convolutional layer can only operate on the local proximity context of each spatial location in the feature map, which does not make good use of the contextual information.

CCNet [52] proposes the Criss-Cross Network to obtain contextual information in a more effective and efficient way, based on which this paper proposes a multidimensional self-attentive feature selection module (MAFM) for remote sensing image generation, which can improve the generator model performance by modeling the attentional feature map and connecting the localized convolutional feature map to a farther range. The data flow is shown in Figure 2a,b, which are two parts in series.

As shown in Figure 2a, the cross-attention module captures contextual information by obtaining the relationship of each point on the feature map with each of its points horizontally and vertically. This vertical and horizontal attention module is more lightweight and uses two cross-attention modules in series for the self-attention calculation on each feature map in the channel domain. The P_i and P_i' weights are shared, and the relationship between each point and all points on that feature map can be obtained via cross-attention twice, which can capture dense global contextual information.

However, it is difficult to capture the relationship between feature maps; therefore, in order to learn the long-distance dependencies within the channel dimension of the feature maps, F'' is passed sequentially through the spatial attention module and the channel attention module. The feature attention selection module shown in Figure 2b is added after Figure 2a to model the importance and interdependence of the feature maps in three dimensions: two in the spatial domain and one in the channel domain. F_{ws} and F_{hs} are the spatial attention maps and F_{cs} is the channel attention map, which are obtained in the same way, the only difference being that the dimensions are different.

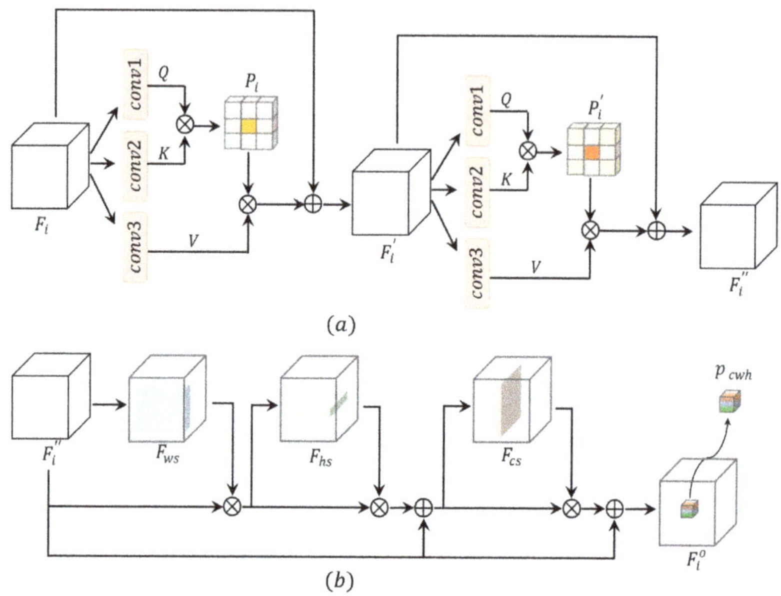

Figure 2. MAFM data flow diagram, (**a**) indicates the cross-attention module, (**b**) indicates the feature attention selection module.

As shown in Figure 3, taking channel attention as an example, F is first reduced to a one-dimensional vector along the channel dimension, using the maximum pooling operation. Then, the importance of each channel feature map is obtained by a 1 × 1 convolutional layer, ReLU activation function, 1 × 1 convolutional layer, and sigmoid activation function in turn. Subsequent to multiplying the features after the spatial attention module, they are summed with F'' after the 1 × 1 convolutional layer and ReLU activation function in turn to obtain the final output feature F_o. F_o contains not only the long-range dependencies of all spatial locations within the whole spatial dimension but also the correlations between channel dimensions and the importance of each feature map under that dimension, which enables the generator to better use the global information to generate high-quality remote sensing images.

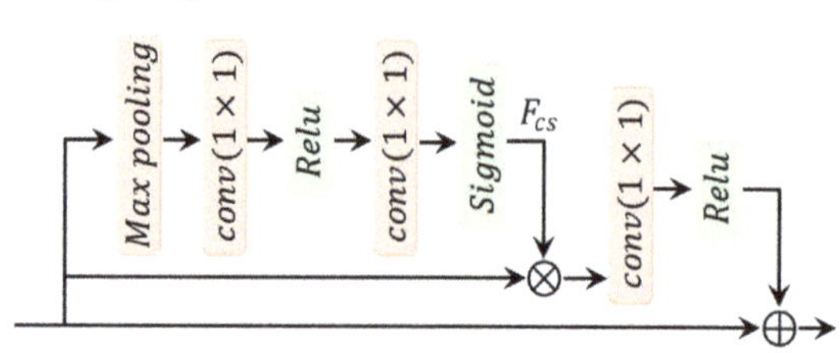

Figure 3. Diagram of feature importance selection process.

3.2. Gradient Structure Information Guidance Model

Since the gradient map reveals clear details of each local region in the sample image [53], the gradient structure information can be used to guide the generation process of the image generation network. However, the sample gradient map extracted by the conventional operator can only reflect one aspect of the sample structure (i.e., the value difference of neighboring pixels), and it is difficult to obtain richer and more general information (e.g., more complex structural textures) in the sample structure.

In this paper, we propose a learnable gradient structure information branching network, GSNet, to generate high-quality remote sensing images with clear structures by extracting the gradient structure information of samples in real remote sensing images and guiding the generator to pay attention to the sample structure information during the generation process.

A structural diagram of the gradient structure information branching network used in this paper is shown in Figure 4. Since most of the area of the gradient map is close to zero and the convolutional neural network can focus more on the spatial relationships of the contours, the network model may capture structural dependencies more easily. To ensure that GSNet can learn the geometric structural information of the samples rather than the deep semantic information of the samples, GSNet does not use too many convolutional layers, and the conv block is a residual block consisting of three layers of convolution. GSNet is divided into three stages to extract the sample image structure tensor, and the spatial size of the structure information feature map is downsampled by a factor of 2 each time and incorporated into the corresponding position of the main structure of the generative network.

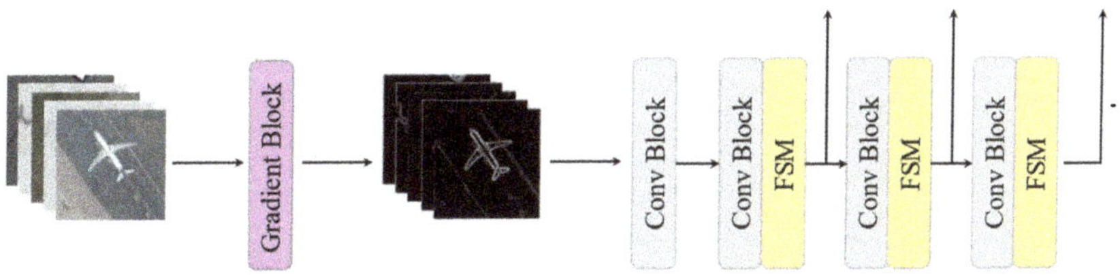

Figure 4. Structural diagram of the gradient structure information branching network.

The gradient block is used to obtain the gradient map of the original image of the sample without considering the gradient direction information but using a convolutional layer with a fixed kernel, as shown in Equation (1). The gradient calculation formula used is shown in Equation (2).

$$K_y = \begin{bmatrix} 0 & -1 & 0 \\ 0 & 0 & 0 \\ 0 & 1 & 0 \end{bmatrix}, \quad K_x = \begin{bmatrix} 0 & 0 & 0 \\ -1 & 0 & 1 \\ 0 & 0 & 0 \end{bmatrix}, \tag{1}$$

$$G_{I'} = \|((f(x+1,y) - f(x-1,y)),\ (f(x,y+1) - f(x,y-1)))\|_2, \tag{2}$$

After obtaining the gradient map, it is first fed into the first convolutional block to extract simple structural features, which consists of a 3 × 3 convolutional layer with a step size of 1 and an instance normalization layer. It is then fed into three modules consisting of a convolution block and a multidimensional feature selection module (FSM) in turn to further extract structural features. The convolutional block consists of three dense residual blocks, each consisting of five 3 × 3 convolutional layers using dense connections and the Leaky-ReLU activation function. The specific structure of the multidimensional feature selection module (FSM) is shown in Figure 5, which uses transposition operations to select features from three dimensions, namely the spatial dimension W, spatial dimension H, and channel dimension C, respectively. Each feature selection module consists of a maximum pooling layer, a 1 × 1 convolutional layer, a ReLU activation function, a 1 × 1 convolutional layer, and a sigmoid activation function.

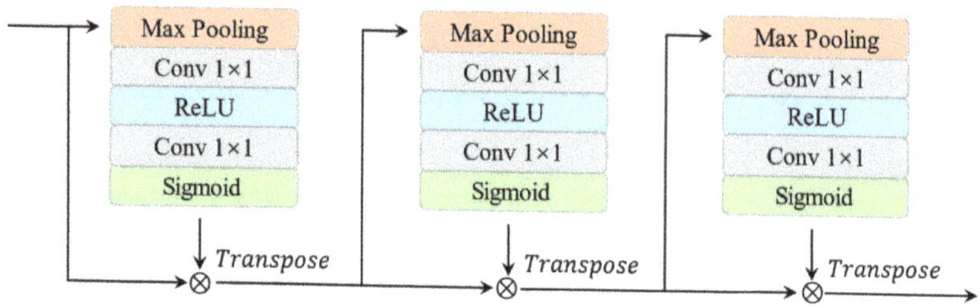

Figure 5. Multidimensional feature selection module.

The structural features at different scales are obtained in the branching network of the gradient structural information, and they are input into the generator as a priori structural information to guide the generation process. Using a learnable attention graph, we assign each part of the input different weights to extract more critical and important information so that the branching network can output structural information with more details. The branching network can effectively extract the structural information of the preserved sample images, allowing the generative network to obtain results with high perceptual quality and less geometric distortion. In addition, the obtained gradient information can highlight the sharpness and structural regions that need more attention, thus implicitly leading to the high-quality generation of remote sensing images.

3.3. *Loss Function*

In order to train the algorithm in this paper, it is necessary to compare the image in the training phase, the reconstructed image, and the original image under certain metrics. We designed a loss function L consisting of three components. It is denoted by the following equation:

$$\mathcal{L}_{total} = \lambda \mathcal{L}_{GAN} + \mu \mathcal{L}_{pl} + \gamma \mathcal{L}_{ms}, \tag{3}$$

where $\mathcal{L}_{GAN}$ is the adversarial loss, $\mathcal{L}_{pl}$ is the perceptual length regularization term, $\mathcal{L}_{ms}$ is the mode seeking regularization term, and λ, μ, and γ are the weight coefficients.

The work in this paper is mainly based on StyleGANv2 for improvement, and thus the loss function in StyleGANv2 is retained, while the mode seeking regularization term is added to improve the diversity of the generated images.

3.3.1. Adversarial Loss

This paper uses the adversarial loss to make the image distribution of the generated image match the image distribution of the real image with the following equation:

$$\min_G \max_D \mathcal{L}(G,\ D) = \mathbb{E}_{x \sim p_{data}(x)}[\log D(x)] + \mathbb{E}_{z \sim p_z(z)}[\log(1 - D(G(z)))], \tag{4}$$

where G tries to generate images that look similar to the real image, and D tries to distinguish the generated images from the real image. The goal of G is to minimize the adversarial loss, while the goal of D is to maximize the adversarial loss.

3.3.2. Path Length Regularization

The perceptual path length (PPL), proposed in StyleGANv2 [25], which measures the degree of feature coupling between different random input vectors, representing the perceptual distance length, is also found to be correlated with the quality of the generated images, and thus the quality of the generated images can be improved using the PPL regularization term.

$$\mathcal{L}_{pl} = \mathbb{E}_{w,\ y \in N(0,1)}(\|J_w^T y\|_2 - \alpha)^2, \tag{5}$$

where $w \in W$, W is the intermediate potential space [35], $w \sim f(z)$, z obeys a normal distribution, J_w denotes the Jacobian matrix, α denotes the exponential moving average, which aims to preserve the expected length of the vector regardless of the vector direction, and y is a random image whose pixel (numerical) intensities obey a normal distribution.

3.3.3. Mode Seeking Regularization

GANs can easily fall into mode collapse, which leads to a loss of diversity in the generated images. Therefore, additional regularization terms need to be introduced to alleviate mode collapse and thus ensure the diversity of the generated images.

When mode collapse occurs, two different random input vectors, z_1 and z_2, are mapped to images I_1 and I_2 with similar patterns, which means that the GAN learns only a few patterns and loses the diversity of the generated images. Therefore, we introduce the mode seeking regularization term to maximize the ratio of the distance between I_1 and I_2 relative to the distance between z_1 and z_2 with the following equation:

$$\mathcal{L}_{ms} = \max_{G}\left(\frac{d_I(G(z_1),\ G(z_2))}{d_z(z_1, z_2)}\right), \tag{6}$$

where z_1 and z_2 are two different random vectors, $G(z_1)$ and $G(z_2)$ denote the corresponding generated images of z_1 and z_2, respectively, and $d_I(\cdot)$ and $d_z(\cdot)$ are distance metrics.

4. Experiments

In order to verify the effectiveness and generalizability of the model proposed in this paper, the model proposed in this paper and other advanced models were tested on two remote sensing datasets and compared. To quantitatively evaluate the performance of the different network models, the IS (Inception Score), FID (Frechet Inception Distance), and KID (Kernel Inception Distance) were used to evaluate the quality of the images generated by the different models. These are widely used metrics for image generation quality evaluation.

In addition, in this paper, the generated remote sensing sample images were tested in practical applications using current commonly used object detection models, including single-object class experiments as well as multi-object class experiments. The object detection models included YOLOv5 [54], YOLOX [55], and Efficientdet [56]. The results of object detection using the DOTA and UCAS-AOD datasets were used to verify the effectiveness of the generated images.

In this paper, we conducted an ablation study, which is presented in Section 4.5, to verify the effectiveness of the gradient structure information guidance model proposed in this paper and explore the effect of adding a multidimensional self-attentive module on the algorithm.

4.1. Evaluation Metrics

The larger the value of the Inception Score (IS) [57], the better the image quality and diversity. Specifically, the generated image x is fed into the Inception [58] classification network, which outputs a 1000-dimensional vector y. It is desired that the entropy $p(y|x)$ of the generated image over the conditional distribution is small, and that the edge distribution $p(y)$ of the generated image over all class probabilities is large. IS is the average KL divergence of these two distributions, and is expressed as follows:

$$IS(x) = exp(\mathbb{E}_x[D_{KL}(p(y|x) \parallel p(y))]) = exp(H(y) - \mathbb{E}_x[H(y|x)]), \tag{7}$$

The FID [59] determines the quality of the generated image by calculating the distance between the feature vectors of the real image and the generated image, which is extracted and calculated using an image classification model (e.g., Inception v3). A lower FID score means that the two sets of images, or the statistics of the two, are more similar. The calculation formula is as follows:

$$FID(r,g) = \|u_r - u_g\|_2^2 + Tr\left(\Sigma_r + \Sigma_g - 2\left(\Sigma_r\Sigma_g\right)^{1/2}\right), \tag{8}$$

where u_r and u_g represent the mean values of the real image and the generated image on a certain layer of features of Inception v3, respectively; Σ_r and Σ_g represent the variance of the real image and the generated image on a certain layer of features, respectively.

The KID [60] is used to evaluate the realism of the generated image, and the lower the KID, the higher the visual similarity between the real image and the generated image. It calculates the square of the maximum mean discrepancy (MMD) of the higher perceptual features of the image (the last layer of the Inception model). The two data distributions of the real and generated images are P_r and P_g, respectively. The KID between the two distributions can be expressed as follows:

$$KID(P_r, P_g) = \mathbb{E}_{x,x' \sim P_r}[\mathcal{K}(x, x')] - 2\mathbb{E}_{x \sim P_r, y \sim P_g}[\mathcal{K}(x, y)] + \mathbb{E}_{y,y' \sim P_g}[\mathcal{K}(y, y')], \tag{9}$$

where $x \sim P_r$ and $y \sim P_g$ denote the high-level perceptual features of the real image and the generated image, respectively, and $\mathcal{K}(\cdot, \cdot)$ denotes the kernel function used for feature transformation:

$$\mathcal{K}(x, y) = \left(\frac{1}{d}x^T y + 1\right)^3, \tag{10}$$

4.2. Datasets

1. DOTA dataset [61,62]: This is a large image dataset for aerial image object detection. The image sources include different sensors and platforms, including Google Earth, the JL-1 satellite, and the GF-2 satellite of the China Resource Satellite Data and Application Center. The dataset consists of a total of 2806 aerial images, each with pixel sizes ranging from 800 × 800 to 4000 × 4000, containing objects of different scales, orientations, and shapes, and the images are annotated by experts using 15 common target categories. In this paper, the images were cropped to a 640 × 640 pixel size for testing.

2. UCAS-AOD dataset [63]: Remote sensing images were collected from Google Earth and used for plane detection. The plane dataset consists of 600 images of 3210 planes, and all the images are carefully selected so that the target directions in the dataset are evenly distributed. In this paper, the images were cropped to a 640 × 640 pixel size for testing.

4.3. Image Generation Comparison Results

In order to evaluate the performance of the algorithm in this paper, the method in this paper was compared with other SOTA image generation methods on two datasets. All comparison procedures and test results of the compared models were obtained from their authors' official websites.

DOTA dataset: The results of the algorithm proposed in this paper and the other algorithms were quantitatively analyzed on the DOTA dataset. The FID, KID, and IS metrics of different image generation algorithms are shown in Table 1. Compared with the other five image generation models, the SGA-GAN algorithm proposed in this paper returned the optimal results for the three metrics, i.e., FID, KID, and IS. Compared with the baseline model StyleGANv2, the SGA-GAN algorithm proposed in this paper reduced the FID value from 85.967 to 72.924, the KID value from 0.059 to 0.047, and the IS value from 4.581 to 4.815. Compared with StyleGAN-XL, the FID and KID of SGA-GAN decreased by 23.927 and 0.003, respectively, and the IS improved by 2.351 on the DOTA dataset. The quantitative comparison in terms of the FID, KID, and IS indicates that the SGA-GAN algorithm proposed in this paper can generate higher-quality remote sensing images.

Table 1. FID, KID, and IS metrics for different methods on the DOTA dataset.

Model	FID	KID	IS
ProGAN [32]	174.600	0.141	4.419
StyleGANv1 [35]	148.882	0.112	4.712
StyleGANv2 [25]	85.967	0.059	4.581
FastGAN [64]	77.918	0.049	4.798
StyleGAN-XL [65]	96.851	0.050	2.464
SGA-GAN (Ours)	72.924	0.047	4.815

The qualitative evaluation of the results of the different algorithms on the DOTA dataset is shown in Figure 6. The remote sensing sample images generated by ProGAN and StyleGANv1 were more seriously distorted, and the partial structures of the wings of some planes in the images were severely distorted. The StyleGANv2, FastGAN, and StyleGAN-XL methods generated a few clearer sample images, but there was also distortion in the partial structures of the wings and tails. In contrast, the method proposed in this paper generated sample images with realistic visual effects, and the image structure was more realistic and reliable.

Figure 6. Comparison chart of sample images generated by different algorithms on the DOTA dataset.

UCAS-AOD dataset: The results of the algorithm proposed in this paper and the other algorithms were quantitatively analyzed on the UCAS-AOD dataset. The FID, KID, and IS metrics of the different image generation algorithms are shown in Table 2. Compared with the other five image generation models, SGA-GAN again returned the optimal results for the three metrics, i.e., FID, KID, and IS. Compared with StyleGANv2, the SGA-GAN algorithm proposed in this paper decreased the FID value from 58.405 to 54.096, the KID value from 0.047 to 0.042, and the IS value from 3.493 to 3.728. Compared to StyleGAN-XL, the FID and KID of SGA-GAN on the UCAS-AOD dataset decreased by 41.464 and 0.009, respectively, and the IS improved by 1.646. The quantitative comparison of the three metrics, i.e., FID, KID, and IS, shows that the SGA-GAN algorithm proposed in this paper can generate higher-quality remote sensing images.

Table 2. FID, KID, and IS metrics for different methods on the UCAS-AOD dataset.

Model	FID	KID	IS
ProGAN	158.0978	0.154	3.285
StyleGANv1	117.276	0.128	3.537
StyleGANv2	58.405	0.047	3.493
FastGAN	63.469	0.049	3.342
StyleGAN-XL	95.560	0.051	2.082
SGA-GAN (Ours)	54.096	0.042	3.728

The qualitative evaluation of the results of the different algorithms on the UCAS-AOD dataset is shown in Figure 7. The target types in the UCAS-AOD dataset are more concentrated with less inter-class variation. ProGAN, StyleGANv1, and StyleGAN-XL generated a small number of sample target images with a clear structure, but most of the images were severely distorted. The StyleGANv2 and FastGAN methods generated clearer sample images, but there was structural distortion. The method proposed in this paper generated more realistic sample images with clearer and more reliable structures.

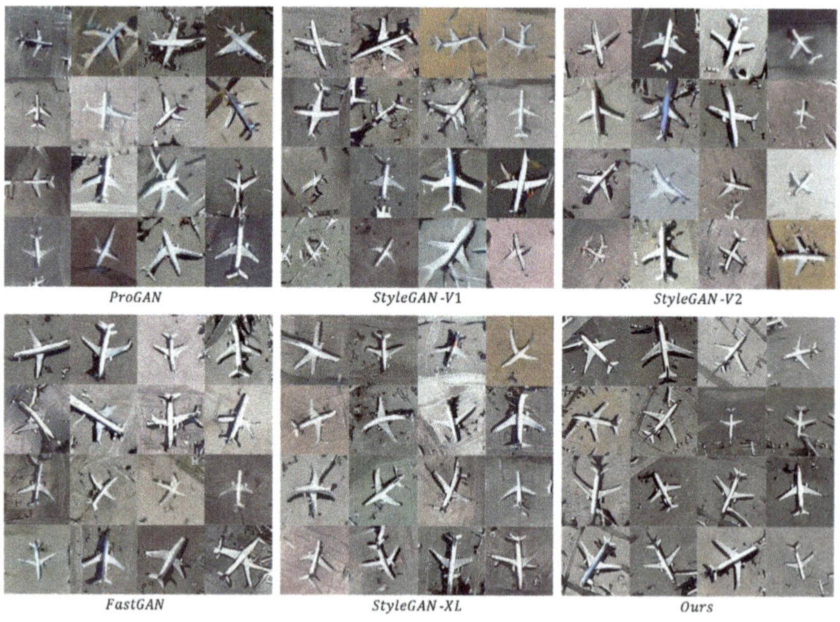

Figure 7. Comparison chart of sample images generated by different algorithms on the UCAS-AOD dataset.

Comparing the results of all algorithms on the DOTA and UCAS-AOD datasets, it can be seen that the algorithms in this paper have good performance metrics and visual effects on both datasets.

4.4. Object Detection Test Results

To evaluate the performance of the algorithm in this paper, target samples generated using the method in this paper were added to the training set of two publicly available aerial remote sensing object detection datasets, while the test set was left unchanged. Comparative experiments were conducted on three different types of generic object detection models, namely YOLOv5 [51], YOLOX [52], and Efficientdet [53], to verify the ability of this paper's method to solve the problems of insufficient image diversity and unbalanced numbers between classes in existing datasets.

4.4.1. Multi-Class Object Data

Adding single-class targets to multi-class target data: The DOTA airborne remote sensing dataset contains 15 common object classes. The number of plane targets is significantly lower than that of ships and vehicles, and the plane targets are of different sizes. We added 1000 samples of plane targets of different sizes generated by the method in this paper to the training set of the original dataset, as shown in Figure 8; the left subfigure (a) shows the image slices of the original dataset, and the right subfigure (b) shows the added generated data to compare the experimental results of object detection.

(a) (b)

Figure 8. Schematic of adding images to the DOTA dataset, (**a**) is the original image, (**b**) is the generated image.

To verify the effectiveness of SGA-GAN in improving the performance of remote sensing object detection, the YOLOv5 and YOLOX detection networks were used to verify the effectiveness on the DOTA dataset. The test results are shown in Table 3.

Table 3. Comparison of object detection results (Single-Class) before and after data augmentation on the DOTA dataset.

Model	Train	All			Plane		
		P (%)	R (%)	mAP (%)	P (%)	R (%)	AP (%)
YOLOv5	DOTA	75.4	67.4	70.6	87.5	84.1	88.5
	DOTA + ORI_{1000}	75.6	67.8	70.9	86.9	84.3	88.6
	DOTA + GEN_{1000}	76.6	67.5	71.6	90.6	85.8	90.2
YOLOX	DOTA	81.3	80.3	84.7	90.8	86.9	91.0
	DOTA + ORI_{1000}	81.5	80.9	85.0	89.3	87.7	91.2
	DOTA + GEN_{1000}	82.1	81.1	85.6	90.1	89.3	92.3

ORI_{1000} represents 1000 plane target images in the original DOTA data with conventional data enhancement (random cropping, scaling, rotation, etc.). GEN_{1000} represents 1000 plane target samples generated using this paper's method, SGA-GAN, stitched into a 640×640-size image and added to the training set. From Table 3, it can be seen that, for YOLOv5, the mAP and AP of plane objects were improved by 1.0% and 1.7%, respectively, after adding the plane target images generated by SGA-GAN. For YOLOX, the mAP and AP of plane objects were improved by 0.9% and 1.3%, respectively, after adding the plane target sample images generated by SGA-GAN. It can also be seen from the table that, compared with the traditional data enhancement methods, the SGA-GAN-generated images had a greater enhancement effect on remote sensing object detection, which can effectively alleviate the inter-class and intra-class imbalance problems existing in the dataset.

Adding multi-class targets to the multi-class target data: A total of 1000 images each of swimming pools, storage tanks, and planes were added to the training dataset of the original dataset, as shown in Figure 9; the left subfigure (a) shows the image slices of the original dataset, and the right subfigure (b) shows the added generated data. The experimental results of multi-class object detection were compared and observed using the YOLOv5 and YOLOX detection networks.

(a) (b)

Figure 9. Schematic of adding multi-class images to the DOTA dataset, (**a**) is the original image, (**b**) is the generated image.

The test results are shown in Table 4, where $ORIm_{1000}$ represents the multi-class target images in the original DOTA data with conventional data enhancement, and $GENm_{1000}$ represents the multi-class target samples generated using the SGA-GAN method proposed in this paper. As can be seen from the table, for YOLOv5, the mAP and AP of the three classes of objects were improved by 2.1%, 1.9%, 0.9%, and 2.4% after adding the target images generated by SGA-GAN. For YOLOX, the mAP and AP of the three classes of objects were improved by 1.7%, 0.9%, 1.7%, and 3.1% after adding the target images generated by SGA-GAN.

Table 4. Comparison of object detection results (multi-class) before and after data augmentation on the DOTA dataset.

Model	Train	All	Plane	Swimming Pool	Storage Tank
		mAP (%)		AP (%)	
YOLOv5	DOTA	70.6	88.5	54.5	77.8
	DOTA + $ORIm_{1000}$	70.9	88.6	54.6	78.2
	DOTA + $GENm_{1000}$	72.7	90.4	55.4	80.2
YOLOX	DOTA	84.7	91.0	61.2	83.6
	DOTA + $ORIm_{1000}$	85.0	91.1	61.8	83.7
	DOTA + $GENm_{1000}$	86.4	92.0	62.9	86.7

4.4.2. Single-Class Data

Using the UCAS-AOD dataset, which contains only a single plane target, 1000 generated plane targets of different sizes were added to the original dataset, as shown in Figure 10, with the original dataset on the left (a) and the added generated data on the right (b), to compare the experimental results of the observed object detection.

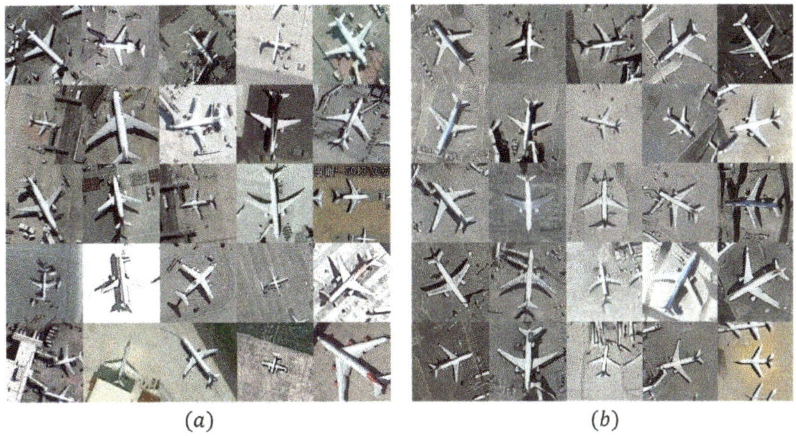

Figure 10. Schematic of the dataset, (a,b) are the original and generated images, respectively.

The test results are shown in Table 5. ORI_{1000} represents 1000 plane target images in the original UCAS-AOD data, and GEN_{1000} represents 1000 plane target samples generated by the method proposed in this paper. The AP of plane objects was significantly improved by 1.6% for YOLOv5, 0.3% for YOLOX, and 1.8% for Efficientdet, after adding the plane objects generated by the method proposed in this paper, which proves that this paper's method can effectively alleviate the problem of the model performance being limited by the size of the dataset.

Table 5. Comparison of object detection results before and after data augmentation on the UCAS-AOD dataset.

Model	Train	Plane		
		P (%)	R (%)	AP (%)
YOLOv5	UCAS-AOD	96.8	92.5	95.5
	UCAS-AOD + ORI_{1000}	96.6	93.1	95.8
	UCAS-AOD + GEN_{1000}	96.8	94.4	97.1
YOLOX	UCAS-AOD	95.1	96.3	96.8
	UCAS-AOD + ORI_{1000}	95.2	96.5	96.7
	UCAS-AOD + GEN_{1000}	95.4	97.3	97.1
EfficientDet	UCAS-AOD	98.8	83.6	91.8
	UCAS-AOD + ORI_{1000}	98.7	85.2	92.6
	UCAS-AOD + GEN_{1000}	98.1	88.3	93.6

4.5. Ablation Experiments

4.5.1. The Effectiveness of Mode Seeking Regularization ($\mathcal{L}_{ms}$)

During the model training, mode seeking regularization ($\mathcal{L}_{ms}$) was removed, and the network was retrained on the DOTA dataset using the same training scheme to verify its effect on the model.

As can be seen from Table 6, the addition of $\mathcal{L}_{ms}$ reduced the FID from 74.421 to 72.924, the KID from 0.049 to 0.047, and the IS from 4.710 to 4.815. Thus, the effectiveness of mode seeking regularization in the image generation process is verified, showing that it can increase the diversity of the generated images.

Table 6. FID, KID, and IS metrics on the DOTA dataset. ($\mathcal{L}_{ms}$ ablation experiments).

Model	FID	KID	IS
SGA-GAN without $\mathcal{L}_{ms}$	74.421	0.049	4.710
SGA-GAN	72.924	0.047	4.815

4.5.2. The Effectiveness of Path Length Regularization ($\mathcal{L}_{pl}$)

During the model training, path length regularization ($\mathcal{L}_{pl}$) was removed, and the network was retrained on the DOTA dataset using the same training scheme to verify its effect on the model.

As can be seen from Table 7, the addition of $\mathcal{L}_{pl}$ reduced the FID from 103.777 to 72.924, the KID from 0.076 to 0.047, and the IS from 4.560 to 4.815. Thus, the effectiveness of path length regularization in the image generation process is verified.

Table 7. FID, KID, and IS metrics on the DOTA dataset. ($\mathcal{L}_{pl}$ ablation experiments).

Model	FID	KID	IS
SGA-GAN without $\mathcal{L}_{pl}$	76.561	0.050	4.761
SGA-GAN	72.924	0.047	4.815

4.5.3. The Effectiveness of the Attention Module (MAFM)

The attention module (MAFM) was removed from the generator structure, and the network was retrained on the DOTA dataset using the same training scheme to verify the effectiveness of the MAFM.

As can be seen from Table 8, the addition of the MAFM reduced the FID from 85.967 to 80.156, the KID from 0.059 to 0.054, and the IS from 4.581 to 4.643, thus verifying the effectiveness of the MAFM in the image generation process.

Table 8. FID, KID, and IS metrics on the DOTA dataset. (MAFM ablation experiments).

Model	FID	KID	IS
Baseline	85.967	0.059	4.581
Baseline + MAFM	80.156	0.054	4.643

4.5.4. Impact of the Gradient Structure Information-Guided Model on the Algorithm (GSNet)

GSNet was removed, and the network was retrained on the DOTA dataset using the same training scheme to verify its effectiveness.

As can be seen from Table 9, the FID was decreased from 80.156 to 72.924, the KID was decreased from 0.054 to 0.047, and the IS was decreased from 4.643 to 4.815 after adding the branching network, which verifies the effectiveness of the gradient structure information guidance in the image generation process.

Table 9. FID, KID, and IS metrics on the DOTA dataset. (GSNet ablation experiments).

Model	FID	KID	IS
SGA-GAN without GSNet	80.156	0.054	4.643
SGA-GAN	72.924	0.047	4.815

5. Conclusions

The distortion phenomenon exists in remote sensing images generated by existing image generation methods. To address this problem, this paper proposed an attention generative adversarial network based on gradient structure information guidance (SGA-GAN) for remote sensing image generation. The gradient structure information extraction branching network can effectively alleviate the structure distortion phenomenon existing in remote sensing image generation and improve the quality of the generated remote sensing images. Comparison experiments were conducted on two remote sensing datasets, namely UCAS-AOD and DOTA. Compared with five other advanced image generation models, the SGA-GAN algorithm proposed in this paper returned the optimal results for the three studied metrics, i.e., FID, KID, and IS, and generated reliable structures and realistic visual sample images. At the same time, experiments were conducted on two datasets using three different object detection models, namely YOLOv5, YOLOX, and EfficientDet. After adding aircraft target images generated by SGA-GAN to the DOTA dataset, the APs of YOLOv5 and YOLOX for aircraft objects were improved by 1.7% and 1.3%, respectively. The comparison experiments show that the algorithm in this paper can effectively alleviate the problems of the insufficient scale of existing remote sensing datasets and the imbalance within and between categories, meaning it can be useful in practical applications.

In the next step, we will jointly train the remote sensing image generation and object detection networks to improve the performance of the model in practical applications and provide data and experimental support for the interpretation of satellite remote sensing images.

Author Contributions: Conceptualization, B.Z. and Z.T.; methodology, B.Z.; software, B.Z. and Y.Y.; investigation, B.Z., K.Z., X.S. and Y.T.; writing—original draft preparation, B.Z.; writing—review and editing, B.Z. and Z.T.; project administration, Q.L.; funding acquisition, Z.T. All authors have read and agreed to the published version of the manuscript.

Funding: This research was funded by the Key Program Project of Science and Technology Innovation of the Chinese Academy of Sciences (no. KGFZD-135-20-03-02) and the Strategic Priority Research Program of the Chinese Academy of Sciences (Grant No. XDA28050401).

Data Availability Statement: Not applicable.

Conflicts of Interest: The authors declare no conflict of interest.

References

1. Christaki, M.; Vasilakos, C.; Papadopoulou, E.-E.; Tataris, G.; Siarkos, I.; Soulakellis, N. Building Change Detection Based on a Gray-Level Co-Occurrence Matrix and Artificial Neural Networks. *Drones* **2022**, *6*, 414. [CrossRef]
2. Pan, X.; Yang, L.; Sun, X.; Yao, J.; Guo, J. Research on the Extraction of Hazard Sources along High-Speed Railways from High-Resolution Remote Sensing Images Based on TE-ResUNet. *Sensors* **2022**, *22*, 3784. [CrossRef] [PubMed]
3. Himeur, Y.; Rimal, B.; Tiwary, A.; Amira, A. Using Artificial Intelligence and Data Fusion for Environmental Monitoring: A Review and Future Perspectives. *Inf. Fusion* **2022**, *86–87*, 44–75. [CrossRef]
4. Zhang, B.; Chen, Z.; Peng, D.; Benediktsson, J.A.; Liu, B.; Zou, L.; Li, J.; Plaza, A. Remotely Sensed Big Data: Evolution in Model Development for Information Extraction [Point of View]. *Proc. IEEE* **2019**, *107*, 2294–2301. [CrossRef]
5. Yan, P.; Liu, X.; Wang, F.; Yue, C.; Wang, X. LOVD: Land Vehicle Detection in Complex Scenes of Optical Remote Sensing Image. *IEEE Trans. Geosci. Remote Sens.* **2022**, *60*, 1–13. [CrossRef]
6. Lin, T.-Y.; Maire, M.; Belongie, S.; Hays, J.; Perona, P.; Ramanan, D.; Dollár, P.; Zitnick, C.L. Microsoft Coco: Common Objects in Context. In Proceedings of the European Conference on Computer Vision, Zurich, Switzerland, 6–12 September 2014; Springer: Berlin/Heidelberg, Germany, 2014; pp. 740–755.
7. Deng, J.; Dong, W.; Socher, R.; Li, L.-J.; Li, K.; Fei-Fei, L. ImageNet: A Large-Scale Hierarchical Image Database. In Proceedings of the CVPR: 2009 IEEE Conference on Computer Vision and Pattern Recognition, Miami, FL, USA, 20–25 June 2009; IEEE: New York, NY, USA, 2009; Volume 1–4, pp. 248–255.
8. Zhang, Y.; Wen, F.; Gao, Z.; Ling, X. A Coarse-to-Fine Framework for Cloud Removal in Remote Sensing Image Sequence. *IEEE Trans. Geosci. Remote Sens.* **2019**, *57*, 5963–5974. [CrossRef]
9. Wang, C.; Wang, Y.; Wang, L.; Hu, Z.; Zhang, S.; Wang, S.; Xiu, W.; Cui, H.; Wang, D.; Li, Q. Volunteered Remote Sensing Data Generation with Air Passengers as Sensors. *Int. J. Digit. Earth* **2021**, *14*, 158–180. [CrossRef]
10. Li, K.; Wan, G.; Cheng, G.; Meng, L.; Han, J. Object Detection in Optical Remote Sensing Images: A Survey and a New Benchmark. *ISPRS-J. Photogramm. Remote Sens.* **2020**, *159*, 296–307. [CrossRef]
11. Li, Y.; Wang, Q.; Luo, X.; Yin, J. Class-Balanced Contrastive Learning for Fine-Grained Airplane Detection. *IEEE Geosci. Remote Sens. Lett.* **2022**, *19*, 1–5. [CrossRef]
12. Leevy, J.L.; Khoshgoftaar, T.M.; Bauder, R.A.; Seliya, N. A Survey on Addressing High-Class Imbalance in Big Data. *J. Big Data* **2018**, *5*, 42. [CrossRef]

13. Wang, Q.; Huang, W.; Xiong, Z.; Li, X. Looking Closer at the Scene: Multiscale Representation Learning for Remote Sensing Image Scene Classification. *IEEE Trans. Neural Netw. Learn. Syst.* **2022**, *33*, 1414–1428. [CrossRef]
14. Chen, Z.; Shang, Y.; Python, A.; Cai, Y.; Yin, J. DB-BlendMask: Decomposed Attention and Balanced BlendMask for Instance Segmentation of High-Resolution Remote Sensing Images. *IEEE Trans. Geosci. Remote Sens.* **2022**, *60*, 1–15. [CrossRef]
15. Miao, W.; Geng, J.; Jiang, W. Multigranularity Decoupling Network with Pseudolabel Selection for Remote Sensing Image Scene Classification. *IEEE Trans. Geosci. Remote Sens.* **2023**, *61*, 1–13. [CrossRef]
16. Zhou, Z.; Zheng, C.; Liu, X.; Tian, Y.; Chen, X.; Chen, X.; Dong, Z. A Dynamic Effective Class Balanced Approach for Remote Sensing Imagery Semantic Segmentation of Imbalanced Data. *Remote Sens.* **2023**, *15*, 1768. [CrossRef]
17. Shorten, C.; Khoshgoftaar, T.M. A Survey on Image Data Augmentation for Deep Learning. *J. Big Data* **2019**, *6*, 60. [CrossRef]
18. Naji, H.A.H.; Li, T.; Xue, Q.; Duan, X. A Hypered Deep-Learning-Based Model of Hyperspectral Images Generation and Classification for Imbalanced Data. *Remote Sens.* **2022**, *14*, 6406. [CrossRef]
19. Ruiz-Ponce, P.; Ortiz-Perez, D.; Garcia-Rodriguez, J.; Kiefer, B. POSEIDON: A Data Augmentation Tool for Small Object Detection Datasets in Maritime Environments. *Sensors* **2023**, *23*, 3691. [CrossRef]
20. Goodfellow, I.; Pouget-Abadie, J.; Mirza, M.; Xu, B.; Warde-Farley, D.; Ozair, S.; Courville, A.; Bengio, Y. Generative Adversarial Networks. *Commun. ACM* **2020**, *63*, 139–144. [CrossRef]
21. Mo, L.; Zhu, Y.; Wang, G.; Yi, X.; Wu, X.; Wu, P. Improved Image Synthesis with Attention Mechanism for Virtual Scenes via UAV Imagery. *Drones* **2023**, *7*, 160. [CrossRef]
22. Jozdani, S.; Chen, D.; Pouliot, D.; Alan Johnson, B. A Review and Meta-Analysis of Generative Adversarial Networks and Their Applications in Remote Sensing. *Int. J. Appl. Earth Obs. Geoinf.* **2022**, *108*, 102734. [CrossRef]
23. Berg, P.; Pham, M.-T.; Courty, N. Self-Supervised Learning for Scene Classification in Remote Sensing: Current State of the Art and Perspectives. *Remote Sens.* **2022**, *14*, 3995. [CrossRef]
24. Gu, S.; Zhang, R.; Luo, H.; Li, M.; Feng, H.; Tang, X. Improved SinGAN Integrated with an Attentional Mechanism for Remote Sensing Image Classification. *Remote Sens.* **2021**, *13*, 1713. [CrossRef]
25. Karras, T.; Laine, S.; Aittala, M.; Hellsten, J.; Lehtinen, J.; Aila, T. Analyzing and Improving the Image Quality of StyleGAN. In Proceedings of the 2020 IEEE/CVF Conference on Computer Vision and Pattern Recognition (CVPR), Seattle, WA, USA, 13–19 June 2020; pp. 8107–8116.
26. Mao, Q.; Lee, H.-Y.; Tseng, H.-Y.; Ma, S.; Yang, M.-H. Mode Seeking Generative Adversarial Networks for Diverse Image Synthesis. In Proceedings of the 2019 IEEE/CVF Conference on Computer Vision and Pattern Recognition (CVPR 2019), Long Beach, CA, USA, 16–17 June 2019; IEEE Computer Soc: Los Alamitos, CA, USA, 2019; pp. 1429–1437.
27. Wang, J.; Li, Y.; Chen, W. UAV Aerial Image Generation of Crucial Components of High-Voltage Transmission Lines Based on Multi-Level Generative Adversarial Network. *Remote Sens.* **2023**, *15*, 1412. [CrossRef]
28. Yu, Y.; Li, X.; Liu, F. Attention GANs: Unsupervised Deep Feature Learning for Aerial Scene Classification. *IEEE Trans. Geosci. Remote Sens.* **2020**, *58*, 519–531. [CrossRef]
29. Zhu, B.; Lv, Q.; Tan, Z. Adaptive Multi-Scale Fusion Blind Deblurred Generative Adversarial Network Method for Sharpening Image Data. *Drones* **2023**, *7*, 96. [CrossRef]
30. Che, T.; Li, Y.; Jacob, A.; Bengio, Y.; Li, W. Mode Regularized Generative Adversarial Networks. *arXiv* **2022**, arXiv:1612.02136.
31. Ghosh, A.; Kulharia, V.; Namboodiri, V.; Torr, P.H.S.; Dokania, P.K. Multi-Agent Diverse Generative Adversarial Networks. In Proceedings of the 2018 IEEE/CVF Conference on Computer Vision and Pattern Recognition (CVPR), Salt Lake City, UT, USA, 18–23 June 2018; IEEE: New York, NY, USA, 2018; pp. 8513–8521.
32. Karras, T.; Aila, T.; Laine, S.; Lehtinen, J. Progressive Growing of GANs for Improved Quality, Stability, and Variation. In Proceedings of the International Conference on Learning Representations, Vancouver, BC, Canada, 30 April–3 May 2018.
33. Zhang, H.; Goodfellow, I.; Metaxas, D.; Odena, A. Self-Attention Generative Adversarial Networks. In Proceedings of the International Conference on Machine Learning; PMLR, Long Beach, CA, USA, 10–15 June 2019; pp. 7354–7363.
34. Brock, A.; Donahue, J.; Simonyan, K. Large Scale GAN Training for High Fidelity Natural Image Synthesis. *arXiv* **2018**, arXiv:1809.11096.
35. Karras, T.; Laine, S.; Aila, T. A Style-Based Generator Architecture for Generative Adversarial Networks. *IEEE Trans. Pattern Anal. Mach. Intell.* **2021**, *43*, 4217–4228. [CrossRef]
36. Ma, D.; Tang, P.; Zhao, L. SiftingGAN: Generating and Sifting Labeled Samples to Improve the Remote Sensing Image Scene Classification Baseline In Vitro. *IEEE Geosci. Remote Sens. Lett.* **2019**, *16*, 1046–1050. [CrossRef]
37. Lin, D.; Fu, K.; Wang, Y.; Xu, G.; Sun, X. MARTA GANs: Unsupervised Representation Learning for Remote Sensing Image Classification. *IEEE Geosci. Remote Sens. Lett.* **2017**, *14*, 2092–2096. [CrossRef]
38. Jiang, K.; Wang, Z.; Yi, P.; Wang, G.; Lu, T.; Jiang, J. Edge-Enhanced GAN for Remote Sensing Image Superresolution. *IEEE Trans. Geosci. Remote Sens.* **2019**, *57*, 5799–5812. [CrossRef]
39. Han, W.; Wang, L.; Feng, R.; Gao, L.; Chen, X.; Deng, Z.; Chen, J.; Liu, P. Sample Generation Based on a Supervised Wasserstein Generative Adversarial Network for High-Resolution Remote-Sensing Scene Classification. *Inf. Sci.* **2020**, *539*, 177–194. [CrossRef]
40. Zhu, D.; Xia, S.; Zhao, J.; Zhou, Y.; Jian, M.; Niu, Q.; Yao, R.; Chen, Y. Diverse Sample Generation with Multi-Branch Conditional Generative Adversarial Network for Remote Sensing Objects Detection. *Neurocomputing* **2020**, *381*, 40–51. [CrossRef]

41. Bello, I.; Zoph, B.; Le, Q.; Vaswani, A.; Shlens, J. Attention Augmented Convolutional Networks. In Proceedings of the 2019 IEEE/CVF International Conference on Computer Vision (ICCV), Seoul, Republic of Korea, 27 October–2 November 2019; pp. 3285–3294.
42. Huang, Z.; Li, Y.; Bai, M.; Wei, Q.; Gu, Q.; Mou, Z.; Zhang, L.; Lei, D. A Multiscale Spatiotemporal Fusion Network Based on an Attention Mechanism. *Remote Sens.* **2023**, *15*, 182. [CrossRef]
43. Hu, J.; Shen, L.; Sun, G. Squeeze-and-Excitation Networks. In Proceedings of the 2018 IEEE/CVF Conference on Computer Vision and Pattern Recognition, Salt Lake City, UT, USA, 18–23 June 2018; pp. 7132–7141.
44. Gao, Z.; Xie, J.; Wang, Q.; Li, P. Global Second-Order Pooling Convolutional Networks. In Proceedings of the IEEE/CVF Conference on Computer Vision and Pattern Recognition, Long Beach, CA, USA, 15–20 June 2019; pp. 3024–3033.
45. Wang, Q.; Wu, B.; Zhu, P.; Li, P.; Zuo, W.; Hu, Q. ECA-Net: Efficient Channel Attention for Deep Convolutional Neural Networks. In Proceedings of the IEEE Conference on Computer Vision and Pattern Recognition (CVPR), Seattle, WA, USA, 13–19 June 2020.
46. Yang, Z.; Zhu, L.; Wu, Y.; Yang, Y. Gated Channel Transformation for Visual Recognition. In Proceedings of the IEEE/CVF Conference on Computer Vision and Pattern Recognition, Seattle, WA, USA, 13–19 June 2020; pp. 11794–11803.
47. Qin, Z.; Zhang, P.; Wu, F.; Li, X. Fcanet: Frequency Channel Attention Networks. In Proceedings of the IEEE/CVF International Conference on Computer Vision, Montreal, BC, Canada, 11–17 October 2021; pp. 783–792.
48. Huang, W.; Deng, Y.; Hui, S.; Wang, J. Adaptive-Attention Completing Network for Remote Sensing Image. *Remote Sens.* **2023**, *15*, 1321. [CrossRef]
49. Wang, C.; Meng, L.; Gao, Q.; Wang, J.; Wang, T.; Liu, X.; Du, F.; Wang, L.; Wang, E. A Lightweight UAV Swarm Detection Method Integrated Attention Mechanism. *Drones* **2023**, *7*, 13. [CrossRef]
50. Cao, H.; Gao, Y.; Cai, W.; Xu, Z.; Li, L. Segmentation Detection Method for Complex Road Cracks Collected by UAV Based on HC-Unet++. *Drones* **2023**, *7*, 189. [CrossRef]
51. Chen, S.; Lan, J.; Liu, H.; Chen, C.; Wang, X. Helmet Wearing Detection of Motorcycle Drivers Using Deep Learning Network with Residual Transformer-Spatial Attention. *Drones* **2022**, *6*, 415. [CrossRef]
52. Huang, Z.; Wang, X.; Wei, Y.; Huang, L.; Shi, H.; Liu, W.; Huang, T.S. CCNet: Criss-Cross Attention for Semantic Segmentation. In Proceedings of the IEEE/CVF International Conference on Computer Vision, Seoul, Republic of Korea, 27 October–2 November 2019. [CrossRef]
53. Zuo, Y.; Wang, H.; Fang, Y.; Huang, X.; Shang, X.; Wu, Q. MIG-Net: Multi-Scale Network Alternatively Guided by Intensity and Gradient Features for Depth Map Super-Resolution. *IEEE Trans. Multimed.* **2022**, *24*, 3506–3519. [CrossRef]
54. Ultralytics. Yolov5. Available online: Https://Github.Com/Ultralytics/Yolov5 (accessed on 28 November 2022).
55. Ge, Z.; Liu, S.; Wang, F.; Li, Z.; Sun, J. Yolox: Exceeding Yolo Series in 2021. *arXiv* **2021**, arXiv:2107.08430.
56. Tan, M.; Pang, R.; Le, Q.V. Efficientdet: Scalable and Efficient Object Detection. In Proceedings of the IEEE/CVF Conference on Computer Vision and Pattern Recognition, Seattle, WA, USA, 13–19 June 2020; pp. 10781–10790.
57. Salimans, T.; Goodfellow, I.; Zaremba, W.; Cheung, V.; Radford, A.; Chen, X.; Chen, X. Improved Techniques for Training GANs. In Proceedings of the Advances in Neural Information Processing Systems, Barcelona, Spain, 5–10 December 2016; Curran Associates, Inc.: Red Hook, NY, USA, 2016; Volume 29.
58. Szegedy, C.; Liu, W.; Jia, Y.; Sermanet, P.; Reed, S.; Anguelov, D.; Erhan, D.; Vanhoucke, V.; Rabinovich, A. Going Deeper with Convolutions. In Proceedings of the 2015 IEEE Conference on Computer Vision and Pattern Recognition (CVPR), Boston, MA, USA, 7–12 June 2015; pp. 1–9.
59. Hensel, M.; Ramsauer, H.; Unterthiner, T.; Nessler, B.; Hochreiter, S. GANs Trained by a Two Time-Scale Update Rule Converge to a Local Nash Equilibrium. In Proceedings of the Advances in Neural Information Processing Systems 30 (NIPS 2017), Long Beach, CA, USA, 4–9 December 2017; Guyon, I., Luxburg, U.V., Bengio, S., Wallach, H., Fergus, R., Vishwanathan, S., Garnett, R., Eds.; Neural Information Processing Systems (NIPS): La Jolla, CA, USA, 2017; Volume 30.
60. Bińkowski, M.; Sutherland, D.J.; Arbel, M.; Gretton, A. Demystifying MMD GANs. *arXiv* **2018**, arXiv:1801.01401.
61. Xia, G.-S.; Bai, X.; Ding, J.; Zhu, Z.; Belongie, S.; Luo, J.; Datcu, M.; Pelillo, M.; Zhang, L. DOTA: A Large-Scale Dataset for Object Detection in Aerial Images. In Proceedings of the 2018 IEEE/CVF Conference on Computer Vision and Pattern Recognition (CVPR), Salt Lake City, UT, USA, 18–23 June 2018; IEEE: New York, NY, USA, 2018; pp. 3974–3983.
62. Ding, J.; Xue, N.; Xia, G.-S.; Bai, X.; Yang, W.; Yang, M.Y.; Belongie, S.; Luo, J.; Datcu, M.; Pelillo, M.; et al. Object Detection in Aerial Images: A Large-Scale Benchmark and Challenges. *IEEE Trans. Pattern Anal. Mach. Intell.* **2022**, *44*, 7778–7796. [CrossRef] [PubMed]
63. Zhu, H.; Chen, X.; Dai, W.; Fu, K.; Ye, Q.; Jiao, J. Orientation Robust Object Detection in Aerial Images Using Deep Convolutional Neural Network. In Proceedings of the 2015 IEEE International Conference on Image Processing (ICIP), Quebec City, QC, Canada, 27–30 September 2015; pp. 3735–3739.

64. Liu, B.; Zhu, Y.; Song, K.; Elgammal, A. Towards Faster and Stabilized Gan Training for High-Fidelity Few-Shot Image Synthesis. In Proceedings of the International Conference on Learning Representations, Addis Ababa, Ethiopia, 26–30 April 2020.
65. Sauer, A.; Schwarz, K.; Geiger, A. Stylegan-Xl: Scaling Stylegan to Large Diverse Datasets. In Proceedings of the ACM SIGGRAPH 2022 Conference Proceedings, Vancouver, BC, Canada, 7–11 August 2022; pp. 1–10.

sensors

MDPI

Article

CE-BART: Cause-and-Effect BART for Visual Commonsense Generation

Junyeong Kim [1], Ji Woo Hong [2], Sunjae Yoon [2] and Chang D. Yoo [2,*]

[1] Department of AI, Chung-Ang University, Seoul 06974, Republic of Korea
[2] School of Electrical Engineering, Korea Advanced Institute of Science and Technology, Daejeon 34141, Republic of Korea
* Correspondence: cd_yoo@kaist.ac.kr; Tel.: +82-42-350-3470

Abstract: "A Picture is worth a thousand words". Given an image, humans are able to deduce various cause-and-effect captions of past, current, and future events beyond the image. The task of visual commonsense generation has the aim of generating three cause-and-effect captions for a given image: (1) what needed to happen before, (2) what is the current intent, and (3) what will happen after. However, this task is challenging for machines, owing to two limitations: existing approaches (1) directly utilize conventional vision–language transformers to learn relationships between input modalities and (2) ignore relations among target cause-and-effect captions, but consider each caption independently. Herein, we propose Cause-and-Effect BART (CE-BART), which is based on (1) a structured graph reasoner that captures intra- and inter-modality relationships among visual and textual representations and (2) a cause-and-effect generator that generates cause-and-effect captions by considering the causal relations among inferences. We demonstrate the validity of CE-BART on the VisualCOMET and AVSD benchmarks. CE-BART achieved SOTA performance on both benchmarks, while an extensive ablation study and qualitative analysis demonstrated the performance gain and improved interpretability.

Keywords: deep learning; visual–language reasoning; visual commonsense generation; video-grounded dialogue; VisualCOMET; AVSD

Citation: Kim, J.; Hong, J.W.; Yoon, S.; Yoo, C.D. CE-BART: Cause-and-Effect BART for Visual Commonsense Generation. *Sensors* **2022**, *22*, 9399. https://doi.org/10.3390/s22239399

Academic Editors: Moulay A. Akhloufi and Mozhdeh Shahbazi

Received: 4 November 2022
Accepted: 29 November 2022
Published: 2 December 2022

1. Introduction

Visual Commonsense Generation (VCG) [1] is a challenging task that requires the generation of commonsense and cause-and-effect captions regarding visual and textual information. To be specific, given a still image and a description of the event shown in that image, the goal is to understand the cause-and-effect relations within the event and generate free-form natural language sentences that describe the inferred past/future events and the present intents of characters in the image. For example, in Figure 1, given the image on the left of a woman approaching a man at a table, the agent generates three kinds of cause-and-effect captions: (1) sometime in the past, she walked into the room and saw a man sitting at the table; (2) the intent of the woman is to talk to the man; (3) sometime in the future, she will sit down at the table and speak with him about a serious topic. While reasoning about the rich dynamic story of the visual scene is easy for humans, it is difficult for machines since it requires a higher-order cognitive-level understanding of the world.

In recent years, several visual reasoning tasks [2–4] were proposed and drew attention in the computer vision and natural language processing communities. To elaborate on a few such proposals, the Visual-Question-Answering (VQA) task defines a question-answering paradigm as a test to measure a machine's reasoning abilities for a given image or video. The Visual Dialog (VisDial) task asks a series of questions in the form of dialogue grounded in an image or video. The Visual-Commonsense-Reasoning (VCR) task further requires the machine to provide a rationale explaining why its answer is correct. While the above visual reasoning tasks are defined as recognition-level understanding and only consider the concepts and relations *within* the provided image or video, VCG focuses on reasoning about the rich cognitive-level dynamic story that goes beyond the directly visible contents by requiring cause-and-effect caption generation. Piaget's cognitive development theory [5] describes the drive of human intelligence to know in two forms, that is as states and transformations, suggesting that people must possess functions to represent both static and transformational aspects of reality. If the former tasks represent reasoning in a stationary situation, then VCG represents reasoning in a transforming situation. Hence, the research on VCG opens the door for a major leap from recognition-level understanding to cognition-level reasoning.

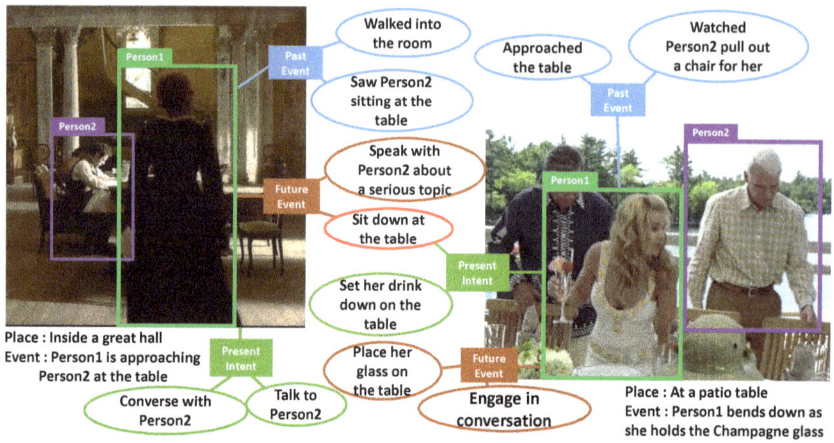

Figure 1. Illustration of visual commonsense generation. Given a person in an image and a corresponding textual event, an agent is required to generate (1) what needed to happen *before*, (2) what is the current *intent* of person, and (3) what will happen *after*.

Only a few works on VCG have been published. Park et al. [1] constructed a benchmark for visual commonsense generation, VisualCOMET, and proposed the baseline method. Xing et al. [6] proposed Knowledge-Enhanced BART (KM-BART), which leverages knowledge from external corpora to pre-train the BART. Previous approaches only operated on the conventional learning scheme of visual and textual information, overlooking the distinctiveness of cause-and-effect generation. Two major limitations of previous approaches are that they (1) directly utilized conventional vision–language transformers to learn relationships between input modalities and (2) ignored relations among target captions, but considered each caption independently. Due to the first limitation, previous approaches ignored the intra- and inter-modality relationships, which have proven to be beneficial to transformer-based generation [7]. Due to the second limitation, the existing models pay no attention to the intrinsic structure of the task or dataset. As the goal of VCG is to generate cause-and-effect captions, it is essential to consider causal relations among each inference for *before*, *intent*, and *after*. While existing approaches consider these inferences as separate cases and are trained independently, we argue that the generation of these three captions should be considered holistically.

In this paper, we address the aforementioned limitations with our novel Cause-and-Effect BART (CE-BART), which is composed of (1) a Structured Graph Reasoner (SGR) and (2) a Cause-and-Effect Generator (CEG). The SGR first builds semantic graphs for each modality to interpret the intra-modality relationships from the spatial or token domain via graph structures, then it captures higher-order semantic relations among graphs (i.e., inter-modality relationships) via tripartite graph attention and strengthens the multimodal graph representations. As SGR comprehends the intra- and inter-relationships interspersed in multimodal representations beforehand, the latter workload of the transformer-based CEG is unburdened, allowing it to focus more on understanding the commonsense and cause-

and-effect inference of the given input. The CEG generates cause-and-effect captions for *before, intent*, and *after* situations. While all existing approaches for visual commonsense generation are trained to generate each cause-and-effect caption separately (i.e., there are no connections between the generation of *before, intent*, and *after*, even for the same image), the proposed CEG infers all three cause-and-effect captions holistically by considering the causal relations. It consists of one transformer encoder for modeling multimodal representations and three transformer decoders each for generating the *before, intent*, and *after* captions. To consider causal relations among cause-and-effect inferences, the decoders for *intent* and *after* are connected to those of *before* and *intent*, respectively. Through causal connections between the three decoders, it can attend to the hidden states of the former decoders, which take the role of *cause* to generate *effect* captions (i.e., the proposed intent/after decoder can attend to not only the hidden states of the transformer encoder, but also the hidden states of the before/intent decoder).

The overall contribution and novelty of this work are summarized as follows: (1) We propose CE-BART, which is a novel transformer-based reasoning pipeline to handle both comprehensive understandings of multimodal input and cause-and-effect caption generation. (2) We extended BART with a graph-based information encoder and to have three decoders in order to address the issues of existing VCG methods. (3) We empirically show that our proposed CE-BART is the state-of-the-art on the VisualCOMET and AVSD benchmarks.

2. Related Works

2.1. Commonsense Reasoning

Commonsense knowledge has attracted a great deal of attention in both the computer vision and natural language communities. Commonsense or causality knowledge refers to the basic level of practical knowledge and reasoning about everyday situations and events commonly shared among most people [6,8]. For example, if the Sun is out, it is unlikely to rain; if we drop a cup, it is likely to break. Such causality knowledge has been shown to be beneficial for many tasks [9,10], and thus, it is essential for machines to learn to understand causality [11].

In the field of natural language processing, several commonsense Knowledge Bases (KBs) have been constructed to help machines better understand commonsense causality. ConceptNet [12] and ATOMIC [13] are widely used commonsense KBs that leverage human annotations to provide high-quality causality knowledge. These KBs are built based on tuples (s, r, o), where s and o are subject and object phrases and r defines the relation between them. Relations in commonsense KBs include *causes, because, before, as a result, ...*, which are essential for learning causality. Bosselut et al. [14] proposed COMET, a transformer-based architecture, for automatic commonsense knowledge base completion. COMET is trained to predict the object o, given subject s and relation r. In the field of computer vision, the Visual Commonsense Reasoning (VCR) task [4] has been proposed, which is a visual question-answering benchmark that requires the machine to provide a rationale explaining why its answer is correct.

2.2. Visual Commonsense Generation

Park et al. [1] proposed the task of visual commonsense generation and a corresponding benchmark, VisualCOMET, which aims to generate cause-and-effect descriptions for a given image and corresponding textual event and place. VisualCOMET is a visual commonsense knowledge base where an image and corresponding textual event and place take the place of the object in ATOMIC. There are only a few works [1,6] dealing with the task of visual commonsense generation. Park et al. [1] first proposed a baseline model based on GPT-2 [15]. The baseline model feeds visual and textual context as inputs and is trained to predict each of the cause-and-effect descriptions. Xing et al. [6] proposed Knowledge-enhanced Multimodal BART (KM-BART), which utilizes a BART [16] to pre-train on large external datasets and leverages knowledge from them. KM-BART was first pre-trained with knowledge-based commonsense generation by leveraging knowledge from COMET [14], attribute and relation prediction using the Visual Genome benchmark [17], and masked language and region modeling using various pre-training benchmarks. It was then fine-tuned on the VisualCOMET benchmark to achieve state-of-the-art performance on the VisualCOMET benchmark. However, we argue that these systems operate on the conventional learning scheme of visual and textual information, overlooking the distinctiveness of the cause-and-effect generation task, and possess two major limitations: (1) conventional vision–language transformers are directly utilized to learn relationships between input modalities; (2) every training example is trained independently without considering the relations with others.

3. Cause-and-Effect BART

First, we provide a formal definition of the visual commonsense generation task [1] as follows. We are given tuples of (v, e, p), consisting of an image v, the event description e, and the place description p. The goal of visual commonsense generation is to generate three cause-and-effect captions corresponding to (1) what needed to happen *before*, (2) what is the current *intent* of the person, and (3) what will happen *after*.

Figure 2 shows a schematic of Cause-and-Effect BART (CE-BART), consisting of a multimodal input encoder, structured graph reasoner, and cause-and-effect generator. The multimodal input encoder first embeds three input modalities into the feature space. Then, the structured graph reasoner captures intra- and inter-modality relationships among the input modalities. Finally, the cause-and-effect generator generates cause-and-effect captions by considering the causal relationships among inferences. All three components of CE-BART are elaborated on in the following subsections.

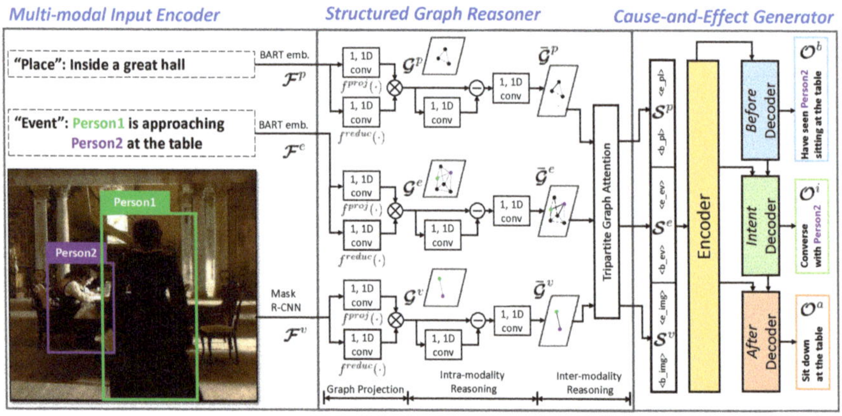

Figure 2. Illustration of Cause-and-Effect BART (CE-BART), which is composed of a multimodal input encoder, structured graph reasoner, and cause-and-effect generator. (1) Multimodal input encoder: we first obtain multimodal features $(\mathcal{F}^v, \mathcal{F}^e, \mathcal{F}^p)$ using pre-trained models. We used mask R-CNN for the visual inputand the BART word-embedding layer for textual inputs. (2) Structured Graph Reasoner (SGR): we then built semantic graphs $(\mathcal{G}^v, \mathcal{G}^e, \mathcal{G}^p)$ and strengthened their representations $(\mathcal{S}^v, \mathcal{S}^e, \mathcal{S}^p)$ by capturing the intra- and inter-modality relations. (3) Cause-and-Effect Generator (CEG): we finally generated cause-and-effect descriptions $(\mathcal{O}^b, \mathcal{O}^i, \mathcal{O}^a)$ using the BART-based transformer architecture, which considers the causal relations among inferences.

3.1. Multimodal Input Encoder

Following the previous work on visual commonsense generation, we used the mask R-CNN [18] to detect the visual person, which extracts $\mathcal{N}^v$ number of appearance features $\mathcal{A} = \{a_i\}_{i=1}^{\mathcal{N}^v}$ and their corresponding location features $\mathcal{B} = \{b_i\}_{i=1}^{\mathcal{N}^v}$. Each location feature $b_i = [x_i, y_i, w_i, h_i]$ represents a spatial coordinate, where $[x_i, y_i]$ denotes the relative coordinate of the top-left point of the bounding box, while $[w_i, h_i]$ denotes the width and height of the box. We calculate the final visual feature as $\mathcal{F}^v = \{v_i\}_{i=1}^{\mathcal{N}^v} \in \mathbb{R}^{\mathcal{N}^v \times d_v}$, where $v_i = w^a a_i + w^b b_i$ and w^a, w^b are learnable weights that embed both features into the visual feature dimension d_v.

There are two types of text for each image (i.e., event e and place p). Each sentence for event and place is fed into the word-embedding layer of the pre-trained BART to be further utilized. We obtain the textual feature as follows: $\mathcal{F}^e = \{e_i\}_{i=1}^{\mathcal{N}^e}$, $\mathcal{F}^p = \{p_i\}_{i=1}^{\mathcal{N}^p}$, where $\mathcal{N}^v, \mathcal{N}^p$ are the number of token features, and $e_i, p_i \in \mathbb{R}^{d_t}$ are the embedding of the i-th token in the event and place, respectively. Further, Table 1 summarizes the descriptions of the important symbols used in this paper.

Table 1. Descriptions of the important symbols used in this paper.

Symbol	Description	Symbol	Description
v	Input image	e	Input event description
p	Input place description	$\mathcal{A}$	Appearance feature
$\mathcal{B}$	Location feature	$\mathcal{F}^v$	Input visual feature
$\mathcal{F}^e$	Input event description feature	$\mathcal{F}^p$	Input place description feature
$\mathcal{G}^v$	Image semantic graph	$\mathcal{G}^e$	Event semantic graph
$\mathcal{G}^p$	Place semantic graph	$\mathcal{S}^v$	Strengthened semantic image feature
$\mathcal{S}^e$	Strengthened semantic event feature	$\mathcal{S}^p$	Strengthened semantic place feature
$\mathcal{O}^b$	Before caption	$\mathcal{O}^i$	Intent caption
$\mathcal{O}^a$	After caption		

3.2. Structured Graph Reasoner

In order to capture the intra-modality relationships from individual modalities (i.e., image, event, and place) and the inter-modality relationships among the input modalities, the structured graph reasoner first builds semantic graphs for each modality: an image semantic graph $\mathcal{G}^v$, an event semantic graph $\mathcal{G}^e$, and a place semantic graph $\mathcal{G}^p$. Motivated by [19], which projects visual features in the spatial domain into the graph domain for relational reasoning over a global context, the structured graph reasoner performs graph convolutions to capture intra-modality relations. It then captures the higher-order semantic relations among graphs (i.e., inter-modality relationships) via tripartite graph attention to strengthen the multimodal graph representations. The final strengthened semantic representations ($\mathcal{S}^v, \mathcal{S}^e, \mathcal{S}^p$) are fed into the following cause-and-effect generator.

For simplicity, we denote the feature representation as $\mathcal{F}^x$ and the semantic graph as $\mathcal{G}^x$ for each modality $x \in \{v, e, p\}$. We first project the feature representation $\mathcal{F}^x$ into semantic graph $\mathcal{G}^x$, which is a lightweight *fully connected* graph. Since we will directly reason over graph nodes, the projection into the graph domain is formulated as a linear combination among input features. A linear combination over input features can be thought of as weighted global pooling or global attention. We divide graph projection into two parts: dimension reduction and graph projection. In dimension reduction, we embed all three modality features into the same small feature space. In graph projection, we compute the weights for weighted global pooling. Finally, embedded features are weighted globally pooled to form the graph node feature:

$$\mathcal{G}^x = f^{proj}(\mathcal{F}^x; W_{f^{proj}}) \times f^{reduc}(\mathcal{F}^x; W_{f^{reduc}}) \in \mathbb{R}^{\mathcal{N}^x \times d_g}, \tag{1}$$

where the dimension reduction function f^{reduc} parameterized by $W_{f^{reduc}}$ projects each feature into the graph feature dimension d_g and the graph projection function f^{proj} parameterized by $W_{f^{proj}}$ produces the weights for linear combination. Here, both $f^{reduc} and f^{proj}$ are 1D convolution layers with a kernel size of 1.

In order to capture intra-modality relations in individual semantic graphs, we utilize graph convolution [20] to update node representations and obtain $\tilde{\mathcal{G}}^x$. We reason over fully connected graphs by learning the edge weights, which model the interactions among globally pooled graph node features of each modality. Given a fully connected graph $\mathcal{G}^x$, graph convolution learns the edge weights that correspond to the correlations between node representations. We divide graph convolution into two parts: channelwise convolution and nodewise convolution. By implementing 2D convolution with two 1D convolutions, graph convolution can be more efficient. A single layer of graph convolution is formulated as:

$$\bar{\mathcal{G}}^x = \Lambda \mathcal{G}^x W_x = ((\mathcal{I} - \mathcal{A}^x)\mathcal{G}^x)W_x, \tag{2}$$

where Λ and $\mathcal{A}^x$ are $\mathcal{N}^x \times \mathcal{N}^x$ an adjacency matrix for diffusing information across nodes of $\mathcal{G}^x$, $W_x \in \mathbb{R}^{d_g \times d_g}$ denotes the state update weight, and $\mathcal{I} \in \mathbb{R}^{\mathcal{N}^x \times \mathcal{N}^x}$ is the identity matrix. Here, the adjacency matrix $\mathcal{A}^x$ is randomly initialized and learned during training, together with W_x, and the identity matrix serves as a shortcut connection. We can implement Equation (2) using two consecutive 1D convolution layers along different directions: channelwise convolution (i.e., modeling $(\mathcal{I} - \mathcal{A}^x)$) and nodewise convolution (i.e., modeling W_x).

Finally, we capture inter-modality relations among the three semantic graphs $(\bar{\mathcal{G}}^v, \bar{\mathcal{G}}^e, \bar{\mathcal{G}}^p)$ via tripartite graph attention and calculate the strengthened semantic representations $(\mathcal{S}^v, \mathcal{S}^e, \mathcal{S}^p)$. We perform graph attention over a *tripartite* graph, which connects all the nodes in individual modalities to all the nodes belonging to the other modalities. By doing so, every node in each modality learns to integrate informative semantics from the other modalities into its representation in order to effectively capture inter-modality relations. First, we concatenate $\bar{\mathcal{G}}^v, \bar{\mathcal{G}}^e, \bar{\mathcal{G}}^p$ along the node axis to make a tripartite graph structure $\bar{\mathcal{G}}^T$, where each node is connected to all other nodes belonging to different modalities: $\bar{\mathcal{G}}^T = [\bar{\mathcal{G}}^v || \bar{\mathcal{G}}^e || \bar{\mathcal{G}}^p] \in \mathbb{R}^{\mathcal{N}^T \times d_g}$. We perform graph attention [21] over $\bar{\mathcal{G}}^T$ to calculate the multi-head attention to capture relations between each node and its neighboring nodes (i.e., inter-modality relations):

$$\mathcal{S}^T = \mathrm{GAT}(\bar{\mathcal{G}}^T), \tag{3}$$

$$\mathcal{S}^v, \mathcal{S}^e, \mathcal{S}^p = slice(\mathcal{S}^T), \tag{4}$$

where the $slice(\cdot)$ operation slices the multimodal representations along the node axis with the corresponding length of each modality. Graph attention effectively models the interaction among graph nodes through the self-attention mechanism inside. Basically, it calculates self-attention over neighboring nodes with the residual connection. Since the proposed tripartite graph attention is performed over the tripartite graph, we can effectively diffuse semantic information to different modalities.

3.3. Cause-and-Effect Generator

A Cause-and-Effect Generator (CEG) is proposed to generate cause-and-effect captions by considering the causal relationships among inferences. It is a sequence-to-sequence transformer architecture that feeds the strengthened semantic graph $(\mathcal{S}^v, \mathcal{S}^e, \mathcal{S}^p)$ and decodes cause-and-effect captions (*before* $\mathcal{O}^b$, *intent* $\mathcal{O}^i$, and *after* $\mathcal{O}^a$) in an autoregressive manner. In contrast to existing approaches that treat the generation of each caption as a separate objective, the cause-and-effect generator infers all three cause-and-effect captions holistically. Formally, we have the function of the CEG f_{CEG} with its parameter $\mathcal{W}_{f_{CEG}}$, whose goal is:

$$\mathcal{O}^b, \mathcal{O}^i, \mathcal{O}^a = f_{CEG}(\mathcal{S}^v, \mathcal{S}^e, \mathcal{S}^p; \mathcal{W}_{f_{CEG}}). \tag{5}$$

3.3.1. Encoder

The encoder of the CEG is based on a multi-layer bidirectional transformer, as in BART and its variant in visual commonsense generation, KM-BART. In contrast to the encoder in KM-BART, whose input sequence starts with one of the special tokens <before>, <intent>, or <after> to indicate to the model which cause-and-effect caption should be generated, the CEG only takes the three sets of semantic graph representations (i.e., $\mathcal{S}^v, \mathcal{S}^e, \mathcal{S}^p$), as it infers all three captions holistically. To inform the encoder about the start and end of different input modalities, we add three sets of special tokens: <b_img>, <e_img> for image embedding $\mathcal{S}^v$; <b_ev>, <e_ev> for event embedding $\mathcal{S}^e$; and <b_pl>, <e_pl> for place embedding $\mathcal{S}^p$.

3.3.2. Decoder

The decoders of the CEG are based on a multi-layer unidirectional transformer as it works in an autoregressive manner during generation. There are a total of three decoders for the CEG—one each for the generation of *before, intent*, and *after* captions. To inform each decoder about the start of generation, we add three special starting tokens for each decoder: <before>, <intent>, and <after>. Further, we add a special end inference token <e_inf> at the end of the target sequence to indicate the stop of a decoding process. During training, we use teacher-forcing [22] to supervise each decoding step, that is ground truth tokens are used as the decoder input. The decoders of the CEG only take a right-shifted target token sequence as the input.

To consider causal relations among cause-and-effect captions, the decoders for *intent* and *after* are connected to those of *before* and *intent*, respectively. Through causal connections between the three decoders, the CEG can attend to the hidden states of the former decoder, which take the role of *cause* to generate *effect* captions (i.e., the proposed *intent/after* decoder can attend to not only the hidden states of the transformer encoder, but also the hidden states of the *before/intent* decoder), as shown in Figure 2. Formally, we divide the function of the CEG f_{CEG} in Equation (5) as an encoder E_{CEG} and a set of decoders D^x_{CEG}, where $x \in \{b, i, a\}$. The conventional approaches [1,6] generate the cause-and-effect captions separately without considering causal relations:

$$\mathcal{O}^b = D_{con}(E_{con}(v, e, p)), \tag{6}$$
$$\mathcal{O}^i = D_{con}(E_{con}(v, e, p)), \tag{7}$$
$$\mathcal{O}^a = D_{con}(E_{con}(v, e, p)), \tag{8}$$

where E_{con} and D_{con} represent the functions of the encoder and decoder in existing approaches that feed the image v, event e, and place p. On the other hand, the proposed CEG has sequential connections among decoders to preserve causal relations among cause-and-effect captions:

$$\mathcal{O}^b = D^b_{CEG}(E_{CEG}(\mathcal{S}^v, \mathcal{S}^e, \mathcal{S}^p)), \tag{9}$$
$$\mathcal{O}^i = D^i_{CEG}(E_{CEG}(\cdot), D^b_{CEG}(\cdot)), \tag{10}$$
$$\mathcal{O}^a = D^a_{CEG}(E_{CEG}(\cdot), D^i_{CEG}(\cdot)). \tag{11}$$

4. Experiments

4.1. Benchmark Dataset

VisualCOMET [1] is a large-scale benchmark dataset for visual commonsense generation and is the only available dataset of its kind at present. It consists of over 1.4 million textual captions of visual commonsense inferences carefully annotated over a diverse set of 59,000 images paired with 139,000 event descriptions. The visual commonsense inferences are divided into 1174k, 146k, and 145k examples for training, validation, and testing, respectively.

Audio Visual Scene-Aware Dialogue (AVSD) [23] provides video, caption, and dialogue history consisting of a series of textual QA pairs, as well as follow-up questions about the video. The goal is to generate a free-form natural language answer to the question. As both tasks share similar input–output relations, CE-BART can be easily applied to video-grounded dialogue and can transfer causal knowledge learned from a visual commonsense generation for better video understanding.

4.2. Metrics

For generative evaluation, we follow the official object metrics for the VisualCOMET and AVSD benchmarks, including BLEU [24], METEOR [25], ROUGE-L [26], and CIDEr [27]. The metrics are formulated to compute the word overlapping between each generated caption and reference caption.

The BLEU score is a basic evaluation method often used in natural language processing. It measures the precision between the generated caption and the reference caption by measuring how much the ordered word pairs overlap through n-grams (1 to 4). Although it has obvious limitations because it lacks consideration of grammatical structures and synonyms, it is still widely used.

The METEOR score computes the weighted F-score, which is the harmonic mean of the precision and recall values based on mapping unigrams, which replaces the simple n-gram precision/recall. The reordering penalty term in METEOR penalizes captions that contain the correct words, but in the wrong order.

ROUGE is based on the recall value and is mostly used for text summarization evaluation. Depending on the sequence used for recall computation, ROUGE can be divided into various types; ROUGE-N, ROUGE-L, ROUGE-W, and ROUGE-S. ROUGE-N is based on the n-gram recall value. For instance, ROUGE-1 calculates the recall based on the matching unigram, and so on. ROUGE-L/W/S are based on the Longest Common Subsequence (LCS), weighted LCS, and skip-bigram co-occurrence statistics, respectively. Instead of only using the recall value, they use the F-score based on the corresponding sequence (e.g., the longest common subsequence between the generated and reference caption for ROUGE-l).

The CIDEr score is based on TF-IDF and is proposed for image captioning evaluation. First, the TF-IDF features are calculated for generated and reference captions based on the n-gram. Then, the CIDEr score is calculated by the cosine similarity between two TF-IDF features.

4.3. Experimental Details

We initialized the cause-and-effect generator with a pre-trained BART-based model with 6 transformer layers in both the encoder and decoder, and a hidden size of 768. For tripartite graph attention in the structured graph reasoner, the number of heads in multi-head attention was set to 8. We trained using 4 NVIDIA Quadro RTX 8000 (48 GB of memory) and Adam optimizer with $\beta_1 = 0.9$ and $\beta_2 = 0.999$. The learning rate was initially set to 0.0001, and the model was trained up to 30 epochs with an effective training batch size of 512. During inference, we adopted a beam search, and for each set of inputs, we decoded *before*, *intent*, and *after* captions sequentially.

4.4. Experimental Results on VisualCOMET

We compared our proposed Cause-and-Effect BART (CE-BART) with state-of-the-art methods on the VisualCOMET benchmark. Table 2 summarizes the experimental results on the VisualCOMET benchmark on both the validation and test splits, since the current state-of-the-art method, KM-BART [6], only provides the results on the validation split. We also provide the results of the ablation study in Table 2 with several variants of CE-BART in order to measure the effectiveness of the proposed key components of CE-BART. All the reported performances in Table 2 are the average values of five independently trained models with different seeds.

Starting from the direct fine-tuning of the BART-based model in VisualCOMET, which showed slightly lower performance than the previous SOTA method KM-BART, every component of the proposed CE-BART demonstrated improved performance on all three metrics. The results of the ablation study suggest that the limitations of existing approaches that we introduced are valid: (1) conventional vision–language transformers are directly utilized to learn relationships between input modalities, and (2) every training example is trained independently without considering relations with others.

Table 2. Comparison with state-of-the-art methods on the VisualCOMET benchmark. Here, "Proj-SGR" denotes the graph projection (i.e., Equation (1)), "Intra-SGR" denotes the intra-modality reasoning (i.e., Equation (2)), and "Inter-SGR" stands for the inter-modality reasoning (i.e., Equation (3)). "CEG" stands for the Cause-and-Effect Generator with three decoders (i.e., Equations (9)–(11)). The best results are in bold.

Methods	Validation Set			Test Set		
	BLEU2	**METEOR**	**CIDEr**	**BLEU2**	**METEOR**	**CIDEr**
Baseline [1]	13.50	11.55	18.27	12.71	11.13	17.36
KM-BART [6]	23.47	15.02	39.76	-	-	-
Variants on CE-BART						
BART-base	22.51	14.73	37.86	-	-	-
+ Proj-SGR	22.47	14.97	38.91	-	-	-
+ Intra-SGR	23.85	15.72	39.59	-	-	-
+ Inter-SGR	25.07	18.24	41.07	-	-	-
+ CEG	28.60	19.32	43.58	-	-	-
CE-BART	**28.60**	**19.32**	**43.58**	**28.14**	**18.91**	**42.64**

The structured graph reasoner is proposed to capture intra- and inter-modality relations among visual and textual representations. The inclusion of graph reasoning led to a 3.21 point boost in the CIDEr metric compared to the BART-based model. Among the components of the structured graph reasoner, intra-modality reasoning provided a 0.68 point gain in CIDEr, and inter-modality reasoning provided a 1.48 point gain. As our inter-modality reasoning module performs multi-head attention over a tripartite graph, whose neighborhood is defined as the nodes of a heterogeneous modality, each node reinforces its representation with the information from other modalities, and thus, it is able to comprehend inter-modality relations effectively. The design of our structured graph reasoner was effective in capturing intra- and inter-modality relations, which is essential in visual commonsense generation.

The cause-and-effect generator is proposed to generate cause-and-effect descriptions holistically by considering the causal relationships among inferences. It improved the CIDEr score by 2.51 points. Through causal connections between three decoders, the cause-and-effect generator looks at the former decoder, which takes the role of cause, to generate an effect description. The design of our cause-and-effect generator was effective in modeling causal relations among generated descriptions, which is essential in a visual commonsense generation.

The results of our ablation study suggest that the proposed CE-BART can effectively capture intra- and inter-modality relationships interspersed in multimodal input representations and can effectively generate cause-and-effect descriptions holistically by considering causal relations through the cause-and-effect generator with causal connections between decoders. In our comparison, CE-BART surpassed the other state-of-the-art methods on both the validation and test splits of the VisualCOMET benchmark. Compared to KM-BART [6], the state-of-the-art method in the validation split, CE-BART obtained a CIDEr score of 43.58, an improvement of nearly 4 points. The CIDEr score of CE-BART (42.64) was nearly double that obtained with the baseline [1], the state-of-the-art method in the test split (17.36). CE-BART also improved the BLEU-2 score by almost 16 points and the METEOR score by more than 7 points.

We also provide an in-depth quantitative analysis of the CEG in Table 3 to show the significance of the dependencies among the three decoders. The motivation behind capturing relations between training samples is to consider the relations among the *before*, *intent*, and *after* descriptions while generating them for an image: the current *intent* is related to the situation *before*, and the situation *after* is related to the current *intent*. As the CEG has connections among decoders, the *intent* decoder can operate by using not only the image, but also the information from the *before* decoder. Similarly, the *after* decoder can generate predictions with the help of the *intent* decoder. We conducted separate evaluations for *before*/*intent*/*after* captions. Without connections among decoders (i.e., without the CEG), *before* prediction showed superior performance to *intent* and *after* predictions. However, with connections among decoders (i.e., with the CEG), we observed a performance improvement in *intent* and *after* predictions, as expected, obtained by considering relations.

Table 3. Analysis conducted on the validation split of VisualCOMET. We provide an analysis of the behavior of the CEG by observing separate performance evaluations with (w CEG) and without the CEG (w/o CEG).

Methods	Before			Intent			After		
	B2	M	C	B2	M	C	B2	M	C
w/o CEG	29.7	20.4	45.1	19.4	15.4	40.7	26.1	18.9	37.7
w CEG	30.9	20.9	45.9	25.5	16.6	42.4	29.6	20.2	41.9

4.5. Experimental Results on AVSD

We conducted additional experiments to validate the generalizability of the proposed CE-BART in other VL tasks. We conducted experiments in the task of video-grounded dialogue, which is a multi-turn question-answering task. The formal definition of video-grounded dialogue is as follows: we are given a video, a dialogue history consisting of a series of textual QA pairs, and a follow-up question about the video, and the goal is to generate a free-form natural language answer. Both tasks share similar input–output relations; therefore, CE-BART can be easily applied to the new task of video-grounded dialogue and can transfer causal knowledge learned from visual commonsense generation for better video understanding. We trained CE-BART for video-grounded dialogue in two settings: (1) CE-BART without pre-training on VisualCOMET and (2) CE-BART with pre-training on VisualCOMET. Through this comparative experiment, we observed that causal information learned from VisualCOMET could help with understanding the video. Table 4 summarizes the results of the AVSD benchmark. We observed that CE-BART showed an improvement over existing methods and achieved SOTA performance in all metrics. By pre-training CE-BART on VisualCOMET, we could further boost the performance, indicating that causal knowledge trained on VisualCOMET can be successfully transferred for better video understanding in AVSD. We were able to validate that our proposed CE-BART can benefit other VL tasks by effectively transferring causal knowledge learned from VisualCOMET.

Table 4. Comparison with state-of-the-art methods on the AVSD benchmark. We compared CE-BART with various state-of-the-art systems on the AVSD benchmark: baseline [28], STSGR [29], MTN [30], MTN-TMT [31], and VX2TEXT [32]. The best results are in **bold**.

Methods	BLEU1	BLEU2	BLEU3	BLEU4	METEOR	ROUGE-L	CIDEr
Baseline [28]	-	-	-	0.078	0.113	0.277	0.727
STSGR [29]	-	-	-	0.133	0.165	0.362	1.272
MTN [30]	0.356	0.242	0.174	0.135	0.165	0.365	1.366
MTN-TMT [31]	-	-	-	0.142	0.171	0.371	1.357
VX2TEXT [32]	0.361	0.260	0.197	0.154	0.178	0.393	1.605
CE-BART	0.364	0.266	0.203	0.158	0.181	0.400	1.681
CE-BART w pre-train	**0.365**	**0.268**	**0.205**	**0.161**	**0.183**	**0.404**	**1.721**

4.6. Qualitative Analysis

Figure 3 shows examples from the test split of VisualCOMET and compares predictions from CE-BART and the baseline. CE-BART successfully utilized the SGR, which captures intra- and inter-modality relationships and is flexible in selecting important information regardless of whether it is textual or visual. In the first example, CE-BART noticed that the text of the father holding his daughter's shoulder was not crucial information and focused more on the image. With this ability to consider and select crucial information, in the second example, CE-BART focused on the key object in the scene and people's interaction centered on that object and avoided stating the obvious. Further, as the CEG was trained to successively generate captions through causal connections, it had the information from the former decoder, which took the role of *cause* to generate the *effect* caption. In the third example, CE-BART was continuous regarding the contents, whereas the baseline model produced a discontinuous caption. In the lower-right example, the baseline model produces overlapping captions, while CE-BART effectively generated the rich dynamic story of the visual scene.

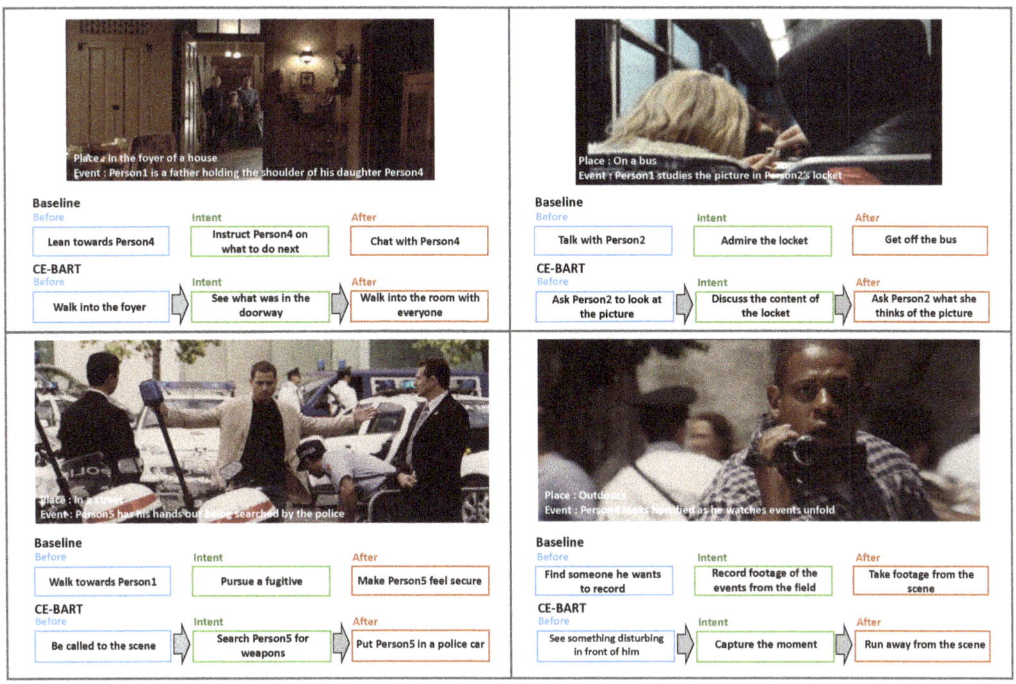

Figure 3. Four examples from the test split of the VisualCOMET benchmark.

5. Limitations

We believe that our proposed CE-BART has several limitations that can be addressed through further experiments in the future. First, its scalability is limited due to the requirement of large GPU resources. We conducted experiments using four NVIDIA Quadro RTX 8000 units (48 GB of memory), which are extremely expensive. Second, its scalability to control the time scale is limited. There is no factor in the current task setting that selects how much of a past/future situation is required. We will further develop our methods to overcome the limitations of the model.

6. Conclusions

We proposed a novel cause-and-effect BART for the task of visual commonsense generation. The proposed CE-BART consists of two major components: (1) a structured graph reasoner and (2) a cause-and-effect generator. The structured graph reasoner builds semantic graphs for individual modalities and strengthens their representations by capturing intra- and inter-modality relations among graph structures. The cause-and-effect generator is a transformer architecture with three decoders, one each for generating before, intent, and after captions. In experiments on the VisualCOMET and AVSD benchmarks, CE-BART achieved new state-of-the-art performance.

Author Contributions: Conceptualization, J.K.; methodology, J.K.; software, J.K., J.W.H. and S.Y.; validation, J.K. and S.Y.; formal analysis, J.K. and C.D.Y.; writing, J.K., J.W.H., S.Y. and C.D.Y.; supervision, C.D.Y. All authors have read and agreed to the published version of the manuscript.

Funding: This research was supported by the Chung-Ang University Research Grants in 2022 and by the Institute of Information and Communications Technology Planning and Evaluation (IITP) Grant through the Ministry of Science and ICT (MSIT), Government of Korea (Development of Previsional Intelligence Based on Long-Term Visual Memory Network) under Grant 2020-0-00004.

Institutional Review Board Statement: Not applicable.

Informed Consent Statement: Not applicable.

Data Availability Statement: Not applicable.

Acknowledgments: The authors would like to thank Jinwoo Park, Jangyoung Jeon, Jungeun Kim, and Zhang Nan for providing help in proofreading this paper and the researchers at ETRI for providing constructive comments on this paper.

Conflicts of Interest: The authors declare no conflicts of interest.

Abbreviations

The following abbreviations are used in this manuscript:

VCG	Visual Commonsense Generation
VQA	Visual Question Answering
VCR	Visual Commonsense Reasoning
CE-BART	Cause-and-Effect BART
SGR	Structured Graph Reasoner
CEG	Cause-and-Effect Generator

References

1. Park, J.S.; Bhagavatula, C.; Mottaghi, R.; Farhadi, A.; Choi, Y. VisualCOMET: Reasoning about the Dynamic Context of a Still Image. In Proceedings of the European Conference on Computer Vision (ECCV), Glasgow, UK, 23–28 August 2020.
2. Antol, S.; Agrawal, A.; Lu, J.; Mitchell, M.; Batra, D.; Lawrence Zitnick, C.; Parikh, D. Vqa: Visual question answering. In Proceedings of the IEEE International Conference on Computer Vision (ICCV), Santiago, Chile , 7–13 December 2015.
3. Das, A.; Kottur, S.; Gupta, K.; Singh, A.; Yadav, D.; Moura, J.M.; Parikh, D.; Batra, D. Visual Dialog. In Proceedings of the IEEE Conference on Computer Vision and Pattern Recognition (CVPR), Honolulu, HI, USA, 21–26 July 2017.
4. Zellers, R.; Bisk, Y.; Farhadi, A.; Choi, Y. From Recognition to Cognition: Visual Commonsense Reasoning. In Proceedings of the IEEE/CVF Conference on Computer Vision and Pattern Recognition (CVPR), Long Beach, CA, USA, 15–20 June 2019.
5. Piaget, J. The role of action in the development of thinking. In *Knowledge and Development*; Springer: Berlin/Heidelberg, Germany, 1977; pp. 17–42.
6. Xing, Y.; Shi, Z.; Meng, Z.; Ma, Y.; Wattenhofer, R. KM-BART: Knowledge Enhanced Multimodal BART for Visual Commonsense Generation. *arXiv* **2021**, arXiv:2101.00419.
7. Kim, J.; Yoon, S.; Kim, D.; Yoo, C.D. Structured Co-reference Graph Attention for Video-grounded Dialogue. In Proceedings of the AAAI Conference on Artificial Intelligence (AAAI), Virtually, 2–9 February 2021.
8. Sap, M.; Shwartz, V.; Bosselut, A.; Choi, Y.; Roth, D. Commonsense Reasoning for Natural Language Processing. In Proceedings of the 58th Annual Meeting of the Association for Computatioonal Linguistics (ACL), Tutorial, Online, 5–10 July 2020.
9. Hashimoto, C.; Torisawa, K.; Kloetzer, J.; Sano, M.; Varga, I.; Oh, J.H.; Kidawara, Y. Toward Future Scenario Generation: Extracting Event Causality Exploiting Semantic Relation, Context, and Association Features. In Proceedings of the 52th Annual Meeting of the Association for Computatioonal Linguistics (ACL), Baltimore, ML, USA, 23–24 June 2014.
10. Ning, Q.; Feng, Z.; Wu, H.; Roth, D. Joint Reasoning for Temporal and Causal Relations. In Proceedings of the 56th Annual Meeting of the Association for Computational Linguistics (ACL), Melbourne, Australia, 15–20 July 2018.
11. Pearl, J.; Mackenzie, D. *The Book of Why: The New Science of Cause and Effect*, 1st ed.; Basic Books, Inc.: New York, NY, USA, 2018.

12. Li, X.; Taheri, A.; Tu, L.; Gimpel, K. Commonsense Knowledge Base Completion. In Proceedings of the 54th Annual Meeting of the Association for Computational Linguistics (ACL), Berlin, Germany, 7–12 August 2016.
13. Sap, M.; Bras, R.L.; Allaway, E.; Bhagavatula, C.; Lourie, N.; Rashkin, H.; Roof, B.; Smith, N.A.; Choi, Y. ATOMIC: An Atlas of Machine Commonsense for If-Then Reasoning. In Proceedings of the 57th Annual Meeting of the Association for Computatioonal Linguistics (ACL), Florence, Italy, 28 July–2 August 2019.
14. Bosselut, A.; Rashkin, H.; Sap, M.; Malaviya, C.; Çelikyilmaz, A.; Choi, Y. COMET: Commonsense Transformers for Automatic Knowledge Graph Construction. In Proceedings of the 57th Annual Meeting of the Association for Computational Linguistics (ACL), Florence, Italy, 28 July–2 August 2019.
15. Radford, A.; Wu, J.; Child, R.; Luan, D.; Amodei, D.; Sutskever, I. Language models are unsupervised multitask learners. *OpenAI Blog* **2019**, *1*, 9.
16. Lewis, M.; Liu, Y.; Goyal, N.; Ghazvininejad, M.; Mohamed, A.; Levy, O.; Stoyanov, V.; Zettlemoyer, L. BART: Denoising Sequence-to-Sequence Pre-training for Natural Language Generation, Translation, and Comprehension. In Proceedings of the 58th Annual Meeting of the Association for Computational Linguistics (ACL), Online, 5–10 July 2020.
17. Krishna, R.; Zhu, Y.; Groth, O.; Johnson, J.; Hata, K.; Kravitz, J.; Chen, S.; Kalantidis, Y.; Li, L.J.; Shamma, D.A.; et al. Visual Genome: Connecting Language and Vision Using Crowdsourced Dense Image Annotations. *arXiv* **2016**, arXiv:1602.07332.
18. He, K.; Gkioxari, G.; Dollar, P.; Girshick, R. Mask R-CNN. In Proceedings of the IEEE International Conference on Computer Vision (ICCV), Venice, Italy, 22–29 October 2017.
19. Chen, Y.; Rohrbach, M.; Yan, Z.; Shuicheng, Y.; Feng, J.; Kalantidis, Y. Graph-Based Global Reasoning Networks. In Proceedings of the IEEE/CVF Conference on Computer Vision and Pattern Recognition (CVPR), Long Beach, CA, USA, 15–20 June 2019.
20. Kipf, T.N.; Welling, M. Semi-Supervised Classification with Graph Convolutional Networks. In Proceedings of the International Conference on Learning Representations (ICLR), Toulon, France, 24–26 April 2017.
21. Veličković, P.; Cucurull, G.; Casanova, A.; Romero, A.; Lio, P.; Bengio, Y. Graph attention networks. *arXiv* **2017**, arXiv:1710.10903.
22. Lamb, A.M.; ALIAS PARTH GOYAL, A.G.; Zhang, Y.; Zhang, S.; Courville, A.C.; Bengio, Y. Professor Forcing: A New Algorithm for Training Recurrent Networks. In Proceedings of the Advances in Neural Information Processing Systems (NIPS), Barcelona, Spain, 5–10 December 2016.
23. Alamri, H.; Cartillier, V.; Das, A.; Wang, J.; Cherian, A.; Essa, I.; Batra, D.; Marks, T.K.; Hori, C.; Anderson, P.;et al. Audio Visual Scene-Aware Dialog. In Proceedings of the IEEE/CVF Conference on Computer Vision and Pattern Recognition (CVPR), Long Beach, CA, USA, 15–20 June 2019.
24. Papineni, K.; Roukos, S.; Ward, T.; Zhu, W.J. Bleu: A Method for Automatic Evaluation of Machine Translation. In Proceedings of 40th Annual Meeting of the Association for Computational Linguistics, Philadelphia, PA, USA, 7–12 July 2002; pp. 311–318.
25. Banerjee, S.; Lavie, A. METEOR: An Automatic Metric for MT Evaluation with Improved Correlation with Human Judgments. In *Proceedings of the ACL Workshop on Intrinsic and Extrinsic Evaluation Measures for Machine Translation and/or Summarization*; Association for Computational Linguistics: Ann Arbor, MI, USA, 2005; pp. 65–72.
26. Lin, C.Y. ROUGE: A Package for Automatic Evaluation of Summaries. In *Proceedings of the Text Summarization Branches Out*; Association for Computational Linguistics: Barcelona, Spain, 2004; pp. 74–81.
27. Vedantam, R.; Lawrence Zitnick, C.; Parikh, D. CIDEr: Consensus-Based Image Description Evaluation. In Proceedings of the IEEE Conference on Computer Vision and Pattern Recognition (CVPR), Boston, MA, USA, 7–12 June 2015.
28. Hori, C.; Alamri, H.; Wang, J.; Wichern, G.; Hori, T.; Cherian, A.; Marks, T.K.; Cartillier, V.; Lopes, R.G.; Das, A.; et al. End-to-end Audio Visual Scene-aware Dialog Using Multimodal Attention-based Video Features. In Proceedings of the IEEE International Conference on Acoustics, Speech and Signal Processing (ICASSP), Brighton, UK, 12–17 May 2019.
29. Geng, S.; Gao, P.; Chatterjee, M.; Hori, C.; Roux, J.; Zhang, Y.; Li, H.; Cherian, A. Dynamic Graph Representation Learning for Video Dialog via Multimodal Shuffled Transformers. In Proceedings of the AAAI Conference on Artificial Intelligence (AAAI), Virtually, 2–9 February 2021.
30. Le, H.; Sahoo, D.; Chen, N.; Hoi, S.C. Multimodal Transformer Networks for End-to-End Video-Grounded Dialogue Systems. In Proceedings of the 57th Annual Meeting of the Association for Computational Linguistics (ACL), Florence, Italy, 28 July–2 August 2019.
31. Li, W.; Jiang, D.; Zou, W.; Li, X. TMT: A Transformer-based Modal Translator for Improving Multimodal Sequence Representations in Audio Visual Scene-aware Dialog. In Proceedings of the Interspeech, Shanghai, China, 25–29 October 2020.
32. Lin, X.; Bertasius, G.; Wang, J.; Chang, S.F.; Parikh, D.; Torresani, L. VX2TEXT: End-to-End Learning of Video-Based Text Generation From Multimodal Inputs. In Proceedings of the IEEE/CVF Conference on Computer Vision and Pattern Recognition (CVPR), Virtual, 19–25 June 2021.

remote sensing

MDPI

Article

Machine-to-Machine Visual Dialoguing with ChatGPT for Enriched Textual Image Description

Riccardo Ricci [1,†], Yakoub Bazi [2,†] and Farid Melgani [1,*]

1 Department of Information Engineering and Computer Science, University of Trento, 38123 Trento, Italy; riccardo.ricci-1@unitn.it

2 Department of Computer Engineering, College of Computer and Information Sciences, King Saud University, Riyadh 4545, Saudi Arabia; ybazi@ksu.edu.sa

* Correspondence: melgani@disi.unitn.it

† These authors contributed equally to this work.

Abstract: Image captioning is a technique that enables the automatic extraction of natural language descriptions about the contents of an image. On the one hand, information in the form of natural language can enhance accessibility by reducing the expertise required to process, analyze, and exploit remote sensing images, while on the other, it provides a direct and general form of communication. However, image captioning is usually restricted to a single sentence, which barely describes the rich semantic information that typically characterizes remote sensing (RS) images. In this paper, we aim to move one step forward by proposing a captioning system that, mimicking human behavior, adopts dialogue as a tool to explore and dig for information, leading to more detailed and comprehensive descriptions of RS scenes. The system relies on a questions–answers scheme fed by a query image and summarizes the dialogue content with ChatGPT. Experiments carried out on two benchmark remote sensing datasets confirm the potential of such an approach in the context of semantic information mining. Strengths and weaknesses are highlighted and discussed, as well as some possible future developments.

Keywords: ChatGPT; image captioning; visual question answering (VQA); visual question generation (VQG); visual dialoguing

Citation: Ricci, R.; Bazi, Y.; Melgani, F. Machine-to-Machine Visual Dialoguing with ChatGPT for Enriched Textual Image Description. *Remote Sens.* **2024**, *16*, 441. https://doi.org/10.3390/rs16030441

Academic Editors: Chiman Kwan, Moulay A. Akhloufi and Mozhdeh Shahbazi

Received: 19 September 2023
Revised: 28 December 2023
Accepted: 19 January 2024
Published: 23 January 2024

1. Introduction

Remote sensing images are characterized by rich and detailed semantics, which makes them difficult and time-consuming to analyze, especially for non-experts. Traditional approaches of automatic information extraction such as scene classification, semantic segmentation, and others can extract very low-level information that, usually, must be refined or reasoned further to translate it into a more useful source of information for the final user. In this context, remote sensing (RS) image captioning aims at bridging the need for a more complete and straightforward source of information by extracting information directly in the form of natural language, which is more interpretable by a human user. Image captioning teaches machines to maximize the probability of generating the ground truth description through end-to-end training of large neural networks. This can be difficult to achieve due to the enormous space of possibilities that image–description couples represent. Furthermore, image captioning often extracts a single sentence, which may be not enough to capture the rich and detailed semantics of (especially) RS images. In this paper, we try to model the image captioning process by proposing a different paradigm that, instead of directly optimizing a set of weights conditioned on a set of supervised training pairs, tackles the description problem in a more human-like way, namely through dialoguing. We argue that dialogue is beneficial in two ways: (1) it decomposes the problem into subproblems represented by various questions, whose solutions can be easier; (2) it enables exploring to a deeper degree the semantic content hidden in an image; and (3) it is less sensitive to the common problem of error accumulation in image captioning [1] in which an error in the prediction of a word will most likely influence all the other words by breaking down the description and so reduces the errors to sub-problems. In particular, we explore different solutions for enriched remote sensing image description under the paradigm of "description through dialogue", providing strengths and weaknesses of each, as well as directions for further research. The paper is organized as follows: in Section 2, after an introduction to the fundamentals of image captioning (IC) in the context of RS images, and various related research areas such as visual question answering (VQA) and visual question generation (VQG), we introduce our proposed paradigm of description through dialogue by bridging these different research areas. Specifically, we outline three different variants of the dialoguing system and the metrics used for assessment. Section 3 reports the results obtained by the various approaches, while Section 4 reports quantitative and qualitative results for all three solutions of the dialoguing system on both datasets, followed by a general discussion about the performance in different contexts. Finally, Section 5 outlines open issues and future directions, concluding the paper.

1.1. Related Research Areas

First, we want to introduce various techniques for the automatic extraction of information from images using natural language. Starting with the most seminal area of generative-based image captioning, additional research areas, namely visual question answering (VQA) and visual question generation (VQG), are also introduced. It is noteworthy that, in the following subsections, the concept of token refers to a generic piece of a text sequence, which can be different depending on the tokenization scheme.

1.1.1. RS Image Captioning (RSIC)

Remote sensing image captioning (RSIC) is the task of automatically describing a remote sensing image using natural language. RSIC is being applied to tackle multimodal image retrieval and change captioning [2] and can be used in other areas such as traffic command or rescue scenarios [3] by providing automatic text or voice descriptions of unmanned aerial vehicle (UAV) images. Generative-based remote sensing image captioning witnessed a similar evolution compared to the natural image counterpart. RSIC commonly relies on an encoder-decoder architecture, in which the encoder is devoted to the extraction of discriminative features from the image, and the decoder, conditioned by the extracted features, is employed to generate a description. For a practical example, the seminal paper [4] introduces an encoder-decoder architecture employing a pre-trained convolutional neural network (CNN) to extract image features and an auto-regressive decoder (RNN) to generate the description word-by-word conditioned on both the image features and the previously predicted words. Soon, building on this first approach, advances in encoder (deeper architectures [5]), decoder (gated recurrent unit (GRU [6], SVM-based decoder [7]), and in the interaction between the text and image features (multilevel features [8], attention mechanism [9–11]), enabled researchers to push further the limits of RSIC performance. Recently, the introduction of transformer architecture led to fully attention-based captioning systems [12]. All image captioning approaches based on deep neural networks model their task as an auto-regressive process, in which the next token is generated conditioned on the image features and the previously predicted tokens. More specifically, given an image–description pair, traditional image captioning algorithms try to maximize the following probability distribution:

$$P(\mathbf{T}|I) = \prod_{k=1}^{L} P(t_k|\mathbf{T}_{1...(k-1)}, I) \tag{1}$$

where $\mathbf{T}$ is a ground-truth description composed of L tokens and I is the vector of image features. Given a collection of image–description pairs, the network is enforced to increase the probability of assigning the correct description to every image in the dataset. Afterward, the trained network associates an input image with the highest fitting sequence of words according to the learned distribution. Such a paradigm has been widely applied in RSIC but presents some limitations. The first is that all datasets for RSIC are composed of images and short descriptions, which barely is enough to describe the wide variety of objects and concepts that characterize RS images. The second is the difficulty in learning such a complex distribution with the relatively small RS datasets that exist in the literature [4,8,13]. The cost associated with the collection and labeling of data for remote sensing image captioning makes it unfeasible to deliver comprehensive and large datasets adequate for an algorithm to reliably learn a good approximation of the true distribution, even if some efforts are being made in this direction [13]. This makes algorithms for remote sensing image captioning less robust and general, thus restricting their applicability.

1.1.2. RS Visual Question Answering (RSVQA)

The branch of visual question-answering research started shortly after image captioning [14]. The task in VQA is to answer user-defined questions by looking at the contents of an image. This enables users to interact with VQA systems to analyze images, retrieving specific information using natural language questions. In the context of remote sensing, a VQA system can be compared to an expert who can provide answers on specific concepts about remote sensing images. VQA methods may be subdivided into two main categories: closed-ended and open-ended answering. Closed-ended answering is defined as a classification task on a closed set of possible answers, modeling a probability distribution as:

$$P(a|\mathbf{Q}, I) \tag{2}$$

where a is one answer between a set of predetermined fixed answers, $\mathbf{Q}$ is the sequence of tokens representing the natural language question, and I is the vector of image features. In open-ended visual question answering, similarly to the task of image captioning, the answer is a sequence of tokens generated in an auto-regressive manner. The probability distribution modeled by open-ended VQA algorithms is reported in Equation (3), which closely resembles that modeled by a captioning system, except for the further conditioning given by the question Q:

$$P(\mathbf{A}|\mathbf{Q}, I) = \prod_{k=1}^{L} P(A_k|A_{1..k-1}, Q, I) \tag{3}$$

RS visual question answering (RSVQA) has been tackled for the first time in [15] using a closed-form answering paradigm. The authors used OpenStreetMap and a template-based question generation process to gather question–answer pairs for RS images, creating the first dataset for remote sensing visual question answering [15]. However, as shown by them, RSVQA suffers from the complexity of remote sensing scenes, for example, when dealing with the counting and numbering of objects. Despite being very recent, the field of RSVQA has witnessed some advancements, from the introduction of the attention mechanism [16], to ways that allow the algorithm to account for subsequent questions [17], to the leveraging of large language models in the context of prompt answering [18] and the adoption of the transformer architecture for both images and text and their fusion [19].

1.1.3. RS Visual Question Generation (RSVQG)

Visual question generation is a novel research area, which aims at automating the questioning process by generating coherent and sound questions for an input image. Visual question generation can be useful, for example, to gather data for visual question answering, where data labeling is very expensive and time-consuming. Other areas where visual question generation can be useful are in the field of child education, interactive lectures, and visual conversations [20]. VQG methods can be divided into three groups, namely rule-based, template-based, and generative-based visual question generation. In rule-based VQG, some specific syntactical rules are used to build questions from image descriptions [21]. Template-based VQG proposes to generate questions using templates and a set of objects, attributes, and relations. Those can be fixed [22] or automatically recognized in the input image using some visual detectors [23]. Generative-based VQG shares similarities with image captioning, in which both are required to generate text in an open-handed way. One peculiarity of generative-based VQG that differentiates it from image captioning is the requirement to generate not just one, but a set of plausible and coherent sequences of text (questions) for the same input image. Different generative-based methods have been envisioned to generate different questions for the same image. For example, in [22], questions are generated that explicitly explore objects detected in the image, further constraining the generation to predefined types of questions. In [24], different questions are created, starting from regional captions and adding the conditioning given by the question type. In [25], multiple question generation is addressed in an end-to-end leveraging the variational autoencoder framework. Despite advances in generative-based visual question generation, the application to remote sensing images is still an open area of research. From what we know, ref. [26] is the only work in the literature exploring generative-based visual question generation for RS images. In [26] the authors propose a transformer-based architecture to generate different plausible questions for RS images. They use questions from the dataset proposed in [15] to train the model in a fully supervised fashion.

2. Materials and Methods

In this work, we propose a system that can extract information from images, condensing it in a descriptive paragraph of natural language. The system building blocks are derived from the previously introduced research areas, bridging visual question generation with visual question answering. Its general idea, depicted in Figure 1, is to establish a machine-to-machine (M2M) visual dialoguing (VD) that requires no (or little) concrete human intervention and that enables information digging from images. The collected information is summarized at the end of the dialogue to generate the final output. This way of approaching the problem is different from standard image captioning. In image captioning, there is no real exploration of the information in an adaptive way, but all the information is already condensed in a single round output description, which can reduce the extraction of information. Under the dialoguing paradigm, the exploration is performed in steps; each question–answer pair allows the machine to explore more of the image content, possibly building on previously extracted knowledge to decide which follow-up questions to ask to maximize the collection of information. To formalize the idea, we envision a system fed by an input image and composed of two machines, one that is devoted to the task of asking questions, and the other that, based on a given question, provides an appropriate answer. This conceptual division is useful to obtain the idea of the actors involved in the process since it is not necessarily required that those be two distinct machines. Indeed, different variants can emerge, for example, with a single machine that provides both questions and answers or with predefined template questions. In this paper, we will explore different preliminary solutions, but as we expect, more can emerge in the future. We will explore some strengths and drawbacks of different strategies, trying to set the path for more exploration of this promising but challenging paradigm.

We present three different solutions to the dialoguing system. The first, called "Open-ended dialogue", is based entirely on the proposal outlined in [27]. We adopt the same configuration, experimenting with zero-shot dialoguing for remote-sensing images. The second, called "Closed-form dialogue", uses a predefined set of questions and generates a fixed dialogue for each image. The third, called "Closed-form dialogue with context", inherits the dialoguing approach of the previous closed-form solution but proposes a way to integrate the context given by the previous questions and answers in the answering process.

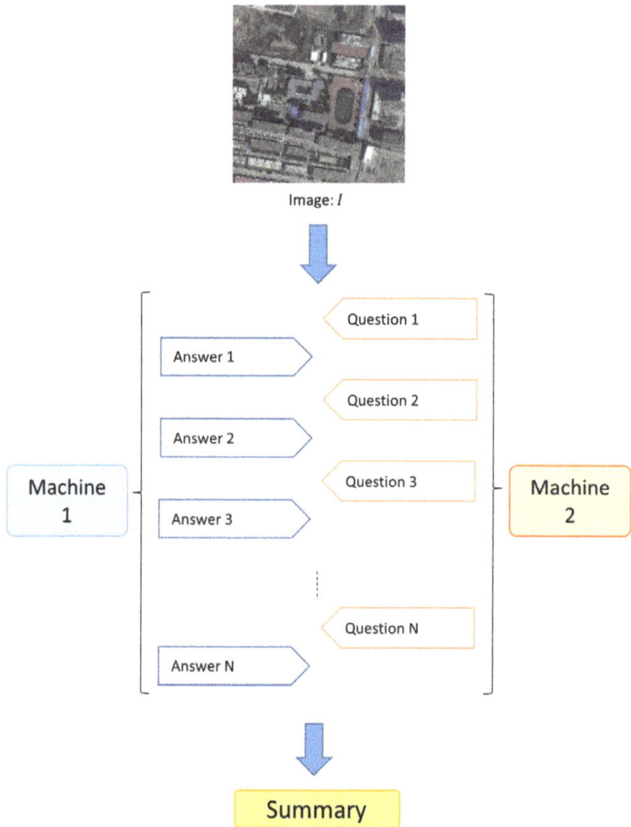

Figure 1. Conceptual representation of the machine-to-machine visual dialoguing (M2M-VD) paradigm.

2.1. Open-Ended Dialogue

In open-ended dialogue, the idea is to have two separate machines, one that produces questions and the other that answers. In this work, we adopt the method proposed in [27], in which Blip-2 [28] is used to answer questions generated by ChatGPT [29]. Blip-2 is a multimodal architecture that bridges the visual and text modalities by connecting a frozen image encoder to a frozen large language model (LLM). The goal is to preserve the reasoning ability of large language models while injecting image understanding by adaptively merging the two modalities through a so-called "query transformer". The alignment between the image encoder and the LLM follows a two-step procedure, using a corpus of 129M of images with corresponding text descriptions. In the first step, alignment of the extracted visual features is performed by connecting the query transformer to the frozen image encoder.

In detail, the authors make use of three pre-training objectives, namely (1) image-text contrastive learning (ITC), (2) image-grounded text generation (ITG), and (3) image-text matching (ITM). ITC aims at enhancing the mutual information between image–text representations of corresponding (positive) pairs while reducing that of non-corresponding (negative) pairs. ITG trains the query transformer to generate texts conditioned on the input image. This objective forces the queries to capture the most meaningful information from the frozen image encoder representation to generate the text. ITM consists of a binary classification of whether the image corresponds to the text (positive pair) or not (negative pair). The prediction is achieved by generating logits' scores from each query vector in the

Query Transformer and averaging to obtain the overall matching result. In the second step, the authors connect the query transformer to a frozen large language model. The goal of this stage is to align the large language model in generating text conditioned on the queries extracted by the query transformer. According to the authors, keeping the models frozen during alignment can mitigate the catastrophic forgetting problem, therefore preserving the ability to perform prompt-based text generation while including conditioning information from the image.

ChatGPT [29] is a fine-tuned version of GPT-3.5, and it belongs to the family of large language models. GPT-3.5 has been pre-trained to perform next token prediction in a self-supervised fashion on a huge amount of textual data collected from the web. Starting from the pre-trained GPT-3.5 checkpoint, in [30], authors use reinforcement learning from human feedback (RLHF) to enforce the instruction following capability. This passage equipped ChatGPT with superior capability in language understanding, generation, interaction, and reasoning.

Figure 2 depicts the block scheme for the open-ended dialogue system. In this framework, the question generation process is dynamically controlled by ChatGPT relying on the previous dialogue history (questions and answers). Specifically, the following steps are performed:

1. A fixed prompt "Describe this image in detail" is used to spark the conversation.
2. Blip-2 answers with a description of the image.
3. ChatGPT generates a question based on the description.
4. Blip-2 answers by looking at the image.
5. ChatGPT, using the context of the first description and the dialoguing history, produces another question to further explore the image contents.
6. Blip-2 answers by looking at the image.

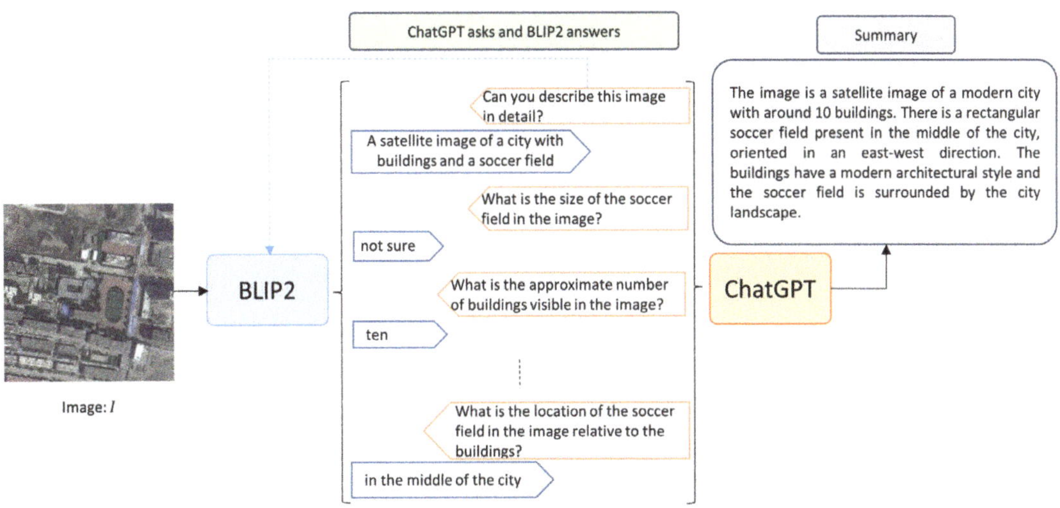

Figure 2. Open-ended dialogue block scheme.

Points (5) and (6) are repeated N times to simulate a dialogue between the questioning machine and the answering machine. In our experiments, we simulated a total of 10 rounds of questions and answers, so $N = 9$. At the end of the dialogue, ChatGPT is also leveraged to summarize the content of the conversation and generate the final description.

2.2. Closed-Form Dialogue

In closed-form dialogue, we propose an alternative to alleviate some of the drawbacks of the open-ended pipeline. The major weakness of our strategy is the generation of follow-up questions, which heavily relies on the first description generated by Blip-2. Therefore, if this is incorrect, the dialogue may evolve completely incorrectly. One way to alleviate this problem is to decouple the question generation from the dialogue history. In closed-form dialogue, a predetermined set of questions is used for every image. Closed-form dialoguing can be useful in situations in which (1) the user needs to collect specific information regarding the scenes and is thus able to define a fixed set of targeted questions or (2) the first description by Blip-2 does not capture enough/correct information about the image to process meaningful subsequent questions. In the second case, the predefined set of questions, if well-designed for the target images, allows for a more consistent extraction of information and thus for more reliable image descriptions. As depicted in Figure 3, in closed-form dialogue, Blip-2 is used to answer the questions listed in the fixed set, one by one. Then, the set of questions and corresponding answers composing the dialoguing history is summarized by ChatGPT to produce a descriptive paragraph.

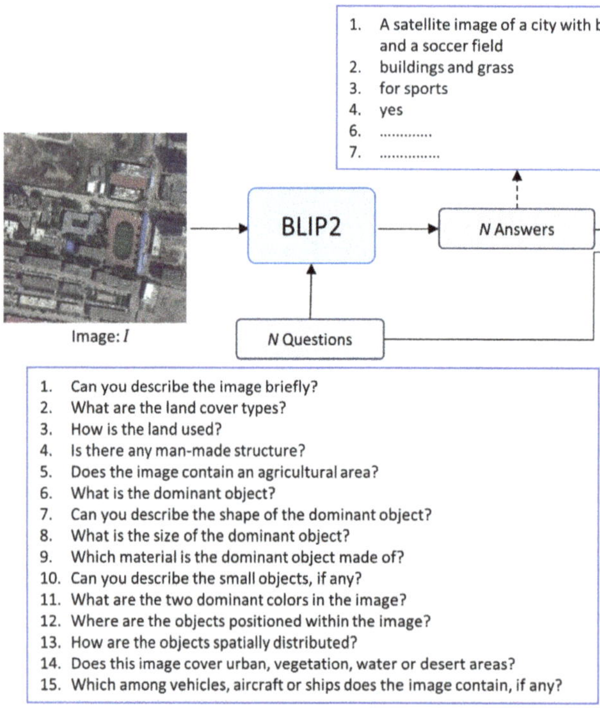

Figure 3. Closed-form dialogue block scheme.

2.3. Closed-Form Dialogue with Context

In closed-form dialogue with context, we try to move one step forward in exploring the interaction between the image and the previous dialogue, and how it can influence the generation of answers for subsequent questions. In Figure 4, we propose an encoder-decoder architecture based on Clip [31] and GPT-2 [32] models. We rely on a ViT-base-16 to encode the image I into a sequence of tokens $f(I)$. Then, we concatenate these tokens with the first question $[Q_1]$ and feed them as input to an auto-regressive distil-GPT-2 [33] decoder to predict the answer $\hat{A}_1$ in an auto-regressive manner. To form the contextual information for the subsequent question Q_2, we concatenate the previous question and its predicted answer as follows $[f(I), Q_1, \hat{A}_1, Q_2]$ to generate the answer $\hat{A}_2$. The process continues until the generation of the last answer $\hat{A_N}$ for the question Q_N with the following contextual information $[f(I), Q_1, \hat{A}_1, Q_2, \ldots, Q_{N-1}, \hat{A_{N-1}}, Q_N]$. After that, we feed the concatenation of all questions and answers to ChatGPT to generate the summary as in the previous methods. It is worth noting that fine-tuning the weights of this encoder-decoder architecture (see Figure 5) is achieved by optimizing the so-called cross-entropy loss on the training set, which can be given for one dialogue related to the image I as follows:

$$L = \frac{1}{N}\sum_{k=1}^{N} Cross\_entropy(A_k, \hat{A}_k) \quad (4)$$

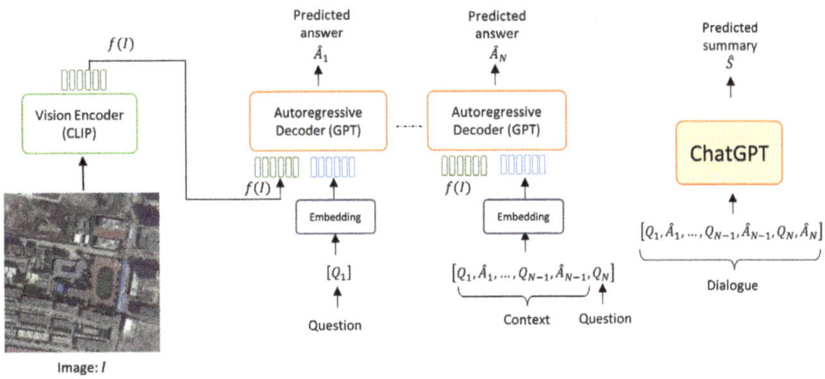

Figure 4. Block scheme of closed-form dialogue with context during inference phase.

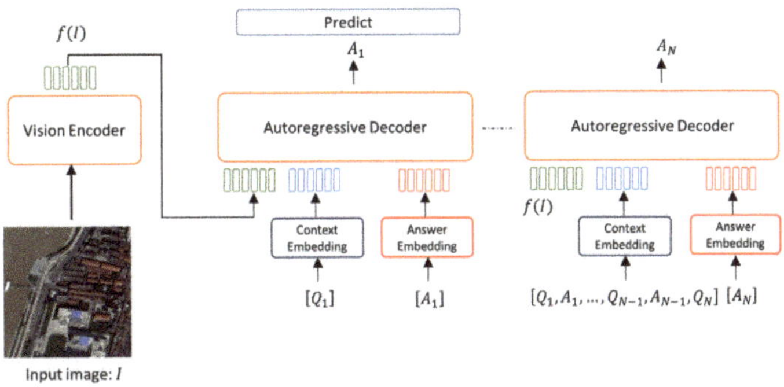

Figure 5. Closed-form dialogue with context during training phase.

2.4. Dataset and Metrics

Experiments have been carried out on two popular datasets for remote sensing image captioning, namely the RSICD dataset [8] and UCM-Captions dataset [4]. The RSICD dataset contains 10921 images of size 224 × 224 with varying spatial resolutions. Each image is annotated with a set of five captions. For some images, replication has been used to artificially augment the number of captions up to five. UCM-Captions includes 2100 aerial images of size 256 × 256 with a spatial resolution of one foot. This dataset is defined based on the UC Merced Land Use dataset, in which each image is associated with 1 of 21 land-use classes. Each image in the UCM-Captions dataset was annotated with five captions as well. Although five captions per image are available, in both datasets, captions of images belonging to the same class are very similar. Some examples of image–text couples for both datasets can be found in Figure 6. As stated in [27,34], assessing the performance in dialogue-based image description is a daunting problem, mainly because of the lack of ground-truth summaries to compare with. As explored in [27], the use of traditional metrics leads to huge drops in performance. In both these studies, authors use human assessment to evaluate the descriptions and the dialogues, which is expensive and time-consuming.

- This square playground is made up of two tennis courts coloured with green.
- Two small courts locate in a large grass.
- Two tennis courts surrounded by green meadows are near to several buildings.
- Two tennis fields are surrounded by green meadows.
- Two tennis courts are surrounded by a large piece of green meadow.

- A viaduct divided into some beige ring area and two gray straight viaduct.
- Viaduct is located in dark green vegetation area.
- A viaduct through the dark green plants and dark brown land.
- A pool with black water inside while surrounded by many dark green plants.
- Many green trees are near a viaduct.

- Lots of boats docked in lines at the harbor.
- Lots of boats docked neatly at the harbor.
- Many boats docked neatly at the harbor and the water is deep blue.
- Many boats docked neatly at the harbor and just a few positions are free.
- Lots of boats docked in lines at the harbor and just a few positions are free.

- There are two airplanes at the airport.
- Two airplanes are stopped at the airport.
- Two white airplanes are stopped at the airport.
- Two white airplanes are stopped at the airport with some cars beside it.
- Two white airplanes are stopped at the airport with some cars beside it.

Figure 6. Examples of images and corresponding captions from the RSICD and UCM-Captions datasets.

Therefore, we choose to experiment with other types of metrics, addressing the strengths and weaknesses of each in evaluating such results. The first is a reference-free metric called CLIPScore [35], another is text-to-image retrieval, and the last is a novel proposal, based on the generation of synthetic images from the summaries and comparison with the ground-truth images. CLIPScore is based on Clip [31], a multimodal deep neural network that learns shared vector representations of text and images. Clip has been trained on 400 M image–text pairs to align the representations of corresponding pairs while misaligning representations of mismatched pairs. In the contrastive learning paradigm used by Clip, "align" means to increase the cosine similarity between image and text representations. Leveraging the pre-trained Clip network, CLIPScore measures image–text compatibility using the following formula:

$$ClipScore(\vec{c}, \vec{v}) = 2.5[max(cosine(\vec{c}, \vec{v}), 0)] \quad (5)$$

where $\vec{c}$ is an n-dimensional vector representation of the text and $\vec{v}$ an n-dimensional vector representation of the image. In the experiments, we adopted the Clip-ViT-L/14 variant, obtaining 512-dimensional vectors. The coefficient of 2.5 is used to stretch the measure in the range $[0-1]$. Higher scores have been shown to correlate with better descriptions of images. To obtain a measure for an entire dataset, the simple average for all samples is used.

For the second metric, we leverage text-to-image retrieval as a prompt of description goodness. The motivation for this choice is that if with a generated text one can isolate the image from which it has been created from the other images in the dataset, this means that the text must be more descriptive and coherent with the image itself. Indeed, suppose that the same description is generated for every image, then the retrieval is purely a matter of chance. The more the descriptions are targeted to the relative image, the more the description can isolate an image from the rest. We choose to experiment with text-to-image retrieval using Clip embeddings. Specifically, we experiment with Clip-ViT-L/14. We will report three recall metrics, R@1, R@5, and R@10, in decreasing order of strictness. More in detail, the recall measure is computed as follows. We have a dataset of N images, and we generate a description for image i. Using Clip-ViT-L/14, we encode all the images in the dataset and the generated description, yielding I and $\vec{t}$. I is a matrix of size $N \times 768$, where $\vec{I_j}$ is the vector representation of the j-th image. $\vec{t}$ is a vector of size 768, representing the text. We then compute the cosine similarity between $\vec{t}$ and $\vec{I_j}$ using Equation (6), for $j = 1 \ldots N$

$$Cosine\_similarity = \frac{\vec{t} \cdot \vec{I_j}}{||\vec{t}|| \times ||\vec{I_j}||} \tag{6}$$

This gives us a vector $\vec{s}$ of size N, where s_j is a scalar corresponding to the cosine similarity between the text and the j-th image in the dataset. We sort $\vec{s}$ in descending order and analyze the indices of the sorted array. If the i-th index of the image we are analyzing is falling within the first m entries, we increase the R@m by 1. Therefore, for example, R@10 is using $m = 10$, R@5 $m = 5$, and so on. We repeat this process for all the images and compute the recall values.

The last measure we want to introduce has never been used in this context. The main idea is that if an image can be faithfully reconstructed from its description, the description is a faithful distillation of the image contents. This new evaluation method, which we call "text-to-image and comparison" (T2IC), is based on generating synthetic images from the extracted descriptions and comparing them with the original images. The higher the similarity, the more compatible and richer the image description. From each generated description, a synthetic image has been generated using a fine-tuned version of stable-diffusion-v1-4 [36]. Fine-tuning has been performed using the union of both UCM and RSICD training set images and ground-truth captions to target the generation of remote-sensing images. We used the AdamW optimizer with a learning rate of 1×10^{-5} , for 20 epochs. The measure is then articulated in two parts. The first works on couples of true and synthetic images, extracting image representations using InceptionV3 [37] and measuring the cosine similarity between them as follows:

$$T2IC = \frac{1}{N} \sum_{k=1}^{N} cosinesim(\vec{i_k}, \vec{\hat{i}_k}) \tag{7}$$

where $\vec{i_k}$ is the n-dimensional feature vector extracted from the $k-th$ original image I_k and $\vec{\hat{i}_k}$ is the n-dimensional feature vector extracted from the $k-th$ synthetic image $\hat{I}_k$. The other is the FID score, a widely used score to compare image generation systems. FID directly compares an entire dataset of original images with a dataset of synthetic images. It does so by extracting image representations with InceptionV3 and fitting two Gaussian distributions on top of the extracted feature, calculating the mean and the covariance matrix of both the original and the synthetic image distribution. Then, the FID score is calculated as the distance between the two Gaussians, measured as:

$$FID = ||\vec{m} - \hat{\vec{m}}||_2^2 + Tr(\mathbf{C} + \hat{\mathbf{C}} - 2(\mathbf{C}\hat{\mathbf{C}})^{\frac{1}{2}}) \tag{8}$$

where $\vec{m}$ and $\hat{\vec{m}}$ are the $2048 - mean$ vector for the distributions of original and synthetic images, respectively, while $\mathbf{C}$ and $\hat{\mathbf{C}}$ are the covariance matrices estimated for the distribution of original and synthetic images, respectively. The lower the FID score, the more the two distributions are similar, hence the synthetic images are similar to the original ones.

2.5. Prompts

Large language models have been shown to have great capability of prompt-following. Prompts are pieces of text that are placed before the request to direct the model to satisfy the request. As both Blip-2 and ChatGPT employ large language models, authors in [27] used specific prompts for both, placing them before question–answer generation.

2.5.1. ChatGPT Prompts

1. I have an image. Ask me questions about the content of this image. Carefully asking me informative questions to maximize your information about this image content. Each time ask one question only without giving an answer. Avoid asking yes/no questions. I'll put my answer beginning with "Answer":
2. Next Question. Avoid asking yes/no questions. Question:
3. Now summarize the information you get in a few sentences. Ignore the questions with answers no or not sure. Don't add information. Don't miss information. Summary:

Prompt (1) is used at the beginning of the dialogue to provide information to ChatGPT about what the user wants and the format in which the user wants the questions to be generated. Prompt (2) is to gather follow-up questions, and it is placed after the previous dialogue history to ask the model to provide another question in the correct format. Prompt (3) is used at the end of the dialogue to ask the model to summarize the dialogue history in a descriptive paragraph.

2.5.2. Blip-2 Prompts

1. Answer given questions. If you are not sure about the answer, say you don't know honestly. Don't imagine any contents that are not in the image.
2. Answer:

Prompt (1) is used to condition Blip-2 to output an answer only if it is sure that it is the correct answer. In [27], the authors show that if this prompt is not used, the model tends to hallucinate concepts more often. Prompt (2) is used for follow-up answers.

3. Results

The results will be split by dataset. We will report quantitative results for each metric. At the end of the section, some qualitative results are reported to further demonstrate the capability of the proposed pipeline.

In Tables 1 and 2, the quantitative results of the three solutions described previously, namely open-ended dialogue (OED), closed-form dialogue (CFD), and closed-form dialogue with context (CFD-C), are reported. We compare the performance obtained using each generated summary by OED, CFD, and CFD-C with the results obtained using the initial answer of Blip-2. In this way, we can assess if dialoguing successfully adds value to the initial description, which is a caption-style piece of text. Inspecting the values of each of the defined metrics, we can see a trend, in which the dialogue obtained by CFD achieves the highest CLIPScore, while dialogue obtained with OED achieves the highest recall scores. This can in part be explained by the different approaches taken by OED and CFD. Indeed, in CFD, the predefined set of questions is not affected by a wrong initial description, because the question generation simply does not rely on it. For this reason, we think that summaries generated by CFD are more reliable than summaries generated with OED in situations in which the first description is inaccurate. On the other hand, the recall measure expects that the summary is more targeted because the more the summary is targeted and not general, the more an image can be differentiated from all the other images. OED obtains higher scores on the recall metrics because it does not rely on a predefined template of questions and has more possibility to target the dialogue to the input image under consideration. Lastly, CFD-C obtains the lowest recall scores, while it places in the middle between CFD and OED according to the CLIPScore. It seems that the context embedded in CFD-C is not able to improve the performance. The training using weak annotations is probably not effective since errors produced in CFD are replicated by CFD-C. We expect that with correct annotations, CFD-C can achieve higher performance by leveraging previous questions and answers to predict the next answer. Some qualitative results for both datasets are shown in Figures 7 and 8, where it can be noticed how the dialogues in the OED method are more focused, with questions that in some cases tend to be too specific and thus receive undefined answers. The CFD and CFD-C methods have more general questions and receive more general answers. In the third example, the network hallucinates some aircrafts, highlighting the necessity of a correct and coherent predefined set of questions. The T2IC score between the original and the images generated from the descriptions depict OED as the best in terms of FID score, which suggests that summaries generated through dialogue are indirectly better than the others since they allow stable-diffusion-v1-4 to reconstruct images that are more similar to the originals. Based on the T2IC score, on the RSICD dataset, it is not possible to identify the best method, with OED and CFD sharing similar scores. We believe that this can be a more truthful conclusion since RSICD is larger than UCM, and so the statistics are more reliable. Despite being very promising, this measure needs further work to be more reliable in this context. Indeed, better image generation algorithms must be used, targeted to the remote sensing scenario. We expect that leveraging image generation from rich and detailed descriptions can provide the foundation for a wider use of this measure in this context. Some examples of real and synthetic images can be found in Figure 9. It can be seen how summaries generated through dialoguing can collect more information about color, number of objects, and spatial distribution.

Table 1. Results on UCM dataset. Bold indicates the best result.

		Blip-2	**MLAT**	**Summary**		
				OED	**CFD**	**CFD-C**
Clipscore		65.2	69.0	66.1	**72.7**	67.8
ViT-L/14	R@1	21.9	4.8	**25.2**	22.9	9.0
	R@5	59.5	20.0	**61.0**	59.5	36.2
	R@10	83.8	37.1	**84.8**	81.4	63.3
T2IC	CosineSim	0.59	0.59	**0.60**	0.59	0.58
	FID	260.68	241.40	**238.71**	248.33	250.28

Table 2. Results on RSICD dataset. Bold indicates the best result.

		Blip-2	**MLAT**	**Summary**		
				OED	**CFD**	**CFD-C**
Clipscore		65.2	71.1	66.1	**72.7**	67.8
ViT-L/14	R@1	9.2	2.56	**11.0**	6.7	3.9
	R@5	29.4	12.53	**32.7**	24.4	17.2
	R@10	44.3	21.68	**49.3**	38.6	28.5
T2IC	CosineSim	0.62	0.62	**0.63**	**0.63**	0.62
	FID	140.02	151.23	**122.03**	122.89	128.65

Figure 7. Examples of dialogues and summaries on RSICD images (**top row**) and UCM-Captions images (**bottom row**).

Figure 8. Examples of dialogues and summaries on RSICD images (**top row**) and UCM-Captions images (**bottom row**).

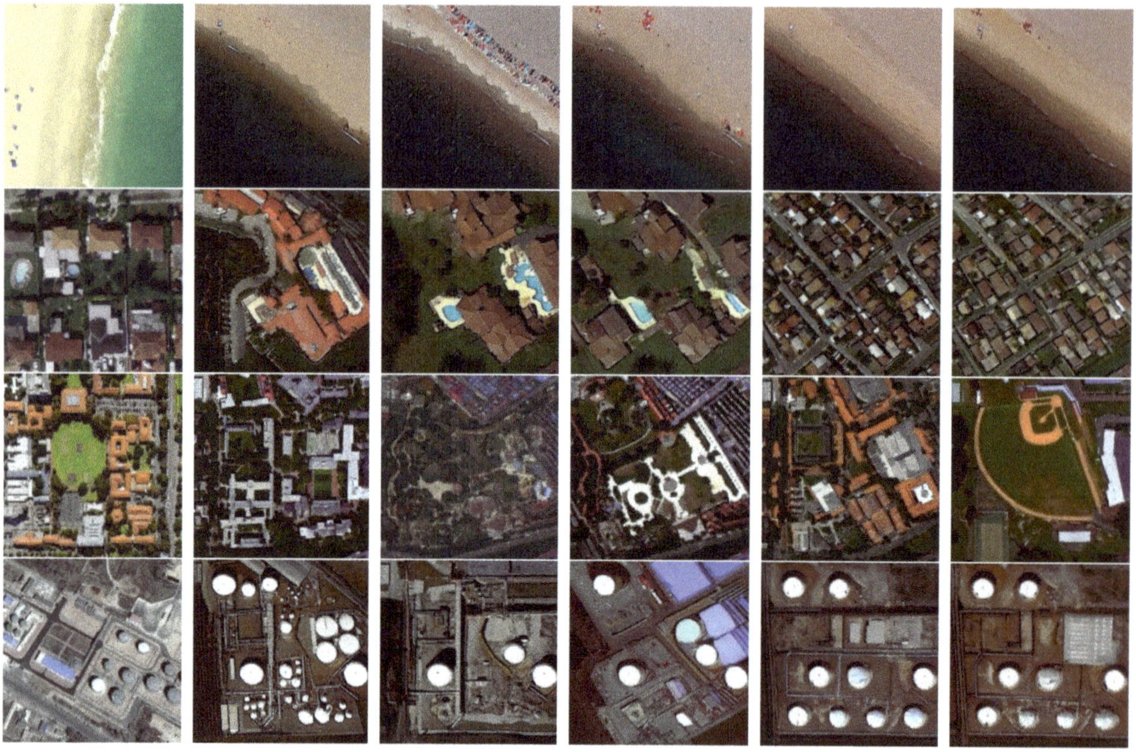

(a) (b) (c) (d) (e) (f)

Figure 9. Original (**a**) and synthetic images generated starting from descriptions generated with different methods. (**b**) MLAT, (**c**) Blip-2, (**d**) OED, (**e**) CFD, and (**f**) CFD-C.

4. Discussion

This paper presents a new approach for remote sensing image description generation that, thanks to a dialogue between two machines, allows it to further dig information from images and convert it into a natural language. The approach is innovative with respect to traditional captioning methods, which instead produce a one-shot description that barely can capture the rich semantics of remote sensing images. Despite the promising results, several weaknesses must be addressed to build a more robust and customized pipeline.

4.1. Input Image Integration

In this work, the conditioning of the dialogue on the input image is limited to the answering process. The questioner, a large language model (ChatGPT), exclusively relies on the textual exchange to build subsequent questions. We noticed that this can severely impact the performance when the initial description from Blip-2 is inaccurate. Despite the closed-ended strategy seeming to cope better with this problem, we argue that a better integration of the visual information in the dialoguing process is of outmost importance and can improve the overall process by making the questions more targeted to the image itself.

4.2. Adaptation of Answerer and Questioner to the RS Context

Both the networks employed in OED and CFD have been used in a zero-shot fashion, leading to criticalities when the network is presented with remote sensing images. In particular, the answerer is the most critical part since it deals with the image. The questioner is not directly affected, since it relies only on the textual information to generate questions, but errors in the answerer indirectly impact the generation of subsequent questions. To alleviate this problem, remote sensing visual question answering datasets can be exploited to refine the algorithm that generates the answers. The questioner, on the other hand, can be fine tuned using ground-truth conversation, opening the path to the creation of a dataset for RS visual dialoguing. However, the expensiveness of this process creates the need to find solutions that lie somewhere between human and automated annotation, such as trying to filter the conversations created by isolating the most coherent and plausible conversations for the remote sensing domain. Lastly, we point out that our proposed CFD-C is not to be considered a fine-tuned version, since it uses the results of CFD as soft labels for training.

4.3. Removal of Uncertain Question–Answer Pairs

With the use of the uncertainty prompt in Blip-2, several questions, especially in the open-ended dialogue, receive "I don't know" or "not sure" answers. Authors in [28] adopted a specific ChatGPT prompt to try to eliminate this redundant and useless information when generating the summary. However, in our context, this approach is not effective, and we argue that authors in [28] did not notice this criticality because they received less answers of this type, being in a natural image scenario, where Blip-2 is more effective. A simpler way to deal with such a situation can be to remove all the question–answer pairs that receive uncertain answers prior to the summarization step.

4.4. Better Predefined Questions for CFD Method

We noticed that the definition of some questions for the CFD method turned out to be a weakness. This is the case, for example, of the question "Which among vehicles, aircraft or ships does the image contain, if any?", that, in situations where none of these objects are present, misleads Blip-2, leading to erroneous outputs. Specifically, for CFD, creating more specific templates can lead to more detailed descriptions, especially in situations where the user already knows which information to extract.

4.5. Evaluation Metrics and Generation of Targeted Datasets

The evaluation strategies adopted in this paper use different kinds of reference-free metrics. In general, these metrics provide indirect measures of the goodness of the generated dialogues. To fully evaluate the consistency of a dialogue directly of the generated description, it would be necessary to have ground-truth summaries generated by human operators. This could give rise to new datasets targeted to the task of richer image description through a paragraph. However, the generation of ground-truth summaries is very costly and time-consuming. As pointed out in subsection B, joint use of human operator and machine-generated data can be useful to speed up the creation of such datasets. At the same time, research on other suitable reference-free metrics must be conducted, as it offers a valuable alternative.

4.6. Customizing the Dialogue and Multimodality

Another aspect that can be worth exploring is the targeting of the dialogue on the need of the user. As explored in this work, large language models adopt prompts to focus their attention on specific outputs needed by the user. This can give rise to multimodal-driven dialoguing, in which the user is also able, through the definition of specific prompts, to direct the attention of the dialoguing process to specific aspects of the image that are worth analyzing. Finding the best way to introduce such conditioning on the dialogue can be a major advancement, since it enables users to interact with the system in a natural and engaging way through natural language. On the other hand, it can provide the user with a more useful and detailed description of the highlighted concepts. Practically, this approach can be seen as an automatic generalization of CFD, in which instead of fixing a priori the questions, the user directs the questioner in producing set of questions targeted to the need of the moment, alleviating the burden of the question definition. Moreover, one may extend the future perspective to scenarios where the user not only may interact with the dialoguing system but feeds it with other sources of information such as ancillary information from geographic information systems.

5. Conclusions

In this paper, we propose three approaches to extract richer textual descriptions of remote sensing images. Our three approaches use machine-to-machine dialogue to extract information sequentially and in small bits. Once the dialogue is concluded, its content is summarized in a paragraph, creating the final description. We evaluate the coherence of our captions using different indirect metrics, where our open-handed dialogue generally obtains the higher scores. Each of the approaches is targeted to a different use case. Where open-handed dialogue can retrieve general descriptions, in certain applications where it is possible to define a set of predefined set of questions, closed-form dialoguing can be the best choice to customize the descriptions.

Author Contributions: Conceptualization, R.R., Y.B. and F.M.; Methodology, R.R. and Y.B.; Validation, R.R. and Y.B.; Writing—original draft, R.R.; Writing—review & editing, Y.B. and F.M.; Supervision, F.M.; Project administration, F.M. All authors have read and agreed to the published version of the manuscript.

Funding: This research received no external funding.

Data Availability Statement: The datasets used in this work are publicly available and can be found https://github.com/201528014227051/RSICD_optimal (accessed on 18 January 2024). Data are contained within the article. We release all our code https://github.com/RicRicci22/M2MVD.git (accessed on 18 January 2024).

Conflicts of Interest: The authors declare no conflicts of interest.

Abbreviations

The following abbreviations are used in this manuscript:

R@1-5-10	Recall value considering the first match, the first five, and the first ten, respectively.
T2IC	Text-to-Image and Comparison
FID	Frechet Inception Distance
OED	Open-Ended Dialogue
CFD	Closed-Form Dialogue
CFD-C	Closed-Form Dialogue with Context

References

1. Wu, L.; Tan, X.; He, D.; Tian, F.; Qin, T.; Lai, J.; Liu, T.Y. Beyond Error Propagation in Neural Machine Translation: Characteristics of Language. *arXiv* **2018**, arXiv:1809.00120.
2. Hoxha, G.; Chouaf, S.; Melgani, F.; Smara, Y. Change Captioning: A New Paradigm for Multitemporal Remote Sensing Image Analysis. *IEEE Trans. Geosci. Remote Sens.* **2022**, *60*, 1–14. [CrossRef]
3. Shi, Z.; Zou, Z. Can a Machine Generate Humanlike Language Descriptions for a Remote Sensing Image? *IEEE Trans. Geosci. Remote Sens.* **2017**, *55*, 3623–3634. [CrossRef]
4. Qu, B.; Li, X.; Tao, D.; Lu, X. Deep semantic understanding of high resolution remote sensing image. In Proceedings of the 2016 International Conference on Computer, Information and Telecommunication Systems (CITS), Kunming, China, 6–8 July 2016 ; pp. 1–5. [CrossRef]
5. He, K.; Zhang, X.; Ren, S.; Sun, J. Deep Residual Learning for Image Recognition. *arXiv* **2015**, arXiv:1512.03385.
6. Cho, K.; van Merrienboer, B.; Gulcehre, C.; Bahdanau, D.; Bougares, F.; Schwenk, H.; Bengio, Y. Learning Phrase Representations using RNN Encoder-Decoder for Statistical Machine Translation. *arXiv* **2014**, arXiv:1406.1078.
7. Hoxha, G.; Melgani, F. A Novel SVM-Based Decoder for Remote Sensing Image Captioning. *IEEE Trans. Geosci. Remote Sens.* **2022**, *60*, 1–14. [CrossRef]
8. Lu, X.; Wang, B.; Zheng, X.; Li, X. Exploring Models and Data for Remote Sensing Image Caption Generation. *IEEE Trans. Geosci. Remote Sens.* **2018**, *56*, 2183–2195. [CrossRef]
9. Huang, W.; Wang, Q.; Li, X. Denoising-Based Multiscale Feature Fusion for Remote Sensing Image Captioning. *IEEE Geosci. Remote Sens. Lett.* **2021**, *18*, 436–440. [CrossRef]
10. Zhang, X.; Wang, X.; Tang, X.; Zhou, H.; Li, C. Description Generation for Remote Sensing Images Using Attribute Attention Mechanism. *Remote Sens.* **2019**, *11*, 612. [CrossRef]

11. Zhang, Z.; Diao, W.; Zhang, W.; Yan, M.; Gao, X.; Sun, X. LAM: Remote Sensing Image Captioning with Label-Attention Mechanism. *Remote Sens.* **2019**, *11*, 2349. [CrossRef]
12. Wang, J.; Chen, Z.; Ma, A.; Zhong, Y. Capformer: Pure Transformer for Remote Sensing Image Caption. In Proceedings of the IGARSS 2022 IEEE International Geoscience and Remote Sensing Symposium, Kuala Lumpur, Malaysia, 17–22 July 2022 ; pp. 7996–7999. [CrossRef]
13. Cheng, Q.; Huang, H.; Xu, Y.; Zhou, Y.; Li, H.; Wang, Z. NWPU-Captions Dataset and MLCA-Net for Remote Sensing Image Captioning. *IEEE Trans. Geosci. Remote Sens.* **2022**, *60*, 1–19. [CrossRef]
14. Antol, S.; Agrawal, A.; Lu, J.; Mitchell, M.; Batra, D.; Zitnick, C.L.; Parikh, D. VQA: Visual Question Answering. In Proceedings of the IEEE International Conference on Computer Vision (ICCV), Santiago, Chile, 7–13 December 2015.
15. Lobry, S.; Marcos, D.; Murray, J.; Tuia, D. RSVQA: Visual Question Answering for Remote Sensing Data. *IEEE Trans. Geosci. Remote Sens.* **2020**, *58*, 8555–8566. [CrossRef]
16. Zheng, X.; Wang, B.; Du, X.; Lu, X. Mutual Attention Inception Network for Remote Sensing Visual Question Answering. *IEEE Trans. Geosci. Remote Sens.* **2022**, *60*, 1–14. [CrossRef]
17. Yuan, Z.; Mou, L.; Wang, Q.; Zhu, X.X. From Easy to Hard: Learning Language-Guided Curriculum for Visual Question Answering on Remote Sensing Data. *IEEE Trans. Geosci. Remote Sens.* **2022**, *60*, 1–11. [CrossRef]
18. Chappuis, C.; Zermatten, V.; Lobry, S.; Le Saux, B.; Tuia, D. Prompt–RSVQA: Prompting visual context to a language model for Remote Sensing Visual Question Answering. In Proceedings of the 2022 IEEE/CVF Conference on Computer Vision and Pattern Recognition Workshops (CVPRW), New Orleans, LA, USA, 19–20 June 2022; pp. 1371–1380. [CrossRef]
19. Bazi, Y.; Rahhal, M.M.A.; Mekhalfi, M.L.; Zuair, M.A.A.; Melgani, F. Bi-Modal Transformer-Based Approach for Visual Question Answering in Remote Sensing Imagery. *IEEE Trans. Geosci. Remote Sens.* **2022**, *60*, 1–11. [CrossRef]
20. Patil, C. Visual Question Generation: The State of the Art. *ACM Comput. Surv.* **2020**, *53*, 1–22. [CrossRef]
21. Ren, M.; Kiros, R.; Zemel, R. Exploring Models and Data for Image Question Answering. *arXiv* **2015**, arXiv:1505.020742.
22. Geman, D.; Geman, S.; Hallonquist, N.; Younes, L. Visual Turing test for computer vision systems. *Proc. Natl. Acad. Sci. USA* **2015**, *112*, 3618–3623. [CrossRef]
23. Yang, J.; Lu, J.; Lee, S.; Batra, D.; Parikh, D. Visual Curiosity: Learning to Ask Questions to Learn Visual Recognition. *arXiv* **2018**, arXiv:1810.00912.
24. Vedd, N.; Wang, Z.; Rei, M.; Miao, Y.; Specia, L. Guiding Visual Question Generation. *arXiv* **2012**, arXiv:2110.08226.
25. Jain, U.; Zhang, Z.; Schwing, A. Creativity: Generating Diverse Questions using Variational Autoencoders. *arXiv* **2017**, arXiv:1704.03493.
26. Bashmal, L.; Bazi, Y.; Melgani, F.; Ricci, R.; Al Rahhal, M.M.; Zuair, M. Visual Question Generation From Remote Sensing Images. *IEEE J. Sel. Top. Appl. Earth Obs. Remote Sens.* **2023**, *16*, 3279–3293. [CrossRef]
27. Zhu, D.; Chen, J.; Haydarov, K.; Shen, X.; Zhang, W.; Elhoseiny, M. ChatGPT Asks, BLIP-2 Answers: Automatic Questioning Towards Enriched Visual Descriptions. *arXiv* **2023**, arXiv:2303.06594.
28. Li, J.; Li, D.; Savarese, S.; Hoi, S. BLIP-2: Bootstrapping Language-Image Pre-training with Frozen Image Encoders and Large Language Models. *arXiv* **2023**, arXiv:2301.12597.
29. Brown, T.B.; Mann, B.; Ryder, N.; Subbiah, M.; Kaplan, J.; Dhariwal, P.; Neelakantan, A.; Shyam, P.; Sastry, G.; Askell, A.; et al. Language Models are Few-Shot Learners. *arXiv* **2020**, arXiv:2005.14165.
30. Ouyang, L.; Wu, J.; Jiang, X.; Almeida, D.; Wainwright, C.L.; Mishkin, P.; Zhang, C.; Agarwal, S.; Slama, K.; Ray, A.; et al. Training language models to follow instructions with human feedback. *arXiv* **2022**, arXiv:2203.02155.
31. Radford, A.; Kim, J.W.; Hallacy, C.; Ramesh, A.; Goh, G.; Agarwal, S.; Sastry, G.; Askell, A.; Mishkin, P.; Clark, J.; et al. Learning Transferable Visual Models From Natural Language Supervision. *arXiv* **2021**, arXiv:2103.00020.
32. Radford, A.; Wu, J.; Child, R.; Luan, D.; Amodei, D.; Sutskever, I. Language Models are Unsupervised Multitask Learners. *OpenAI Blog* **2019**, *1*, 9.
33. Huggingface: Distilgpt2. Available online: https://huggingface.co/distilgpt2 (accessed on 14 April 2022).
34. Yang, Y.; Li, Y.; Fermuller, C.; Aloimonos, Y. Neural Self Talk: Image Understanding via Continuous Questioning and Answering. *arXiv* **2015**, arXiv:1512.03460.
35. Hessel, J.; Holtzman, A.; Forbes, M.; Bras, R.L.; Choi, Y. CLIPScore: A Reference-free Evaluation Metric for Image Captioning. *arXiv* **2022**, arXiv:2104.08718
36. Huggingface: CompVis/Stable-Diffusion-v1-4. Available online: https://huggingface.co/CompVis/stable-diffusion-v1-4 (accessed on 14 April 2022).
37. Szegedy, C.; Vanhoucke, V.; Ioffe, S.; Shlens, J.; Wojna, Z. Rethinking the Inception Architecture for Computer Vision. *arXiv* **2015**, arXiv:1512.00567.

sensors

Article

Dual-Scale Doppler Attention for Human Identification

Sunjae Yoon [1], Dahyun Kim [1], Ji Woo Hong [1], Junyeong Kim [2] and Chang D. Yoo [1,*]

[1] School of Electrical Engineering, Korea Advanced Institute of Science and Technology, Daejeon 34141, Korea
[2] Department of AI, Chung-Ang University, Seoul 06974, Korea
* Correspondence: cd_yoo@kaist.ac.kr; Tel.: +82-10-3774-1007

Abstract: This paper considers a Deep Convolutional Neural Network (DCNN) with an attention mechanism referred to as Dual-Scale Doppler Attention (DSDA) for human identification given a micro-Doppler (MD) signature induced as input. The MD signature includes unique gait characteristics by different sized body parts moving, as arms and legs move rapidly, while the torso moves slowly. Each person is identified based on his/her unique gait characteristic in the MD signature. DSDA provides attention at different time-frequency resolutions to cater to different MD components composed of both fast-varying and steady. Through this, DSDA can capture the unique gait characteristic of each person used for human identification. We demonstrate the validity of DSDA on a recently published benchmark dataset, IDRad. The empirical results show that the proposed DSDA outperforms previous methods, using a qualitative analysis interpretability on MD signatures.

Keywords: deep learning; human identification; micro-Doppler radar; fine-grained feature analysis

Citation: Yoon, S.; Kim, D.; Hong, J.W.; Kim, J.; Yoo, C.D. Dual-Scale Doppler Attention for Human Identification. *Sensors* **2022**, *22*, 6363. https://doi.org/10.3390/s22176363

Academic Editors: Moulay A. Akhloufi and Mozhdeh Shahbazi

Received: 26 July 2022
Accepted: 20 August 2022
Published: 24 August 2022

1. Introduction

Human Identification (HI) serves as an essential building block for many personal identification services including surveillance, security and identification systems. In general, HI adopts visual video as it has information that can be easily understood. However, HI through visual video can be problematic under low lighting conditions and with the privacy infringement issues [1]. As an alternative to the use of cameras, radar devices bypass these problems. Radar is a detection system that measures the distance, direction, angle, and speed of a target by emitting electromagnetic waves from a device and analyzing electromagnetic waves that are reflected and returned from an object. Radar can operate under low light conditions, and its tendency to bend around obstacles makes it suitable for identification in obscured environments [2–4]. Moreover, radar information is relatively safe regarding privacy, as people cannot directly interpret information obtained by radar. Furthermore, radar also has the advantage of observing far-distance targets. Overall, radar is a more robust sensor for HI than visual video. However, since it is difficult to observe the iris, voice, and face using radar, this paper adopts gait instead of aforementioned biometrics. Gait can be observed from a distance, and unlike the biometrics (iris, voice, and face), it is behavior over a certain period of time, so the security is relatively high. These advantages are consistent with the reason for using radar-based HI instead of video-based HI.

For this radar-based HI, the micro-Doppler (MD) signature has been a popular choice. As shown in Figure 1, it sequentially records Doppler effects in electromagnetic signals of moving targets [5], and the superposition of these Doppler signals is summarized to make MD signatures, where it holds granularity to specify their information. In the methodology, conventional machine learning (ML) algorithms have attempted to analyze the statistics of the MD signatures [6–9]. Unfortunately, ML has several drawbacks as it is based on heuristic feature extraction and a low capacity model. Recent Deep Convolutional Neural Network (DCNN), via capturing spatial relationships and comprehending high-level spatial features, overcomes these limitations and has revolutionized many applications including radar-based human identification and motion recognition. Motion recognition is easily conducted by training DCNN to recognize the common patterns among identical motions. However, human identification [10,11] should identify the unique characteristics of each person in an uncontrolled scenario, which requires more fine-grained understanding than motion recognition.

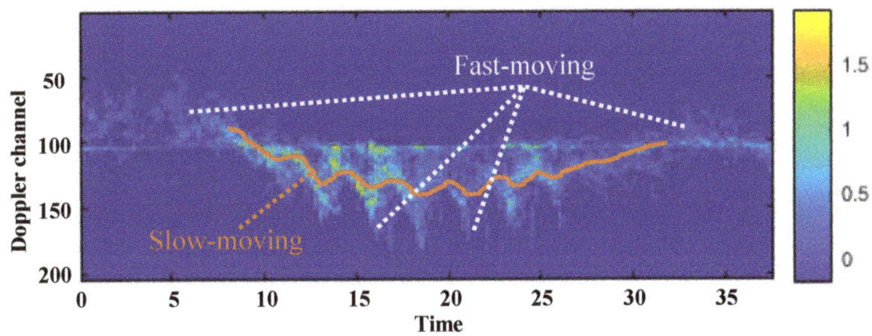

Figure 1. Micro-Doppler signature of human walking. When a person walks, the arms and legs generate fast-moving signatures represented by a white line, and the torso generates slow-moving signatures indicated by an orange line.

For these fine-grained understandings, we exploit that MD signatures from different body parts such as arms, legs, and torso display different signal characteristics. Each of us has a unique gait characteristic distinguished by different sized body parts moving and swinging in a distinctive pattern, as the arms and legs move rapidly, while the torso moves slowly. Our empirical analysis validates that more than 95% of MD signatures on the radar dataset (Validation is performed on the IDRad dataset) include this distinguishable gait characteristic. Therefore, we utilize these unique gait characteristics from the MD signatures to identify humans. For the detailed explanation of the gait characteristics of MD signatures, as shown in Figure 1, MD signatures have fast and slow-moving components. The MD signatures recorded the Doppler effects along the time, where the orange curve shows a small amplitude wave denoting slow-moving features, and the signal that spreads around the orange curve makes a high-amplitude wave representing fast-moving features.

Under this observation, our proposed Dual-Scale Doppler Attention (DSDA) is composed of Temporal Window Attention (TWA), which attends fast-moving MD signatures via building temporal windows for MD signatures, and Holistic Window Attention (HWA), which attends slow-moving MD signatures via building holistic windows. To perform TWA and HWA, we define a common attention module defined as Window Attention (WA), which conducts self-attention among windows by calculating their similarities. In the overall pipeline, TWA extracts fast-moving features by generating multiple temporal windows and applying WA among them. HWA generates slow-moving features by subtracting the attended fast-moving features from original signals, and WA again performs self-attention between subtracted and original signals. With these attended fast and slow-moving features, DSDA extracts unique characteristics of each person in fast and slow moving, and identifies human identity. Using the IDRad [10] dataset, we validate state-of-the-art performances on human identification tasks and show the interpretability of the proposed DSDA.

2. Related Works

2.1. Doppler Radar Systems for Human Identification

Doppler radar, which uses a single-tone radio wave, has been frequently used for human identification [6,10,12,13]. By the Doppler effect, the received frequency f_r of the moving target is shifted away from the transmitted frequency f_t, and the Doppler frequency is defined by subtracting f_t from f_r as:

$$f_r = f_t(1 + v/c)/(1 - v/c), \quad (1)$$

$$f_d = f_r - f_t = 2vf_t/(c - v), \quad (2)$$

where c is the speed of light, and v is the radial speed of the moving target. By capturing the Doppler shift, it is able to detect the human motions [5,14,15]. Moreover, frequency-modulated continuous-wave (FMWC) radar is commonly used for short-range multiple targets detection by generating a Doppler map within a certain range [10,16,17]. Vandersmissen et al. [10] utilized low-power 77 GHz FMCW radar for person identification and constructed the IDentification with Radar (IDRad) data set. IDRad is a micro-Doppler map received from several people walking around spontaneously in any possible direction. Our proposed DSDA is experimented on IDRad and validates its interpretability on MD signatures.

2.2. Deep Learning for Micro-Doppler Signatures

As the use of radar-based systems increases, several applications of MD signatures using deep learning have emerged [18–21]. Kim et al. [22] first applied a neural network for human motion recognition on MD signatures and showed the applicability of deep learning on radar signals. After that, several deep learning techniques have been applied for the radar-based motion recognition, including large-scale pre-training [23] and recurrent neural network [24]. Lin et al. [25] proposed iterative CNN followed with random forests in MD signatures which showed performance boost. Park et al. [23] utilized DCNN pre-trained with a large-scale image classification dataset, ImageNet [26], which presented the connectivity between radar and computer vision. Furthermore, Wang et al. [24] used a recurrent neural network (RNN) to detect dynamic gestures with a short-range radar-based sensor, Google's Soli. Recent studies have been conducted via performing human identification (HI) on an MD signature, which requires the understanding of unique characteristics in a single person. In detail, MD signatures from heartbeat signals are utilized for HI [11,27]. Henceforth, MD signatures on gait characteristics of humans [10] are used for this HI, which is more challenging as they are performed in an uncontrolled scenario where a target is allowed to walk around in a free and spontaneous way. Cao et al. [27] primarily applied DCNN to MD signatures for HI. Vandersmissen et al. [10] also used the DCNN and constructed public dataset IDRad for HI, which contributed to subsequent research. Although several methods have been proposed for the aforementioned tasks, they do not fully utilize the details of MD signatures induced from moving human body parts. Therefore, we propose DSDA, which can exclusively recognize unique signals generated by human walking.

2.3. Recent Radar-Based Human Identification Analysis

Radar systems have mainly been applied on the radar-based human identifications [10,11,27–29]. We also compare these previous works to our research in terms of the differences, advantages and disadvantages. The work [28] is performed on detecting humans in specific conditions (i.e., short-range through-wall and long-range foliage penetration). The difference between this study and ours is that it was carried out to find people under specific conditions, but our proposed DSDAs are more contributing on human identification from general human behaviors including human walking motions, arm and body movements. In terms of the method, they applied SVM for human detection. However, our study also contain SVM methods and makes more experimental contributions (SVM, CNN, RNN, and Attention model). The advantage of this study, we think, is that it defines specific tasks well and suggests their solutions. However, the disadvantage is that it is too task-specific and reduces its applicability for other research.

The work [29] holds the commonality with our work in that it performs human identification through the recognition of gait characteristics in MD signatures. However, this work mainly focused on open-set feature analysis, which means how the model can do better when the 'unknown class' exists in the inference, but our research is mainly about sequential feature pattern analysis; thus, our experimental contributions are more relying on sequential pattern analysis and its solutions. The advantage of the work [29] is novel problem definition for establishing the generality of human identification, but the disadvantage is that the feature analysis is insufficient in that the MD signatures contain sequential information.

The works in [11,27] are also holding commonality with our work in that they perform feature pattern analysis (limbs, torso, heart beat). However, the difference is that our model introduces an attention method to better recognize sequential feature patterns. As shown in Table 1, in the paper, our initial attempts also include the CNN models such as [10,11,27], but we confirmed that the RNN-based model should perform better. Therefore, our further studies are focused on designing sequential feature processing model (RNN and DSDA). The advantages of works [11,27] are specific feature pattern analysis, and we speculate that the disadvantages are a lack of concern for a model that can recognize the feature pattern well because of reliance on the popular CNN model.

Table 1. Window Encoder Specifications for Temporal Window Attention.

Layer	Kernel Size	Stride	# of Filters	Data Shape
INPUT				$(150 \times 205 \times 1)$
Temporal Window				$(3 \times 50 \times 205 \times 1)$
Conv 1	(3,3)	(1,1)	8	$(3 \times 50 \times 205 \times 8)$
ELU 1				$(3 \times 50 \times 205 \times 8)$
MAXPooL 1	(2,3)	(2,3)		$(3 \times 25 \times 68 \times 8)$
Conv 2	(3,3)	(1,1)	16	$(3 \times 25 \times 68 \times 16)$
ELU 2				$(3 \times 25 \times 68 \times 16)$
MAXPooL 2	(2,3)	(2,3)		$(3 \times 12 \times 22 \times 16)$
Conv 3	(3,3)	(1,1)	32	$(3 \times 12 \times 22 \times 32)$
ELU 3				$(3 \times 12 \times 22 \times 32)$
MAXPooL 3	(3,1)	(3,1)		$(3 \times 4 \times 22 \times 32)$
Conv 4	(3,3)	(1,1)	64	$(3 \times 4 \times 22 \times 64)$
ELU 4				$(3 \times 4 \times 22 \times 64)$
MAXPooL 4	(1,5)	(1,5)		$(3 \times 4 \times 4 \times 64)$
Pooling				(3×1024)

For the work [10], our proposed DSDA is validated on the same dataset (IDRad) released in [10]. However, the simple convolutional neural network used in [10] is limited in understanding sequential patterns in the radar continuity; thus, we more focused on a method that can accurately recognize sequential information of the radar sequence. In this respect, we have performed several experimental contributions including Recurrent Neural Network models and an Attention model. Finally, we proposed a Dual-Scale Doppler Attention technique for recognizing fast and slow-moving patterns that are prominently present in continuous signals, where we speculate that this makes the methodological differences comparing to the work [10]. The advantages in the work [10] exist in the contributions from dataset release and task proposal, but for the disadvantages, we guess that there is a lack of contribution on how to better understand the features pattern in sequential MD signatures.

3. Method

3.1. Generating Micro-Doppler Signatures

Our proposed Dual-Scale Doppler Attention (DSDA) is validated on the Micro-Doppler signatures (MD signatures); we first explain how to generate the MD signatures. To make 45 time stamps of MD signatures, we first build a single time stamp MD signature and connect 45 stamps in a row. As shown in Figure 2, the single time stamp MD signature is obtained from a single range-Doppler map. We integrate the range-Doppler map (e.g., the dimension for the range-Doppler map is 256 × R, where 256 is the Doppler channel and R is the number of discrete range bins) along the range axis, which constructs the single MD signature (e.g., 256 × 1).

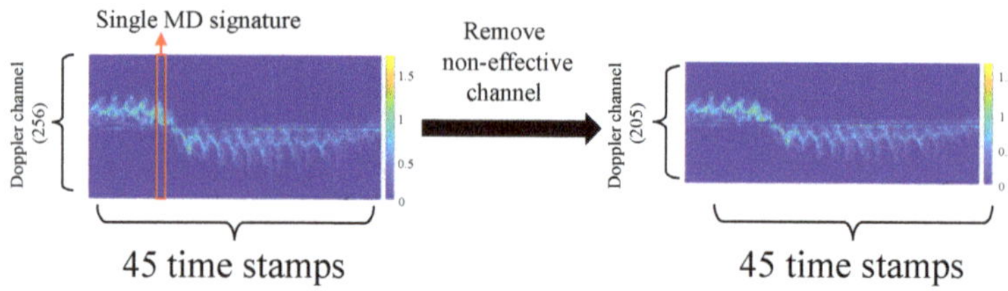

Figure 2. Illustration of Generating Micro-Doppler signature for Dual-Scale Doppler Attention.

This single MD signature contains the Doppler-shift information, where this Doppler-shift value denotes whether the target is moving closer or farther away (e.g., 129–256 channels contain the Doppler shift by the target moving closer and 1–128 channels contain the Doppler shift by the target moving farther away). Following the [10], some channels (i.e., 127–129 channels are zero-Doppler effective, and Doppler channels at both ends are too noisy) are not helpful to identify the human. We also remove them and finally can generate 205 × 45 time stamps MD signatures.

3.2. Model Overview

MD signatures are provided for identifying the human identity [10,11,27–31]. DSDA takes MD signature $R \in \mathbb{R}^{C \times T}$ consisting of the number of channels C and the number of time stamps T ($C = 205$, $T = 45$). The time stamp can be modulated according to the size of MD signatures. As shown in Figure 3, the MD signature is a time-evolving sequence representing the micro-Doppler effect of a person moving.

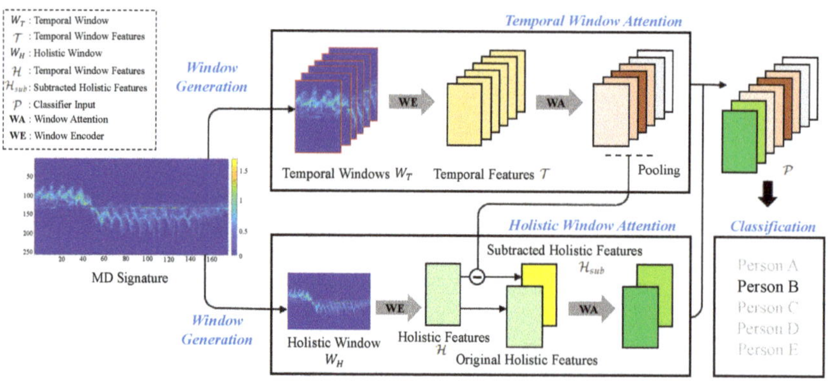

Figure 3. Illustration of Dual-Scale Doppler Attention (DSDA) for human identification. DSDA is composed of the following components: (1) Window Generation for designing temporal window and holistic window, (2) Temporal Window Attention for observing fast-moving features of arms and legs, (3) Holistic Window Attention for recognizing slow-moving features of the torso, and (4) Classification for classifying final targets.

Figure 3 gives a schematic of DSDA consisting of Window Generation, Temporal Window Attention (TWA) and Holistic Window Attention (HWA). Window Generation provides dual-scale windows: a temporal window and holistic window. The temporal window W_T is focused on recognizing fast-moving features extracted by MD signatures so that it is uniformly divided with a temporal sliding window of stride S. The temporal window size is $C \times L$; as such, the total number of temporal windows is N ($S = 5, L = 25$ and $N = 5$). The holistic window W_H is focused on recognizing the slow-moving signal and extracted by an original MD signature. For the Window Encoder (WE), it embeds temporal windows and holistic window into d-dimensional feature representation through a series of Conv-ELU-MaxPooling layers. The encoded temporal features $\mathcal{T}$ and holistic features $\mathcal{H}$ are defined as:

$$\mathcal{T} = \text{LN}(\text{MaxPool}(\text{ELU}(\text{Conv}(W_T)))) \in \mathbb{R}^{N\times d}, \quad (3)$$

$$\mathcal{H} = \text{LN}(\text{MaxPool}(\text{ELU}(\text{Conv}(W_H)))) \in \mathbb{R}^{1\times d}, \quad (4)$$

where LN denotes the layer normalization [32] and the Exponential Linear Unit (ELU) [33] is nonlinearity operation. We present detailed specifications of the Window Encoder used in TWA and HWA in Tables 1 and 2. The number of temporal windows is considered as N = 3 in Table 1 (i.e., this is designed to help understand the Window Encoder's structure used in the Temporal Window Attention). The following sub-sections will explain the details of the remaining model components.

Table 2. Window Encoder Specifications for Holistic Window Attention.

Layer	Kernel Size	Stride	# of Filters	Data Shape
INPUT				(150 × 205 × 1)
Holistic Window				(150 × 205 × 1)
Conv 1	(3,3)	(1,1)	8	(150 × 205 × 8)
ELU 1				(150 × 205 × 8)
MAXPooL 1	(6,3)	(6,3)		(25 × 68 × 8)
Conv 2	(3,3)	(1,1)	16	(25 × 68 × 16)
ELU 2				(25 × 68 × 16)
MAXPooL 2	(2,3)	(2,3)		(12 × 22 × 16)
Conv 3	(3,3)	(1,1)	32	(12 × 22 × 32)
ELU 3				(12 × 22 × 32)
MAXPooL 3	(3,1)	(3,1)		(4 × 22 × 32)
Conv 4	(3,3)	(1,1)	64	(4 × 22 × 64)
ELU 4				(4 × 22 × 64)
MAXPooL 4	(1,5)	(1,5)		(4 × 4 × 64)
Pooling				(1 × 1024)

3.3. Window Attention

Our proposed basic attention unit of the TWA and HWA is referred to as Window Attention (WA), which calculates the attention weights among multi-element features. Given $X = [\mathbf{x}_1 \dots \mathbf{x}_I]^T \in \mathbb{R}^{I\times d}$ and $Y = [\mathbf{y}_1 \dots \mathbf{y}_J]^T \in \mathbb{R}^{J\times d}$, the WA $W(X,Y) : \mathbb{R}^{I\times d} \times \mathbb{R}^{J\times d} \rightarrow \mathbb{R}^{I\times d}$ attends X using Y, which is defined as follows:

$$\text{w}(\mathbf{x}_i, Y) = \Sigma_{j=1}^{J}(\text{ELU}(W_1\mathbf{x}_i)\text{ELU}(W_2\mathbf{y}_j)^T), \quad (5)$$

$$W(X,Y) = \text{Softmax}([\text{w}(\mathbf{x}_1, Y) \ \cdots \ \text{w}(\mathbf{x}_I, Y)])X, \quad (6)$$

where $w(\mathbf{x}_i, Y) : \mathbb{R}^d \times \mathbb{R}^{J\times d} \rightarrow \mathbb{R}$ and $W_1, W_2 \in \mathbb{R}^{d\times d}$ are learnable parameters. (Our experiments also validate the attention method in the Section 4.4) Using the ELU nonlinearity, WA tries to preserve common signatures between input X and Y windows. Based on the WA, the following TWA and HWA are defined and attend their moving signatures. We also provide the process of Window Attention in Figure 4, where WA is performed on two window features. For the TWA, the two types of window are temporal windows and, for the HWA, the two types are holistic windows.

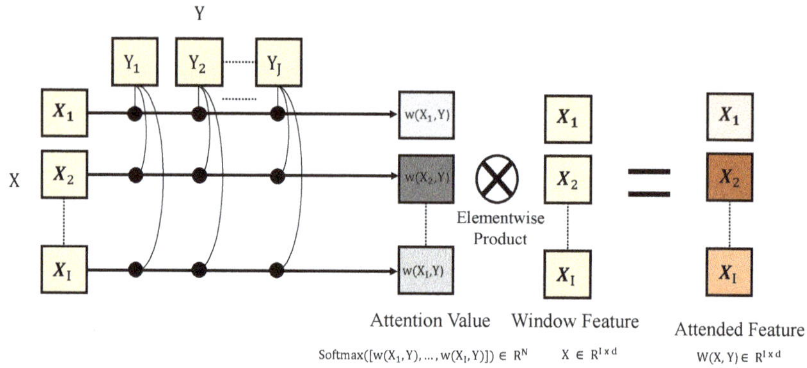

Figure 4. Illustration of Window Attention. The WA is performed on TWA and HWA according to the window features.

3.4. Temporal Window Attention

We observe that the fast-moving features of arms and legs are unique depending on the person's height and walking pattern. To preserve and highlight these features, the Temporal Window Attention (TWA) takes multiple temporal windows as inputs and applies WA as follows:

$$\mathcal{T} \leftarrow \mathrm{LN}(W(\mathcal{T}, \mathcal{T}) + \mathcal{T}), \tag{7}$$

$$\mathcal{T}_{avg} = \mathrm{AvgPool}(\mathcal{T}) \in \mathbb{R}^{1\times d}. \tag{8}$$

where AvgPool() is the average pooling function. TWA produces two types of features. One is the original TWA output $\mathcal{T}$ in Equation (7), which contributes to classifying the person identification, and the other is the average of $\mathcal{T}$ over N temporal windows $\mathcal{T}_{avg}$ in Equation (8). This average of temporal window attended features $\mathcal{T}_{avg}$ is treated to have the characteristic of the overall fast-moving features. Furthermore, $\mathcal{T}_{avg}$ is used to remove fast-moving information from the original holistic features in the following Holistic Window Attention.

3.5. Holistic Window Attention

The Holistic Window Attention (HWA) is designed to recognize slow-moving signatures such as the torso. Different from TWA, HWA takes two types of features as input. One is original holistic features $\mathcal{H}$, and the other is subtracted holistic features $\mathcal{H}_{sub}$ obtained by subtracting the $\mathcal{T}_{avg}$ from $\mathcal{H}$ in Equation (9). Semantically, the $\mathcal{H}$ can include slow and fast-moving features, and the $\mathcal{H}_{sub}$ includes only the slow-moving features by removing characteristics of overall fast-moving features. These two holistic features are concatenated, and HWA attends the common slow-moving signals using the WA in Equations (10) and (11):

$$\mathcal{H}_{sub} = \mathcal{H} - \mathcal{T}_{avg}, \tag{9}$$

$$\mathcal{H}^* = [\mathcal{H}\mathcal{H}_{sub}] \in \mathbb{R}^{2\times d}, \tag{10}$$

$$\mathcal{H}^* \leftarrow \mathrm{LN}(W(\mathcal{H}^*, \mathcal{H}^*) + \mathcal{H}^*). \tag{11}$$

where $[\cdot\cdot]$ is the concatenation operation. Holistic features $\mathcal{H}^*$ through HWA represent the slow-moving feature in the MD signature and are utilized to classify the targets.

3.6. Classification

In classification, we use two aforementioned temporal and holistic features for classifying targets. These two attended temporal and holistic features are concatenated and transmitted to the final classifier as follows:

$$\mathcal{P} = [\mathcal{T}\mathcal{H}^*] \in \mathbb{R}^{(N+2)\times d}, \tag{12}$$

$$\mathcal{P}^* = \text{Flat}(\mathcal{P}) \in \mathbb{R}^{((N+2)\times d)}, \tag{13}$$

$$\mathcal{C} = W_c(\text{Dropout}(\text{ELU}(W_p\mathcal{P}^*))) \in \mathbb{R}^{C_T}, \tag{14}$$

$$p = (\text{Softmax}(\mathcal{C})) \tag{15}$$

where Flat() is a function that converts 2D data into 1D in order to apply the CNN data type to the fully connected neural network, Equation (14) is the last Dropout–ELU–Linear block in the classifier, $W_p \in \mathbb{R}^{128\times((N+2)\times d)}$ and $W_c \in \mathbb{R}^{C_T\times 128}$ are the learnable parameters, $C_T = 5$ (IDRad dataset contain five different targets made by five different people.) is the number of targets and $p = \{p[1], p[2], p[3], p[4], p[5]\}$ is inferred prediction distribution values. In the inference, prediction is performed using argmax function on the p. We also add some specifications of the classifier network in Table 3. The (3 + 2) in Table 3 denotes concatenation between TWA and HWA, where HWA includes (2 × 1024) features from the holistic feature (1 × 1024) in Table 2 and the subtracted feature (1 × 1024). TWA includes (3 × 1024) features in Table 3.

Table 3. Classifier detailed specifications.

Layer	Kernel SIZE	Stride	# of Filters	Data Shape
INPUT				$((3+2)\times 1024)$
Pooling	(3,3)	(1,1)	8	(5120)
Linear				(128)
ELU	(3,3)	(1,1)	16	
Dropout 4	(1,5)	(1,5)		
Linear				5

3.7. Training Loss

The entire model is trained in an end-to-end manner using cross-entropy loss $\mathcal{L}_{HI}(y, p)$, where the y is the ground-truth target information and the p is the prediction of human identification from DSDA, as shown below:

$$\mathcal{L}_{HI}(y, p) = -log(p[y]) \tag{16}$$

4. Experiments

4.1. IDRad Dataset

As the MD signature dataset for HI, we use IDRad [10] using FMCW radar, which records range-Doppler maps with a speed of around 15 FPS. The IDRad dataset contains 95,650 frames of 20 min for a training set and 22,535 frames of 5 min for a test set. One frame contains a Doppler frequency channel along the range axis. To construct micro-Doppler signatures, the IDRad dataset integrates a range-Doppler map along the range axis and connects the integrated Doppler signals in temporal order to make Doppler-time maps as previously stated [10]. Here, one time stamp includes 256 Doppler channels, and one MD signature is composed of 45 time stamps. For a fair comparison, we also performed the same preprocessing as [10] and used the default input of 205 × 45 MD signatures, which translates to 205 Doppler channels and 45 time stamps. Regarding the details of preprocessing, among 256 Doppler channels, 127∼129 Doppler channels representing static objects are removed and 24 Doppler channels in the top and bottom of the Doppler axis in MD signatures are also removed, because they are a too high speed range for humans to demonstrate. The IDRad dataset films the movements of five subjects whose age ranges are from 23 to 32 years old, weight ranges are from 60 to 90 kg and their height ranges are from 178 to 185 cm. Five subjects are able to move freely within a certain range of filmed room, and they are able to freely walk, run, or stop in various directions.

4.2. Experimental Details

The dimension of the hidden layer is set to $d = 1024$. For DSDA, we use three blocks of the Conv-ELU-MaxPooling layer in the Window Encoder and four blocks of the Dropout [34]-ELU-Linear layer in the classifier. Our model can be easily implemented with six layers of convolutional neural networks and trained on NVIDIA TITAN V (12 GB of memory) GPU with an Adam optimizer [35] with $\beta_1 = 0.9, \beta_2 = 0.98$ and $\epsilon = 10^{-9}$. For all experiments, we select the batch size of 64, a dropout rate of 0.2, and the model is trained up to 15 epochs. From this condition, we do not perform any hyperparameter fine tunning.

The overall evaluation of DSDA is performed using error rate, where the error rate is calculated as: 'error rate' $= 100 \times (\text{number of incorrect predictions})/(\text{total samples})$. The total sample can be composed of a validation set and test set.

4.3. Experimental Results

Table 4 summarizes the experimental results on the IDRad Dataset where we compare DSDA with several recent methods. Considering the temporal properties, we use 10 s MD signatures as an input (150 time stamps), which is different from the input of 3 s MD signatures (45 time stamps) in [10]. For the fair comparison, we also measured the baseline with the same sizes of the input MD signatures, where the baseline with 10 s MD signatures is reproduced using their public code and reported as 'Baseline (150 time stamp)' in Table 4. The extended input size of the baseline also shows slight effectiveness, but it still remained in the variation of original performances reported in the paper [10]. To validate the deep learning-based model on this task, we first built a PCA model with SVM classifier, which shows the classical classifying performance with a machine learning algorithm. Comparing to the performances of the PCA model, the deep learning-based models (i.e., baseline, LSTM-based model, DSDA) are giving superior performances. Although there are sequential data in the IDRad dataset, there have not been sequential models to perform human identification. Thus, for these sequential radar image data, we also consider a sequential RNN (Recurrent Neural Network) model on this task. We devised a Long Short-Term Memory (LSTM) [36,37] based model. The LSTM model shows better performances compared to the baseline built with a CNN structure, which explains the necessity of sequential understanding for radar data. The recent success of Transformer [38–40], our proposed DSDA, is based on the attention mechanism and also utilizes the specifications in radar (fast-moving and slow-moving features). DSDA achieves state-of-the-art performance against all methods with a large margin. The results indicate that extracting slow and fast-moving features and attending them improve the interpretability of MD signatures.

Table 4. Error rate (%) on the IDRad Dataset.

Methods	Validation	Test
Baseline [10]	24.70	21.54
Baseline [10] (reproduced on 150 time stamp)	22.42	20.37
PCA SVM	37.67	32.91
LSTM based model	22.83	18.42
DSDA	**10.87**	**8.65**

4.4. Ablation Study

We experiment with several variants of DSDA to measure the effectiveness of the proposed key components. The first block of Table 5 is full DSDA with three multi-scale windows composed of strides $s = \{5, 10, 20\}$ and windows $L = \{35, 25, 25\}$. The second block of Table 5 provides ablation results of TWA and HWA. Since TWA and HWA have influenced performance improvements, both fast-moving features and slow-moving features imply target features. Especially as TWA boosts performance significantly, we can see that unique information, which identifies a person, is inherent in fast-moving features. The third block of Table 5 provides ablation results on the scale of temporal windows in TWA. Window size and stride between windows influence the extraction of the proper characteristic features. We have confirmed that the best performance is achieved when the stride s is 5 and window size L is 35 for single fixed window. The fourth block of Table 5 provides ablation results with multi-scale windows. Here, we use multi-scale temporal windows of various sizes for TWA via selecting several windows in the third block of Table 5. The M is the number of different types of windows. When $M = 3$, we select the several combinations of three windows, where the best performances are shown with the windows composed of strides $s = \{5, 10, 20\}$ and windows $L = \{35, 25, 25\}$. We also validate for $M = 4$ via adding one more window on top of the best-performance condition in $M = 3$. If all the results are not improved, then $M = 3$ is the best-performance condition, so we report the averaging results on $M = 4$. To adopt multi-scale temporal windows, TWA is applied to each window scale, and HWA uses several subtracted holistic features obtained by multi-scale temporal windows. We consider that several moving features extracted by multi-scale temporal windows make it suitable to capture a person's unique characteristics.

Table 5. Ablation study on model variants of DSDA on the validation split of IDRad.

Model Variants	Error Rate (%)
Full DSDA	10.87
w/o TWA	24.32
w/o HWA	23.94
w/ stride $s = 10$, window $L = 25$	12.99
w/ stride $s = 20$, window $L = 25$	13.71
w/ stride $s = 5$, window $L = 15$	14.73
w/ stride $s = 10$, window $L = 15$	14.80
w/ stride $s = 5$, window $L = 35$	11.93
w/ stride $s = 10$, window $L = 35$	13.45
Multi-Scale (M = 3)	10.87
Multi-Scale (M = 4)	12.79

Table 6 presents the ablation on the attention methods for $\mathrm{ELU}(W_1\mathbf{x}_i)\mathrm{ELU}(W_2\mathbf{y}_j)^T$ in Equation (5). We experimented with three attention methods that highlight common features. Based on the results, we selected $A * B$ for the final DSDA. Here, the operation; denotes concatenation with a d-dimensional axis. To follow the dimensional condition, we add more of an embedding matrix $W_{add} \in \mathbb{R}^{d\times 1}$ for $A + B$ and embedding matrix $W_{cat} \in \mathbb{R}^{2d\times 1}$ for $A; B$. Our empirical experiments for more attention methods are in the variance of $A + B$ and $A; B$, and they do not give further performance gain.

Table 7 represents the Kappa Index Analysis of our proposed DSDA with Baseline [10]. 'TRUE' means 'correct' on the target, and 'FALSE' means 'incorrect' on the target. To calculate the kappa value $K = (1 - Pa)/(Pa - Pc)$, where Pa is the observational probability of agreement and Pc is the hypothetical expected probability of agreement. Pa is obtained as $Pa = (1082 + 122)/1490 = 0.808$ and Pc is obtained as $Pc = (1328/1490) \times (1122/1490) + (162/1490) \times (368/1490) = 0.698$. Thus, the kappa value is obtained as $K = 0.36$. Therefore, according to the kappa value analysis, our proposed DSDA decision performance has 'fair' strength of agreement with the baseline [10]. We consider this is because DSDA includes improved performance (10.8% error rate) compared to the baseline (24.7% error rate), which contributes to the disagreement with baseline in the case the baseline performs an incorrect decision on the target.

Table 6. Comparison of attention methods in WA on the validation split of IDRad.

	Attention Method for Calculating Similarity	Error Rate (%)
$A * B$	$w(\mathbf{x}_i, Y) = \sum_{j=1}^{J}(\text{ELU}(W_1\mathbf{x}_i)\text{ELU}(W_2\mathbf{y}_j)^T)$	10.87
$A + B$	$w(\mathbf{x}_i, Y) = \sum_{j=1}^{J}([\text{ELU}(W_1\mathbf{x}_i) + \text{ELU}(W_2\mathbf{y}_j)^T]W_{add})$	11.28
$A; B$	$w(\mathbf{x}_i, Y) = \sum_{j=1}^{J}([\text{ELU}(W_1\mathbf{x}_i); \text{ELU}(W_2\mathbf{y}_j)^T]W_{cat})$	11.02

Table 7. Kappa Index Analysis of DSDA on the validation split of IDRad.

Human Identification		Baseline [10]		Total
		True	False	
DSDA	True	1082	246	1328
	False	40	122	162
Total		1122	368	1490

4.5. Qualitative Results

Figure 5 visualizes attention weights for temporal windows. For the input MD signature of 150 time stamps, attention weights are represented as a bar chart in the Figure 5. Here, temporal windows adopt 45 time stamps and 15 strides. Although the multi-scale windows performed well in this human identification task, we select the single window that can help easily understand how the attention weights of TWA are formed and these weights identify the fast-moving information of the MD signatures. Thus, we can find out which windows have been highlighted through the attention weight. The MD signatures corresponding to three windows (i.e., 1, 2, 3) from the left contain fast-moving features, and also, the signatures corresponding to two windows (i.e., 7, 8) from the right contain fast-moving features. Our analysis on synchronized video identifies that these fast-moving features are from the human walking. The other signatures (i.e., MD signatures from windows 4, 5, 6) are formed when the humans stop moving and change the direction of walking. The attention weights in TWA are highlighted on these windows containing fast-moving features such as walking; however, the small weights are given on slow-moving features such as changing direction or non-moving. Therefore, it is confirmed that the TWA is properly trained to recognize the fast-moving information from the MD signatures.

In Figure 6, we build a confusion matrix of DSDA. The vertical axis denotes the ground-truth category and the horizontal axis denotes the predicted category. For all targets, DSDA perform 100% accuracy on target 3 with the highest accuracy and performs 72% accuracy on target 4. We consider the reason why DSDA's prediction is the lowest on target 4, where the target 4, target 1 and target 5 show similar movements and have a relatively similar body shape. This gives the challenge in identifying their characteristics. To qualitatively compare the confusion matrix with the baseline [10], we give the confusion matrix predicted from the baseline (We obtain confusion matrices from fully trained baseline and DSDA.) in the (b) of Figure 6. Comparing to the baseline, the performance improvement can be confirmed in all targets for our proposed DSDA, and it was confirmed that both the baseline and DSDA showed excellent performance in target 3. However, it is also notable that our DSDA is more improved in target 4, where this is because DSDA is more robust to distinguish fine-grained information in the MD signatures.

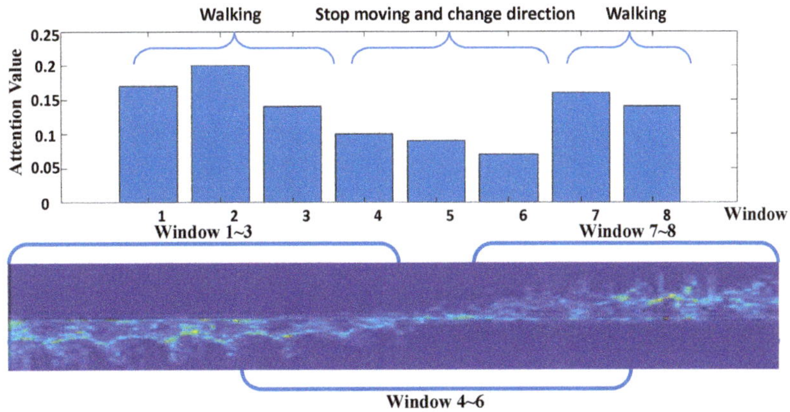

Figure 5. Attention weights according to temporal windows in TWA. The 8 windows are generated via traversing MD signatures with a sliding window composed of 45 time stamps and 15 strides.

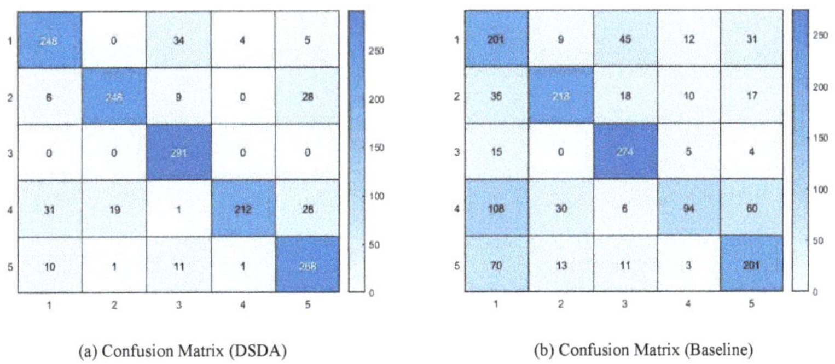

Figure 6. (**a**) Confusion matrix (5 × 5) of human identification from DSDA and (**b**) Confusion matrix (5 × 5) of human identification from baseline [10] on IDRad validation split. Vertical axis denotes ground-truth type and horizontal axis denotes predicted type.

In Figure 7, we also perform a target-wise confusion matrix to calculate the sensitivity and specificity according to the targets. The average sensitivity is 0.88, and the average specificity is 0.97. Our DSDA is effective in the specificity, which means our model is highly sensible on the negative targets. It is also notable that DSDA performs 100% on target 3, which implies that the model finds suitable feature space that can classify target 3, and also, the model localizes other target features on the proper feature spaces.

In Figure 8, we perform efficiency analysis. The attention mechanism in DSDA does not require many memory resources (The resources that are required are joint space embedding matrices.) but also performs early saturation of training error rate. As shown in orange curve of Figure 8, the error rate on the IDRad training dataset converges more early than the blue curve of the baseline, which denotes that the window attention mechanism promotes weights in the network to be sensible on this identification task. After epoch 15, two curves from the baseline and DSDA are converged and become saturated.

In Figure 9, we also perform dataset sensitivity analysis according to the ratio of training dataset usage. Our proposed DSDA gives robustness until 60% usage of the training dataset by keeping the error rate below 30%, and then, the error rate increases. However, the baseline shows the robustness until 80% usage of the training dataset, which shows that our model learns more efficiently on the dataset and is less sensitive to dataset scarcity.

DSDA Target-wise Confusion Matrix																			
Target 1		GT		Target 2		GT		Target 3		GT		Target 4		GT		Target 5		GT	
		T	F			T	F			T	F			T	F			T	F
DSDA	T	248	43	DSDA	T	248	43	DSDA	T	291	0	DSDA	T	212	79	DSDA	T	268	23
	F	47	1117		F	20	1144		F	55	1109		F	5	1159		F	61	1103
SE	**0.841**	**SP**	**0.963**	**SE**	**0.925**	**SP**	**0.964**	**SE**	**0.841**	**SP**	**1**	**SE**	**0.977**	**SP**	**0.936**	**SE**	**0.815**	**SP**	**0.98**

Figure 7. Target-wise Confusion Matrix to calculate sensitivity (SE) and specificity (SP) between DSDA and the ground-truth (GT) on the validation split of the IDRad Dataset (T: True, F: False).

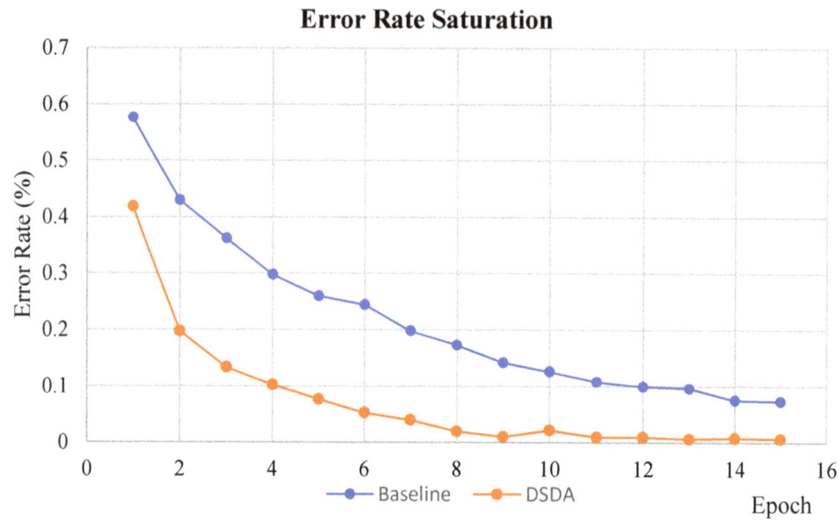

Figure 8. Error rate saturation along the epoch on the training dataset. The blue curve shows the saturation of baseline [10] and the orange curve shows the saturation of the proposed DSDA.

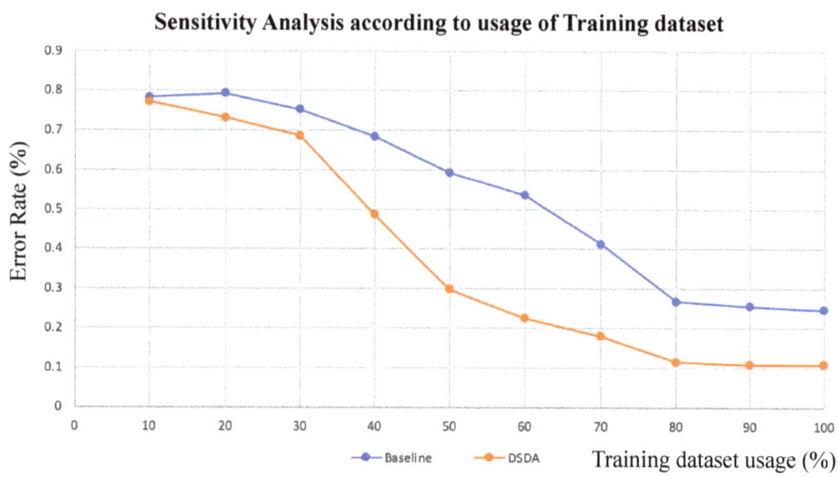

Figure 9. Sensitivity analysis of model performances according to the training dataset usage. The blue curve shows baseline [10], and the orange curve shows the proposed DSDA.

In Figure 10, to verify the practicality of the proposed model, we extended the HI task into a human localization task, where the MD signatures are extended to the longer size (i.e., 450 time stamps), and we perform time stamp wise human identifications. This can be thought of as human localization in MD signatures, which is more challenging and requires a fine-grained understanding of MD signatures. MD signatures are randomly selected in the validation set. The prediction of each stamp is performed on the center of 150 time stamp windows. DSDA traverses all the MD signatures and predicts every time stamp, which shows its predictions in the below Figure 10, where the y-axis denotes the probability of each target. Thus, five distributions corresponding to five targets are generated, and the above bar shows the ground-truth target for every signature. It is clearly confirmed that the distributions matched with the ground-truth target are highlighted for the given MD signatures. This denotes that DSDA is available to perform fine-grained human identification and is properly extended to the human localization task in radar signals.

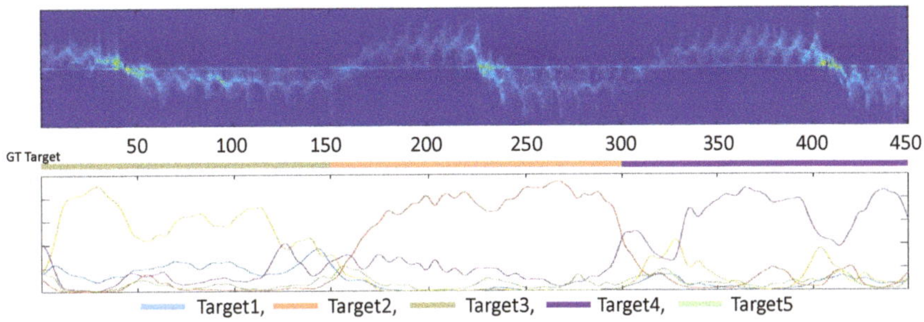

Figure 10. Extended experiment on MD signatures. The below distribution denotes time stamp wise target predictions, where this represents the availability of the human localization task of DSDA.

5. Limitation

We would like to present two limitations from two different perspectives. The first limitation exists in terms of task. Our current experiments are mainly performed under a human identification task. However, as shown in Figure 10 of the paper, we found that this task can be extended up to the human detection task in the radar sequence by stiching MD signatures and localizing the human in them. Our further studies will include this extended analysis and possibilities of human detection. The second limitation is the lack of datasets for this study. We validate DSDA on the IDRad dataset. To build a general radar-based model, we should also validate more different radar datasets. In this respect, our further research will focus on how to establish model generality on radar signals.

6. Future Work

Our future works are three-fold. The first is that we will annotate the radar dataset to be suitable localization tasks. As shown in Figure 10, we further evaluate our DSDA in terms of human identification and localization tasks on the radar dataset, where DSDA shows enough applications on these tasks. The second is that we will modify the window. Currently, our window operates on the sliding window mechanism, but we will consider more diverse windowing methods. The third is that we will validate our DSDA on another radar dataset to be a general attention module for radar understanding.

7. Conclusions

We propose Dual-Scale Doppler Attention (DSDA) for human identification. DSDA adopts Temporal Window Attention (TWA) that attends fast-moving MD signatures of arms and legs and Holistic Window Attention (HWA) that attends slow-moving MD signatures of the torso. Our experimental results on the IDRad dataset show state-of-the-art performance and the qualitative results validate the efficiency of our proposed module. The model analysis including a sensitivity and confusion matrix shows DSDA's robustness. Extended experiments incorporating radar-based human detection tasks show the flexibility of DSDA. From these experiments, our future works are more obvious and substantial.

Author Contributions: Conceptualization, S.Y.; methodology, S.Y., D.K. and J.K.; software, S.Y. and D.K.; validation, S.Y. and D.K.; formal analysis, S.Y. and C.D.Y.; writing S.Y., D.K., J.W.H., J.K. and C.D.Y.; supervision, C.D.Y. All authors have read and agreed to the published version of the manuscript.

Funding: This workwas partly supported by the Institute for Information & Communications Technology Promotion (IITP) grant funded by the Korea government (MSIT) (No. 2021-0-01381, Development of Causal AI through Video Understanding) and partly supported by Institute of Information & Communications Technology Planning & Evaluation (IITP) grant funded by the Korea government (MSIT) (No. 2022-0-00951, Development of Uncertainty-Aware Agents Learning by Asking Questions).

Institutional Review Board Statement: Not applicable.

Informed Consent Statement: Not applicable.

Data Availability Statement: Not applicable.

Acknowledgments: The authors would like to thank the experimental contributions from researcher Junehyune Park and Jaewook Park in LIG Nex1, where they perform additional experiments in this paper to validate model generality including target-wise accuracy analysis via confusion matrix and model efficiency analysis.

Conflicts of Interest: The authors declare no conflict of interest.

Sample Availability: Samples of the compounds ... are available from the authors.

Abbreviations

The following abbreviations are used in this manuscript:

DCNN	Deep Convolution Neural Network
DSDA	Dual-Scale Doppler Attention
HWA	Holistic Window Attention
IDRad	IDentification with Radar
MD	Micro-Doppler
ML	Machine Learning
TWA	Temporal Window Attention
WE	Window Encoder
WA	Window Attention

References

1. Cunningham, S.J.; Masoodian, M.; Adams, A. Privacy issues for online personal photograph collections. *J. Theor. Appl. Electron. Commer. Res.* **2010**, *5*, 26–40. [CrossRef]
2. Kang, D.; Kum, D. Camera and radar sensor fusion for robust vehicle localization via vehicle part localization . *IEEE Access* **2020**, *8*, 75223–75236. [CrossRef]
3. Bai, J.; Li, S.; Huang, L.; Chen, H. Robust detection and tracking method for moving object based on radar and camera data fusion. *IEEE Sens. J.* **2021**, *21*, 10761–10774. [CrossRef]
4. Yang, B.; Guo, R.; Liang, M.; Casas, S.; Urtasun, R. Radarnet: Exploiting radar for robust perception of dynamic objects. In Proceedings of the European Conference on Computer Vision, Glasgow, UK, 23–28 August 2020; Springer: Cham, Switzerland, 2020.
5. Chen, V.C.; Li, F.; Ho, S.S.; Wechsler, H. Micro-Doppler effect in radar: Phenomenon, model, and simulation study. *IEEE Trans. Aerosp. Electron. Syst.* **2006**, *42*, 2–21. [CrossRef]
6. Kim, Y.; Ling, H. Human activity classification based on micro-Doppler signatures using a support vector machine. *IEEE Trans. Geosci. Remote Sens.* **2009**, *47*, 1328–1337.
7. Zhou, Z.; Cao, Z.; Pi, Y. Dynamic gesture recognition with a terahertz radar based on range profile sequences and Doppler signatures. *Sensors* **2017**, *18*, 10. [CrossRef] [PubMed]
8. Fioranelli, F.; Ritchie, M.; Griffiths, H. Classification of unarmed/armed personnel using the NetRAD multistatic radar for micro-Doppler and singular value decomposition features. *IEEE Geosci. Remote Sens. Lett.* **2015**, *12*, 1933–1937. [CrossRef]
9. Kim, Y.; Ha, S.; Kwon, J. Human detection using Doppler radar based on physical characteristics of targets. *IEEE Geosci. Remote Sens. Lett.* **2014**, *12*, 289–293. [CrossRef]
10. Vermissen, B.; Knudde, N.; Jalalv, A.; Couckuyt, I.; Bourdoux, A.; De Neve, W.; Dhaene, T. Indoor person identification using a low-power FMCW radar. *IEEE Trans. Geosci. Remote Sens.* **2018**, *56*, 3941–3952. [CrossRef]

11. Cao, P.; Xia, W.; Li, Y. Heart id: Human identification based on radar micro-doppler signatures of the heart using deep Learning. *Remote Sens.* **2019**, *11*, 1220. [CrossRef]
12. Kim, Y.; Moon, T. Human detection and activity classification based on micro-Doppler signatures using deep convolutional neural networks. *IEEE Geosci. Remote Sens. Lett.* **2015**, *13*, 8–12. [CrossRef]
13. Le H.T.; Phung, S.L.; Bouzerdoum, A.; Tivive, F.H. Human motion classification with micro-Doppler radar and Bayesian-optimized convolutional neural networks. In Proceedings of the 2018 IEEE International Conference on Acoustics, Speech and Signal Processing (ICASSP), Calgary, AB, Canada, 15–20 April 2018; pp. 2961–2965.
14. Gong, T.; Cheng, Y.; Li, X.; Chen, D. Micromotion detection of moving and spinning object based on rotational Doppler shift. *IEEE Microw. Wirel. Compon. Lett.* **2018**, *28*, 843–845. [CrossRef]
15. Seddon, N.; Bearpark, T. Observation of the inverse Doppler effect. *Science* **2003**, *302*, 1537–1540. [CrossRef] [PubMed]
16. Winkler, V. Range Doppler detection for automotive FMCW radars. In Proceedings of the 2007 European Radar Conference, Waltham, MA, USA, 17–20 April 2007; pp. 166–169.
17. Lin, J., Jr.; Li, Y.P.; Hsu, W.C.; Lee, T.S. Design of an FMCW radar baseband signal processing system for automotive application. *SpringerPlus* **2016**, *5*, 42. [CrossRef] [PubMed]
18. Li, X.; He, Y.; Jing, X. A survey of deep learning-based human activity recognition in radar. *Remote Sens.* **2019**, *11*, 1068. [CrossRef]
19. Lee, D.; Park, H.; Moon, T.; Kim, Y. Continual learning of micro-Doppler signature-based human activity classification. *IEEE Geosci. Remote Sens. Lett.* **2021**, *19*, 1–5. [CrossRef]
20. Martinez, J.; Vossiek, M. Deep learning-based segmentation for the extraction of micro-doppler signatures. In Proceedings of the 2018 15th European Radar Conference (EuRAD), Madrid, Spain, 26–28 September 2018; pp. 190–193. [CrossRef]
21. Abdulatif, S.; Wei, Q.; Aziz, F.; Kleiner, B.; Schneider, U. Micro-doppler based human-robot classification using ensemble and deep learning approaches. In Proceedings of the 2018 IEEE Radar Conference (RadarConf18), Oklahoma City, OK, USA, 23–27 April 2018.
22. Kim, Y.; Ling, H. Human activity classification based on micro-Doppler signatures using an artificial neural network. In Proceedings of the 2008 IEEE Antennas and Propagation Society International Symposium, San Diego, CA, USA, 5–11 July 2008; pp. 1–4.
23. Park, J.; Javier, R.J.; Moon, T.; Kim, Y. Micro-Doppler based classification of human aquatic activities via transfer learning of convolutional neural networks. *Sensors* **2016**, *16*, 1990. [CrossRef]
24. Wang, S.; Song, J.; Lien, J.; Poupyrev, I.; Hilliges, O. Interacting with soli: Exploring fine-grained dynamic gesture recognition in the radio-frequency spectrum. In Proceedings of the 29th Annual Symposium on User Interface Software and Technology, Tokyo, Japan, 16–19 October 2016; pp. 851–860.
25. Lin, Y.; Le Kernec, J.; Yang, S.; Fioranelli, F.; Romain, O.; Zhao, Z. Human activity classification with radar: Optimization and noise robustness with iterative convolutional neural networks followed with random forests. *IEEE Sens. J.* **2018**, *18*, 9669–9681. [CrossRef]
26. Krizhevsky, A.; Sutskever, I.; Hinton, G.E. Imagenet classification with deep convolutional neural networks. *Adv. Neural Inf. Process. Syst.* **2012**, *25*, 1097–1105. [CrossRef]
27. Cao, P.; Xia, W.; Ye, M.; Zhang, J.; Zhou, J. Radar-ID: Human identification based on radar micro-Doppler signatures using deep convolutional neural networks. *IET Radar Sonar Navig.* **2018**, *12*, 729–734. [CrossRef]
28. Narayanan, R.M.; Smith, S.; Gallagher, K.A. A multifrequency radar system for detecting humans and characterizing human activities for short-range through-wall and long-range foliage penetration applications. *Int. J. Microw. Sci. Technol.* **2014**, *2014*. [CrossRef]
29. Ni, Z.; Huang, B. Open-set human identification based on gait radar micro-Doppler signatures. *IEEE Sens. J.* **2021**, *21*, 8226–8233. [CrossRef]
30. Kwon, J.; Kwak, N. Human detection by neural networks using a low-cost short-range Doppler radar sensor. In Proceedings of the 2017 IEEE Radar Conference (RadarConf), Seattle, WA, USA, 8–12 May 2017.
31. Qiao, X.; Shan, T.; Tao, R. Human identification based on radar micro-Doppler signatures separation. *Electron. Lett.* **2020**, *56*, 195–196. [CrossRef]
32. Ba, J.L.; Kiros, J.R.; Hinton, G.E. Layer normalization. *arXiv* **2016**, arXiv:1607.06450.
33. Clevert, D.A.; Unterthiner, T.; Hochreiter, S. Fast and accurate deep network learning by exponential linear units (elus). *arXiv* **2015**, arXiv:1511.07289.
34. Srivastava, N.; Hinton, G.; Krizhevsky, A.; Sutskever, I.; Salakhutdinov, R. Dropout: A simple way to prevent neural networks from overfitting. *J. Mach. Learn. Res.* **2014**, *15*, 1929–1958.
35. Kingma, D.P.; Ba, J. Adam: A method for stochastic optimization. *arXiv* **2014**, arXiv:1412.6980.
36. Zhang, Z.; Tian, Z.; Zhou, M. Latern: Dynamic continuous hand gesture recognition using FMCW radar sensor. *IEEE Sens. J.* **2018**, *18*, 3278–3289. [CrossRef]
37. Hochreiter, S.; Schmidhuber, J. Long short-term memory. *Neural Comput.* **1997**, *9*, 1735–1780. [CrossRef]
38. Vaswani, A.; Shazeer, N.; Parmar, N.; Uszkoreit, J.; Jones, L.; Gomez, A.N.; Kaiser, Ł.; Polosukhin, I. Attention is all you need. *Adv. Neural Inf. Process. Syst.* **2017**, *30*. Available online: https://proceedings.neurips.cc/paper/2017/file/3f5ee243547dee91fbd053c1c4a845aa-Paper.pdf (accessed on 10 August 2022).

39. Ma, M.; Yoon, S.; Kim, J.; Lee, Y.; Kang, S.; Yoo, C.D. Vlanet: Video-language alignment network for weakly-supervised video moment retrieval. In Proceedings of the European Conference on Computer Vision, Glasgow, UK, 23–28 August 2020; Springer: Cham, Switzerland, 2020.
40. Kim, J.; Ma, M.; Pham, T.; Kim, K.; Yoo, C.D. Modality shifting attention network for multi-modal video question answering. In Proceedings of the IEEE/CVF Conference on Computer Vision and Pattern Recognition, Seattle, WA, USA, 14–19 June 2020.

sensors

MDPI

Article

Sensor Data Prediction in Missile Flight Tests

Sang-Gyu Ryu [1,2], Jae Jin Jeong [3] and David Hyunchul Shim [1,*]

1 Department of Electrical Engineering, Korea Advanced Institute of Science and Technology (KAIST), 291, Daehak-ro, Yuseong-gu, Daejeon 34141, Republic of Korea
2 The 1st R&D Institute, Agency for Defense Development (ADD), 160, Bugyuseong-daero 488 Beon-gil, Yuseong-gu, Daejeon 34060, Republic of Korea
3 Department of Electronic Engineering, Kumoh National Institute of Technology, 61, Daehak-ro, Gumi-si 39177, Republic of Korea
* Correspondence: hcshim@kaist.ac.kr

Abstract: Sensor data from missile flights are highly valuable, as a test requires considerable resources, but some sensors may be detached or fail to collect data. Remotely acquired missile sensor data are incomplete, and the correlations between the missile data are complex, which results in the prediction of sensor data being difficult. This article proposes a deep learning-based prediction network combined with the wavelet analysis method. The proposed network includes an imputer network and a prediction network. In the imputer network, the data are decomposed using wavelet transform, and the generative adversarial networks assist the decomposed data in reproducing the detailed information. The prediction network consists of long short-term memory with an attention and dilation network for accurate prediction. In the test, the actual sensor data from missile flights were used. For the performance evaluation, the test was conducted from the data with no missing values to the data with five different missing rates. The test results showed that the proposed system predicts the missile sensor most accurately in all cases. In the frequency analysis, the proposed system has similar frequency responses to the actual sensors and showed that the proposed system accurately predicted the sensors in both tendency and frequency aspects.

Keywords: sensor data prediction; time series prediction; deep learning; long short-term memory; recurrent neural network; missing data imputation; wavelet reconstruction

Citation: Ryu, S.-G.; Jeong, J.J.; Shim, D.H. Sensor Data Prediction in Missile Flight Tests. *Sensors* **2022**, *22*, 9410. https://doi.org/10.3390/s22239410

Academic Editor: Moulay A. Akhloufi and Mozhdeh Shahbazi

Received: 23 October 2022
Accepted: 28 November 2022
Published: 2 December 2022

1. Introduction

Prediction of time series sensor data is becoming increasingly important as sensors proliferate and are used in almost all applications. Predicting sensor data can reduce costs and make up for compromised or faulty sensors, so the needs are increasing in many areas. Examples are the prediction of temperature in the industry [1], health-related applications [2], and the pandemic of COVID-19 [3].

In missile development, sensor data obtained from flight tests are valuable, as a test requires many resources and members of the workforce. Further, the data can also be used in the analysis for troubleshooting, performance improvement, and missile design. In a flight test, unexpected situations may occur, and the measured sensor data may be compromised. At times, sensors are removed after verification of missile design, but sensor data may be needed to solve problems. The prediction algorithm can reproduce corrupted or removed sensor data in these cases.

For the prediction of sensor data, model-based methods can reproduce the data with high accuracy and high sensitivity when the environments are limited (in the laboratory) and when additional human endeavors or apparatus exist in advance [4]. However, the real-world environment varies, and a complex model should be prepared by trial and error to reproduce it. In the case of using multiple sensors, as in missile tests, there must be many tests and human resources to get accurate models for multiple sensors. In preparation for the case of sensor data loss, it is not cost-effective to produce all the models, and the prediction of missile sensor data should be achieved by utilizing the previous data, which is the data-driven method.

For these reasons, data-driven methods utilize statistical techniques or deep learning-based achievements in predicting various sensor data. Traditional data-driven methods are the auto regressive moving average model (ARMA) and the auto regressive integrated moving average model (ARIMA). These models, however, have difficulty achieving high performance in the prediction of real-world sensors due to the complexity, irregularity, randomness, and nonlinearity of the data [5]. For better performance with these methods, the data need expert adjustments (e.g., deleting errors, normalizing the data, adding/mixing/deleting some data vectors, and interpolations), and the results depend on the expert's ability.

These problems can be solved using a deep learning approach. In the past decade, deep learning-based prediction algorithms with various network structures have been studied, and the data can be predicted more accurately with a deep learning network. With a deep learning network, data can be normalized and adjusted with little human intervention (Table 1). Based on these advantages and characteristics of the deep learning approach, we designed an effective deep learning-based network to predict the sensor data in missile flight tests.

Table 1. Categories of sensor prediction methods.

		Advantage	Disadvantage
Test-based modelling		• Computationally compact	• Time consuming in analyzing and modeling • Needs multiple tests in a limited environment setting
Data driven	Traditional (ARMA, ARIMA)	• Can predict data after the test	• Difficulty in achieving high performance in a complex system • Need expert's preprocessing
	Deep learning -based method	• Effective and accurate with little human effort • Various methods can be combined in the network	• Requires computing power for training the large model

To analyze data characteristics, many decomposition methods were developed to extract various components (e.g., seasonal, trend, and abrupt components) from the non-stationary time series for improved predictability and interpretability [6]. Among the methods, wavelet transform has become a widely used tool for signal analysis as a temporal-frequency representor. We combined wavelet transform, the powerful data analysis method, with a deep learning network to assist in the prediction of sensor data. Related work and details of the implementation follow.

2. Related Work

In our missile flight data, missing values exist due to loss of communication or corruption. Data imputation algorithms can help predict the output sensor data by making up for the missing values. In Section 2.1, previous work on data imputation algorithms is introduced. In Section 2.2, wavelet decomposition and reconstruction for data analysis are described, and generative adversarial network, which is an additional helper for an imputer network, is explained in Section 2.3. In Section 2.4, long short-term memory for the prediction network is explained.

2.1. Data Imputation

Imputation is widely used, as missing values can significantly affect performance [7], so missing filling algorithms have been introduced, e.g., k-nearest neighbor (kNN), tree-based, ensemble methods (bagging and boosting algorithm), and deep learning-based methods.

The kNN-based method imputes the missing data with the closest values or classes by computing the k number of the nearest normal data with the specific distance metric. In applications of kNN-based imputation, missing species-level forest biomass data are imputed [8]. The authors tested six distance metrics and fifteen k values to find optimal parameters in imputing biomass. The kNN-based methods are simple, but the performance varies depending on the parameters (distance metric, k value), and the proper parameters are needed to achieve high performance.

Tree-based methods were also used in imputation. Single tree-based algorithms have trouble with generalization and overfitting, so ensemble algorithms such as bagging and boosting have been introduced, resulting in better performance [9]. Among the bagging algorithms, random forest (RF) randomly selects features, makes a decision tree model, and iterates it to make multiple decision trees. The resultant decision trees are aggregated in the last step, and an improved tree-based imputation model is generated.

Boosting algorithms train and combine the learners by considering the importance of the data. In boosting algorithms, adaptive boosting (AdaBoost) arranges multiple learners sequentially by calculating the importance of learners through the Gini impurity measure. Each learner only considers a different single feature or combination of features in the data and is trained iteratively. In training, the "weight of the data observations" (concept of boosting) is adjusted through the error measure of the result from the previous iteration. The higher the weight of data observations, the larger the data are reflected in training.

Gradient boosting machine (GBM) trains the learners to minimize the difference (loss) between the true value and the output of learners. It weights the observation of the data by calculating the gradient of the loss. Generally, it updates the learners to the negative direction of the gradient as the loss metric is a mean squared error, but the loss metric can vary. In updating, the concept of learning rate is applied to prevent overfitting; however, the algorithm can still be overfitted.

Extreme gradient boosting (XGB) adds more complicated concepts to prevent overfitting, and a regularization term is one example. The regularization term considers the number of each tree node and node score, so it makes the tree not have too large a number of tree nodes and node scores by adding it to the loss function. Through multiple techniques, XGB has been successfully applied to the imputation of relatively large and complex datasets [10].

Various imputation algorithms have been developed and applied to many areas, but parameter tuning is essential for acceptable performance [11]. An example is that AdaBoost showed inferior performance to the simple kNN algorithm when the parameters were not optimized [12]. For stable performance, many researchers have studied and set the rules to select the effective parameters, but the algorithms still have many parameters to tune for high performance, such as the number of nodes, the number of trees, and the number of features. Further, thorough dependence consideration between the features is needed to handle a more complex dataset. The algorithms noted above consider the dependencies between the sensors less for computation simplicity, and deep learning-based imputation is a better approach to high dimensionality features, severe nonlinearities, and unconventional data [13].

For these reasons, there have been many deep learning-based approaches for imputation using various deep learning structures, such as convolution neural network (CNN) [14], recurrent neural network [15], and the generative adversarial network (GAN) [16]. In our problems, deep learning-based imputation is implemented, as the data are complex with high dimensional (the number of channels is over two hundred), nonlinear, and unconventional problems.

2.2. Wavelet Analysis

Wavelet transform (WT), as a time-frequency analysis method, has been successfully applied to an extraordinary range of time series applications. Wavelet filters exist over a finite time limit and are better suited for interpreting seasonal, trend, and abrupt components [6,17].

WT decomposes information into approximated and detailed parts. The approximated parts include low-frequency information, and the detailed parts include high frequency information. The approximated parts can be decomposed continuously, and the low-frequency information can be subdivided and analyzed more closely. Both the approximated and detailed parts are obtained using wavelet coefficients $a_{j,k}$, below.

$$\psi_{j,k}(t) = 2^{-j/2}\psi(2^{-j}t - k) \tag{1}$$

$$x(t) =< a_{j,k}, \psi_{j,k}(t) >= \sum_{j,k} a_{j,k} \cdot \overline{\psi}_{j,k}(t) \tag{2}$$

$$a_{j,k} =< x(t), \psi_{j,k}(t) >= \int x(t) \cdot \overline{\psi}_{j,k}(t)dt \tag{3}$$

where $\psi(t)$ is a kernel function, j is the dilation (number of scale), k is a translation, and $x(t)$ is the original signal.

In WT, signals can be decomposed with the various kernel functions (e.g., Haar, Meyer, Morlet, Daubechies, Symlet, Coiflet). Each wavelet kernel has its characteristics, and some kernel functions can be more useful to the specific signal analysis. In electrocardiogram signal denoising, four different kinds of wavelet kernels (Haar, Daubechies, Symlet, and Coiflet) were used in wavelet decomposition, and the Symlet kernel showed the best performance [17].

Further, there are various wavelet analysis methods, such as continuous/discrete way, multi-resolution analysis, wavelet packet decomposition, wavelet reconstruction, maximal overlap discrete wavelet transform, and so on. This variety of kernel functions and types are the main advantage of WT [6].

In the various methods, the wavelet reconstruction method selectively analyzes multi-frequencies by repeatedly decomposing time series data. The approximated parts can be decomposed again by discrete wavelet transform (DWT), so data can be analyzed in multi-frequencies. The decomposed parts can be reconstructed by inverse discrete wavelet transform (IDWT), and specific parts can be weighted and inverse transformed for rejecting or weakening the specific parts. Figure 1 shows the wavelet reconstruction method with a wavelet level of two.

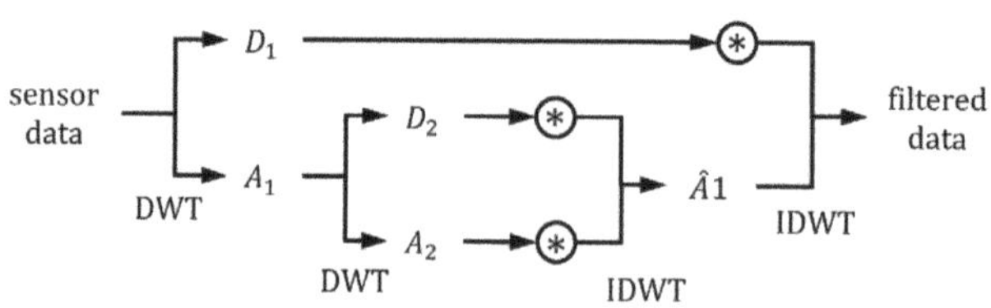

Figure 1. Structure of the wavelet reconstruction method. The convolution notation (asterisk with circle) represents the weighting process of the specific decomposed part.

The wavelet analysis methods can improve performance when used with other methods. Time series data are forecasted more accurately when the wavelet transform was combined with ARMA and ARIMA compared to the results when only ARMA and ARIMA were used [18], and defects were effectively detected in steel images with the wavelet reconstruction method and particle swarm optimization [19].

The effectiveness of WT also has been combined with the powerful performance of the deep learning approach. Wavelet decomposition and CNNs were used to analyze returned waveform data for the prediction of water level [20]. In the image processing area, deep learning networks were applied to the wavelet transformed image for image inpainting [21], super-resolution [22], and brain tumor detection [23]. We also utilized the effectiveness of multi-frequency analysis in WT and the powerful performance of deep learning networks. The details of the implementation are stated in Section 3.2.

2.3. GAN

GAN was inspired by the game where a generator and a discriminator compete with each other. The generator imitates the real value to deceive the discriminator, and the discriminator detects the generator as fake [24]. The generator network (**G**) is trained to generate a value similar to the real value to fool the discriminator, and it is represented as minimizing $log(1 - \mathbf{D}(\mathbf{G}(\mathbf{z})))$. The discriminator network (**D**) is trained to discern the real value as real and the generated output as fake, and it is represented as maximizing $log(\mathbf{D}(\mathbf{x}))$. Simultaneous training of the generator and discriminator networks are represented below:

$$\underset{\mathbf{G}}{min}\,\underset{\mathbf{D}}{max}\, V(\mathbf{D}, \mathbf{G}) = \mathbb{E}_{x \sim p_{data}(\mathbf{x})}[log(\mathbf{D}(\mathbf{x}))] + \mathbb{E}_{z \sim p_z(\mathbf{z})}[log(1 - \mathbf{D}(\mathbf{G}(\mathbf{z})))] \tag{4}$$

where $p_{data}(\mathbf{x})$ is the real data distribution, $p_z(\mathbf{z})$ is the noise distribution, $log(\cdot)$ is cross-entropy, and $\mathbf{D}(\cdot)$ and $\mathbf{G}(\cdot)$ are the results of the discriminator and generator network, respectively.

For training GAN, each network is trained one by one. The generator network is fixed when the discriminator is training, and vice versa. In the process, the generator network can generate output very close to the real data due to the discriminator network. This characteristic makes the GAN useful in many applications, such as imputation [16], texture conversion [25], image inpainting [26], and fault diagnosis [27]. GAN tends to recover high frequency information as it focuses on detailed information to fool the discriminator [21,25], resulting in improved performance in many applications.

However, the generator and discriminator networks are trained separately, so learning stability and mode collapse are issues in GAN [28]. As the networks are trained one by one, GAN has inherent training loss instability, and mode collapse can occur when the generator focuses on the local distribution. The causes of these issues can vary, and there have been many kinds of research to handle these problems, from reengineering network architectures to new loss functions [29].

Among them, Wasserstein GAN with gradient penalty (WGAN-GP) improved the performance and stability by using Earth Mover's distance and penalizing the norm of the discriminator gradient [30]. In this paper, we utilized WGAN-GP in recovering the detailed information by applying the adversarial loss to the decomposed data.

2.4. Long Short-Term Memory (LSTM)

Recurrent neural networks were introduced to consider time dependencies and relationships by connecting their units recurrently. The early stage of the recurrent network was recursive neural networks (RNNs). RNNs have been applied to time series applications [31], but they have problems with gradients exploding and vanishing when updating the networks. Gradient exploding can be solved by regularizing the weights. However, the gradient vanishing problem cannot be solved, as the gradient of the farther data becomes smaller in backpropagation. The limitation of these problems in learning long-term data motivated the different recurrent structures.

LSTM overcomes gradient vanishing and exploding problems by introducing the concept of forgetting [32]. The dependencies between the current input and the previous state are calculated as the weights through the sigmoid function. The weights, the dependency measure of how much the input is reflected in the cell state and affects the results, make differences in the learning process and result in forgetting less critical information and maintaining critical information.

More specifically, an LSTM has internal parameters named cell state (C) and hidden state (h) as in Figure 2. The cell state preserves the previous information by weighting the previous cell state (C_{t-1}) with the forget weight (f_t) and reflects the input to the cell state through the newly estimated cell state ($\tilde{C}_t$) from weighting it with the input weight

(i_t). The cell state is updated not by multiplying but by weighting and adding, so the long-term information is preserved (multiplication in applying long-term information results in gradient vanishing in backpropagation).

The hidden state, which contains important information for predicting the current time, are filtered parameters from the cell state. The hidden state is updated by the tanh of the cell state (C_t) and weighting with output weight (o_t), and the output of the LSTM is the cell state and hidden state.

$$f_t = \sigma(W_f \cdot [h_{t-1}, x_t] + b_f) \tag{5}$$

$$i_t = \sigma(W_i \cdot [h_{t-1}, x_t] + b_i) \tag{6}$$

$$\tilde{C}_t = tanh(W_C \cdot [h_{t-1}, x_t] + b_C) \tag{7}$$

$$C_t = f_t \odot C_{t-1} + i_t \odot \tilde{C}_t \tag{8}$$

$$o_t = \sigma(W_o \cdot [h_{t-1}, x_t] + b_o) \tag{9}$$

$$h_t = o_t \odot tanh(C_t) \tag{10}$$

where x_t is the network input, h_t is the hidden layer, C is the cell state with the subscript t and $t-1$ denoting time instant, $\odot$ is element-wise multiplication, σ is the sigmoid function, and W and b are the weight and bias with the subscript f, i, C, o meaning the forget, input, cell, and output.

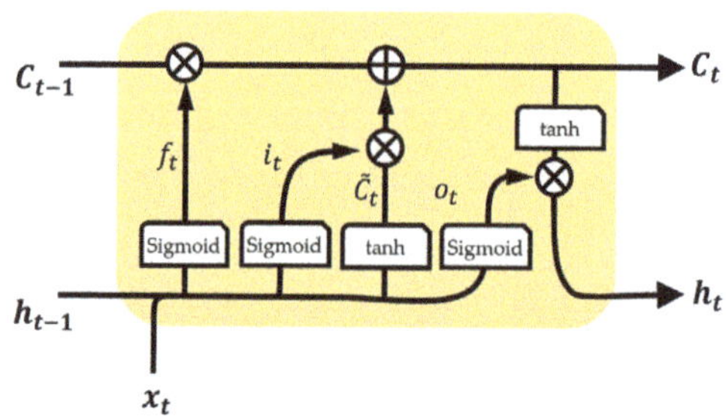

Figure 2. Structure of the fundamental LSTM network.

The property of long-term memory retention made LSTM to be applied in natural language translation [33], time series prediction [3,5,32], time series classification [34], and many other applications. The prediction performance of the simple recurrent network is inferior to the statistical ARIMA model [35,36], so most applications of LSTM was combined with other deep learning structures and techniques, e.g., fully-connected network, convolution neural network, autoencoder [32,33], squeeze-and-excitation block [34], and attention block [5].

Gated recurrent unit (GRU) is a simplified version of LSTM [37]. It is considered an efficient alternative to LSTM with a reduced number of parameters and a simpler structure [38], and it is unclear which network is better as it differs from application to application; LSTM is slightly superior to GRU in speech recognition [39]; GRU is superior to LSTM in control of a dynamic system [40]. Generally, however, LSTM is known as more fitted to complex data [41]. In our case, the data are more complex than the other applications, and we utilized the LSTM network with the attention mechanism and other deep learning techniques.

3. Missile System and the Proposed Network

3.1. Missile Data

In a missile system, various kinds of data are acquired by the remote acquisition system or the other relaying systems (Figure 3). Missile data consist of two parts: sensor data and intra-communication data. Sensor data are pressures, inner/outer temperatures of the missile, strain gauge readings from the missile body, acceleration, voltage, current sensors, etc. Intra-missile communication data include angles of fins, control-related data, missile attitude information, position-related information, the status of the equipment, etc. These sensor data and intra-communication data are used in controlling the missile guidance or analyzing the flight test result.

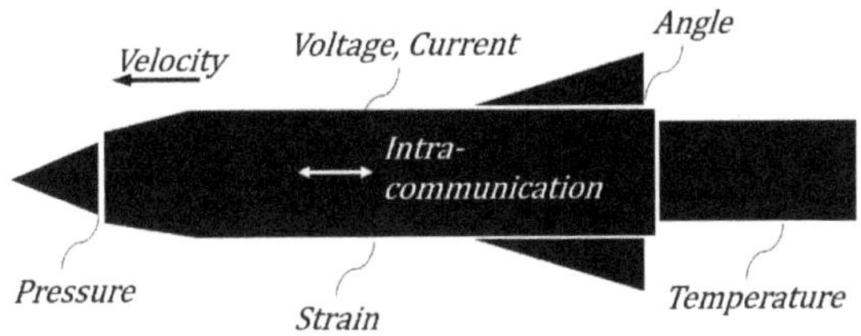

Figure 3. Types of data in a missile system.

Each sensor or communication datum is digitized with a different sampling rate, and the time standard of the sensor or intra-communication data is different, so the data are out of sync. As a result, many data points are marked as not a number (NaN) or redundant values. Further, some data are missing when they are not received by the ground station. In Table 2, the time intervals of the data are not uniform, and some values are redundant or NaN.

Table 2. An example of missile data (sensor and intra-communication).

Time	Data1	Data2	Data3	Data4	Data5	Data6	Data7	Data8	Data9	Data10	Data11
50.433 s	0.9940	2.7940	0.9830	10070	−1.423	827.42	23.816	22.346	NaN	41.806	24.032
50.437 s	NaN	NaN	NaN	NaN	NaN	NaN	NaN	NaN	45.2868	NaN	NaN
50.438 s	0.9940	2.7940	0.9830	10070	−1.423	827.42	23.780	23.346	NaN	41.792	24
50.442 s	NaN	NaN	NaN	NaN	NaN	NaN	NaN	NaN	45.2868	NaN	NaN
50.443 s	0.9940	2.7940	0.9830	10070	−1.423	827.42	23.780	22.325	NaN	41.792	24
50.447 s	NaN	NaN	NaN	NaN	NaN	NaN	NaN	NaN	45.0913	NaN	NaN
50.448 s	0.9940	2.7940	0.9830	10070	−1.423	827.42	23.754	22.325	NaN	41.758	23.983
50.452 s	NaN	NaN	NaN	NaN	NaN	NaN	NaN	NaN	45.2868	NaN	NaN
50.453 s	0.9940	2.7930	0.9830	10070	−1.423	827.42	23.754	22.308	NaN	41.758	23.983
50.457 s	NaN	NaN	NaN	NaN	NaN	NaN	NaN	NaN	45.2868	NaN	NaN
50.458 s	0.9940	2.7930	0.9830	10070	−1.423	827.42	23.733	22.308	NaN	41..734	23.974
50.462 s	NaN	NaN	NaN	NaN	NaN	NaN	NaN	NaN	45.6778	NaN	NaN
50.463 s	1	2.7930	0.9830	10070	−1.423	826.91	23.733	22.284	NaN	41.734	23.974
50.467 s	NaN	NaN	NaN	NaN	NaN	NaN	NaN	NaN	45.0913	NaN	NaN
50.468 s	1	2.7930	0.9830	10082	−1.423	826.91	23.716	22.284	NaN	41.711	23.923

Figure 4 is the display of acquired missile flight data. The data seem simple and clear in the full view, but the magnified views (Figure 4b,c) show complex and noisy values with missing or redundant intervals. The missing data occur completely randomly and are sometimes continued for a random interval in case of communication loss. The complication and incompleteness in the missile data make it difficult to predict the sensor data accurately.

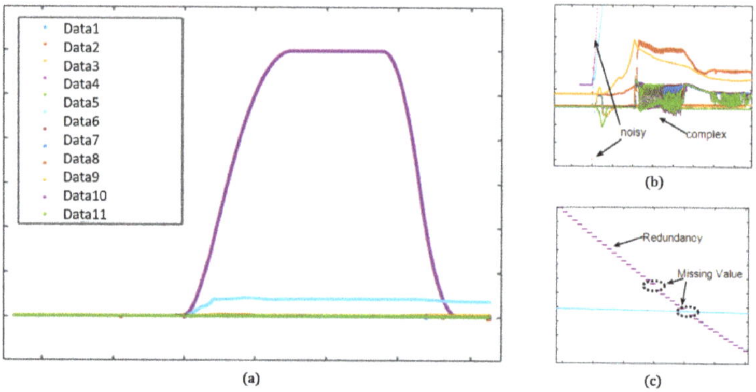

Figure 4. Missile data plot: (**a**) full view; (**b**) magnified plot with noisy and complex values; (**c**) magnified plot with redundant and missing values.

The proposed network predicts the missing sensor data utilizing the acquired dataset. For a clear explanation, examples of the missile dataset are shown in Figure 5. For the flight dataset (FT), the length of the lines in each FT shows the different test times, which means the test settings are different from test to test, as noted above. In sensor data (s), the superscript means flight dataset number with the maximum number of flight data L, and the subscript means sensor number with the maximum sensor number ns. The sampling rates can be different from sensor to sensor, and they are depicted as the different styles of the lines and dots, with blanks representing missing data. When all the sensor data are missing in a specific flight dataset ($s_{2,2}$), the proposed system can reproduce the missing sensor by using other flight datasets besides the FT_2.

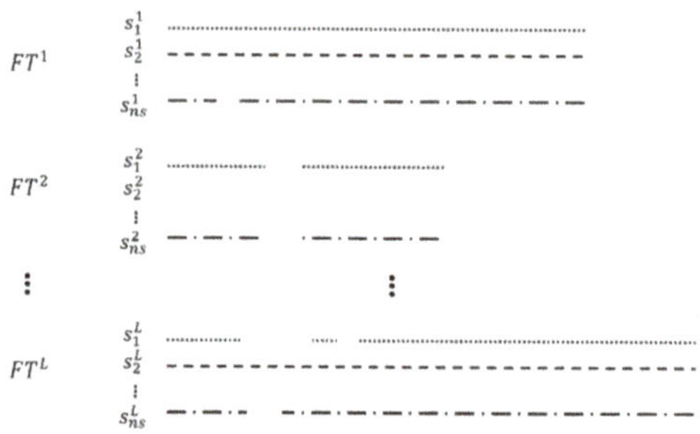

Figure 5. Examples of the missile dataset.

3.2. Network Architecture

The sensor data from missile flight tests are complex, noisy, and have many redundant values, uncertain dependencies, and a large amount of data, as shown above. Although various deep learning and signal analysis methods have been implemented for prediction, there is a limit to predicting the sensor value due to the aforementioned characteristics. To accurately predict the sensor data, we combined various analysis methods and deep learning structures; wavelet is utilized for decomposing the complex sensor data into multiple characteristics; GAN helps to handle missing or erroneous values by recovering high frequency information; LSTM is used as it is suitable for complex time series data; various deep learning techniques, such as dilation convolution and attention, are used for extracting important information in the large datasets.

Network architectures are shown in Figure 6. The input data are decomposed by DWT to analyze the time series data effectively. The imputer network imputes the missing values for each decomposed part separately, as the data's characteristics differ for the decomposed part per wavelet level and translation (approximated and detailed parts).

The imputed decomposed data are inverse transformed by IDWT to make imputed data. The imputed data are then fed into the attention-based LSTM predictor. The attention mechanism is applied to both the feature and the temporal dimensions. The proposed structure is straightforward, as the imputer and predictor networks are based on the deep learning network. The details of each network are explained more closely below.

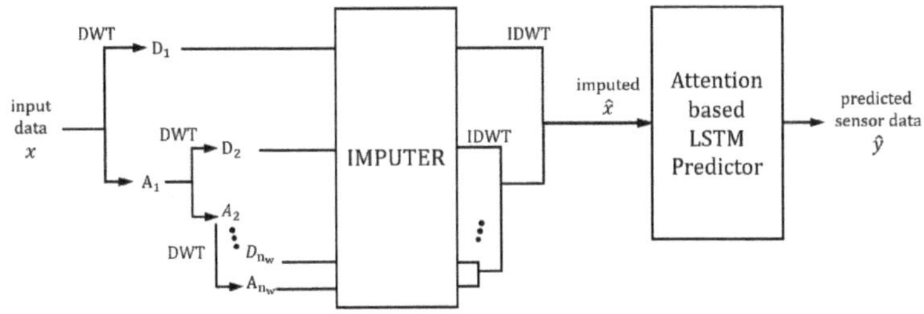

Figure 6. The architecture of the proposed network.

The detailed structure of the imputer network is shown in Figure 7. It is effective to combine the proper analysis method with a deep learning network, and it can enhance performance. In visual odometry, the movement characteristics in images are effectively analyzed by optical flow, and the geometrical analysis combined with deep learning networks [42] improve performance. Similarly, the wavelet analysis was combined with a deep learning network in image inpainting and showed improved performance [21]. The proposed network also utilized the wavelet analysis method for analyzing the time series data effectively.

In the imputer network, the input data are decomposed (D_1, D_2, A_2) and estimated, respectively, as the characteristics of each decomposed part are different. The imputing network is regarded as a part of the generator in the GAN structure. For the discriminator, we constructed the discriminator for each decomposed part to reflect the different characteristics of each decomposed part. The missing mask is estimated $(\hat{M})$ using the input data, and the estimated mask assists the imputation. The convolution network is designed with dilation and gated convolution with residual techniques to use the information of both temporal and sensor dimensions. For imputing the detailed parts (D_1, D_2), the data are adjusted by scaling (γ) and bias (β), as the detailed parts are sparse [21]. The imputed parts are inversely transformed (IDWT), and the imputed data $(\hat{x})$ is computed.

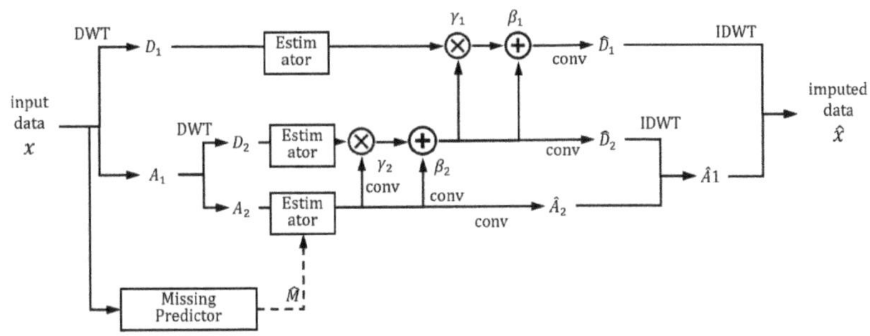

Figure 7. Details of the proposed imputer network.

The output of the imputer network is then input into the attention-based prediction network (Figure 8). The encoder–decoder structure of the prediction network can squeeze the abundant sensor information in the encoder and use the compressed context in the decoder for prediction. To predict the output sensor more accurately, the LSTM needs to concentrate on the important information in the imputed data $(\hat{x})$, and we utilized an attention mechanism. In the attention, the hidden state, cell state of the LSTM, and input data are used for selecting the important information adaptively [43]. The attention network includes the combination of fully connected layers, hyperbolic tangent, and softmax activation network to calculate attention weights, which are multiplied with the input to make attended input.

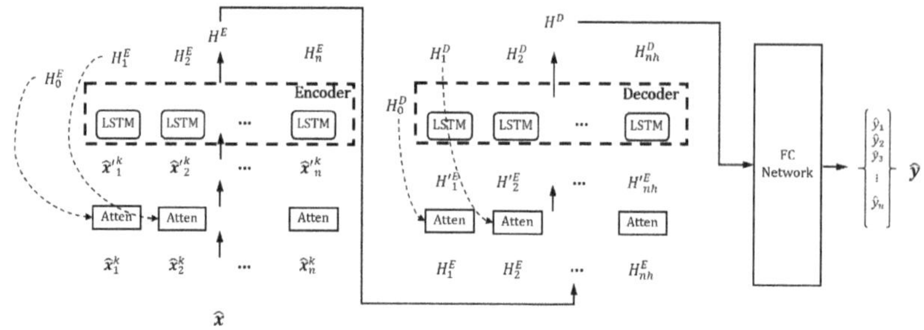

Figure 8. Details of the proposed attention-based LSTM predictor network.

The attended input ($\hat{x}'$) is then fed into the encoder LSTM network, and the encoder calculates the cell state and hidden state. The decoder also uses the attention mechanism in the temporal dimension for the attended hidden state of the encoder (H'^E), and the hidden state of the decoder is calculated. Finally, the output of the decoder is fully connected with the concatenation of the encoder and decoder hidden state (the description is skipped in the figure for clarification) and outputs the estimated sensor data $\hat{y}$.

3.3. Loss Function

For the loss metric, the L1 norm is used, as it better learns the tendency. The MSE loss exaggerates the difference in the spiky noises, and the network learns the erroneous values. It also reproduces noise values and degrades the prediction of normal values. In the loss, the results in the process of the imputer network, the imputer network's output, and the prediction network's output are considered. The masks are estimated to distinguish whether the value is a missing value, which is used in the imputer network. The mask values are binary, and the mask loss is computed using the binary cross-entropy loss.

$$L_M = -\frac{1}{CL}\sum_i^C\sum_j^L[M_{GT}^{ij}log(\hat{M}^{ij}) + (1 - M_{GT}^{ij})log(1 - \hat{M}^{ij})] \tag{11}$$

where C is the number of the input channel and L is the length of the data.

We take into account the estimation of the decomposed parts through wavelet loss. The wavelet loss (L_W) considers the imputed approximated part and the detailed parts. The imputed approximated part ($\hat{A}$) and the imputed detailed parts ($\hat{D}_n$) are compared with true values through the L1 norm as below.

$$L_W = \frac{2^{n_w}}{CL}||\hat{A} - A_{GT}||_1 + \sum_{n=1}^{n_w}\frac{2^n}{CL}||\hat{D}_n - D_{n_{GT}}||_1 \tag{12}$$

where n_w denotes the maximum number of wavelet decomposition levels.

To ensure the imputer network correctly recovered the input value, we consider imputed loss (L_I) by comparing the imputed value ($\hat{x}$) with true input (x_{GT}). Similarly, the final prediction output is considered through the regression loss (L_R) using the L1 norm of the predicted ($\hat{y}$) and true sensor output (y_{GT}).

$$L_I = \frac{1}{CL}||\hat{x} - x_{GT}||_1 \tag{13}$$

$$L_R = \frac{1}{L}||\hat{y} - y_{GT}||_1 \tag{14}$$

We adopt WGAN-GP for better performance and higher training stability, and the discriminator loss function follows the original paper [30]. We set discriminators for each decomposed part, and the discriminator loss is the summation of the loss of each approximated and detailed part. In training, we trained the discriminator five times per the one training of the generator, as specified in the paper. For each decomposed part, the discriminator is constructed, and the discriminator loss (L_D) and the adversarial (L_{Adv}) loss is formulated as below.

$$L_{Adv} = -\mathbb{E}[D_A(\hat{A})] - \sum_{n=1}^{n_w} \mathbb{E}[D_{D_n}(\hat{D}_n)] \tag{15}$$

$$\begin{aligned} L_D = \mathbb{E}[D_A(A_{GT})] \quad &- \mathbb{E}[D_A(\hat{A})] + \lambda\mathbb{E}[(||\nabla D_A(\tilde{A})|| - 1)^2] + \\ \sum_{n=1}^{n_w} [\mathbb{E}[D_{D_n}(D_{n_{GT}})] &- \mathbb{E}[D_{D_n}(\hat{D}_n)] + \lambda\mathbb{E}[(||\nabla D_{D_n}(\tilde{D}_n)|| - 1)^2]] \end{aligned} \tag{16}$$

where $\mathbf{D}$ is the discriminator of each decomposed part with the subscript D and A meaning detailed and approximated, $\mathbb{E}[\cdot]$ implies expectation, $\tilde{D}_n = \varepsilon D_{n_{GT}} + (1-\varepsilon)\hat{D}_n$, $\tilde{A} = \varepsilon A_{GT} + (1-\varepsilon)\hat{A}$ with $\varepsilon \sim U[0,1]$, and λ is set to ten.

In the proposed network, the generator can be regarded as the whole network, including the imputer network and the prediction network. Therefore, the generator loss is expanded to the addition of the adversarial loss and the other losses with the balancing terms. In the test, the balancing terms were set to one as the losses were normalized with the data size. The full objective is the maximization of the discriminator loss and the minimization of the generator loss.

$$L_G = L_{Adv} + \lambda_M L_M + \lambda_W L_W + \lambda_I L_I + \lambda_R L_R \tag{17}$$

4. Test Result

4.1. Test Setup

The proposed algorithm is trained and tested with five sets of flight data. Each flight dataset is unique, as the flight tests are conducted in different settings and environments (trajectories, speed, control, etc.). Therefore, the datasets for the training and the testing should be separated. In the five flight datasets, we used four datasets in training and one in the test. In preparation for training, the batches are generated through a sliding window and multiple batches are trained simultaneously. In the flight test dataset, if there exist ns sensors and k data length, batches are generated through the sliding window process, as shown in the left part of the Figure 9. In each batch, the x are inputted to the network sequentially from x_1 to x_w, as in the right part of the Figure 9 (b_j represents an arbitrary batch). In total, we used 230 of sensor data and about 130k data length.

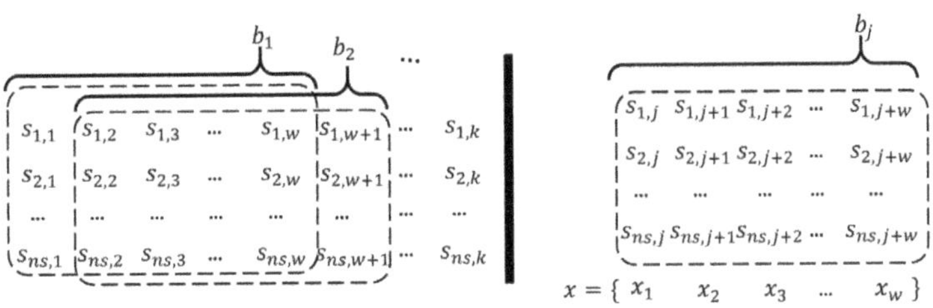

Figure 9. Preparation of input data.

All the compared methods used the same batch size and window size of the data for the preparation of the input. Parameters related to the training and testing are stated in Table 3. For the learning rate, inverse sqrt scheduling is used.

Table 3. Parameters in training and testing.

Model	Parameters
No. iterations	200
Learning rate	0.001
Optimizer	Adam optimizer
Batch size	512
Window size	128
No. LSTM layer	2
No. hidden layer	64
Dropout in LSTM	0.4

Generally, the categories of missing values are divided as missing completely at random (MCAR), missing at random (MAR), and missing not at random (MNAR). In the missile dataset, missing values can occur in communication loss, sensor error, and random error, as explained below. The probability of missing is completely random in these cases, and we assumed the missing values as MCAR.

In the flight data, the causes of missing can be divided into three cases as below, and the three cases are explained in Figure 10.

1. Communication loss: Communication loss can occur at a random time for random time intervals, and all the data are lost. The missing values are generated at a random time (*cl*) for random time intervals (f).
2. Sensor error: Sensor error can be caused by contact problems or corruption of sensors, and the specific sensor data are missing. The missing values are generated for the random sensor data for a random time (*se*) for random time intervals (k).
3. Random error: Random error can be caused by random noise or an unknown reason, and the missing values are generated for a random feature and random time.

Most of the missing values in the three cases are caused by communication loss, and we only considered the first case in this paper. The missingness is defined as the missing rate, which is the percentage of the number of missing values over the number of normal values. The missing values are generated completely randomly, and the missing rate varies slightly between the missing generations, but we marked the average value, as the difference is very small and negligible. In the test, we tested with five missing rates for the performance evaluation.

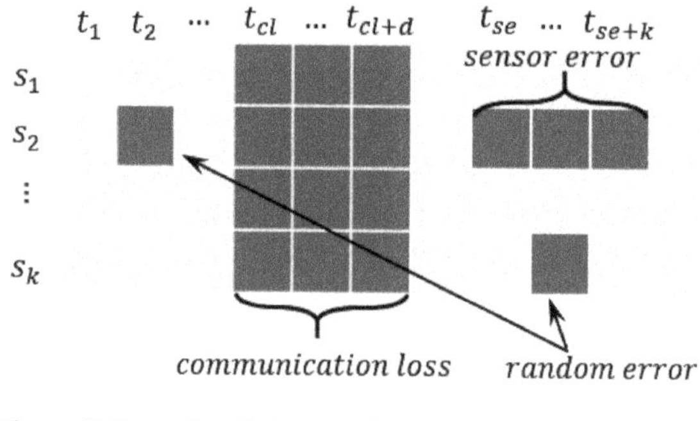

Figure 10. Examples of missing values cases in the dataset.

To evaluate the prediction performance, we compared the results with the four commonly used statistical metrics, e.g., mean absolute error (MAE), mean squared error (MSE), root mean squared error (RMSE), and mean absolute percentage error (MAPE).

4.2. Performance Evaluation

For the performance evaluation, the proposed network is compared with three different deep learning prediction models. In the different models, the basic LSTM model consists of multiple dilations, fully connected layers, and LSTM networks. Wavelet analysis with LSTM (WLSTM) consists of wavelet reconstruction method, dilation, and LSTM networks. Attention LSTM (ALSTM) is an encoder–decoder LSTM with attention techniques. The proposed method combined the wavelet method with attention LSTM with normalization and GAN architecture to reconstruct the detailed information of the wavelet decomposed data. For a fair comparison, we set the same network parameters for the LSTM used in all models of the tests, such as the number of layers and the number of hidden layers.

4.2.1. Quantitative Evaluation

First, we tested the networks with clean data (without missing values). Some sensors can be predicted relatively easily due to simple statistical characteristics, and some cannot. Therefore, the networks were tested for various sensors, and the test results are the averaged values of several test results from different sensors. The comparison result (Table 4) shows that the proposed method predicts the sensor most accurately for all four metrics. Basic LSTM shows inferior results compared to the other models, because it fails to predict some sensor values. It results in the model having large variations in metrics, but the proposed method shows very stable predictability in all sensors. The results of basic LSTM are omitted in the subsequent results, as it fails to predict the sensor value, so the comparison is meaningless.

Table 4. Quantitative comparison without missing data.

Model	MAE	MSE	RMSE	MAPE
Basic LSTM	0.454	0.555	0.676	5.184
WLSTM	0.075	0.022	0.126	1.290
ALSTM	0.041	0.073	0.215	0.951
Proposed Method	0.035	0.014	0.077	0.489

Second, we tested the methods with the five different missing rates (Table 5). In comparison, MAE, MSE, and RMSE metrics were used. It is obvious that as the missing rate increases, the prediction performance degrades. When the missing rate is low, the performance is degraded slightly because the prediction performance is only slightly affected. In the resulting table, some metrics show similar results when the missing rate is low, and the differences are shown more clearly when the missing rate is high. In all cases, the proposed method shows the best results for all the missing rates.

Table 5. Quantitative comparison with different missing rates. (The top, middle, and bottom represent MAE, MSE, and RMSE values, respectively).

Model	Missing Rate				
	1.15%	2.3%	4.6%	9%	17%
WLSTM	0.067 0.022 0.127	0.071 0.023 0.133	0.066 0.023 0.128	0.081 0.029 0.147	0.099 0.041 0.179
ALSTM	0.045 0.081 0.226	0.049 0.078 0.227	0.056 0.071 0.227	0.070 0.081 0.241	0.097 0.088 0.263
Proposed Method	0.037 0.016 0.091	0.039 0.017 0.097	0.044 0.021 0.115	0.053 0.025 0.129	0.070 0.039 0.163

4.2.2. Qualitative Evaluation

The proposed network is compared in both the time and frequency domains in the qualitative evaluation. The graphs plot only the proposed method and the best results of other methods for simple comparison. The overview of the prediction results shows that both the proposed and other methods predict the sensor values very close to the real sensor value, ground truth (GT). Both methods predict abrupt and gradual changes relatively well. In addition, in the prediction results, the spiky noise is less than that of the GT, and the spiky noises are alleviated. In the magnified view, however, the proposed network predicted the sensor more accurately and stably than the best of the others (ALSTM). In Figure 11, periodic noises are included in the result from ALSTM, which were caused by the spiky noises. On the other hand, the proposed method is less affected by the spiky noises, as it analyzes data using wavelet transform, so it predicted the sensor data robustly. Sensor A and sensor B represent the arbitrary sensors. In Figure 12, GT shows more repeated spiky noises compared to Figure 11. ALSTM is affected by excessive spiky noises, and, consequently, many spiky noises are generated. However, the proposed predictor follows the tendency of the GT, and the spiky noises are rejected by utilizing the wavelet analysis.

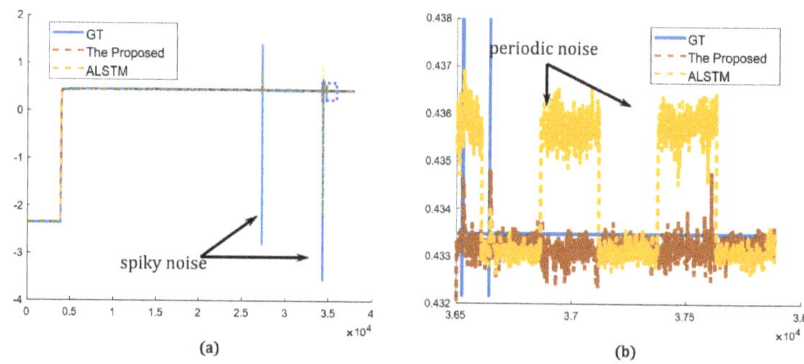

Figure 11. Prediction results in time domain (sensor A): (**a**) overview of the proposed and the best in other methods compared to the GT; (**b**) detailed view of the blue dotted rectangle of (**a**).

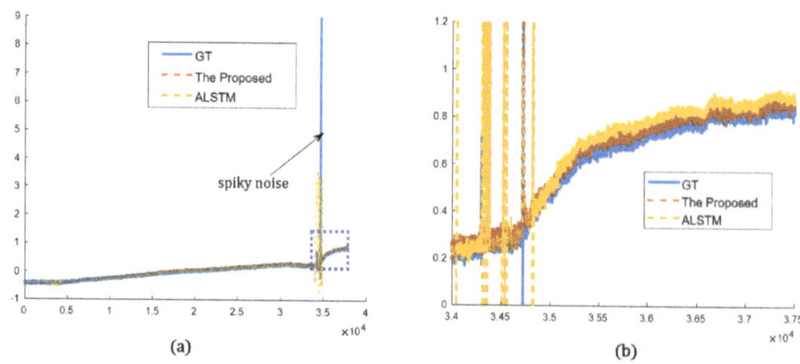

Figure 12. Prediction results in time domain (sensor B): (**a**) overview of the proposed and the best of other methods compared to the GT; (**b**) detailed view of the blue dotted rectangle of (**a**).

In the frequency analysis, the prediction results were transformed by Fourier transform to compare the frequency characteristics of the sensor data. As can be seen, the differences are more clear. In Figure 13, ALSTM follows the frequency tendency of the GT, but it has oscillations or errors compared to the GT. In contrast, the proposed method shows frequency characteristics very close to the GT. It better predicted the true sensor in both frequency tendency and characteristics. In Figure 14, both the proposed method and ALSTM seem to predict the frequency characteristics well. However, the magnified view shows that the ALSTM predicts high frequencies much larger than the GT, and it is caused by the overproduced spiky noises. The proposed method suppresses the spiky noises in the prediction, and the frequency characteristics are very close to the GT, even in the magnified view.

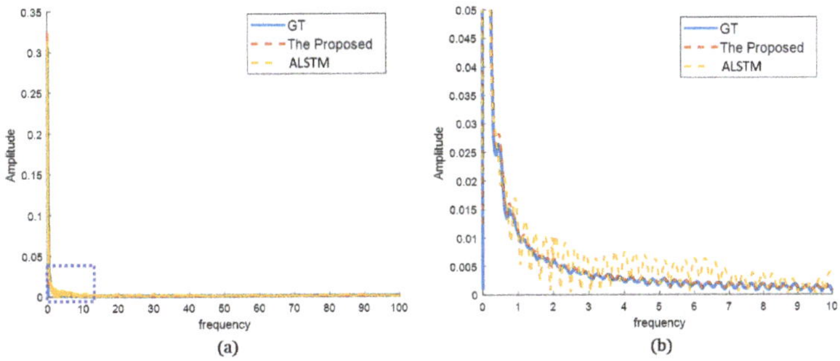

Figure 13. Prediction results in frequency domain (sensor A): (**a**) overview of the proposed and the best of the other methods compared to the GT; (**b**) detailed view of the blue dotted rectangle of (**a**).

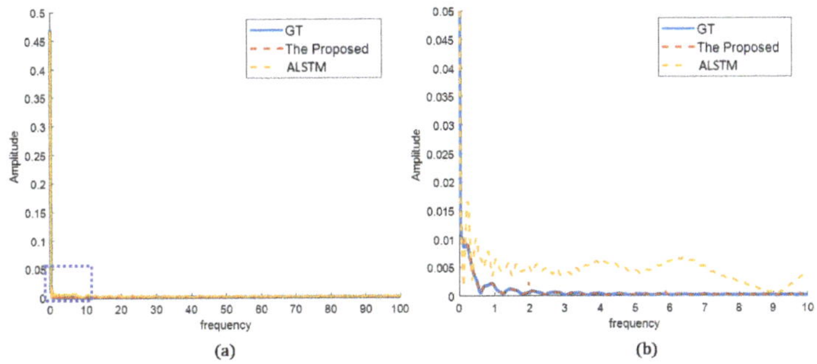

Figure 14. Prediction results in frequency domain (sensor B): (**a**) overview of the proposed and the best of the other methods compared to the GT; (**b**) detailed view of the blue dotted rectangle of (**a**).

As noted above, the proposed method utilizes wavelet analysis and GAN architecture. The wavelet analysis interprets frequency information more precisely, and GAN architecture can generate detailed information in competing between the generator and the discriminator. It results in the predicted sensor data being closer to the true sensor values.

5. Conclusions

This paper proposes a deep learning-based prediction network for sensor data from missile flight tests. The missile datasets are highly different and independent from test to test, as environment and planning differ for each flight. Further, data are incomplete due to communication errors or loss, different sampling rates/time standards, and the number of sensor data is over two hundred with complex correlations. Therefore, the incompleteness, complexity, and independencies among datasets make it difficult to predict the data.

The proposed network utilizes wavelet decomposition in analyzing the sensor data and reconstructs the incomplete data using the recent GAN architectures with better training stability and performance. The interpretability of time series data with wavelet analysis and the advantage of GAN helps to handle incomplete and complex data accurately. The final outputs are predicted by focusing on the important information using the LSTM network with an attention mechanism. The proposed network combined appropriate network structures considering the wavelet technique suitable for analyzing temporal data and the characteristics of the state-of-the-art deep learning technology, and the new network structure can predict sensors accurately.

The test result shows that the proposed network can effectively predict the missile sensor data in quantitative and qualitative aspects. The accuracy and stability of the proposed network outperform the previous methods in the MAE, MSE, RMSE, and MAPE metrics. The qualitative evaluation showed that the proposed network predicted the target sensor well in statistical and frequency analysis. The proposed network can be applied to reproduce corrupted or broken sensors in flight tests when unexpected situations occur.

Author Contributions: Conceptualization, S.-G.R.; methodology, S.-G.R. and J.J.J.; software, S.-G.R.; validation, S.-G.R. and J.J.J.; formal analysis, S.-G.R.; investigation, J.J.J.; resources, S.-G.R.; data curation, S.-G.R.; writing—original draft preparation, S.-G.R.; writing—review and editing, J.J.J. and D.H.S.; visualization, J.J.J.; supervision, D.H.S.; project administration, D.H.S. All authors have read and agreed to the published version of the manuscript.

Funding: This research received no external funding.

Institutional Review Board Statement: Not applicable

Data Availability Statement: Not applicable

Conflicts of Interest: The authors declare no conflict of interest.

References

1. Leon-Medina, J. X.; Camacho, J.; Gutierrez-Osorio, C.; Salomón, J. E.; Rueda, B.; Vargas, W.; Sofrony, J.; Restrepo-Calle F.; Tibaduiza, D., T. Temperature Prediction Using Multivariate Time Series Deep Learning in the Lining of an Electric Arc Furnace for Ferronickel Production. *Sensors* **2021**, *21*, 6894. [CrossRef] [PubMed]
2. Macias, E.; Boquet, G.; Serrano, J.; Vicario, J. L.; Ibeas J.; Morel, A. Novel imputing method and deep learning techniques for early prediction of sepsis in intensive care units. In Proceedings of the 2019 Computing in Cardiology, Singapore, 8–11 September 2019; pp. 1–4.
3. Nikparvar, B.; Rahman, M.; Hatami, F.; Thill, J. C. Spatio-temporal prediction of the COVID-19 pandemic in US counties: Modeling with a deep LSTM neural network. *Sci. Rep.* **2021**, *11*, 21715. [CrossRef] [PubMed]
4. Han, L.; Yu, C.; Xiao, K.; Zhao, X. A new method of mixed gas identification based on a convolutional neural network for time series classification. *Sensors* **2019**, *19*, 1960. [CrossRef] [PubMed]
5. Li, Y.; Zhu, Z.; Kong, D.; Han, H.; Zhao, Y. EA-LSTM: Evolutionary attention-based LSTM for time series prediction. *Knowl.-Based Syst.* **2019**, *19*, 1960. [CrossRef]
6. Rhif, M.; Ben Abbes, A.; Farah, I. R.; Martínez, B.; Sang, Y. Wavelet transform application for/in non-stationary time-series analysis: A review. *Appl. Sci.* **2019**, *9*, 1345. [CrossRef]
7. Kim, T.; Ko, W.; Kim, J. Analysis and Impact Evaluation of Missing Data Imputation in Day-ahead PV Generation Forecasting. *Appl. Sci.* **2019**, *9*, 204. [CrossRef]
8. Fu, Y.; He, H. S.; Hawbaker, T. J.; Henne, P. D.; Zhu, Z.; Larsen, D. R. Evaluating k-Nearest Neighbor (k NN) Imputation Models for Species-Level Aboveground Forest Biomass Mapping in Northeast China. *Remote Sens.* **2019**, *11*, 2005. [CrossRef]
9. Valdiviezo, H. C.; Van Aelst, S. Tree-based prediction on incomplete data using imputation or surrogate decisions. *J. Inf. Sci.* **2015**, *3131*, 163–181. [CrossRef]
10. Zhang, X.; Yan, C.; Gao, C.; Malin, B.; Chen, Y. XGBoost imputation for time series data. In Proceedings of the 2019 IEEE International Conference on Healthcare Informatics, Xi'an, China, 10–13 June 2019; pp. 1–3.
11. Krithiga, R.;Ilavarasan, E. Hyperparameter tuning of AdaBoost algorithm for social spammer identification. *Int. J. Pervasive Comput. Commun.* **2021**, *17*, 462–482.
12. Mikhchi, A.; Honarvar, M.; Kashan, N. E. J.; Zerehdaran, S.; Aminafshar, M. Analyses and comparison of K-nearest neighbour and AdaBoost algorithms for genotype imputation. *Res. Vet. Sci.* **2017**, *5*, 295–299.
13. Lall, R.; Robinson, T. The MIDAS touch: Accurate and scalable missing-data imputation with deep learning. *Political Anal.* **2022**, *30*, 179–196. [CrossRef]
14. Zhuang, Y.; Ke, R.; Wang, Y. Innovative method for traffic data imputation based on convolutional neural network. *IET Intell. Transp. Syst.* **2019**, *13*, 605–613. [CrossRef]
15. Cao, W.; Wang, D.; Li, J.,; Zhou, H.; Li, L.;Li, Y. Brits: Bidirectional recurrent imputation for time series. In Proceedings of the 32nd Conference on Neural Information Processing Systems (NeurIPS 2018), Montréal, QC, Canada, 2–8 December 2018; p. 31.
16. Yoon, J.; Jordon, J.; Schaar, M. Gain: Missing data imputation using generative adversarial nets. In Proceedings of the 35th International Conference on Machine Learning, Stockholm, Sweden, 10–15 July 2018; pp. 5689–5698.
17. Nibhanupudi, S.; Youssif, R.; Purdy, C. Data-specific signal denoising using wavelets, with applications to ECG data. In Proceedings of the 2004 47th Midwest Symposium on Circuits and Systems, Hiroshima, Japan, 25–28 July 2004; pp. 219–222.
18. Joo, T. W.; Kim, S. B. Time series forecasting based on wavelet filtering. *Expert Syst. Appl.* **2015**, *42*, 3868–3874. [CrossRef]
19. Ryu, S.; Koo G.;Kim S. W. An Adaptive Selection of Filter Parameters: Defect Detection in Steel Image Using Wavelet Reconstruction Method. *ISIJ Int.* **2020**, *60*, 1703–1713. [CrossRef]
20. Memarian S. O.; Asgari, J.; Amiri-Simkooei, A. Wavelet decomposition and deep learning of altimetry waveform retracking for Lake Urmia water level survey. *Mar. Georesources Geotechnol.* **2022**, *40*, 361–369. [CrossRef]

21. Yu, Y.;Zhan, F.;Lu, S.;Pan, J.;Ma, F.;Xie, X.; Miao, C. Wavefill: A wavelet-based generation network for image inpainting. In Proceedings of the IEEE/CVF International Conference on Computer Vision, Montreal, QC, Canada, 11–17 October 2021; pp. 14114–14123.
22. Guo, T.; Seyed Mousavi, H.; Huu Vu, T.; Monga, V. Deep wavelet prediction for image super-resolution. In Proceedings of the IEEE Conference on Computer Vision and Pattern Recognition Workshops, Honolulu, HI, USA, 21–26 July 2017; pp. 104–113.
23. Sarhan, A. M. Brain tumor classification in magnetic resonance images using deep learning and wavelet transform. *J. Biomed. Eng.* **2020**, *133*, 102. [CrossRef]
24. Goodfellow, I.; Pouget-Abadie, J.; Mirza, M., Xu, B.; Warde-Farley, D.; Ozair, S.;Aaron C.; Bengio, Y. Generative adversarial nets. Advances in neural information processing systems. *Adv. Neural Inf. Process. Syst.* **2014**, *27*.
25. Xian, W.; Sangkloy, P.; Agrawal, V.; Raj, A.; Lu, J.; Fang, C.; Yu, F.; Hays, J. Texturegan: Controlling deep image synthesis with texture patches. In Proceedings of the IEEE Conference on Computer Vision and Pattern Recognition, Salt Lake City, UT, USA, 18–23 June 2018; pp. 8456–8465.
26. Demir, U.; Unal, G. Patch-based image inpainting with generative adversarial networks. In Proceedings of the Conference on Computer Vision and Pattern Recognition, Salt Lake City, UT, USA, 18–23 June 2018.
27. Zhang, W.; Li, X.; Jia, X. D.; Ma, H.; Luo, Z.; Li, X. Machinery fault diagnosis with imbalanced data using deep generative adversarial networks. *Measurement* **2020**, *152*, 107377. [CrossRef]
28. Arjovsky, M.; Chintala, S.; Bottou, L. Wasserstein generative adversarial networks. In Proceedings of the International Conference on Machine Learning, Sydney, Australia, 6–11 August 2017; pp. 214–223.
29. Saxena, D.; Cao, J. Generative adversarial networks (GANs) challenges, solutions, and future directions. *ACM Comput. Surv.* **2021**, *54*, 1–42. [CrossRef]
30. Gulrajani, I.; Ahmed, F.; Arjovsky, M.; Dumoulin, V.; Courville, A. C. Improved training of wasserstein gans. *Adv. Neural Inf. Process. Syst.* **2017**, *30*.
31. Connor, J. T.; Martin, R. D.; Atlas, L. E. Recurrent neural networks and robust time series prediction. *IEEE Trans. Neural Netw. Learn. Syst.* **1994**, *5*, 240–254. [CrossRef]
32. Hochreiter, S.; Schmidhuber, J. Long short-term memory. *Neural Comput.* **1997**, *9*, 1735–1780. [CrossRef] [PubMed]
33. Tiwari, G.; Sharma, A.; Sahotra, A.; Kapoor, R. English-Hindi neural machine translation-LSTM seq2seq and ConvS2S. In Proceedings of the 2020 International Conference on Communication and Signal Processing, Shanghai, China, 12–15 September 2020; pp. 871–875.
34. Karim, F.; Majumdar, S.; Darabi, H.; Harford, S. Multivariate LSTM-FCNs for time series classification. *Neural Netw.* **2019**, *116*, 237–245. [CrossRef] [PubMed]
35. Liu, X.; Lin, Z.; Feng, Z. Short-term offshore wind speed forecast by seasonal ARIMA-A comparison against GRU and LSTM. *Energy* **2021**, *227*, 120492. [CrossRef]
36. Yamak, P. T.; Yujian, L.; Gadosey, P. K. A comparison between arima, lstm, and gru for time series forecasting. In Proceedings of the 2019 2nd International Conference on Algorithms, Computing and Artificial Intelligence, New York, NY, USA, 20–22 December 2019; pp. 49–55.
37. Chung, J.; Gulcehre, C.; Cho, K.; Bengio, Y. Empirical evaluation of gated recurrent neural networks on sequence modeling. *Neural and Evol. Comput.* **2014**, arXiv:1412.3555.
38. Shewalkar, A. Performance evaluation of deep neural networks applied to speech recognition: RNN, LSTM and GRU. *J. Artif. Intell. Soft Comput. Res.* **2019**, *9*, 235–245. [CrossRef]
39. Shewalkar, A.N. Comparison of rnn, lstm and gru on speech recognition data. Masters Thesis, North Dakota State University, Fargo, ND, USA, 2018.
40. Yang, S.; Yu, X.; Zhou, Y. Lstm and gru neural network performance comparison study: Taking yelp review dataset as an example. In Proceedings of the 2020 International Workshop on Electronic Communication and Artificial Intelligence, Shanghai, China, 12–14 June 2020; pp. 98–101.
41. Zarzycki, K.; Ławryńczuk, M. LSTM and GRU neural networks as models of dynamical processes used in predictive control: A comparison of models developed for two chemical reactors. *Sensors* **2021**, *21*, 5625. [CrossRef]
42. Zhan, H.; Weerasekera, C. S.; Bian, J. W.; Reid, I. Visual odometry revisited: What should be learnt? In Proceedings of the 2020 IEEE International Conference on Robotics and Automation, Virtual Conference, 1 June–31 August 2020; pp. 4203–4210.
43. Qin, Y.; Song, D.; Chen, H.; Cheng, W.; Jiang, G.; Cottrell, G. A dual-stage attention-based recurrent neural network for time series prediction. *Mach. Learn.* **2017**, arXiv:1704.02971.

remote sensing

MDPI

Article

Automatic Laboratory Martian Rock and Mineral Classification Using Highly-Discriminative Representation Derived from Spectral Signatures

Juntao Yang [1,2,3], Zhizhong Kang [2,3,4,*], Ze Yang [2,3,4], Juan Xie [2,3,4], Bin Xue [5], Jianfeng Yang [5] and Jinyou Tao [5]

1 College of Geodesy and Geomatics, Shandong University of Science and Technology, Qingdao 266590, China
2 School of Land Science and Technology, China University of Geosciences, No. 29 Xueyuan Road, Haidian District, Beijing 100083, China
3 Subcenter of International Cooperation and Research on Lunar and Planetary Exploration, Center of Space Exploration, Ministry of Education of The People's Republic of China, No. 29 Xueyuan Road, Haidian District, Beijing 100083, China
4 Lunar and Planetary Remote Sensing Exploration Research Center, China University of Geosciences, No. 29 Xueyuan Road, Haidian District, Beijing 100083, China
5 Xi'an Institute of Optics and Precision Mechanics, Chinese Academy of Sciences, Xi'an 710119, China
* Correspondence: zzkang@cugb.edu.cn

Abstract: The optical properties of rocks and minerals provide a reliable way to measure their chemical and mineralogical composition due to the specific reflection behaviors, which is also the key insight behind most automatic identification and classification approaches. However, the inter-category spectral similarity poses a great challenge to the automatic identification and classification tasks because of the diversity of rocks and minerals. Therefore, this paper develops a recognition and classification approach of rocks and minerals using the highly discriminative representation derived from their raw spectral signatures. More specifically, a transformer-based classification approach integrated with category-aware contrastive learning is constructed and trained in an end-to-end manner, which would force instances of the same category to remain close-by while pushing instances of a dissimilar category far apart in the high-dimensional feature space, in order to produce the highly discriminative feature representation of the rocks and minerals. From both qualitative and quantitative views, experiments are conducted on the laboratory sample dataset with 30 types of rocks and minerals shared from the National Mineral Rock and Fossil Specimens Resource Center, and the spectral information of the laboratory rocks and minerals is captured using a multi-spectral sensor, with a duplicated payload of the counterpart onboard the Zhurong rover. Quantitative results demonstrate that the developed approach can effectively distinguish 30 types of rocks and minerals, with a high overall accuracy of 96.92%. Furthermore, the developed approach is remarkably superior to other existing methods, with average differences of 4.75% in the overall accuracy. Furthermore, we also visualized the derived highly discriminative features of different types of rocks and minerals by projecting them onto a two-dimensional map, where the same categories tend to be modeled by nearby locations and the dissimilar categories by distant locations with high probability. It can be observed that, compared with those in the raw spectral feature space, the clusters are formed better in the derived highly discriminative feature space, which further confirms the promising representation capability.

Keywords: identification and classification; transformer; highly discriminative representation; contrastive learning; multi-spectral sensor; Mars exploration

Citation: Yang, J.; Kang, Z.; Yang, Z.; Xie, J.; Xue, B.; Yang, J.; Tao, J. Automatic Laboratory Martian Rock and Mineral Classification Using Highly-Discriminative Representation Derived from Spectral Signatures. *Remote Sens.* **2022**, *14*, 5070. https://doi.org/10.3390/rs14205070

Academic Editors: Moulay A. Akhloufi and Mozhdeh Shahbazi

Received: 26 August 2022
Accepted: 10 October 2022
Published: 11 October 2022

1. Introduction

Rocks and minerals are one of the major planetary surface features. Since minerals are stable under the known ranges of temperature and pressure, a rock, made of specific minerals, can be used to identify and deduce the climate or weathering when the rocks were being formed and deposited, based on individual mineral stability ranges and the presence or absence of equilibrium between them [1]. Obviously, the characterization of the chemical and mineralogical composition of rocks and minerals, as well as their alteration over time, are highly complementary to the investigation of the climatic and geologic history of the planetary bodies [2–5]. With the development of deep space exploration, spectroscopic systems (e.g., Raman, near-infrared spectroscopy), as one of most important scientific payloads onboard the orbiters, landers, and rovers, provide a powerful tool for the investigation and characterization of the chemical and mineralogical nature of planetary surfaces [6–9], since the reflection behavior of rocks and minerals is primarily determined by their compositions and physical properties [4].

Accurate determination of the presence of rocks and minerals from spectroscopic data is a fundamental task of planetary exploration and science [10]. Viviano et al. [11] corrected spectral reflectance at key wavelengths in a compact reconnaissance imaging spectrometer for Mars (CRISM) to produce the revised spectral parameters, which revealed an increasingly diverse suite of minerals present on the Martian surface. Jain and Chauhan [12] adopted a spectral reflectance-based mineral detection method to reveal the presence of minerals from CRISM data, which hinted of the aqueous history of the Noachian period on Mars. Fox et al. [13] detected Fe^{3+}-Mg^{2+} smectite deposits in Marathon Valley using five CRISM observations. Núñez et al. [14] obtained the spectral information of gully sediments from CRISM data and used them to identify a variety of mineralogies for constraining the gully-forming processes. Xue et al. [15] identified the category of minerals and mapped their distributions from CRISM near-infrared spectral data in the Martian Gale and Nili Fossae regions. Amador et al. [16] used factor analysis and target transformation methods to produce the global distribution of minerals associated with serpentinization from CRISM spectral data. Lin et al. [17] jointly identified the hydrated mineral deposits and determined their abundances with a large endmember library using CRISM near-infrared data. These aforementioned identification methods generally use simple mathematical functions to judge whether specific spectral features are absent or not for achieving the mineralogical analysis. While these existing classification methods are effective for the manual analysis, the spectral artifacts and noises might affect their performance. Thus, automatic identification and classification of rocks and minerals based on their spectral information has been an active issue to date.

Automatic rock and mineral classification based on their spectral signatures is greatly challenging due to the inter-category spectral similarity [18] and lots of machine learning-based methods, generally consisting of feature representation and recognition systems, have been proposed using spectral information, such as Tsallis entropy [19], fractal dimension [20], Sato's maximum Lyapunov exponent [21], color histogram-based features [22], etc. To further enhance the descriptive capability of the feature representation, several feature selection and optimization strategies, including information gain ranking [23], principal component analysis [24], genetic optimization [24], and K-means [25] have also been developed to autonomously learn the high-level feature representation. Following this, some recognition and classification systems of rocks and samples [18,22,25–29] have then focused on fast and reliable identification without human intervention.

Recently, deep learning technology has achieved great successes and has attempted to deal with the classification and recognition issues because of its inspirational performance in the field of computer vision. As a state-of-the-art deep learning model, the transformer provides a strong representation capability to model and encode heterogeneous interactions [30], and its tremendous success in the language domain has led researchers to investigate its adaptation to computer vision [31]. Hence, this work concentrates on using it to generate the highly discriminative representation of rocks and minerals from their raw spectral signatures. Pascual [32] explored the superior capability of convolutional neural network for classifying different types of rock images, especially in the natural scenes. Saranathan and Parente [33] developed a generative adversarial network-based feature representation learning method to map mineral signatures of interest across the CRISM image database. To deal with the problem of limited samples, Li et al. [18] adopted the idea of transfer learning to fine-tune the super-parameters of the pre-trained convolutional neural network.

To date, some institutes and organizations have been also devoted to establishing the laboratory spectral database of rocks and minerals (e.g., the Berlin Emissivity Database, United States Geological Survey (USGS) spectral library version 7) to study their spectral behaviors [5,34–38]. In this way, the reflectance spectra of terrestrial rocks and minerals, covering at the range of visible to near-infrared wavelengths, are explored and analyzed in laboratory conditions, which is beneficial to further supporting the research of planetary origin and geological evolution [4,27,39,40]. Similarly, this work captures the spectral information of the laboratory rocks and minerals using the multi-spectral sensor, which duplicates the payload of its counterpart onboard the Zhurong rover [41], and focuses on investigating the highly discriminative feature representations from the raw spectral information for the classification task.

Although there have been numerous presented studies related to detection and classification [2,25,29,42–45], automatic rock and mineral classification remains greatly challenging due to the following aspects. First, according to the literature statistics, the majority of existing classification approaches tended to only use a few types of rocks and minerals, where the number of the category in these work is generally less than 10, to verify their own robustness and effectiveness [25,29,43]. As a matter of fact, the increasing number of categories might result in the severe inter-category similarity, which will pose a challenge in the task of accurate classification. In this case, a dataset containing as many types of rocks and minerals as possible is very necessary for verifying the generalization of the classification approaches. Second, although lots of feature extraction and selection strategies have been developed for describing the spectral and texture information of rocks and minerals [25,29,43], the inter-category similarity might still affect the descriptive capability and reduce the classification quality [46] as the number of categories increases. Obviously, a highly discriminative feature representation plays a significant role in the classification task to enhance the intra-category similarity and enlarge the inter-category variants.

To address these challenges mentioned above, we develop a rock and mineral classification approach using the highly discriminative representation derived from their original spectral signatures. Moreover, experiments are carried out on the multi-spectral data captured from 30 types of rocks and minerals for both qualitatively and quantitatively evaluating the robustness and reliability of the developed approach. Our main contributions in this work are as follows.

(1) To efficiently achieve the classification task, we design a transformer-based classification approach for generating the highly discriminative feature representation of both rocks and minerals, where the inter-category representation variant is enlarged and the intra-category representation similarity is aggregated;
(2) A category-aware contrastive learning is integrated within the developed transformer-based classification approach. In this case, the super-parameters of the whole network are learned and trained in an end-to-end multi-task manner. Consequently, the remarkable distinctions among different types of rock and minerals occur in their high-dimensional feature space;
(3) We demonstrate the reliability and robustness of the developed approach on a dataset containing rocks and minerals with complicated categories. It is of significance for the investigation of the developed approach's generalization ability.

The rest of this paper is organized as follows. Section 2 describes the experimental data and the developed classification approach in detail. Section 3 presents the experimental results and analysis for evaluating the developed classification approach both quantitatively and qualitatively. Section 4 discusses the sensitivity of parameters and analyzes the strengths and limitations of the developed classification approach. This paper concludes with a summary of future research considerations in Section 5.

2. Materials and Methods

In this section, we will introduce the proposed method in detail. This section is organized as follows. First, the details of data acquisition are described. Second, we introduce the overview of the proposed method. Subsequently, we present other detailed implementation of the proposed method.

2.1. Data Acquisition

When the number of categories is large, the probability of classification error becomes relatively large. This is because each category is surrounded by a large number of neighboring categories [47], which results in severe inter-category spectral similarity. Consequently, a sample dataset of rocks and minerals with as many categories as possible is acquired from the National Mineral Rock and Fossil Specimens Resource Center. The National Mineral Rock and Fossil Specimens Resource Center is devoted to the digitization and sharing of China's rock mineral specimens, offering more than 170,000 state-owned specimens with scientific value, including minerals, rocks, and fossils. As a data provider, its goal is to promote the understanding of geo-resources and to provide resources for academia, education, and scientific popularization in the field of geoscience through collecting, organizing, and sharing rock mineral specimen resources.

As for the implemented experiments in this work, we reviewed the relevant literature about the distribution of rocks and minerals on the Martian surface and selected 30 types of rocks and minerals, e.g., hydrous minerals. All the used rocks and minerals are from the National Mineral Rock and Fossil Specimens Resource Center, including muscovite (Mu), hematite (He), calcite (Ca), galena (Gal), kaolinite (Ka), talcum Ta), pyrite (Pyr), chalcopyrite (Cha), gray tiemannite (GT), stibnite (St), gabbro (Gab), quartz (Qu), chlorite (Chl), serpentine (Se), smaltite (Sm), tennantite (Te), gypsum (Gy), graphite (Gr), crystal (Cr), suanite (Sua), aragonite (Ar), basalt (Bas), fluorite (Fl), psilomelane (Ps), saponite (Sa), goethite (Go), barite (Bar), sulfur (Sul), biotite vermiculite (BV), and pyroxenite (Pr). Moreover, we also enrich the number of categories of rocks and minerals as much as possible in the sample dataset, although a few rocks and minerals on the list are very uncommon and even unlikely to be found on Mars. By establishing a complicated sample dataset, the generalization of the proposed method can be better evaluated.

Generally, most rocks and minerals can be characterized and classified by their unique physical properties (e.g., hardness, luster, color, cleavage, fracture). For instance, barite can be reliably recognized based on its three directions of right-angle cleavage and the sugary appearance. Gypsum is characterized by its softness and its three directions of unequal cleavage. Biotite vermiculite is almost always much darker in color than muscovite. Talcum has a greasy feel. Psilomelane occurs as botryoidal and stalactitic masses with a smooth shining surface and submetallic luster. Quartz is characterized by its glassy luster, conchoidal fracture, and crystal form. The most obvious physical properties of serpentine are its green color, patterned appearance, and slippery feel. Stibnite typically forms coarse, irregular masses or radiating sprays of needlelike crystals, and its distinguishing characterization is easy fusibility, a bladed habit, perfect cleavage in one direction, lead-gray color, and soft black streaks. Figure 1 illustrates examples of different types of both rocks and minerals. It is obvious that similar mineralogy or appearance characteristics among some types of rocks and minerals make them very challenging to differentiate via direct visual interpretation. For example, both aragonite and calcite with the same formula are tabular, prismatic, or needlelike, often with steep pyramidal or chisel-shaped ends, and can form columnar or spreading aggregates. Basalt consists mainly of plagioclase and pyroxene minerals similar to gabbro, but the former is fine-grained, and the latter is coarse-grained. The yellow color and metallic luster are the most obvious physical properties of chalcopyrite, which shows a similar appearance to pyrite. Chlorite is usually green in color, has a foliated appearance, inelastic cleavage, and an oily to soapy feel, but its variable chemical composition makes it a difficult specimen to. Galena can be easily recognized by its metallic cleavage planes, lead gray and silvery color, and black streaks, which are often associated with sphalerite, calcite, and fluorite. Hematite has an extremely variable appearance, from earthy to submetallic to metallic luster, but it always produces a reddish streak. Smaltite crystallizes in the cubic system with the same hemihedral symmetry as pyrite.

Figure 1. Instances of different types of rocks and samples in the laboratory. For abbreviation, muscovite (Mu), hematite (He), calcite (Ca), galena (Gal), kaolinite (Ka), talcum (Ta), pyrite (Pyr), chalcopyrite (Cha), gray tiemannite (GT), stibnite (St), gabbro (Gab), quartz (Qu), chlorite (Chl), serpentine (Se), smaltite (Sm), tennantite (Te), gypsum (Gy), graphite (Gr), crystal (Cr), suanite (Sua), aragonite (Ar), basalt (Bas), fluorite (Fl), psilomelane (Ps), saponite (Sa), goethite (Go), barite (Bar), sulfur (Sul), biotite vermiculite (BV), and pyroxenite (Pr).

The reflection behavior of rocks and minerals primarily depends on their own mineralogical compositions and physical properties [4]. To capture the reflectance spectra of rocks and minerals in this work, the eight-band multi-spectral camera, which is the alternative payload of its counterpart onboard the Zhurong rover, is used as the multi-spectral sensor [9], and it is able to provide spectral data at the following wavelengths: 480 nm, 525 nm, 650 nm, 700 nm, 800 nm, 900 nm, 950 nm, and 1000 nm. The specifications of the used multispectral camera are listed in Table 1. In our implementation, we only use the multi-spectral camera to capture the eight-band multispectral images of all rock/mineral samples at approximately 12:12 pm, during a period of bright sunshine. Solar elevation angle and the shooting angle of camera is approximately 60° and 37°, respectively. All rock/mineral samples are placed on the ground, and the vertical height of camera from the ground is 1.8 m. As a result, the size of the captured image corresponding to each rock or mineral is $H \times W \times 8$, where H denotes the image height, W denotes the image width.

Table 1. Characteristics of the multispectral camera onboard the Zhurong rover.

Parameters	Values
Number of channels	9 (panchromatic and multispectral)
Weight	1.2 kg
Geometric resolution	2048 × 2048
Radiometric resolution	10 bits
Pixel size	5.5 μm
Focal length	50 mm
Imaging distance	[1.5 m, ∞)

Because Spectralon's optical properties make it ideal as a reference surface in remote sensing and spectroscopy, we used Spectralon as a near-Lambertian reference standard. The images of both the standard reference Spectralon and rocks/minerals are captured concurrently, which guarantees the consistency of observation conditions. After capturing the multi-spectral image of all samples, we compute the reflectance using the following Equation (1) due to the very simple and intuitive computation method [48]:

$$R = \frac{G_{\text{m}}}{G_{\text{ref}}} \cdot R_{\text{ref}} \tag{1}$$

where G_{m} denotes the average gray values of the query sample, G_{ref} denotes the average gray values of the standard reference spectralon, R_{ref} denotes the reflectance coefficient of the standard reference Spectralon that is generally known and measured in the laboratory, and R denotes the computed reflectance of the query sample.

Figure 2 demonstrates some typical examples of the spectral curves corresponding to rocks and minerals, and shows the changes in the band position and shape of compositions because of the unique spectral signatures due to their own physical characteristics. In the real world, small variations in the composition of rocks and minerals might occur, which often causes shifts in the position and shape of absorption bands in the spectrum. We put the spectral curves of the used (partial) samples together into one subfigure due to the similar spectral characteristics. It can be observed that these shifts in the position and shape of absorption bands might result in the similarity and even the overlap of spectral curves, i.e., severe inter-category similarity, such as in the case of chalcopyrite and fluorite, or with pyrite and saponite.

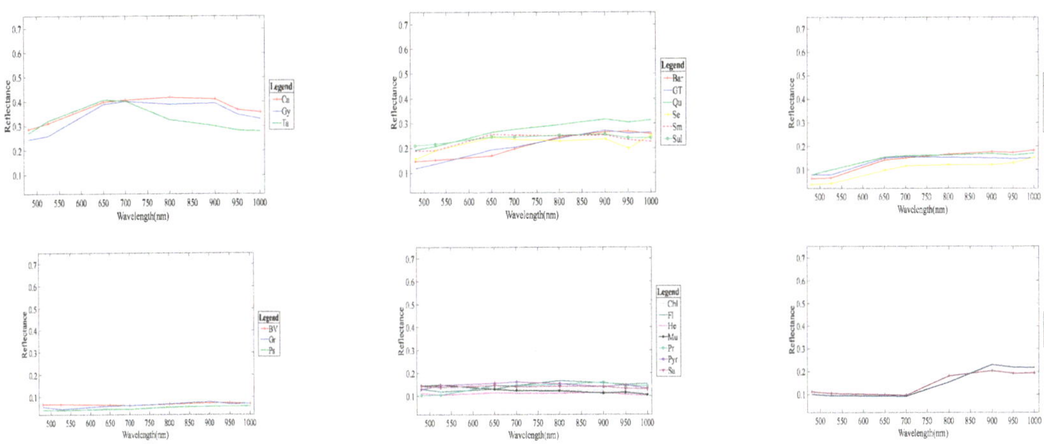

Figure 2. Illustration of spectral curves corresponding to rocks and minerals. For abbreviation, muscovite (Mu), hematite (He), calcite (Ca), galena (Gal), kaolinite (Ka), talcum (Ta), pyrite (Pyr), chalcopyrite (Cha), gray tiemannite (GT), stibnite (St), gabbro (Gab), quartz (Qu), chlorite (Chl), serpentine (Se), smaltite (Sm), tennantite (Te), gypsum (Gy), graphite (Gr), crystal (Cr), suanite (Sua), aragonite (Ar), basalt (Bas), fluorite (Fl), psilomelane (Ps), saponite (Sa), goethite (Go), barite (Bar), sulfur (Sul), biotite vermiculite (BV), and pyroxenite (Pr).

2.2. The Developed Classification Approach of Rocks and Minerals

It is common knowledge that the reflection behaviors of rocks and minerals depend on their own physical characteristics, which is the key insight behind most existing rock and mineral classification approaches. Despite their unique spectral signatures, the inter-category spectral similarity among different types of rocks and minerals might still reduce the classification quality as the number of categories increases. Therefore, using a transformer encoder as the backbone, we integrate it with contrastive learning and develop a classification method of rocks and minerals based on the highly discriminative representation derived from the original spectral signatures. Figure 3 indicates the pipeline of the developed classification method, which consists of a transformer-based feature encoder module and multi-task loss function for optimization. The former is able to remarkably enhance the descriptive capability of the derived feature representation [31] while the latter is to cause the feature representations of similar categories to aggerate with each other while the feature representations of dissimilar categories separate from each other. In our developed approach, an image patch $x \in R^{h \times w \times d}$ (where h is the height, w is the weight, and d is the channels; in this work, $d = 8$) is first operated via the simple linear flatten method, serving as the input of the transformer-based feature encoder module.

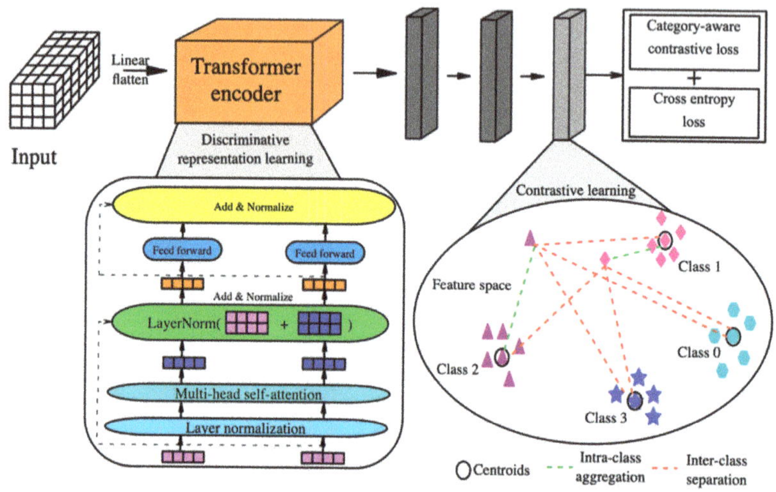

Figure 3. A pipeline showing the developed rock and mineral classification approach.

2.2.1. Transformer-Based Feature Encoder Module

The vision transformer as a deep learning model shows a remarkable capability to capture rich dependency information between variables [49]. We explore and investigate the adaptation of the vision transformer for basic visual feature extraction in order to generate the highly discriminative feature representation. It is found that, unlike the convolutional neural networks (CNNs) that gradually expand the field of view by repeatedly "convoluting" the information around the kernel layer by layer, the transformer-based method uses the stacked multi-head attention module that allows its strong ability to model the long-range dependencies. Recently, the vision transformer has adopted the attention mechanism and achieved promising results for image classification. Generally, an input image is first converted into a sequence of tokens by dividing it with a certain patch size and then linearly projecting each patch into tokens. Then, when a sequence of tokens is passed into a vision transformer model, attention weights are calculated between every token simultaneously. That is to say, the attention weight α_{ij} of token z_j with respect to token z_i is learned, which suggests the relevant information to each token. After calculating the α_{ij} value for all i and j pairs, we update each token z_i to z_i' using a weighted sum of all tokens followed by a nonlinear ReLU layer. This is defined in the following Equations (2)–(4):

$$\alpha_{ij} = \text{softmax}\left(\frac{(W^q z_i)\left(W^k z_j\right)}{\sqrt{d}}\right) \quad (2)$$

$$\overline{z_i} = \sum_{j=1}^{h \times w} \alpha_{ij} W^v z_j \quad (3)$$

$$z_i' = \text{ReLU}(\overline{z_i} W^r + b_1) W^o + b_2 \quad (4)$$

where d denotes the dimension of the key vector, W^k denotes the key weight matrix, W^q denotes the query weight matrix, W^v denotes the value weight matrix, W^r and W^o are the transformation matrices, and b_1 and b_2 are the bias terms.

One set of (W^k, W^q, W^v) matrices is called an attention head, and each layer in a vision transformer model has multiple attention heads, a module for attention mechanisms which runs through an attention mechanism several times in parallel. While each attention head attends to the tokens that are relevant to each token, the model can do this for different definitions of "relevance" with multiple attention heads.

As shown in the subfigure about the highly discriminative representation learning in Figure 3, a transformer-based feature encoder module is composed of a sequence of blocks where each block contains the multi-head attention module. Following this, three successive feedforward networks are used to produce the highly discriminative feature representation for the final label predictions. Empirically, the size of three successive fully connected layers is set to $h \times w \times d$, 360, and the number of classes, respectively. As a result, the output of the whole network, i.e., the last fully-connected layer, is considered as the feature representation of rock and mineral derived from the developed method.

2.2.2. Multi-Task Loss Function for Optimization

The super-parameters within the whole network are learned for mapping a set of the feature representations of rocks and minerals to a set of categories from massive high-quality labeled training data. General speaking, the problem of learning is cast as a search or optimization problem, which navigates the space of possible sets of super-parameters within the whole network in order to make good or good enough predictions. In the context of an optimization process, the function used to evaluate a candidate solution (i.e., a set of super-parameters) is referred to as the loss function, and the value calculated by the loss function is referred to simply as "loss". Typically, a deep learning model is learned using the stochastic gradient descent optimization algorithm, and super-parameters are updated using the backpropagation of error algorithm so that the next evaluation reduces the error, which means that we are searching for a candidate solution that has the lowest score. Here, we primarily discuss how to design an effective loss function, which jointly optimizes both cross entropy loss and category-aware contrastive loss. It enables us to guide the developed model to move towards convergence via measuring the difference between the predicted output and the ground truth. As a consequence, a highly discriminative feature representation is generated from the learned model.

Due to its fast convergence, the cross-entropy loss L_{cls} is used to guarantee accurate classification of the categories, as defined in the following Equation (5):

$$L_{\text{cls}} = \sum_{i=0}^{C} \{y_i \log(\hat{y}_i) + (1 - y_i) \log(1 - \hat{y}_i)\} \tag{5}$$

where C denotes the total number of categories, $\hat{y}_i$ denotes the ith true category of the training samples, and y_i denotes the the ith predicted category from our developed model.

In addition to the cross-entropy loss, the class separation in the latent feature space would also be an ideal characteristic to discriminate among different types of rocks and minerals. Therefore, we define the category-aware contrastive loss, as defined in Equation (6). The right subfigure in Figure 3 gives an explicit description about the role of category-aware contrastive loss. Intuitively, the goal of the category-aware contrastive loss is to force instances of the same category to remain close-by while pushing ones in the dissimilar category far apart in their latent feature space.

More specifically, we consider the set of categories as $K = \{1, 2, \ldots, C\} \subset N^+$, where N^+ represents the set of positive integers. For each category $i \in K$, a mean feature representation $\boldsymbol{p}_i$ is represented and maintained as the specific formed cluster in the latent feature space, which makes up a set of category-specific mean feature representations, namely $\boldsymbol{P} = \{p_1, \ldots, p_C\}$, for computing the category-aware contrastive loss. During the training procedure, the mean of feature representation for each category makes up a set of category-specific mean feature representations, namely $\boldsymbol{P} = \{\boldsymbol{p}_1, \ldots, \boldsymbol{p}_C\}$, for computing the category-aware contrastive loss. Since the super-parameters of the whole network is learned in an end-to-end manner, the mean feature representation corresponding to each category would gradually evolve at the training time. Inspired by the contrastive clustering method [50], a queue $\boldsymbol{q}_i$ with a fixed length is maintained for the i th category to store its associated feature representation. As a result, a feature representation store $\boldsymbol{F}_{\text{store}} = \{\boldsymbol{q}_1, \ldots, \boldsymbol{q}_C\}$ stores the category-specific feature representations in the corresponding queues. After every I_p iterations, a set of new category-specific mean feature representation $\boldsymbol{P}_{\text{new}}$ is calculated. Following this, the existing set of category-specific mean feature representations, namely $\boldsymbol{P}$, is updated by weighting $\boldsymbol{P}$ and $\boldsymbol{P}_{\text{new}}$ with a momentum constant η, as defined in Equation (6). In our implementation, η is set to 0.99, while I_p is set to 2000.

$$\boldsymbol{P} = \eta \boldsymbol{P} + (1-\eta)\boldsymbol{P}_{\text{new}} \tag{6}$$

After obtaining the set of category-specific mean feature representations, let $\boldsymbol{f}_c$ denote a feature representation produced by an intermediate layer of the used transformer-based feature encoder module, for an instance of category c. To force instances of the same category to remain close-by while pushing instances of the dissimilar category far apart in the high-dimensional feature space, the category-aware contrastive loss is defined as follows:

$$L_{\text{cont}}(\boldsymbol{f}_c) = \sum_{i=0}^{C} l(\boldsymbol{f}_c, \boldsymbol{p}_i) \tag{7}$$

$$l(\boldsymbol{f}_c, \boldsymbol{p}_i) = \begin{cases} D(\boldsymbol{f}_c, \boldsymbol{p}_i), i = c \\ \max\{0, \Delta - D(\boldsymbol{f}_c, \boldsymbol{p}_i)\}, \text{otherwise} \end{cases} \tag{8}$$

where $D(\cdot)$ denotes any distance function (e.g., Euclidean, cosine), and Δ denotes how close a similar and dissimilar instance can be. In our implementation, the value of Δ is empirically set to 2.0.

Finally, we simultaneously learn the shared super-parameters through the backpropagation algorithm [51] for optimization to provide better generalization of the developed classification approach [52]. The total loss used in this work is defined as a weighted sum of both the category-aware contrastive loss and the cross-entropy loss as the following Equation (9):

$$\text{Loss}_{\text{total}} = \lambda \times L_{\text{cont}} + L_{\text{cls}} \tag{9}$$

where λ is experimentally set to 0.01, L_{cont} denotes the category-aware contrastive loss, and L_{cls} denotes the cross-entropy loss by measuring the cross entropy between the ground truth and the predicted output of our developed model. By minimizing the total loss, our developed classification approach is progressively trained for optimization until convergence.

3. Experimentation and Analysis

To evaluate the reliability and robustness of the developed classification method, in this section, we performed both qualitative and quantitative analysis on the multi-spectral images of 30 types of rocks and minerals from the National Mineral Rock and Fossil Specimens Resource Center. First, a brief description of evaluation criteria is given. Then, we offer a detailed experimental setting. Finally, we qualitatively and quantitatively analyze the performance of category classification results, where the experiments are conducted using the optimal parameters obtained and discussed in Section 5.

3.1. Evaluation Criteria

To measure the classification quality of rocks, we conduct the evaluation solution using precision (Pr), recall (Re), $F_{1-\text{score}}$ and overall accuracy (OA), as defined in Equations (10)–(13). The recall represents a measure of completeness, the precision denotes a measure of correctness, and the $F_{1-\text{score}}$, also called balanced F_{score}, is a weighted average of the precision and recall; overall accuracy denotes the sum of the true positives plus true negatives divided by the total number of queried individuals. The equations are as follows:

$$Precision = \frac{TP}{TP + FP} \tag{10}$$

$$Recall = \frac{TP}{TP + FN} \tag{11}$$

$$F_{1-\text{score}} = \frac{2 \times precision \times recall}{precision + recall} \tag{12}$$

$$Overall\ accuracy = \frac{TP + FN}{TP + TN + FP + FN} \tag{13}$$

where TP denotes the number of positive objects that are correctly determined as positive ones, TN denotes the number of positive objects that are correctly determined as negative ones, FN denotes the number of negative objects that are incorrectly classified as negative ones, and FP denotes the number of negative objects that are incorrectly determined as positive ones.

3.2. Implementation Details

Computer configuration in the experiment comprised two NVIDIA GeForce RTX 3080Ti GPUs, 128 Gb of memory, Ubuntu 20.04 operating system, Cuda10.0, and Cudnn7.5. Our developed classification approach to generating the highly discriminative features is constructed using Pytorch library [53] and trained in an end-to-end manner. We use Adam for the optimizer with betas $= (0.9, 0.999)$ and weight decay = 1×10^{-6}. We train the developed approach with a batch size of 64, an initial learning rate of 10^{-4}, and maintain a more optimal learning rate throughout the training procedure using a cosine annealing scheduler. We use a dropout with $p = 0.1$ for regularization. The number of heads in the multi-head attention modules is 8. We use a $L = 3$ layer transformer with a residual layer around each embedding update and layer normalization.

With regards to our laboratory sample dataset, we imaged the 30 rocks and minerals in it. To compare the spectra of rocks and minerals as well as to reduce the spectral noises, the image patches with fixed-size coverage are obtained. For example, in our implementation, the image patch is a group of pixels $9 \times 9 \times 8$ if its size is set to 9×9, which will serve as the input for the developed transformer-based classification approach in Section 3.3. The effect of the size of image patch on the classification performance will be discussed in Section 4.1. In our implementation, we randomly selected some of the image patches as training, and then used the rest as validation and test to classify the rocks and minerals. The ratio of training, validation, and test split is 60%, 10%, and 30%, respectively. We count out the number of image patches in the training set, validation set, and test set; Table 2 lists the detailed statistics.

Table 2. The statistics of image patches in training set, validation set and test set. For abbreviation, muscovite (Mu), hematite (He), calcite (Ca), galena (Gal), kaolinite (Ka), talcum (Ta), pyrite (Pyr), chalcopyrite (Cha), gray tiemannite (GT), stibnite (St), gabbro (Gab), quartz (Qu), chlorite (Chl), serpentine (Se), smaltite (Sm), tennantite (Te), gypsum (Gy), graphite (Gr), crystal (Cr), suanite (Sua), aragonite (Ar), basalt (Bas), fluorite (Fl), psilomelane (Ps), saponite (Sa), goethite (Go), barite (Bar), sulfur (Sul), biotite vermiculite (BV), and pyroxenite (Pr).

Category	Mu	He	Ca	Gal	Ka	Ta	Pyr	Cha	GT	St	Gab
Training set	3011	5508	3149	4019	3667	3486	3360	3894	3824	4272	6044
Validation set	502	918	525	670	612	581	560	649	639	712	1008
Test set	1506	2754	2575	2009	1834	1743	1680	1947	1917	2136	3022
Category	Qu	Chl	Se	Sm	Te	Gy	Gr	Cr	Sua	Ar	Bas
Training set	4350	4882	3131	3686	5186	3378	3311	4043	3272	3900	4100
Validation set	725	814	522	614	864	563	552	674	545	650	683
Test set	2175	2441	1566	1843	2593	1689	1655	2022	1637	1950	2050
Category	Fl	Ps	Sa	Go	Bar	Sul	BV	Pr	Sum	In Total	
Training set	3840	3378	5280	4572	4050	4356	6038	4708	128,045		
Validation set	640	563	880	762	675	726	1006	785	21,344	214,421	
Test set	1920	1689	2640	2286	2025	2178	3020	2355	65,032		

3.3. *Rock Classification Results*

The classification performance relies on the descriptive ability of the generated feature representation of rocks and minerals to some extent. In our implementation, we conduct the classification task to distinguish 30 types of rock and mineral samples. Table 3 lists rock classification results generated by the developed method in precision, recall, and $F_{1-\mathrm{score}}$ per class, as well as overall accuracy. From the quantitative perspective, we can observe that the developed method can effectively distinguish 30 types of rock and minerals samples, with a high overall accuracy of 96.92% and an average $F_{1-\mathrm{score}}$ of 97.11%. It is worth noting that some types of rocks and minerals, such as muscovite, talcum, pyrite and so on, even achieve 100% in their $F_{1-\mathrm{score}}$. Experimental results suggest that the feature representation derived from the developed method in this paper is highly discriminative for carrying out the classification task, as summarized in Table 3. That is to say, the generated feature representation shows the capability to reliably and robustly describe the differences among different types of rocks and minerals from the quantitative perspective.

Inevitably, there are some cases of incorrect classification. From the classification results derived from the trained optimal model, it is observed that the sulfur is easily misclassified into crystal, which results in the low recall. Furthermore, the low precision and low recall which occur between chlorite and saponite is primarily due to the mixed classification. For the visualization comparison and analysis (as shown in Figure 4), there exists largely-overlapping regions between sulfur and crystal, as well as between chlorite and saponite in raw spectral signature signatures, which illustrates the high similarity between them. This is one of most important factors responsible for low precision or recall in the classification results.

Table 3. Classification results generated by the developed method in precision, recall and $F_{1-\text{score}}$ per class as well as overall accuracy (the unit: %). For abbreviation, muscovite (Mu), hematite (He), calcite (Ca), galena (Gal), kaolinite (Ka), talcum (Ta), pyrite (Pyr), chalcopyrite (Cha), gray tiemannite (GT), stibnite (St), gabbro (Gab), quartz (Qu), chlorite (Chl), serpentine (Se), smaltite (Sm), tennantite (Te), gypsum (Gy), graphite (Gr), crystal (Cr), suanite (Sua), aragonite (Ar), basalt (Bas), fluorite (Fl), psilomelane (Ps), saponite (Sa), goethite (Go), barite (Bar), sulfur (Sul), biotite vermiculite (BV), and pyroxenite (Pr).

Category	Mu	He	Ca	Gal	Ka	Ta	Pyr	Cha	GT	St	Gab
Precision	100.0	96.62	97.07	93.84	99.89	100.0	100.0	100.0	100.0	92.54	99.93
Recall	100.0	99.93	98.98	97.95%	100.0	100.0	100.0	100.0	100.0	100.0	99.70
F1-score	100.0	98.25	98.01	95.86	99.94	100.0	100.0	100.0	100.0	96.13	99.81
Category	Qu	Chl	Se	Sm	Te	Gy	Gr	Cr	Sua	Ar	Bas
Precision	99.95	89.58	99.23	100.0	95.22	100.0	100.0	81.53	99.51	100.0	100.0
Recall	100.0	82.09	99.93	100.0	100.0	100.0	100.0	100.0	100.0	100.0	100.0
F1-score	99.97	85.67	99.58	100.0	97.55	100.0	100.0	89.82	99.76	100.0	100.0
Category	Fl	Ps	Sa	Go	Bar	Sul	BV	Pr	Overall accuracy		
Precision	100.0	99.12	96.35	100.0	100.0	99.29	93.34	88.28			
Recall	100.0	100.0	100.0	99.65	100.0	58.40	85.03	98.59	96.92%		
F1-score	100.0	99.55	98.14	99.82	100.0	73.54	88.99	93.16			

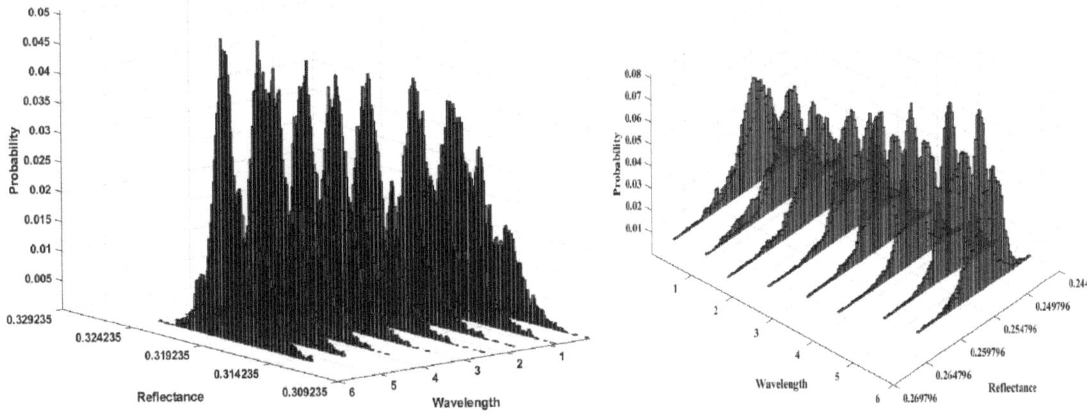

(a) (b)

Figure 4. Spectral analysis of some wrong classification examples. The bars are rendered in different colors according to different categories. (**a**) Crystal–sulfur. (**b**) chlorite–saponite.

3.4. T-SNE Visualization in the Discriminative Feature Space

In this work, the key insight behind the developed method is to make a remarkable distinction among different types of rocks and minerals in their high-dimensional feature space. The T-distributed stochastic neighbor embedding (T-SNE) technique [54] is a dimensionality reduction technique, whose goal is to project high-dimensional data into a two-dimensional map space. In the two-dimensional map space, the same categories tend to be modeled by nearby locations and the dissimilar categories by distant locations with high probability. That is to say, the feature representation is highly discriminative if the clusters belonging to the same categories are formed well. For a more detailed description of the T-SNE technique, please refer to Appendix A. To qualitatively analyze the performance of the highly discriminative feature representation derived from the transformer-based classification method, in this subsection, we visualize the quality of formed clusters in their high-dimensional feature space using T-SNE visualization. Theoretically, the number of formed clusters within the two-dimensional visualization maps should be associated with the number of categories in the rock and mineral dataset. In our implementation, the output of the developed method, i.e., the last fully-connected layer, is considered as the derived feature representation describing both rocks and minerals and serves as the input of the scikit-learn T-SNE package [55] for visualization. For comparison, we project the original spectral signatures and the feature representation derived from the developed method into the two-dimensional map space for visualization. Figure 5 demonstrates a visualization comparison between the original spectral signatures and the feature representation derived from the developed method using the T-SNE technique. As shown in Figure 5a, there occur severe overlapping regions among different types of rocks in the two-dimensional visualization maps. It can be concluded that inter-category spectral similarity among different types of rocks and minerals might appear, which would result in the degradation of the classification performance. Compared with those from the original spectral signatures, as shown in Figure 5b, the clusters are formed well from the derived feature representation for different type of both rocks and minerals, which suggests the highly discriminative representation capability.

(a) (b)

Figure 5. Visualization comparison using the T-SNE technique. (**a**) Original spectral signatures; (**b**) highly discriminative features derived from the developed method.

4. Discussion

In this section, the parameter sensitivity of the developed classification approach is provided, primarily including the size of image patches, the number of transformer layers, the number of transformer heads, and the integration of category-aware contrastive loss. Additionally, we compare the proposed method with other methods. Afterwards, both the strengths and limitations of the developed rock classification method are briefly analyzed and discussed.

4.1. *Effect of the Size of Image Patches on Classification Results*

To compare the spectra of rocks and minerals for automatic classification, in our implementation, spectra were extracted from the image patches with fixed-size coverage. In this subsection, we set different sizes of image patches, including $5 \times 5, 7 \times 7, 9 \times 9$, and 11×11, to discuss the effect of the size of image patch on the classification results. Table 4 lists the effect of the size of image patches on classification results, which demonstrate the fluctuation of classification results from 95.96% to 96.92%. As mentioned in Section 2.1, small changes in the compositions of rocks and minerals cause shifts in the position and shape of absorption bands in the spectrum. With the increasing size, the image patches with fixed-size coverage alleviate the effect of composition variants, which enhance the classification performance to some extent, although it brings computing burdens. Taking the accuracy and computational cost into consideration, we set the image patch size to 9×9 as the optimal parameters in our implementation.

Table 4. Effect of the size of image patches on classification results in $F_{1-score}$ per class and overall accuracy (the unit: %). For abbreviation, muscovite (Mu), hematite (He), calcite (Ca), galena (Gal), kaolinite (Ka), talcum (Ta), pyrite (Pyr), chalcopyrite (Cha), gray tiemannite (GT), stibnite (St), gabbro (Gab), quartz (Qu), chlorite (Chl), serpentine (Se), smaltite (Sm), tennantite (Te), gypsum (Gy), graphite (Gr), crystal (Cr), suanite (Sua), aragonite (Ar), basalt (Bas), fluorite (Fl), psilomelane (Ps), saponite (Sa), goethite (Go), barite (Bar), sulfur (Sul), biotite vermiculite (BV), and pyroxenite (Pr). The best performance is highlighted in **BOLD**.

Patch Size	Mu	He	Ca	Gal	Ka	Ta	Pyr	Cha	GT	St	Gab
Size = 5	95.18	95.86	92.49	94.19	98.66	99.97	100.0	99.80	99.44	97.45	98.33
Size = 7	99.94	97.57	92.23	95.21	99.65	100.0	100.0	100.0	99.93	97.04	99.31
Size = 9	**100.0**	**98.25**	**98.01**	95.86	**99.94**	**100.0**	**100.0**	**100.0**	**100.0**	96.13	**99.81**
Size = 11	100.0	97.97	87.88	**97.47**	99.88	100.0	100.0	99.95	99.73	**98.15**	97.77
Patch size	Qu	Chl	Se	Sm	Te	Gy	Gr	Cr	Sua	Ar	Bas
Size = 5	99.58	81.66	98.54	99.66	91.09	99.91	99.34	92.01	99.54	99.83	99.77
Size = 7	99.56	84.49	99.56	100.0	96.81	99.97	99.27	89.97	99.70	99.97	99.97
Size = 9	**99.97**	85.67	**99.58**	**100.0**	**97.55**	**100.0**	**100.0**	89.82	99.76	**100.0**	**100.0**
Size = 11	99.33	**86.88**	96.55	99.93	96.99	100.0	100.0	**99.21**	**100.0**	99.96	99.94
Patch size	Fl	Ps	Sa	Go	Bar	Sul	BV	Pr	Overall accuracy		
Size = 5	100.0	98.88	97.17	99.32	99.98	67.59	85.33	93.13	95.96		
Size = 7	100.0	99.18	96.87	99.74	100.0	67.27	**89.58**	92.51	96.48		
Size = 9	**100.0**	99.55	**98.14**	99.82	**100.0**	**73.54**	88.99	**93.16**	**96.92**		
Size = 11	100.0	**99.93**	96.36	**100.0**	100.0	68.77	89.33	87.46	96.83		

4.2. Effect of the Number of Transformer Layers on Classification Results

In our implementation, we extract the highly discriminative feature representation from image patches encoded by a series of transformer layers. To evaluate the effect of the number of transformer layers, we set it with steps of 1 from 1 to 7. As with deeper convolutional neural networks, the developed method is capable of perfectly fitting training data, and also performed well on test data as the number of transformer layers increased. Figure 6 demonstrates the experimental results. When the number of Transformer layers rises from 1 to 3, the overall accuracy fluctuates by approximately 0.77%. However, the increase in the number of transformer layers also makes the model convergence very difficult due to the gradient vanishing, which might degrade the classification performance when the number of transformer layers increases from 3 to 7.

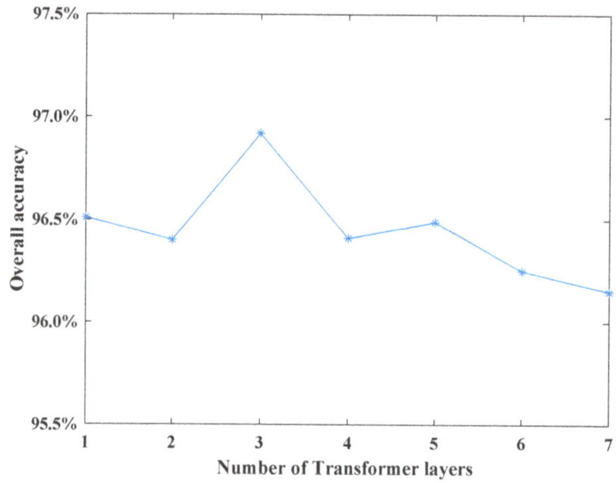

Figure 6. Effect of the number of transformer layers on classification results.

4.3. *Effect of the Number of Transformer Heads on Classification Results*

In addition to the number of transformer layers, the developed transformer-based classification method aggregates the inter-channel and intra-channel information through the multi-head self-attention mechanism. Thus, the accuracy might depend on the number of transformer heads for the developed transformer-based classification method. To demonstrate the effect of the number of transformer heads on classification results, we set it to 1, 2, 4, 8, respectively, since the number of multi-spectral bands must be guaranteed to be divisible. In this way, the multi-spectral bands are split across the multiple attention heads so that each can process them independently, while the correlations are extracted and aggregated for each attention head. From the experimental results shown in Figure 7, we can conclude that the accuracy of classification is easily affected by the number of transformer heads, with difference in the overall accuracy of more than 1.01%. This enables the transformer-based feature encoder module to capture richer interpretations of the multi-spectral bands.

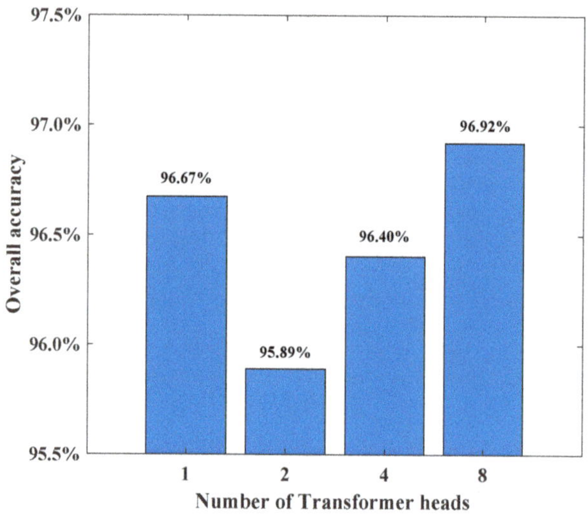

Figure 7. Effect of the number of transformer heads on classification results.

4.4. Effect of Category-Aware Contrastive Loss on Classification Results

To generate the highly discriminative feature representation, we combine the cross-entropy loss with the category-aware contrastive loss to optimize a multi-task loss. In this way, we attempt to force instances of the same category to remain close-by while pushing instances of dissimilar category far apart by introducing the category-aware contrastive loss, as shown in Figure 3. Table 5 compares the results between no contrastive loss and the developed method in $F_{1-\text{score}}$ per class, and the overall accuracy. From the comparative results, an improvement of approximately 0.72% can be observed, which suggests that the generated feature representation becomes more discriminative than that without the category-aware contrastive loss. Hence, we can draw a conclusion that the category-aware contrastive loss is beneficial to enhancing the descriptive capability of the generated feature representation.

Table 5. Effect of category-aware contrastive loss on classification results in $F_{1-\text{score}}$ per class (%). For abbreviation, muscovite (Mu), hematite (He), calcite (Ca), galena (Gal), kaolinite (Ka), talcum (Ta), pyrite (Pyr), chalcopyrite (Cha), gray tiemannite (GT), stibnite (St), gabbro (Gab), quartz (Qu), chlorite (Chl), serpentine (Se), smaltite (Sm), tennantite (Te), gypsum (Gy), graphite (Gr), crystal (Cr), suanite (Sua), aragonite (Ar), basalt (Bas), fluorite (Fl), psilomelane (Ps), saponite (Sa), goethite (Go), barite (Bar), sulfur (Sul), biotite vermiculite (BV), and pyroxenite (Pr). The best performance is highlighted in **BOLD**.

Methods	Mu	He	Ca	Gal	Ka	Ta	Pyr	Cha	GT	St	Gab
No contrastive loss	100.0	97.57	91.21	**95.87**	98.84	100.0	100.0	100.0	100.0	96.34	99.46
Proposed method	**100.0**	**98.25**	**98.01**	95.86	**99.94**	**100.0**	**100.0**	**100.0**	**100.0**	**96.1**	**99.81**
Methods	Qu	Chl	Se	Sm	Te	Gy	Gr	Cr	Sua	Ar	Bas
No contrastive loss	99.97	80.76	99.45	100.0	95.85	100.0	100.0	**93.03**	**99.87**	99.97	99.97
Proposed method	**99.97**	**85.67**	**99.58**	**100.0**	**97.55**	**100.0**	**100.0**	89.82	99.76	**100.0**	**100.0**
Methods	Fl	Ps	Sa	Go	Bar	Sul	BV	Pr	Average	Overall accuracy	
No contrastive loss	100.0	**99.97**	94.70	**99.91**	100.0	68.77	**89.14**	91.49	96.40	96.25	
Proposed method	**100.0**	99.55	**98.14**	99.82	**100.0**	**73.54**	88.99	**93.16**	**97.12**	**96.92**	

4.5. Comparisons with Other Methods

To further evaluate the descriptive performance of the developed method, we also compare it with other methods which implement the frequently-used classifiers, such as the decision tree [56], random forest [57], and support vector machine (SVM) [58], based on the raw spectral signatures of rock samples. Furthermore, we compare our results with other neural networks [59], namely ConvNet, in this work. In our implementation, only one simple convolutional neural network is selected and used because the small image patches in this work are the input of the developed method which cannot be conducted on more complex architectures with many pooling operations. Table 6 lists the comparisons of the classification results between different methods. For the experimental results, we can observe that the overall accuracy and average $F_{1-\text{score}}$ derived from the developed method exceeds those of other methods, with average differences of 4.75% and 4.97%, respectively. Moreover, although the developed method achieves a lower $F_{1-\text{score}}$ for crystal than other methods, the developed method remarkably improves the classification performance, especially for identifying hematite, chalcopyrite, chlorite, serpentine, smaltite, saponite, etc., which further confirms the superior discriminative ability of the produced feature representation.

Table 6. Comparisons with other methods in $F_{1-score}$ per class and overall accuracy (%). For abbreviation, muscovite (Mu), hematite (He), calcite (Ca), galena (Gal), kaolinite (Ka), talcum (Ta), pyrite (Pyr), chalcopyrite (Cha), gray tiemannite (GT), stibnite (St), gabbro (Gab), quartz (Qu), chlorite (Chl), serpentine (Se), smaltite (Sm), tennantite (Te), gypsum (Gy), graphite (Gr), crystal (Cr), suanite (Sua), aragonite (Ar), basalt (Bas), fluorite (Fl), psilomelane (Ps), saponite (Sa), goethite (Go), barite (Bar), sulfur (Sul), biotite vermiculite (BV), and pyroxenite (Pr). The best performance is highlighted in **BOLD** fonts.

Methods	Mu	He	Ca	Gal	Ka	Ta	Pyr	Cha	GT	St	Gab
Decision tree [56]	95.98	85.82	90.06	90.86	99.20	95.20	99.97	87.30	97.62	95.06	86.29
Random forest [57]	97.98	88.83	95.60	91.39	99.91	98.44	100.0	88.82	99.47	94.72	90.52
SVM [58]	99.70	88.44	97.99	95.68	98.86	98.78	99.97	88.62	99.63	**99.09**	96.17
ConvNet [59]	99.33	91.66	93.26	94.96	97.39	99.45	100.0	87.03	99.76	97.62	98.10
Developed method	**100.0**	**98.25**	**98.01**	**95.86**	**99.94**	**100.0**	**100.0**	**100.0**	**100.0**	96.13	**99.81**
Methods	Qu	Chl	Se	Sm	Te	Gy	Gr	Cr	Sua	Ar	Bas
Decision tree [56]	99.15	57.40	62.62	85.76	90.48	99.20	97.58	95.95	87.53	96.26	92.66
Random forest [57]	99.38	65.13	70.33	82.94	91.69	100.0	99.16	**98.92**	93.77	96.27	96.28
SVM [58]	99.70	59.31	77.72	72.96	83.12	100.0	97.66	97.84	93.84	99.82	98.77
ConvNet [59]	**100.0**	74.56	76.19	99.53	96.26	100.0	99.84	98.38	99.31	99.58	99.72
Developed method	99.97	**85.67**	**99.58**	**100.0**	**97.55**	**100.0**	**100.0**	89.82	**99.76**	**100.0**	**100.0**
Methods	Fl	Ps	Sa	Go	Bar	Sul	BV	Pr	Average	Overall accuracy	
Decision tree [56]	98.23	96.93	86.79	97.42	98.99	69.38	80.17	91.37	90.24	90.22	
Random forest [57]	99.89	98.39	91.67	97.62	99.38	71.56	83.98	94.00	92.53	92.41	
SVM	99.94	95.74	87.25	98.55	100.0	30.72	77.18	90.77	90.79	90.79	
ConvNet [59]	99.03	**99.59**	95.29	98.55	98.15	**91.88**	79.34	**98.32**	95.04	95.24	
Developed method	**100.0**	99.55	**98.14**	**99.82**	**100.0**	73.54	**88.99**	93.16	**97.12**	**96.92**	

5. Summary and Outlook

To address the challenges due to the inter-category spectral similarity, we present a category recognition approach of both rocks and minerals using the discriminative representation derived from its spectral signatures. The developed method combines a transformer-based recognition approach with category-aware contrastive learning, which is trained in an end-to-end multi-task manner. The advantages of our proposed approach are as follows: (1) Different from convolutional neural network, the transformer shows its strong ability to model long-range dependencies. (2) We defined a category-aware contrastive loss, which would force instances of the same category to remain close-by while pushing instances of the dissimilar category far apart. Therefore, the derived highly discriminative feature representation from the developed approach is beneficial to enhancing the descriptive capability and alleviating the inter-category spectral similarity for the classification task. Furthermore, we establish a rock sample database with 30 types of rocks and then carry out the experimental analysis both quantitatively and qualitatively. Experimental results confirm the robustness and reliability of the developed method. It can be concluded that the developed method can effectively distinguish 30 types of rock samples, with a high overall accuracy of 96.92%. Additionally, the overall accuracy and average $F_{1-score}$ derived from the developed method exceed those of other common methods, with average differences of 5.78% and 5.93%, respectively. It is well-known that the more rock categories, the more severe the inter-category spectral similarity. The establishment of rock databases with more types of rock samples will enable us to test the generalization of the developed method in our future work. Additionally, for the developed method in this work, the category-aware contrastive loss and cross-entropy loss is aggregated, and the weight is set empirically. In our future research, an adaptive weighting solution is required for the performance stability of the method.

Author Contributions: Methodology, J.Y. (Juntao Yang) and Z.K.; validation, Z.Y., B.X., J.Y. (Jianfeng Yang) and J.T.; visualization, J.X.; formal analysis, J.Y. (Juntao Yang) and Z.K.; supervision, J.Y. (Juntao Yang) and Z.K.; writing—original draft, J.Y. (Juntao Yang) and Z.Y.; writing—review and editing, J.Y. (Juntao Yang) and Z.K. All authors have read and agreed to the published version of the manuscript.

Funding: This research was jointly supported by the National Key Research and Development Program of China (No. 2019YFE0123300), National Natural Science Foundation of China (No. 41872207), Civil Aerospace Technology Advance Research Project of National Defense Science and Engineering (No. D020103), Beijing Science and Technology Project (No. Z191100004319001), the European Union's Horizon 2020 research and innovation programme under grant agreement No. 871149, and Introduction plan of high-end foreign experts (No. G2021025006L).

Acknowledgments: We also would like to thank the National Mineral Rock and Fossil Specimens Resource Center (http://www.nimrf.net.cn (accessed on 25 August 2022)) for the share and free use of their rock samples. The data that support the findings of this work are available from the corresponding author upon the reasonable request.

Conflicts of Interest: The authors declare no conflict of interest.

Appendix A. T-Distributed Stochastic Neighbor Embedding Visualization Technique

The T-distributed stochastic neighbor embedding (T-SNE) technique [54] is a non-linear dimensionality reduction technique for projecting high-dimensional data onto a two-dimensional map space for visualization, where the same categories tend to be modeled by nearby locations and the dissimilar categories by distant locations with high probability. The T-SNE technique consists of two stages, namely the construction of a probability distribution over pairs of high-dimensional data, and the definition of a similar probability distribution over the data in the two-dimensional map space.

Given a set of N high-dimensional data $x_1, \ldots, x_N$, the probabilities p_{ij} that are proportional to the similarity of data x_i and x_j are calculated as defined in the following Equation (A1):

$$p_{j|i} = \frac{\exp(-||x_i - x_j||^2/2\sigma_i^2)}{\sum_{k\neq i}\exp(-||x_i - x_k||^2/2\sigma_i^2)} \tag{A1}$$

where σ_i denotes the bandwidth of the Gaussian kernels set using a predefined perplexity. In our implementation, $p_{i|i}$ is set to 0, and $\sum_j p_{j|i} = 1$ for all i. Define $p_{ij} = \frac{(p_{j|i}+p_{i|j})}{2N}$, then $p_{ij} = p_{ji}$, $p_{ii} = 0$, and $\sum_{i,j} p_{ij} = 1$.

Following this, a two-dimensional map $y_1, \ldots, y_N$, is learned to represent the similarities as well as possible. The similarities q_{ij} between y_i and y_j is defined as follows:

$$q_{ij} = \frac{\left(1+||y_i - y_j||^2\right)^{-1}}{\sum_k \sum_{l\neq k}\left(1+||y_k - y_l||^2\right)^{-1}} \tag{A2}$$

The locations of the point y_i in the two-dimensional map is determined by minimizing the Kullback–Leibler (KL) divergence between the distribution P and Q for the visualization.

References

1. Blake, D.; Bristow, T.; Sarrazin, P.; Zacny, K. In-Situ Mineralogical Analysis of the Venus Surface using X-ray Diffraction. *Bull. Am. Astron. Soc.* **2021**, *53*, 018. [CrossRef]
2. Yang, J.; Kang, Z. A Gradient-Region Constrained Level Set Method for Autonomous Rock Detection from Mars Rover Image. In Proceedings of the International Archives of the Photogrammetry, Remote Sensing and Spatial Information Sciences, Enschede, The Netherlands, 10–14 June 2019.
3. Christensen, P.R.; Bandfield, J.; Clark, R.N.; Edgett, K.; Hamilton, V.E.; Hoefen, T.; Kieffer, H.H.; Kuzmin, R.O.; Lane, M.D.; Malin, M.C.; et al. Detection of crystalline hematite mineralization on Mars by the Thermal Emission Spectrometer: Evidence for near-surface water. *J. Geophys. Res. Earth Surf.* **2000**, *105*, 9623–9642. [CrossRef]
4. Kodikara, G.R.; McHenry, L.J. Machine learning approaches for classifying lunar soils. *Icarus* **2020**, *345*, 113719. [CrossRef]
5. Alemanno, G.; Maturilli, A.; D'Amore, M.; Helbert, J. A new laboratory emissivity and reflectance spectral library for the interpretation of Mars thermal infrared spectral data. *Icarus* **2021**, *368*, 114622. [CrossRef]

6. Anderson, R.C.; Jandura, L.; Okon, A.B.; Sunshine, D.; Roumeliotis, C.; Beegle, L.; A Hurowitz, J.; Kennedy, B.P.; Limonadi, D.; McCloskey, S.; et al. Collecting Samples in Gale Crater, Mars; an Overview of the Mars Science Laboratory Sample Acquisition, Sample Processing and Handling System. *Space Sci. Rev.* **2012**, *170*, 57–75. [CrossRef]
7. Caudill, C.M.; Pontefract, A.J.; Osinski, G.R.; Tornabene, L.L.; Pilles, E.A.; Battler, M.; Francis, R.; Godin, E.; Galofre, A.G.; Haltigin, T.; et al. CanMars mission Science Team operational results: Implications for operations and the sample selection process for Mars Sample Return (MSR). *Planet. Space Sci.* **2019**, *172*, 43–56. [CrossRef]
8. Osinski, G.R.; Battler, M.; Caudill, C.M.; Francis, R.; Haltigin, T.; Hipkin, V.J.; Kerrigan, M.; Pilles, E.A.; Pontefract, A.; Tornabene, L.L.; et al. The CanMars Mars Sample Return analogue mission. *Planet. Space Sci.* **2019**, *166*, 110–130. [CrossRef]
9. Wan, W.; Wang, C.; Li, C.; Wei, Y. China's first mission to Mars. *Nat. Astron.* **2020**, *4*, 721. [CrossRef]
10. Bishop, J.L.; Bell, J.F., III; Moersch, J.E. *Remote Compositional Analysis: Techniques for Understanding Spectroscopy, Mineralogy, and Geochemistry of Planetary Surfaces*; Cambridge U. Press: Cambridge, UK, 2020.
11. Viviano, C.E.; Seelos, F.P.; Murchie, S.L.; Kahn, E.G.; Seelos, K.D.; Taylor, H.W.; Taylor, K.; Ehlmann, B.L.; Wiseman, S.M.; Mustard, J.F.; et al. Revised CRISM spectral parameters and summary products based on the currently detected mineral diversity on Mars. *J. Geophys. Res.-Planet* **2014**, *119*, 1403–1431. [CrossRef]
12. Jain, N.; Chauhan, P. Study of phyllosilicates and carbonates from the Capri Chasma region of Valles Marineris on Mars based on Mars Reconnaissance Orbiter-Compact Reconnaissance Imaging Spectrometer for Mars (MRO-CRISM) observations. *Icarus* **2015**, *250*, 7–17. [CrossRef]
13. Fox, V.K.; Arvidson, R.E.; Guinness, E.A.; McLennan, S.M.; Catalano, J.G.; Murchie, S.L.; Powell, K.E. Smectite deposits in Marathon Valley, Endeavour Crater, Mars, identified using CRISM hyperspectral reflectance data. *Geophys. Res. Lett.* **2016**, *43*, 4885–4892. [CrossRef]
14. Núñez, J.I.; Barnouin, O.S.; Murchie, S.L.; Seelos, F.P.; McGovern, J.A.; Seelos, K.D.; Buczkowski, D.L. New insights into gully formation on Mars: Constraints from composition as seen by MRO/CRISM. *Geophys. Res. Lett.* **2016**, *43*, 8893–8902. [CrossRef]
15. Xue, Y.; Yang, Y.; Yu, L. Mineral composition of the Martian Gale and Nili Fossae regions from Mars Reconnaissance Orbiter CRISM images. *Planet. Space Sci.* **2018**, *163*, 97–105. [CrossRef]
16. Amador, E.S.; Bandfield, J.L.; Thomas, N.H. A search for minerals associated with serpentinization across Mars using CRISM spectral data. *Icarus* **2018**, *311*, 113–134. [CrossRef]
17. Lin, H.; Mustard, J.F.; Zhang, X. A methodology for quantitative analysis of hydrated minerals on Mars with large endmember library using CRISM near-infrared data. *Planet. Space Sci.* **2018**, *165*, 124–136. [CrossRef]
18. Li, J.; Zhang, L.; Wu, Z.; Ling, Z.; Cao, X.; Guo, K.; Yan, F. Autonomous Martian rock image classification based on transfer deep learning methods. *Earth Sci. Inform.* **2020**, *13*, 951–963. [CrossRef]
19. De Albuquerque, M.P.; Esquef, I.; Mello, A.G. Image thresholding using Tsallis entropy. *Pattern Recognit. Lett.* **2004**, *25*, 1059–1065. [CrossRef]
20. Ribas, L.C.; Gonçalves, D.N.; Orué, J.P.M.; Gonçalves, W.N. Fractal dimension of maximum response filters applied to texture analysis. *Pattern Recognit. Lett.* **2015**, *65*, 116–123. [CrossRef]
21. Sato, S.; Sano, M.; Sawada, Y. Practical methods of measuring the generalized dimension and the largest Lyapunov exponent in high dimensional chaotic systems. *Prog. Theor. Phys.* **1987**, *77*, 1–5. [CrossRef]
22. Patel, A.K.; Chatterjee, S. Computer vision-based limestone rock-type classification using probabilistic neural network. *Geosci. Front.* **2016**, *7*, 53–60. [CrossRef]
23. Shang, C.; Barnes, D. Support vector machine-based classification of rock texture images aided by efficient feature selection. In Proceedings of the 2012 International Joint Conference on Neural Networks (IJCNN), Brisbane, QLD, Australia, 10–15 June 2012.
24. Valentín, M.B.; De Bom, C.R.; de Albuquerque, M.P.; Faria, E.L.; Correia, M.; Surmas, R. On a method for Rock Classification using Textural Features and Genetic Optimization. *arXiv* **2016**, arXiv:1607.01679. [CrossRef]
25. Shu, L.; McIsaac, K.; Osinski, G.R.; Francis, R. Unsupervised feature learning for autonomous rock image classification. *Comput. Geosci.* **2017**, *106*, 10–17. [CrossRef]
26. Singh, N.; Singh, T.N.; Tiwary, A.; Sarkar, K.M. Textural identification of basaltic rock mass using image processing and neural network. *Comput. Geosci.* **2010**, *14*, 301–310. [CrossRef]
27. Ishikawa, S.T.; Gulick, V.C. An automated mineral classifier using Raman spectra. *Comput. Geosci.* **2013**, *54*, 259–268. [CrossRef]
28. Sharif, H.; Ralchenko, M.; Samson, C.; Ellery, A. Autonomous rock classification using Bayesian image analysis for Rover-based planetary exploration. *Comput. Geosci.* **2015**, *83*, 153–167. [CrossRef]
29. Díaz, G.F.; Ortiz, J.M.; Silva, J.F.; Lobos, R.A.; Egaña, F. Variogram-Based Descriptors for Comparison and Classification of Rock Texture Images. *Math. Geosci.* **2020**, *52*, 451–476. [CrossRef]
30. Parmar, N.; Vaswani, A.; Uszkoreit, J.; Kaiser, L.; Shazeer, N.; Ku, A.; Tran, D. Image transformer. In Proceedings of the International Conference on Machine Learning, PMLR, Stockholm, Sweden, 10–15 July 2018; Volume 80, pp. 4055–4064.
31. Dosovitskiy, A.; Beyer, L.; Kolesnikov, A.; Weissenborn, D.; Zhai, X.; Unterthiner, T.; Dehghani, M.; Minderer, M.; Heigold, G.; Gelly, S. An image is worth 16x16 words: Transformers for image recognition at scale. *arXiv* **2020**, arXiv:2010.11929.
32. Pascual, A.D. Autonomous and Real Time Rock Image Classification using Convolutional Neural Networks. Electronic Thesis and Dissertation Repository. Master's Thesis, The University of Western Ontario, London, ON, Canada, 2019. Available online: https://ir.lib.uwo.ca/etd/6059 (accessed on 25 July 2022).

33. Saranathan, A.M.; Parente, M. Adversarial feature learning for improved mineral mapping of CRISM data. *Icarus* **2021**, *355*, 114107. [CrossRef]
34. Grove, C.; Hook, S.; Paylor, I. Compilation of Laboratory Reflectance Spectra of 160 Minerals, 0.4 to 2.5 Micrometers. In *Jet Propulsion Laboratory*; NASA: Washington, DC, USA, 1992.
35. Kruse, F. Artificial Intelligence for Geologic Mapping with Imaging Spectrometers. 1993. Available online: https://ntrs.nasa.gov/citations/19930008790 (accessed on 25 August 2022).
36. Baldridge, A.M.; Hook, S.J.; Grove, C.I.; Rivera, G. The ASTER spectral library version 2.0. *Remote Sens. Environ.* **2009**, *113*, 711–715. [CrossRef]
37. Meerdink, S.K.; Hook, S.J.; Roberts, D.A.; Abbott, E.A. The ECOSTRESS spectral library version 1.0. *Remote Sens. Environ.* **2019**, *230*, 111196. [CrossRef]
38. Xie, B.; Wu, L.; Mao, W.; Zhou, S.; Liu, S. An Open Integrated Rock Spectral Library (RockSL) for a Global Sharing and Matching Service. *Minerals* **2022**, *12*, 118. [CrossRef]
39. Christensen, P.R.; Bandfield, J.; Smith, M.D.; Hamilton, V.E.; Clark, R.N. Identification of a basaltic component on the Martian surface from Thermal Emission Spectrometer data. *J. Geophys. Res. Earth Surf.* **2000**, *105*, 9609–9621. [CrossRef]
40. Michalski, J.R.; Niles, P.B.; Glotch, T.D.; Cuadros, J. Infrared Spectral Evidence for K-Metasomatism of Volcanic Rocks on Mars. *Geophys. Res. Lett.* **2021**, *48*, e2021GL093882. [CrossRef]
41. Mallapaty, S. What China's mars rover will do next. *Nature* **2021**, *593*, 323–324. [CrossRef]
42. Li, R.; Di, K.; Howard, A.B.; Matthies, L.; Wang, J.; Agarwal, S. Rock modeling and matching for autonomous long-range Mars rover localization. *J. Field Robot.* **2007**, *24*, 187–203. [CrossRef]
43. Baykan, N.A.; Yılmaz, N. Mineral identification using color spaces and artificial neural networks. *Comput. Geosci.* **2010**, *36*, 91–97. [CrossRef]
44. Xiao, X.; Cui, H.; Yao, M.; Tian, Y. Autonomous rock detection on mars through region contrast. *Adv. Space Res.* **2017**, *60*, 626–635. [CrossRef]
45. Xiao, X.; Cui, H.; Yao, M.; Fu, Y.; Qi, W. Auto rock detection via sparse-based background modeling for mars rover. In Proceedings of the 2018 IEEE Congress on Evolutionary Computation (CEC), Rio de Janeiro, Brazil, 8–13 July 2018.
46. Zhang, J.; Mei, K.; Zheng, Y.; Fan, J. Learning multi-layer coarse-to-fine representations for large-scale image classification. *Pattern Recognit.* **2019**, *91*, 175–189. [CrossRef]
47. Fukunaga, K.; Flick, T.E. Classification error for a very large number of classes. *IEEE Trans. Pattern Anal. Mach. Intell.* **1984**, *6*, 779–788. [CrossRef]
48. Reid, R.J.; Smith, P.H.; Lemmon, M.; Tanner, R.; Burkland, M.; Wegryn, E.; Weinberg, J.; Marcialis, R.; Britt, D.T.; Thomas, N.; et al. Imager for Mars Pathfinder (IMP) image calibration. *J. Geophys. Res.-Planet* **1999**, *104*, 8907–8925. [CrossRef]
49. Lanchantin, J.; Wang, T.; Ordonez, V.; Qi, Y. General Multi-label Image Classification with Transformers. In Proceedings of the IEEE/CVF Conference on Computer Vision and Pattern Recognition, Virtual, 19–25 June 2021.
50. Joseph, K.; Khan, S.; Khan, F.S.; Balasubramanian, V.N. Towards open world object detection. In Proceedings of the IEEE/CVF Conference on Computer Vision and Pattern Recognition, virtual, 19–25 June 2021.
51. Leonard, J.; Kramer, M.A. Improvement of the backpropagation algorithm for training neural networks. *Comput. Chem. Eng.* **1990**, *14*, 337–341. [CrossRef]
52. Caruana, R. Multitask learning. *Mach. Learn.* **1997**, *28*, 41–75. [CrossRef]
53. Paszke, A.; Gross, S.; Massa, F.; Lerer, A.; Bradbury, J.; Chanan, G.; Killeen, T.; Lin, Z.; Gimelshein, N.; Antiga, L. Pytorch: An imperative style, high-performance deep learning library. *Adv. Neural Inf. Process. Syst.* **2019**, *32*, 8026–8037.
54. Van der Maaten, L.; Hinton, G. Visualizing data using t-SNE. *J. Mach. Learn. Res.* **2008**, *9*, 2579–2605.
55. Pedregosa, F.; Varoquaux, G.; Gramfort, A.; Michel, V.; Thirion, B.; Grisel, O.; Blondel, M.; Prettenhofer, P.; Weiss, R.; Dubourg, V.; et al. Scikit-learn: Machine learning in Python. *J. Mach. Learn. Res.* **2011**, *12*, 2825–2830.
56. Zhang, J.; Li, J.; Hu, Y.; Zhou, J.Y. The identification method of igneous rock lithology based on data mining technology. *Adv. Mater. Res.* **2012**, *466*, 65–69. [CrossRef]
57. Masoumi, F.; Eslamkish, T.; Abkar, A.A.; Honarmand, M.; Harris, J.R. Integration of spectral, thermal, and textural features of ASTER data using Random Forests classification for lithological mapping. *J. Afr. Earth Sci.* **2017**, *129*, 445–457. [CrossRef]
58. Chatterjee, S. Vision-based rock-type classification of limestone using multi-class support vector machine. *Appl. Intell.* **2013**, *39*, 14–27. [CrossRef]
59. El-Sawy, A.; El-Bakry, H.; Loey, M. CNN for handwritten arabic digits recognition based on LeNet-5. In Proceedings of the International Conference on Advanced Intelligent Systems and Informatics, Cairo, Egypt, 24–26 October 2016; Springer: Cham, Switzerland, 2016; pp. 566–575.

remote sensing

MDPI

Technical Note

Modeling the Global Relationship via the Point Cloud Transformer for the Terrain Filtering of Airborne LiDAR Data

Libo Cheng [1], Rui Hao [2], Zhibo Cheng [2], Taifeng Li [2], Tengxiao Wang [2], Wenlong Lu [2], Yulin Ding [1] and Han Hu [1,*]

[1] Faculty of Geosciences and Environmental Engineering, Southwest Jiaotong University, Chengdu 610031, China; chenglibo123@my.swjtu.edu.cn (L.C.); dingyulin@swjtu.edu.cn (Y.D.)
[2] China Academy of Railway Sciences Corporation Limited, Beijing 100081, China; haorui@rails.cn (R.H.); chengzhibo@rails.cn (Z.C.); litaifeng@rails.cn (T.L.); wangtengxiao@rails.cn (T.W.); luwenlong@railcs.cn (W.L.)
* Correspondence: han.hu@swjtu.edu.cn

Abstract: Due to the irregularity and complexity of ground and non-ground objects, filtering non-ground data from airborne LiDAR point clouds to create Digital Elevation Models (DEMs) remains a longstanding and unresolved challenge. Recent advancements in deep learning have offered effective solutions for understanding three-dimensional semantic scenes. However, existing studies lack the capability to model global semantic relationships and fail to integrate global and local semantic information effectively, which are crucial for the ground filtering of point cloud data, especially for larger objects. This study focuses on ground filtering challenges in large scenes and introduces an elevation offset-attention (E-OA) module, which considers global semantic features and integrates them into existing network frameworks. The performance of this module has been validated on three classic benchmark models (RandLA-Net, point transformer, and PointMeta-L). It was compared with two traditional filtering methods and the advanced CDFormer model. Additionally, the E-OA module was compared with three state-of-the-art attention frameworks. Experiments were conducted on two distinct data sources. The results show that our proposed E-OA module improves the filtering performance of all three benchmark models across both data sources, with a maximum improvement of 6.15%. The performance of models was enhanced with the E-OA module, consistently exceeding that of traditional methods and all competing attention frameworks. The proposed E-OA module can serve as a plug-and-play component, compatible with existing networks featuring local feature extraction capabilities.

Keywords: terrain filtering; point clouds; semantic segmentation; deep learning

Citation: Cheng, L.; Hao, R.; Cheng, Z.; Li, T.; Wang, T.; Lu, W.; Ding, Y.; Hu, H. Modeling the Global Relationship via the Point Cloud Transformer for the Terrain Filtering of Airborne LiDAR Data. *Remote Sens.* **2023**, *15*, 5434. https://doi.org/10.3390/rs15235434

Academic Editors: Moulay A. Akhloufi and Mozhdeh Shahbazi

Received: 1 October 2023
Revised: 8 November 2023
Accepted: 11 November 2023
Published: 21 November 2023

1. Introduction

Digital Elevation Models (DEMs) are digital models used to describe the height and shape of a terrain surface and have been widely used in various fields such as surveying, hydrology, meteorology, geological hazards [1,2], soil, engineering construction, and communication [3–5]. Airborne LiDAR [6], as one of the main methods for obtaining DEMs [7–9], has become irreplaceable in obtaining large-scale and high-precision geographical scene data due to its high efficiency, strong penetration, and other characteristics. However, in LiDAR data, filtering out non-ground points is crucial for building a DEM. Therefore, correctly distinguishing between ground and non-ground information in airborne LiDAR data has become an important and necessary issue.

Filtering methods based on airborne LiDAR data can mainly be categorized as traditional and machine learning methods. Due to the complex spatial structure and diverse morphology of terrains, the main idea of traditional filtering methods is to construct terrain models at different scales and gradually refine the terrain to achieve better representation of terrain features [10–13]. However, they are often affected by factors such as threshold settings, data transformation losses, and operator errors. Machine learning methods achieve terrain filtering through feature vector construction and spatial distribution features. Classic classifiers include random forests [14,15], support vector machines [16–18], and conditional random fields [19,20], but they suffer from drawbacks such as complicated parameter settings, complicated feature selection, and high memory consumption. Recently, deep learning has been applied to point cloud ground filtering tasks for its strong feature learning ability and has shown significant effectiveness [21], but there are still issues when dealing with large-scale airborne LiDAR point cloud scenes, restricting its development in terrain reconstruction work. The following describes the factors limiting the application of deep learning in point cloud terrain filtering:

(1) **Limited receptive field for point cloud network**. Point clouds are discrete entities [22], and point cloud neural networks require explicit storage of neighborhood information [23] to enable the network to perceive its surrounding environment and perform operations akin to convolution [24]. However, storing excessive neighborhood information can exacerbate the computational complexity of the network's forward propagation, limiting the layers and depth of the network and consequently leading to a smaller receptive field. When storing less neighborhood information and increasing the network depth, issues such as network gradient vanishing [25] and overfitting [26] may arise. This apparent contradiction is a major factor constraining model development.

(2) **Drawbacks of Local Information**. Each point in a point cloud is associated with its neighbors, and neglecting global information limits the network's understanding of the overall structure and higher-level semantic information [27]. Moreover, local information is highly susceptible to occlusion or errors, and ignoring global information may lead to difficulties in identifying local noise [28]. Local information cannot fully reflect the overall shape and properties of the point cloud, resulting in an incomplete representation of information and limiting the stability of model training [29], as illustrated in Figure 1. When processing complex scenes, both local and global information must be taken into account [30].

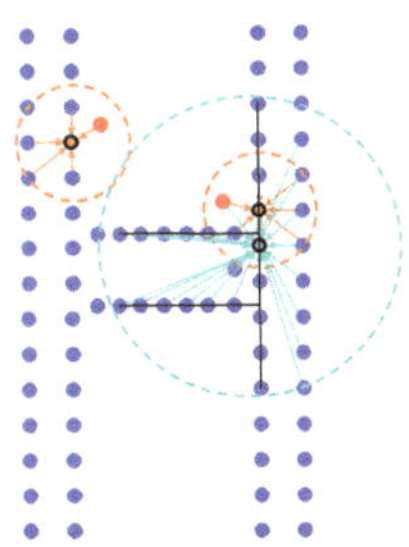

Figure 1. Illustration of Local and Global Receptive Fields: Blue points, black points, and red points respectively represent the outline of the letter H, the center of the receptive field, and noise. The circle denotes the two ranges of receptive fields. Expanding the receptive field enables the model to easily eliminate noise by identifying other similar components of the H outline, such as corners and straight edges.

In this study, we introduce a novel global attention module named Elevation Offset-Attention (E-OA), aimed at addressing the limited receptive field issue in local attention mechanisms. The core concept involves constructing global relationships among point clouds following local information processing within the network, thereby enhancing the network's long-range perceptual capabilities. Specifically, we have developed a novel attention mechanism based on the transformer [31] and integrated it after each layer of the encoder to establish global modeling relationships. The applicability of the E-OA module was validated on three state-of-the-art baseline models, and it was compared with advanced attention mechanisms. Experiments were conducted and analyzed on two datasets.

In summary, this paper makes the following contributions: (1) We developed the E-OA, a global modeling module specifically tailored for ground filtering of airborne LiDAR point cloud data. It leverages the powerful global modeling capabilities of the transformer to learn semantic relationships in large-scale geospatial scenes. (2) We guide the model beyond its local "comfort zone" by introducing global semantics in addition to local relationship modeling. This approach helps establish connections between different local areas, encouraging the model to learn both intra-class and inter-class relationships. (3) The E-AO module is versatile and can be applied as a plug-and-play component to any point cloud model, adapting to various datasets. The remainder of this paper is organized as follows: Section 2 reviews the related work. Section 3 outlines the methodology. Section 4 presents the experimental datasets and provides both qualitative and quantitative assessments. Section 5 discusses the findings. Section 6 concludes the paper.

2. Related Works

2.1. Point-Based Deep Learning

Point-based deep learning has emerged as a prominent area of research in 3D computer vision and robotics, focusing on the analysis and processing of 3D data represented in point cloud form. Point clouds, acquired through methods such as 3D scanning, LIDAR, or stereo vision, have an unstructured data format that challenges traditional deep learning techniques due to its non-grid nature.

To overcome these challenges, various point-based deep learning approaches have been developed. Notable among these are PointNet [22], PointNet++ [32], DGCNN [33], and RandLA-Net [34]. PointNet stands out as the first neural network capable of processing point clouds of arbitrary sizes, achieving significant results in object classification and semantic segmentation. PointNet++ advanced this technology by enhancing feature learning and hierarchical aggregation, thereby addressing more complex point cloud scenarios. DGCNN introduced edge convolution operations, effectively capturing local geometric structures, and demonstrated robust performance in shape classification and object detection. RandLA-Net further increased the efficiency and accuracy of point-based methods by implementing adaptive sampling and spatial segmentation techniques. PointCNN [23] contributed with an X-Conv to better grasp local spatial correlations. In a similar vein, KPConv [35] introduced a kernel convolution method that employs an arbitrary number of kernel points, making it more flexible than traditional fixed convolutions.

These advancements have showcased the potent capabilities of point cloud deep learning in a variety of tasks, such as object classification using point clouds [36], segmentation [37], and detection [38]. Furthermore, these developments have propelled the progress in 3D perception, opening up vast potential in fields like autonomous driving, augmented reality, and robotics.

2.2. Point Cloud Attention Mechanism

The attention mechanism, central to the transformer architecture, is adept at modeling long-range dependencies. Initially introduced in the field of Natural Language Processing (NLP) to model the interrelationships between words in sentences, the transformer quickly became a dominant algorithm in this domain [31]. This modeling capability was later adapted for image processing, capturing the relationships between pixels with significant outcomes [39].

It is this robust ability to model relationships that positions the attention mechanism as a powerful tool for processing point cloud data [40,41]. Coordinates and features of point clouds can be encoded for model training purposes. Transformers in point cloud applications are categorized into local [42] and global transformers [43]. The point transformer [42] employs local transformers to replace MLP modules in PointNet++, enhancing local feature extraction. LFT-Net [44] utilizes transformers for capturing local fine-grained features, with Trans-pool aggregating local information into a global context for semantic segmentation. However, these methods fall short in their ability to facilitate information exchange between global contexts. The point cloud transformer [45] proposes a global transformer network based on PointNet. It initially maps the input point cloud coordinates to a high-dimensional space, followed by processing through four stacked global transformer networks to learn global semantic relationships. Yet, it lacks the capacity to capture local fine-grained semantics. A recent approach, ref. [46], facilitates interaction between local and global information in point cloud segmentation. It identifies axis representatives in each neighborhood group to interact with information from all other axes, reducing computational costs while enhancing performance. However, its application scenarios are somewhat limited.

3. Methods

In this section, we introduce the methodology. Firstly, we summarize the structure of existing point cloud deep learning frameworks; then, we introduce the E-OA module proposed in this paper. Finally, for a fair comparison, we standardize the training strategies of the models involved in the comparison.

3.1. Overall Network Framework

Deep learning frameworks for point clouds encompass five core components as illustrated in Figure 2: input embedding (L_i), an encoding layer (L_e), a middle layer (L_s), a decoding layer (L_d), and a classification layer (L_c). The type, quantity, and presence of each component can be variably adjusted. Within the input embedding (L_i), the features of the input points, denoted as $F \in \mathbb{R}^{N \times C}$, are mapped to a high-dimensional space [45]. In the encoding layer (L_e), neighborhood searches are performed to identify the K neighbors $\{P_i^1, P_i^2, P_i^3, \cdots, P_i^K\}$ for each sampled point P_i. The positional relationship between P_i and its neighbors $\{P_i^1, P_i^2, P_i^3, \cdots, P_i^K\}$ establishes implicit edges between them [33]. The neighborhood aggregation function ϕ aggregates the neighbor features $F_i = \{F_i^1, F_i^2, F_i^3, \cdots, F_i^K\}$ onto the sampled points, represented as $\hat{F}$. The encoding layer (L_e) is constructed by stacking multiple ϕ layers and downsampling layers φ to reduce the number of points in the point cloud while mapping the features to a higher-dimensional space. The middle layer (L_s) is designed to enhance the model's learning capacity. The decoding layer (L_d) upsamples the number of points using the KNN interpolation algorithm [32] and decodes the features. Lastly, the classification layer (L_c) decodes the features into a probability distribution representing the number of categories within the point cloud,

$$\mathcal{I} = \varphi\Big(\phi_{k/r}\Big(\mathrm{F} \in \mathbb{R}^{\mathrm{N} \times \mathrm{C}}\Big)\Big), \tag{1}$$

where $\mathcal{I}$ represents the encoding result obtained after stacking a single aggregation layer ϕ and a downsampling layer φ. The final encoding outcome is computed by stacking several layers of $\mathcal{I}$ during the encoding phase. N and C are point clouds and features, respectively.

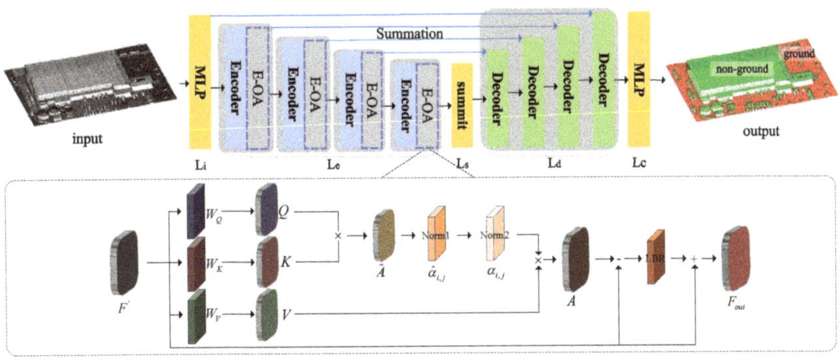

Figure 2. Model Framework Structure.

3.2. Elevation Offset-Attention

Self-Attention Mechanism. The success of attention mechanisms in both the NLP [31,47,48] and image domains [49–51] is attributed to their capability to model long-range dependencies. In the context of point cloud data, the attention mechanism has been recognized as an effective method to capture the correlations within unordered point cloud sequences [45]. According to [31], given input features F, we first compute Query, Key, and Value matrices, $Q = \{q_i\}_{i=1}^{i=S}$, $K = \{k_i\}_{i=1}^{i=S}$, $V = \{v_i\}_{i=1}^{i=S}$, as

$$\begin{aligned} Q &= W_Q \cdot F, \\ K &= W_K \cdot F, \\ V &= W_V \cdot F, \end{aligned} \tag{2}$$

where W_Q, W_K, W_V are three learnable weight matrices. The self-attention mechanism can be described as

$$\begin{aligned} F_{sa} &= (\alpha)_{i,j} \cdot V \\ &= A \cdot V \\ &= \delta \tilde{A} \cdot V \\ &= \delta(Q \cdot K^T) \cdot V, \end{aligned} \tag{3}$$

where the output from self-attention is denoted as F_{sa}. The adjacency weight matrix, $\tilde{A}$, is obtained by calculating the dot product of matrices Q and K^{T}. δ is employed to normalize $\tilde{A}$. Within δ, $\tilde{A}$ is scaled using ${}^1/\sqrt{d_k}$, where $\sqrt{d_k}$ is determined by the feature dimensions of K. Next, the Softmax is computed, resulting in A as the normalized weight. $(\alpha)_{i,j}$ represents an alternative form of A, indicating the similarity level between the i-th Q and the j-th row of K. By applying A to V, the attention outcome is produced.

The features of F_{sa} are remapped by LBR and the resultant features are added to the input features F, facilitating the direct use of self-attention. The output is denoted as F_{out}. LBR refers to the linear, normalization, and activation operations, respectively, as follows:

$$F_{out} = LBR(F_{sa}) + F. \tag{4}$$

Elevation Offset-Attention. Differing from Equation (4), the subtraction operation between the input features F and F_{sa} is computed, with the resulting values replacing the original position of F_{sa}, denoted as $\bar{F}_{sa}$. Consequently, the overall E-OA framework can be described by the following Equation (5):

$$\begin{aligned} F_{out} &= LBR(\bar{F}_{sa}) + F \\ &= LBR(F - F_{sa}) + F. \end{aligned} \tag{5}$$

The normalization operation δ was modified. As indicated by Equations (1)–(3), the form of the adjacency weight matrix $\tilde{A}$ is $(*, N, N)$. Firstly, Softmax was applied to the column vectors of the $N \times N$ matrix. Following this, the L_1 normalization operation was applied to the row vectors of the $N \times N$ matrix, consistent with the procedure in self-attention, as shown in Figure 2. The dual normalization is illustrated as follows:

$$(\tilde{\alpha})_{i,j} = Q \cdot K_k^T \tag{6}$$

$$\hat{\alpha}_{i,j} = \exp(\hat{\alpha}_{i,j}) \,/\, \sum_k \exp\left(\hat{\alpha}_{k,j}\right) \tag{7}$$

$$\alpha_{i,j} = \hat{\alpha}_{i,j} / \sum_k \hat{\alpha}_{i,k} \tag{8}$$

Regarding the interpretation of F in Equation (5), F encompasses basic spatial coordinates (p) and other attribute features f. In this study, the elevation e is utilized as an independent feature. The feature utilization is detailed as follows:

$$F_{out} = LBR(F(p,e,f)) - F_{sa}) + F(p,e,f), \tag{9}$$

3.3. Computational Complexity

A schematic representation of the E-OA module's operation is depicted in Figure 3. It accepts the output features F' from each encoding layer $\mathcal{I}$ and establishes relationships for each F'. Given an input point count of N for $\mathcal{I}$ and a scale factor of S, the corresponding point count for F' is $M = N/S$.

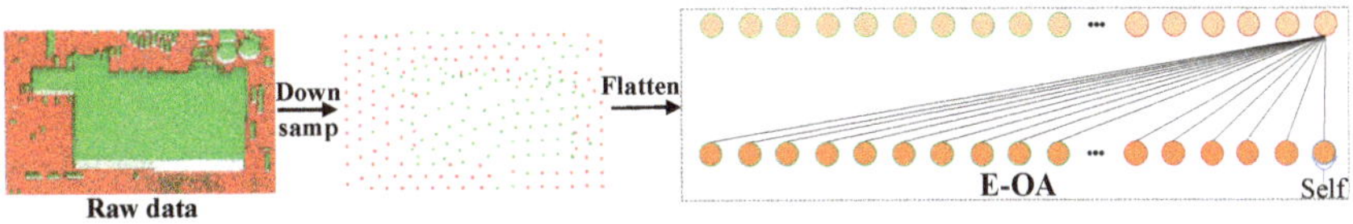

Figure 3. E-OA module builds associations between points after downsampling.

Assuming the E-OA module is applied to the input set of N points, the computational complexity is $\mathcal{O}(N^2)$. This is undoubtedly prohibitive for data batches containing tens of thousands of point clouds. The computational complexity of attention after the first $\mathcal{I}$ layer is $\mathcal{O}(M^2)$, while for the n-th $\mathcal{I}$ layer, it is $\mathcal{O}\left((M/(nS))^2\right)$. When directly computed, the complexity is $\mathcal{O}(N^2) \times n$. Given the high complexity of computing inter-point attention and the significance of the encoder in a model, the E-OA module is employed after $\mathcal{I}$.

3.4. Training Method

In order to compare the impact of E-OA on the performance of different point cloud deep learning frameworks, it is essential to standardize all performance-influencing strategies other than the model framework.

Unified Data Downsampling Strategy. The original data are downsampled using a grid-based downsampling method; during the model training process, the farthest point sampling method is employed for downsampling.

Unified Data Augmentation Strategy. To enhance the effectiveness of the data used [52], we adopted four typical data augmentation methods: (1) Centering. (2) Random rotation around the X-axis or Z-axis within a specified range. This simulates the different observation effects caused by different observation angles. (3) Random scaling, which simulates the different observation effects caused by different distances from the point cloud. (4) Random displacement and drop, which simulates the situation of sensor damage.

4. Results

In this section, the experimental data, evaluation metrics, and experimental results are primarily introduced.

4.1. Train Dataset

To validate the effectiveness of the proposed method, experiments were conducted on two datasets: OpenGF and Navarre. In addition to the coordinates and elevation information, both datasets individually contain intensity and RGB information as input features.

The OpenGF dataset spans four distinct regions situated in four different countries, covering an aggregate area of 47.7 km^2 and encompassing 542.1 MB of point cloud data. It offers a diverse range of terrain scenes, namely urban, suburban, rural, and mountainous. The urban scene is characterized by densely packed large buildings; the suburban terrain presents a mix of undulating and rugged landscapes; rural settings feature sparsely scattered buildings; and the mountainous region is marked by steep slopes and prevalent dense vegetation. The OpenGF data have been meticulously segmented into ground and non-ground classes and were partitioned into tiles measuring 500 × 500 km^2 each. Comprehensive details can be found in [53].

The Navarre dataset, sourced from the Government of Navarra https://cartotecayfototeca.navarra.es/ (accessed on 15 October 2022), utilizes data obtained from the ALS60 linear scan sensors dated between 2011 and 2012 for experimental evaluation. With a point density of 1 point/m^2, the data have been semi-automatically classified and further divided into tiles of 2 × 2 km^2 each. For the study, two distinctive scenes were chosen, leveraging satellite imagery as a reference: (1) urban settings, encompassing buildings of varying dimensions along with expansive factories, and (2) mountainous terrain, characterized by features like mountains, cliffs, valleys, and woodlands. The labels have been restructured, deriving from the official classification, to distinguish between ground and non-ground segments. In total, the curated dataset comprises 71 individual tiles.

4.2. Test Dataset

The OpenGF and Navarre datasets each contain two test sets. Within the OpenGF dataset, Test I encompasses three distinct terrains: a village, a mountainous region, and a smaller urban setting. Test II presents an urban landscape, notably featuring a structure that exceeds 200 m in both length and width. A detailed description of this can be found in [53] and will not be elaborated further here. For the Navarre dataset, two test sets were meticulously designed. Test I captures features such as high-altitude cliffs, profound gorges, and pronounced inclines, as illustrated in Figure 4a. Conversely, Test II showcases an urban environment dominated by a building measuring 265 m in length and 70 m in width, as depicted in Figure 4b. The presence of large-scale structures and canyons poses significant challenges to the performance of the model. A comprehensive breakdown of these test sets is provided in Table 1.

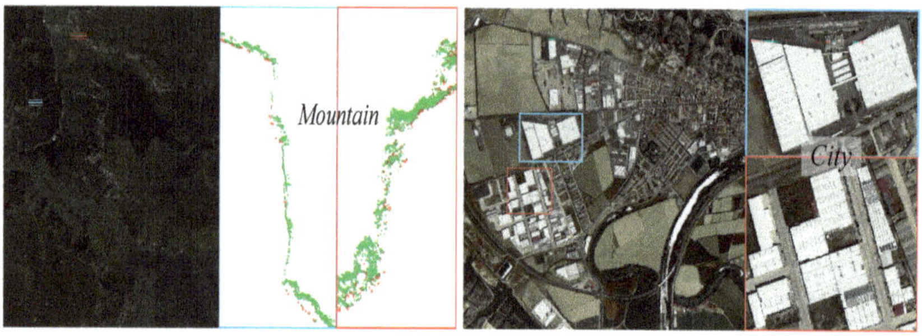

(**a**) Mountainous area. (**b**) Urban area.

Figure 4. Navarre test set. Blue boxes and red boxes distinguish different features on the same terrain.

Table 1. Test set parameters.

Dataset	Coverage	Density (Points/m^2)	Non-Ground (Points)	Ground (Points)
OpenGF Test I	6.6 km^2	6	10,431,224	5,332,664
OpenGF Test II	1.1 km^2	14	962,357	752,242
Navarre Test I	2 km^2	1	7,137,467	976,293
Navarre Test II	2 km^2	1	1,267,144	2,331,908

4.3. Experimental Parameters and Evaluation Indicators

Experiments were executed on a consistent server configuration, featuring an Intel(R) Xeon(R) CPU E5-2680 v4 @ 2.40 GHz, 251 GB of RAM, and an NVIDIA RTX 3090 GPU.

Training Parameters. Uniform training parameters were employed across all models. The initial learning rate was fixed at 0.01, and a cosine annealing approach was adopted for learning rate adjustments. The network was trained for 500 epochs using the AdamW optimizer, with a consistent batch size of 4 and each batch comprising 4×10^4 points.

Evaluation Metrics. In this study, metric evaluations were derived from the union and intersection ratios across all categories, as well as for individual categories. These ratios were denoted as $mIOU$ for overall recognition accuracy, IOU_1 for non-ground recognition accuracy, and IOU_2 for ground recognition accuracy. The accuracy was also used to denote the overall recognition accuracy.

$$\begin{aligned} IOU_1 &= \frac{TP}{(TP+FN+FP)} \\ IOU_2 &= \frac{TN}{(FN+FP+TN)} \\ mIOU &= \frac{IOU_1+IOU_2}{2} \\ Accuracy &= \frac{(TP+TN)}{(TP+FN+FP+TN)} \end{aligned} \tag{10}$$

where the variables TP, FN, FP, and TN denote the number of correctly classified non-ground points, the number of ground points erroneously classified as non-ground, the number of non-ground points incorrectly classified as ground, and the number of correctly classified ground points, respectively.

4.4. Validation Analysis on OpenGF Test Set

Experiments were conducted using two traditional methods, PMF [54] and CSF [55]; three conventional point cloud deep learning methods, RandLA-Net [34], point transformer [42], and PointMeta-L [56]; and one pure transformer model, CDFormer [57].

The experiment focused on the ground filtering performance of models, particularly on IOU_2. The results are displayed in Table 2. Both traditional methods exhibited a poor ground classification accuracy, indicating inadequate ground filtering results (the results are from [53]). As per Section 3, we integrated the E-OA module into three conventional point cloud frameworks, leading to an overall enhancement in the model performance. In Test I, which covers mountainous areas, the enhanced point transformer demonstrated the best filtering effect, with an IOU_2 increase of 0.81% compared to the baseline model. The most significant improvement was observed for PointMeta-L, with an increase of 2.58% in the IOU_2 value; however, improvements in RandLA-Net were not substantial. The corresponding results are depicted in Figure 5b,c,e,f). Due to the only marginal improvement in RandLA-Net, it is not distinctly visible in the figure and is therefore not listed here. In Test II, the enhanced versions of all three conventional point cloud frameworks showed performance improvements, with increases of 1.04%, 3.05%, and 1.83%, respectively. Among these, the RandLA-Net model with the E-OA module excelled in filtering large buildings, as shown in Figure 5h,l. The improved versions of point transformer and PointMeta-L showed slight increases in IOU_2 values, but a lower filtering effectiveness, possibly due to their framework designs, as seen in Figure 5i,j,m,n). Notably, the filtering ability of RandLA-Net in urban areas with large buildings was discussed in [53], and our experimental outcomes reaffirmed this. CDFormer, an attention-based framework, underperformed compared to the improved baseline models in mountainous regions and was less effective than the enhanced RandLA-Net and PointMeta-L in urban settings. This underscores the necessity of the aim of our experiments: how to better utilize attention mechanisms.

4.5. Validation Analysis on the Navarre Test Set

Table 3 provides the quantitative assessment results on the Navarre test dataset. The parameters of the two traditional methods were fine-tuned to their optimum settings, yet yielded the poorest filtering performance. In the mountainous region (Test I), the improved versions of the three baseline models exhibited increases in their IOU_2 values by 6.15%, 0.58%, and 0.91%, respectively. In urban areas (Test II), the enhancements in IOU_2 values for the improved versions of RandLA-Net and PointMeta-L were 0.96% and 1.25%, respectively, while the enhanced point transformer showed no improvement, mirroring its performance on the OpenGF dataset. Due to the minimal improvement in the point transformer in Test I and its unsatisfactory performance on Test II, only the classification results of RandLA-Net and PointMeta are shown in Figure 6, with areas of significant changes highlighted using black circles or rectangles. It is important to note that due to erroneous annotations in overlapping areas of the Navarre dataset, there are stripe artifacts in the classification results in Figure 6g–l, but these do not affect the comparison of all models' performances. The advanced CDFormer framework did not exhibit a ground filtering performance comparable to point cloud frameworks equipped with the E-OA attention mechanism in both mountainous and urban areas.

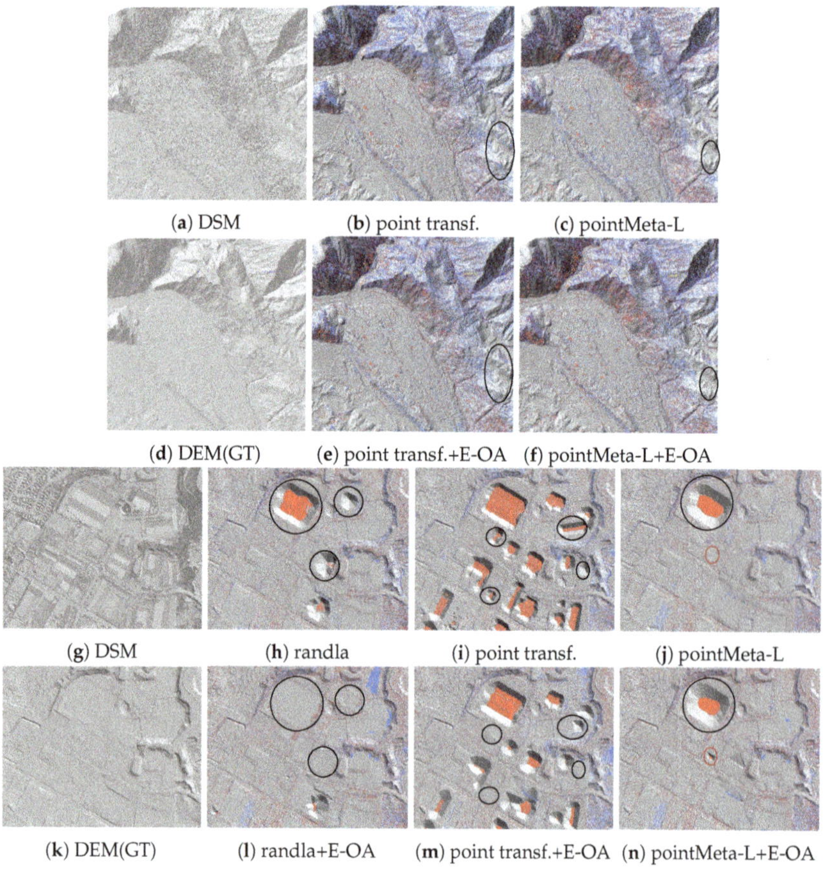

(**a**) DSM (**b**) point transf. (**c**) pointMeta-L

(**d**) DEM(GT) (**e**) point transf.+E-OA (**f**) pointMeta-L+E-OA

(**g**) DSM (**h**) randla (**i**) point transf. (**j**) pointMeta-L

(**k**) DEM(GT) (**l**) randla+E-OA (**m**) point transf.+E-OA (**n**) pointMeta-L+E-OA

Figure 5. Comparative analysis of three baselines and improvements on OpenGF Test Set. DEM (GT) is the real surface. Red is for NG points wrongly classified and blue is for GR points wrongly classified.

Table 2. The quantitative evaluation results on the OpenGF test set.

Method	Test I				Test II			
	Ac	$mIOU$	IoU_1	IoU_2	Ac	$mIOU$	IoU_1	IoU_2
PMF	0.9063	0.8242	0.8522	0.7962	0.8656	0.7606	0.7850	0.7361
CSF	0.9306	0.8691	0.8817	0.8564	0.8934	0.8073	0.8108	0.8038
RandLA-Net	0.9766	0.9487	0.9657	0.9318	0.9533	0.9101	0.918	0.9022
RandLA-Net+E-OA	0.9761	0.9485	0.9650	**0.9319**	**0.9596**	**0.9213**	**0.9301**	**0.9126**
Point transformer	0.9743	0.9436	0.9625	0.9247	0.9144	0.8468	0.8503	0.8433
Point transformer+E-OA	**0.9770**	**0.9495**	**0.9662**	**0.9328**	**0.9392**	**0.8845**	**0.8951**	**0.8738**
PointMeta-L	0.9661	0.9259	0.9510	0.9008	0.9486	0.9010	0.9116	0.8903
PointMeta-L+E-OA	**0.9748**	**0.9448**	**0.9631**	**0.9266**	**0.9578**	**0.9182**	**0.9262**	**0.9102**
CDFormer	0.9717	0.9382	0.9586	0.9179	0.9421	0.8897	0.8988	0.8806

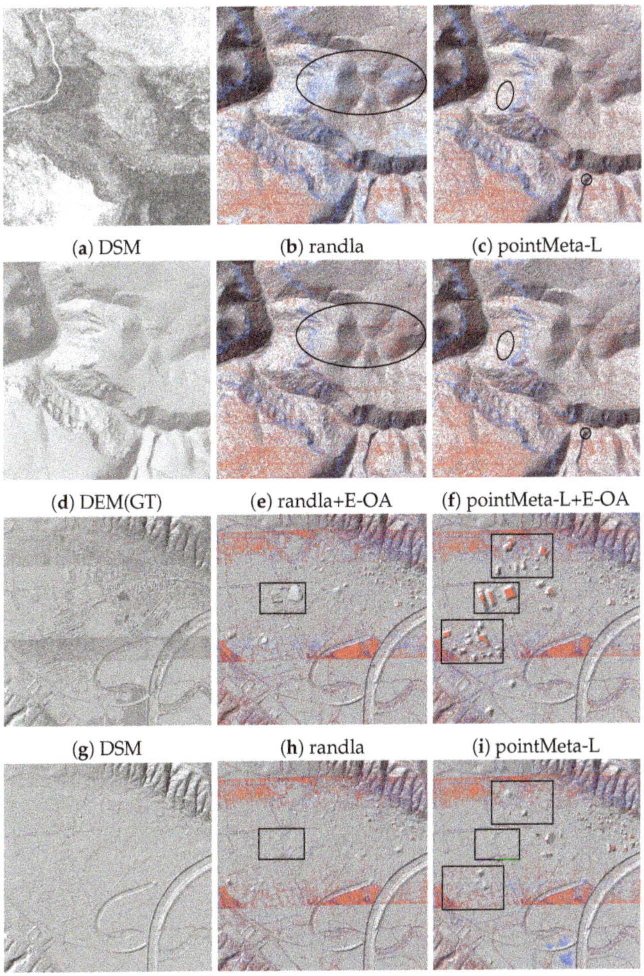

Figure 6. Comparative analysis of three baselines and improvements on the Navarre test set. DEM (GT) is the real surface. Red is for NG points wrongly classified and blue is for GR points wrongly classified.

Table 3. The quantitative evaluation results on the Navarre test set.

Method	Test I				Test II			
	Ac	$mIOU$	IoU_1	IoU_2	Ac	$mIOU$	IoU_1	IoU_2
PMF	0.8512	0.6290	0.8328	0.4253	0.8543	0.6414	0.4729	0.8100
CSF	0.8105	0.5278	0.7970	0.2585	0.8186	0.6549	0.5403	0.7695
RandLA-Net	0.9461	0.7836	0.9408	0.6264	0.9443	0.8899	0.8675	0.9124
RandLA-Net+E-OA	**0.9519**	**0.8170**	**0.9462**	**0.6879**	**0.9504**	**0.9012**	**0.8804**	**0.9220**
Point transformer	0.9487	0.8106	0.9425	0.6786	0.951	0.9023	0.8818	0.9227
Point transformer+E-OA	**0.9509**	**0.8147**	**0.9450**	**0.6844**	0.9489	0.8985	0.8779	0.9191
PointMeta-L	0.9386	0.7765	0.9314	0.6216	0.9293	0.8623	0.8343	0.8903
PointMeta-L+E-OA	**0.9408**	**0.7852**	**0.9339**	**0.6365**	**0.9374**	**0.8758**	**0.8487**	**0.9028**
CDFormer	0.9288	0.7518	0.9210	0.5825	0.8903	0.7740	0.6939	0.8540

5. Discussion

This section presents ablation studies conducted to evaluate the proposed method. The impact of grid sampling size and global scene size on experimental results is discussed; additionally, comparisons are made with three current global modeling frameworks. Furthermore, the effective receptive field is visualized to fully substantiate the reliability of the method proposed in this paper.

5.1. Impact of Different Grid Subsampling Sizes

To assess the impact of varying grid sampling sizes on point cloud classification outcomes, the RandLA-Net network was selected for comparative experiments on the OpenGF Test II dataset. As verified in Section 4.4, the RandLA-Net network demonstrates a robust classification performance for large buildings. OpenGF Test II, which includes a significant building structure, aptly illustrates the results of these comparative experiments. Four distinct grid sampling sizes were employed, 0.25 m, 0.5 m, 0.75 m, and 1 m, with the experimental outcomes presented in Table 4 and Figure 7. It was observed that different grid sampling sizes exert varied impacts on building classification. Specifically, a decrease in grid size led to a reduction in classification accuracy. Notably, the RandLA-Net model, when equipped with global semantic information, consistently maintained effective results in filtering large buildings on the ground at a grid sampling size of 0.5 m. In other words, integrating local and global information within the RandLA-Net model ensures commendable performance, provided the grid size exceeds 0.5 m. Furthermore, we recommend a maximum sampling grid of 1 m, as larger sampling grids would result in the loss of local terrain information.

Table 4. Impact of sampling rate results.

Downsampling (m)	Ac	$mIOU$	IoU_1	IoU_2
0.25	0.9281	0.8659	0.8639	0.8678
0.5	0.9415	0.8895	0.8926	0.8863
0.75	0.9522	0.9082	0.9154	0.9010
1.0	**0.9596**	**0.9213**	**0.9301**	**0.9126**

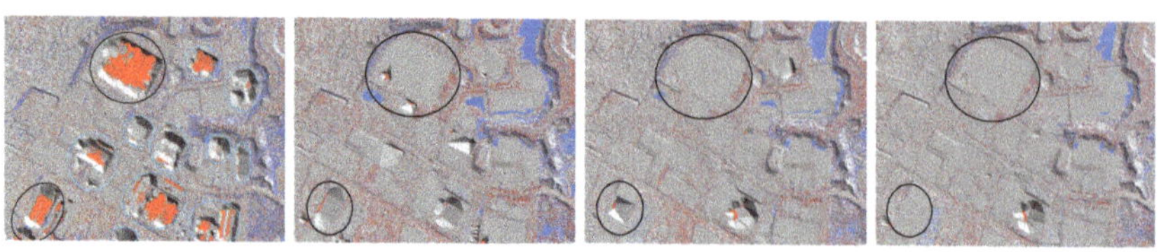

(**a**) 0.25 m (**b**) 0.5 m (**c**) 0.75 m (**d**) 1.0 m

Figure 7. The impact of three different grid sampling sizes on classification results.

5.2. Impact of Global Scene Size

When analyzing the impact of the number of points in each global scene on the model's classification results, it is recognized that loading all point clouds from a tile into the model at once is impractical. In deep learning, batch loading for inference is a feasible method, yet the number of point clouds in a single batch can affect the model's classification results. Hence, we compared and analyzed cases with point cloud numbers in a single batch at 5×10^3, 1×10^4, and 2×10^4, as illustrated in Figure 8. Experiments were conducted using an improved version of RandLA-Net, with the results presented in Table 5. When the number of points is low, such as 5×10^3, the model's classification accuracy is insufficient. With an increase in the point cloud number to 1×10^4, the classification results still show inadequacies. However, when the point cloud number further rises to 2×10^4, it is observed that most ground buildings, especially buildings, are effectively filtered out. This analysis indicates that more scene information provides additional background context, aiding the E-OA module in driving the network to learn more semantic information, thereby enhancing the network's filtering capability.

Table 5. Impact of scene size on filtering performance.

Points	Ac	$mIOU$	IoU_1	IoU_2
2×10^4	**0.9610**	**0.9240**	**0.9324**	**0.9156**
1×10^4	0.9572	0.9170	0.9254	0.9087
5×10^3	0.9502	0.9043	0.9128	0.8958

(**a**) 5×10^3 point (**b**) 1×10^4 point (**c**) 2×10^4 point

Figure 8. The impact of different point cloud quantities on classification results in global scenes.

5.3. Comparison of Different Attention Mechanisms

To further substantiate the superiority of the E-OA module in handling large-scale point cloud data, several state-of-the-art attention mechanisms were compared. External attention [58] utilizes a compact token set (S) as an external memory unit to interact with the input for deriving attention weights, with a computational complexity of $\mathcal{O}(SN)$. In this study, S was set to 512. APPT [46] is a framework that utilizes proxy points to reduce computational complexity. It initially aggregates information from each group of neighbors to the sampled center, i.e., the proxy point, and subsequently calculates attention weights between the proxy point and all original points, resulting in a complexity of $\mathcal{O}(MN)$, where M is determined by the downsampling rate. CDFormer [57], another framework seeking to lower computational complexity via proxy points, initially computes attention weights within each set of neighbors and consolidates neighbor information onto proxy points, with a complexity of $\mathcal{O}(Mk^2)$. Subsequently, it computes the weights for each group of proxy points with a complexity of $\mathcal{O}(Mk^2)$, and finally distributes the proxy point information back to the initial input points with a complexity of $\mathcal{O}(Nk^2)$. In comparison, the computational complexity of the E-OA module is less than that of APPT, but greater than the others. Table 6 below presents the results of embedding the aforementioned attention mechanisms into the RandLA-Net model at the same position as the E-OA module and conducting experiments on OpenGF's test I and II. The experimental results demonstrate that the E-OA module exhibits clear advantages in processing large spatial point clouds.

Table 6. Comparison of attention performance.

Method	Test I				Test II			
	Ac	*mIOU*	IoU_1	IoU_2	*Ac*	*mIOU*	IoU_1	IoU_2
RandLA-Net + External	0.9701	0.9431	0.9582	0.9280	0.9563	0.9169	0.9230	0.9108
RandLA-Net + APPT	0.9696	0.9419	0.9596	0.9242	0.9513	0.9144	0.9233	0.9055
RandLA-Net + CDFormer	0.9534	0.8993	0.9337	0.8648	0.8945	0.8086	0.8192	0.7979
RandLA-Net + E-OA	**0.9761**	**0.9485**	**0.9650**	**0.9319**	**0.9596**	**0.9213**	**0.9301**	**0.9126**

5.4. Effective Receptive Field Analysis

The Effective Receptive Field (ERF) represents the distribution of high-density point clouds contributing to the network model, serving as an efficient means to evaluate the performance of the global module. In this paper, we conduct experiments on OpenGF test I and qualitatively compare the baseline and improved versions of the RandLA-Net, as illustrated in Figures 9 and 10. The baseline RandLA-Net primarily focuses on the local areas of the point cloud, whereas the improved version captures a considerable amount of useful information. To further substantiate the efficacy of our proposed method, we present statistical values of the ERFs for 625 points of interest in Figure 10. We analyze the ERFs of all points in the first decoding layer, as the weights of this layer have been propagated through the transformer layer and encompass an appropriate quantity of point clouds. We quantified the number of ERFs considered effective and discovered that models with global capabilities achieve higher attention coverage on the rooftops of buildings, which further validates the effectiveness of our approach.

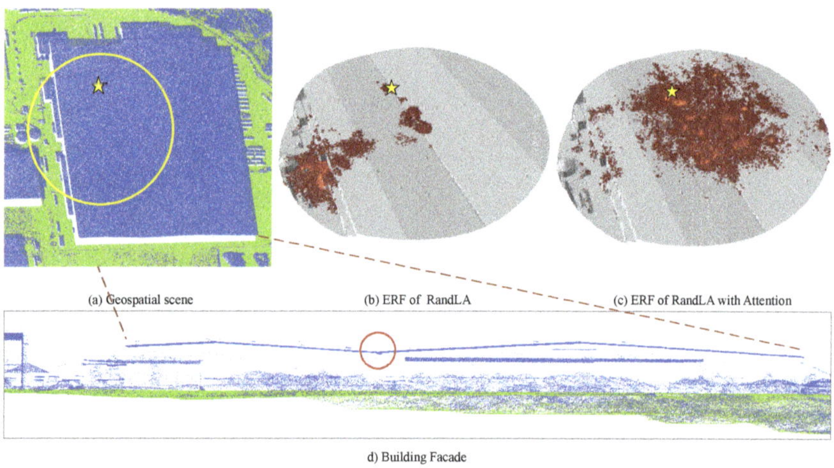

Figure 9. Baselines and improved ERF of RandLA. (**a**) Spatial scene. Yellow circles represent the global scene size in a single batch; asterisks are points of interest, located in the roof depression, whose oblique view is shown in the red circle in (**d**). (**b**) ERF of RandLA. (**c**) ERF of RandLA with attention. Red and bright areas indicate high contributions. (**d**) Oblique view of building facades.

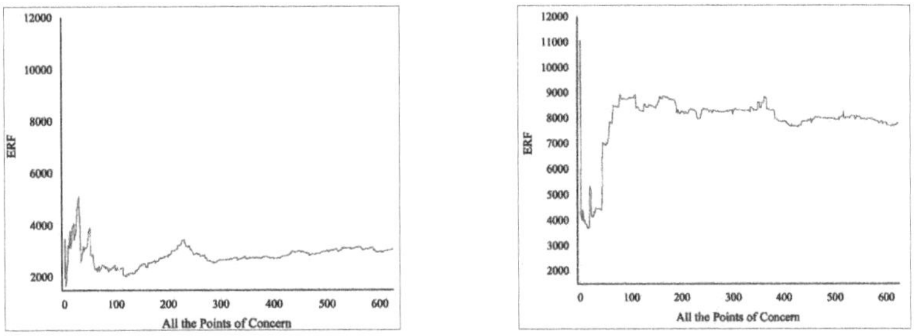

(a) ERF of RandLA (b) ERF of RandLA with Attention

Figure 10. Summary of the ERF of all feature points in the first decoding layer of statistics.

6. Conclusions

In this paper, we address the challenges of modeling geographical relationships based on large-scale airborne LiDAR point cloud data. We critically analyzed the limitations of existing traditional and deep-learning-based terrain filtering methods from point clouds, highlighting the significance of constructing remote relationships in airborne point cloud terrain filtering and integrating global and local modeling for experimental analysis. A novel global semantic capture module, E-OA, is introduced, designed to enhance model performance by combining it with existing networks that capture local information. For a fair comparison, we restructured three classic point cloud deep neural networks, standardized training strategies, and incorporated the E-OA module into these frameworks for comparative experiments. These experiments were conducted on two datasets, each characterized by different data features.

Given that each point in a point cloud can be considered as an individual semantic unit, directly inputting tens of thousands of points into the model for global relationship construction proved to be impractical. Hence, we analyzed computational complexities and decided to integrate the E-OA module into the encoding layers of the base frameworks, placing it after each downsampling layer. This approach not only reduced the computational complexity but also ensured that the points involved in the computation carried local information. The experimental results demonstrated that the E-OA module enhanced the performance of the benchmark models to varying degrees across different data sources. For urban areas, the RandLA-Net equipped with the E-OA module excelled, while the point transformer with the E-OA module was superior in mountainous regions. By introducing the E-OA module into the benchmark models, the models were encouraged to move beyond their local "comfort zones".

In future work, we aim to continue exploring the design principles of different frameworks to elucidate how various forms of global attention impact the network's terrain filtering performance. Additionally, we plan to further investigate compact global modeling frameworks for integration into airborne devices, enhancing their efficiency and applicability in terrain filtering tasks.

Author Contributions: Conceptualization, L.C., R.H. and Z.C.; Methodology, L.C., R.H., Z.C., H.H. and T.L.; Validation: L.C., R.H., Z.C., T.W., W.L., Y.D. and H.H.; Formal analysis, L.C., R.H., Z.C., T.L., T.W., W.L., Y.D. and H.H.; Investigation, L.C., R.H., Z.C., T.L., T.W., W.L., Y.D. and H.H.; Data curation, L.C., R.H., Z.C., T.L., T.W., W.L., Y.D. and H.H.; Writing—original draft preparation, L.C., R.H., Z.C., T.L., T.W., W.L., Y.D. and H.H.; Writing—review and editing, L.C., R.H., Z.C., T.L., T.W., W.L., Y.D. and H.H.; Visualization, L.C., R.H., Z.C. and T.L.; Supervision, L.C., R.H., Z.C. and T.L.; Project administration, L.C., R.H., Z.C. and T.L.; Funding acquisition, H.H. All authors have read and agreed to the published version of the manuscript.

Funding: This work was supported in part by the National Key Research and Development Program of China (Project No. 2022YFF0904400) and the National Natural Science Foundation of China (Project No. 42071355, 42230102, green41871291).

Institutional Review Board Statement: Not applicable.

Informed Consent Statement: Not applicable.

Data Availability Statement: We used two publicly available datasets, respectively, from: https://github.com/Nathan-UW/OpenGF and https://cartotecayfototeca.navarra.es/ (accessed on 15 October 2022).

Conflicts of Interest: Authors Libo Cheng, Yulin Ding, and Han Hu were affiliated with the Faculty of Geosciences and Environmental Engineering at Southwest Jiaotong University. Authors Rui Hao, Zhibo Cheng, Taifeng Li, Tengxiao Wang, and Wenlong Lu were affiliated with China Academy of Railway Sciences Corporation Limited. The remaining authors declare that the research was conducted in the absence of any commercial or financial relationships that could be construed as a potential conflict of interest.

References

1. Du, L.; McCarty, G.W.; Zhang, X.; Lang, M.W.; Vanderhoof, M.K.; Li, X.; Huang, C.; Lee, S.; Zou, Z. Mapping Forested Wetland Inundation in the Delmarva Peninsula, USA Using Deep Convolutional Neural Networks. *Remote Sens.* **2020**, *12*, 644. [CrossRef]
2. Li, D.; Tang, X.; Tu, Z.; Fang, C.; Ju, Y. Automatic Detection of Forested Landslides: A Case Study in Jiuzhaigou County, China. *Remote Sens.* **2023**, *15*, 3850. [CrossRef]
3. Chen, Z.; Ye, F.; Fu, W. The influence of DEM spatial resolution on landslide susceptibility mapping in the Baxie River basin, NW China. *Nat. Hazards* **2020**, *101*, 853–877. [CrossRef]
4. Kakavas, M.P.; Nikolakopoulos, K.G. Digital Elevation Models of Rockfalls and Landslides: A Review and Meta-Analysis. *Geosciences* **2021**, *11*, 256. [CrossRef]
5. McClean, F.; Dawson, R.; Kilsby, C. Implications of Using Global Digital Elevation Models for Flood Risk Analysis in Cities. *Water Resour. Res.* **2020**, *56*, e2020WR028241. [CrossRef]
6. Ozendi, M.; Akca, D. A point cloud filtering method based on anisotropic error model. *Photogramm. Rec.* **2023**, *38*, 1–38. [CrossRef]
7. Gökgöz, T.; Baker, M.K.M. Large Scale Landform Mapping Using Lidar DEM. *ISPRS Int. J. Geo-Inf.* **2015**, *4*, 1336–1345. [CrossRef]
8. O'Banion, M.S.; Olsen, M.J.; Hollenbeck, J.P.; Wright, W.C. Data Gap Classification for Terrestrial Laser Scanning-Derived Digital Elevation Models. *ISPRS Int. J. Geo-Inf.* **2020**, *9*, 749. [CrossRef]
9. Buján, S.; González-Ferreiro, E.M.; Cordero, M.; Miranda, D. PpC: A new method to reduce the density of lidar data. Does it affect the DEM accuracy? *Photogramm. Rec.* **2019**, *34*, 304–329. [CrossRef]
10. Susaki, J. Adaptive Slope Filtering of Airborne LiDAR Data in Urban Areas for Digital Terrain Model (DTM) Generation. *Remote Sens.* **2012**, *4*, 1804–1819. [CrossRef]
11. Evans, J.S.; Hudak, A.T. A Multiscale Curvature Algorithm for Classifying Discrete Return LiDAR in Forested Environments. *IEEE T. Geosci. Remote* **2007**, *45*, 1029–1038. [CrossRef]
12. Yan, W.Y. Scan Line Void Filling of Airborne LiDAR Point Clouds for Hydroflattening DEM. *IEEE J.-STARS* **2021**, *14*, 6426–6437. [CrossRef]
13. Pan, Z.; Tang, J.; Tjahjadi, T.; Wu, Z.; Xiao, X. A Novel Rapid Method for Viewshed Computation on DEM through Max-Pooling and Min-Expected Height. *ISPRS Int. J. Geo-Inf.* **2020**, *9*, 633. [CrossRef]
14. Medeiros, S.C.; Bobinsky, J.S.; Abdelwahab, K. Locality of Topographic Ground Truth Data for Salt Marsh Lidar DEM Elevation Bias Mitigation. *IEEE J-STARS* **2022**, *15*, 5766–5775. [CrossRef]
15. Cao, S.; Pan, S.; Guan, H. Random forest-based land-use classification using multispectral LiDAR data. *Bull. Surv. Mapp.* **2019**, *11*, 79–84. [CrossRef]
16. Wu, J.; Liu, L. Automatic DEM generation from aerial lidar data using multiscale support vector machines. *Mippr 2011 Remote Sens. Image Process. Geogr. Inf. Syst. Other Appl.* **2011**, *8006*, 63–69. [CrossRef]
17. Lodha, S.; Kreps, E.; Helmbold, D.; Fitzpatrick, D. Aerial LiDAR data classification using support vector machines (SVM). In Proceedings of the Third International Symposium on 3D Data Processing, Visualization, and Transmission, Chapel Hill, NC, USA, 14–16 June 2006; pp. 567–574. [CrossRef]
18. Wu, J.; Guo, N.; Liu, R.; Xu, G. Aerial Lidar Data Classification Using Weighted Support Vector Machines. *Int. Conf. Digit. Image Process.* **2013**, *38*, 1–5. [CrossRef]
19. Niemeyer, J.; Wegner, J.D.; Mallet, C.; Rottensteiner, F.; Soergel, U. Conditional Random Fields for Urban Scene Classification with Full Waveform LiDAR Data. *Photogramm. Image Anal. ISPRS Conf.* **2011**, *6952*, 233–244. [CrossRef]
20. Zheng, Y.; Cao, Z.G. Classification method for aerial LiDAR data based on Markov random field. *Electron. Lett.* **2011**, *47*, 934. [CrossRef]
21. Luo, W.; Ma, H.; Yuan, J.; Zhang, L.; Ma, H.; Cai, Z.; Zhou, W. High-Accuracy Filtering of Forest Scenes Based on Full-Waveform LiDAR Data and Hyperspectral Images. *Remote Sens.* **2023**, *15*, 3499. [CrossRef]
22. Charles, R.Q.; Su, H.; Kaichun, M.; Guibas, L.J. PointNet: Deep Learning on Point Sets for 3D Classification and Segmentation. In Proceedings of the IEEE Conference on Computer Vision and Pattern Recognition (CVPR), Honolulu, HI, USA, 21–26 July 2017; pp. 77–85. [CrossRef]
23. Li, Y.; Bu, R.; Sun, M.; Wu, W.; Di, X.; Chen, B. PointCNN: Convolution on x-transformed points. In Proceedings of the Advances in Neural Information Processing Systems, Montreal, QC, Canada, 2–8 December 2018; pp. 820–830. [CrossRef]

24. Kim, Y. Convolutional Neural Networks for Sentence Classification. In Proceedings of the 2014 Conference on Empirical Methods in Natural Language Processing (EMNLP), Doha, Qatar, 25–29 October 2014; Association for Computational Linguistics: Doha, Qatar, 2014; pp. 1746–1751. [CrossRef]
25. Liu, Z.; Sun, M.; Zhou, T.; Huang, G.; Darrell, T. Rethinking the value of network pruning. In Proceedings of the Seventh International Conference on Learning Representations (ICLR), New Orleans, LA, USA, 6–9 May 2019; pp. 2818–2826. [CrossRef]
26. Lu, Z.; Xu, C.; Du, B.; Ishida, T.; Zhang, L.; Sugiyama, M. LocalDrop: A Hybrid Regularization for Deep Neural Networks. *IEEE T. Pattern Anal.* **2021**, *44*, 3590–3601. [CrossRef] [PubMed]
27. Lei, H.; Akhtar, N.; Mian, A. Spherical Convolutional Neural Network for 3D Point Clouds. In Proceedings of the IEEE/CVF Conference on Computer Vision and Pattern Recognition, Long Beach, CA, USA, 15–20 June 2019; pp. 9631–9640. [CrossRef]
28. Fu, K.; Liu, S.; Luo, X.; Wang, M. Robust Point Cloud Registration Framework Based on Deep Graph Matching. In Proceedings of the IEEE Conference on Computer Vision and Pattern Recognition, CVPR 2021, Virtual, 19–25 June 2021; IEEE: New York, NY, USA, 2021; pp. 8893–8902. [CrossRef]
29. Yan, X.; Zheng, C.; Li, Z.; Wang, S.; Cui, S. PointASNL: Robust Point Clouds Processing Using Nonlocal Neural Networks with Adaptive Sampling. In Proceedings of the IEEE/CVF Conference on Computer Vision and Pattern Recognition, Seattle, WA, USA, 13–19 June 2020; pp. 5588–5597. [CrossRef]
30. Pan, X.; Xia, Z.; Song, S.; Li, L.E.; Huang, G. 3d object detection with pointformer. In Proceedings of the IEEE/CVF Conference on Computer Vision and Pattern Recognition, Nashville, TN, USA, 20–25 June 2021; pp. 7463–7472. [CrossRef]
31. Vaswani, A.; Shazeer, N.; Parmar, N.; Uszkoreit, J.; Jones, L.; Gomez, A.N.; Kaiser, L.; Polosukhin, I. Attention is All You Need. *NIPS* **2017**, *30*, 1–15. [CrossRef]
32. Qi, C.R.; Yi, L.; Su, H.; Guibas, L.J. PointNet++: Deep hierarchical feature learning on point sets in a metric space. *NIPS* **2017**, *30*, 1–14. [CrossRef]
33. Wu, B.; Liu, Y.; Lang, B.; Huang, L. DGCNN: Disordered Graph Convolutional Neural Network Based on the Gaussian Mixture Model. *Neurocomputing* **2017**, *321*, 346–356. [CrossRef]
34. Hu, Q.; Yang, B.; Xie, L.; Rosa, S.; Guo, Y.; Wang, Z.; Trigoni, N.; Markham, A. RandLA-Net: Efficient Semantic Segmentation of Large-Scale Point Clouds. In Proceedings of the IEEE Conference on Computer Vision and Pattern Recognition (CVPR), Seattle, WA, USA, 13–19 June 2020; pp. 11105–11114. [CrossRef]
35. Thomas, H.; Qi, C.R.; Deschaud, J.E.; Marcotegui, B.; Goulette, F.; Guibas, L. KPConv: Flexible and Deformable Convolution for Point Clouds. In Proceedings of the 2019 IEEE/CVF International Conference on Computer Vision (ICCV), Seoul, Republic of Korea, 27 October–2 November 2019; pp. 6410–6419. [CrossRef]
36. Su, H.; Jampani, V.; Sun, D.; Maji, S.; Kalogerakis, E.; Yang, M.; Kautz, J. SPLATNet: Sparse Lattice Networks for Point Cloud Processing. In Proceedings of the 2018 IEEE Conference on Computer Vision and Pattern Recognition (CVPR 2018), Salt Lake City, UT, USA, 18–22 June 2018; pp. 2530–2539. [CrossRef]
37. You, Y.; Lou, Y.; Liu, Q.; Tai, Y.; Ma, L.; Lu, C.; Wang, W. Pointwise Rotation-Invariant Network with Adaptive Sampling and 3D Spherical Voxel Convolution. *CVPR* **2019**, *34*, 12717–12724. [CrossRef]
38. Shi, S.; Guo, C.; Jiang, L.; Wang, Z.; Shi, J.; Wang, X.; Li, H. PV-RCNN: Point-Voxel Feature Set Abstraction for 3D Object Detection. In Proceedings of the IEEE/CVF International Conference on Computer Vision (ICCV), Seoul, Republic of Korea, 27 October–2 November 2019; pp. 10529–10538. [CrossRef]
39. Parmar, N.; Vaswani, A.; Uszkoreit, J.; Kaiser, L.; Shazeer, N.; Ku, A.; Tran, D. Image transformer. In Proceedings of the International Conference on Machine Learning PMLR, Stockholm, Sweden, 10–15 July 2018; pp. 4055–4064. [CrossRef]
40. Huang, S.; Chen, Y.; Jia, J.; Wang, L. Multi-View Transformer for 3D Visual Grounding. In Proceedings of the IEEE/CVF Conference on Computer Vision and Pattern Recognition (CVPR), New Orleans, LA, USA, 18–24 June 2022; pp. 15524–15533. [CrossRef]
41. Hui, L.; Wang, L.; Tang, L.; Lan, K.; Xie, J.; Yang, J. 3D Siamese Transformer Network for Single Object Tracking on Point Clouds. *ECCV* **2022**, *13662*, 293–310. [CrossRef]
42. Zhao, H. Point transformer. In Proceedings of the IEEE/CVF International Conference on Computer Vision, Montreal, QC, Canada, 11–17 October 2021; pp. 16259–16268. [CrossRef]
43. Lu, D.; Xie, Q.; Xu, L.; Li, J. 3DCTN: 3D Convolution-Transformer Network for Point Cloud Classification. *IEEE T. Intell. Transp.* **2022**, *23*, 24854–24865. [CrossRef]
44. Gao, Y.; Liu, X.; Li, J.; Fang, Z.; Jiang, X.; Huq, K.M.S. LFT-Net: Local feature transformer network for point clouds analysis. *IEEE Trans. Intell. Transp. Syst.* **2022**, *24*, 2158–2168. [CrossRef]
45. Guo, M.; Cai, J.; Liu, Z.; Mu, T.; Martin, R.R.; Hu, S. PCT: Point cloud transformer. *Comput. Vis. Media* **2021**, *7*, 187–199. [CrossRef]
46. Li, H.; Zheng, T.; Chi, Z.; Yang, Z.; Wang, W.; Wu, B.; Lin, B.; Cai, D. APPT: Asymmetric Parallel Point Transformer for 3D Point Cloud Understanding. *arXiv* **2023**, arXiv:2303.17815. [CrossRef]
47. Devlin, J.; Chang, M.W.; Lee, K.; Toutanova, K. BERT: Pre-training of Deep Bidirectional Transformers for Language Understanding. In Proceedings of the Conference of the North American Chapter of the Association for Computational Linguistics: Human Language Technologies, Minneapolis, MN, USA, 2–7 June 2019; pp. 4171–4186. [CrossRef]
48. Yang, Z.; Dai, Z.; Yang, Y.; Carbonell, J.; Salakhutdinov, R.; Le, V.Q. XLNet: Generalized Autoregressive Pretraining for Language Understanding. *arXiv* **2019**, arXiv:1906.08237. [CrossRef].

49. Fu, J.; Liu, J.; Jiang, J.; Li, Y.; Bao, Y.; Lu, H. Scene segmentation with dual relation-aware attention network. *IEEE Trans. Neural Netw. Learn. Syst.* **2020**, *32*, 2547–2560. [CrossRef]
50. Hu, J.; Shen, L.; Sun, G. Squeeze-and-Excitation Networks. In Proceedings of the IEEE Conference on Computer Vision and Pattern Recognition, Salt Lake City, UT, USA, 18–22 June 2018; pp. 7132–7141. [CrossRef]
51. Wang, X.; Girshick, R.; Gupta, A.; He, K. Non-local neural networks. In Proceedings of the IEEE Conference on Computer Vision and Pattern Recognition (CVPR), Salt Lake City, UT, USA, 18–22 June 2018; pp. 7794–7803. [CrossRef]
52. Xiao, C.; Wachs, J.; Xiao, C.; Wachs, J. Triangle-net: Towards robustness in point cloud learning. In Proceedings of the IEEE/CVF Winter Conference on Applications of Computer Vision, Waikoloa, HI, USA, 3–8 January 2021; pp. 826–835. [CrossRef]
53. Qin, N.; Tan, W.; Ma, L.; Zhang, D.; Li, J. OpenGF: An Ultra-Large-Scale Ground Filtering Dataset Built upon Open ALS Point Clouds around the World. In Proceedings of the IEEE Computer Society Conference on Computer Vision and Pattern Recognition Workshops, Nashville, TN, USA, 19–25 June 2021. [CrossRef]
54. Zhang, K.; Chen, S.; Whitman, D.; Shyu, M.; Yan, J.; Zhang, C. A progressive morphological filter for removing nonground measurements from airborne LIDAR data. *IEEE T. Geosci. Remote* **2003**, *41*, 872–882. [CrossRef]
55. Zhang, W.; Qi, J.; Wan, P.; Wang, H.; Xie, D.; Wang, X.; Yan, G. An Easy-to-Use Airborne LiDAR Data Filtering Method Based on Cloth Simulation. *Remote Sens.* **2016**, *8*, 501. [CrossRef]
56. Lin, H.; Zheng, X.; Li, L.; Chao, F.; Wang, S.; Wang, Y.; Tian, Y.; Ji, R. Meta Architecure for Point Cloud Analysis. In Proceedings of the 2022 IEEE/CVF Conference on Computer Vision and Pattern Recognition (CVPR), New Orleans, LA, USA, 19–23 June 2022; pp. 17682–17691. [CrossRef]
57. Qiu, H.; Yu, b.; Tao, D. Collect-and-Distribute Transformer for 3D Point Cloud Analysis. *arXiv* **2023**, arXiv:2306.01257. [CrossRef]
58. Guo, M.; Liu, Z.; Mu, T.; Hu, S. Beyond Self-Attention: External Attention Using Two Linear Layers for Visual Tasks. *IEEE T. Pattern Anal.* **2023**, *45*, 5436–5447. [CrossRef] [PubMed]

remote sensing

MDPI

Article

DVST: Deformable Voxel Set Transformer for 3D Object Detection from Point Clouds

Yaqian Ning [1], Jie Cao [1,2,*], Chun Bao [1] and Qun Hao [1,2,3]

[1] School of Optics and Photonics, Beijing Institute of Technology, Beijing 100081, China; ningyq@bit.edu.cn (Y.N.); baochun@bit.edu.cn (C.B.); qhao@bit.edu.cn (Q.H.)
[2] Yangtze Delta Region Academy, Beijing Institute of Technology, Jiaxing 314003, China
[3] School of Opto-Electronic Engineering, Changchun University of Science and Technology, Changchun 130022, China
* Correspondence: caojie@bit.edu.cn

Abstract: The use of a transformer backbone in LiDAR point-cloud-based models for 3D object detection has recently gained significant interest. The larger receptive field of the transformer backbone improves its representation capability but also results in excessive attention being given to background regions. To solve this problem, we propose a novel approach called deformable voxel set attention, which we utilized to create a deformable voxel set transformer (DVST) backbone for 3D object detection from point clouds. The DVST aims to efficaciously integrate the flexible receptive field of the deformable mechanism and the powerful context modeling capability of the transformer. Specifically, we introduce the deformable mechanism into voxel-based set attention to selectively transfer candidate keys and values of foreground queries to important regions. An offset generation module was designed to learn the offsets of the foreground queries. Furthermore, a globally responsive convolutional feed-forward network with residual connection is presented to capture global feature interactions in hidden space. We verified the validity of the DVST on the KITTI and Waymo open datasets by constructing single-stage and two-stage models. The findings indicated that the DVST enhanced the average precision of the baseline model while preserving computational efficiency, achieving a performance comparable to state-of-the-art methods.

Keywords: 3D object detection; deformable mechanism; transformer; point clouds

Citation: Ning, Y.; Cao, J.; Bao, C.; Hao, Q. DVST: Deformable Voxel Set Transformer for 3D Object Detection from Point Clouds. *Remote Sens.* **2023**, *15*, 5612. https://doi.org/10.3390/rs15235612

Academic Editor: Mohammad Awrangjeb

Received: 18 October 2023
Revised: 17 November 2023
Accepted: 1 December 2023
Published: 3 December 2023

1. Introduction

Three-dimensional object detection based on LiDAR point clouds has attracted considerable attention, primarily for its applications in autonomous driving, virtual reality, and robotics [1,2]. There is immense potential in improving the performance of 3D object detection, both in industry and academia. Various methods for detecting 3D objects based on point clouds can be classified into point-based and voxel-based methods.

Point-based methods [3–6], such as PointNet and its variant models, take raw point clouds as inputs and extract key point feature representations by iterative sampling and group aggregation operations in point set abstraction modules, which are then used to predict object bounding boxes. Alternatively, voxel-based methods [7–9] divide the input raw point cloud into a regular grid to obtain a discrete voxel representation, which is processed using a deep convolutional neural network. Point-based methods can achieve precise spatial information from point clouds through larger receptive fields, enabling accurate positioning. However, these methods incur higher computational costs due to using point set abstraction [10]. Conversely, voxel-based methods prioritize computational efficiency but inevitably sacrifice information and reduce the precision of fine-grained localization [11]. This study focuses on voxel-based methods that aim to facilitate the development of 3D object detection from point clouds.

Voxel-based 3D object detection models frequently rely on 3D sparse convolutional networks [12] to extract voxel features. Although this method is computationally efficient, its limited receptive field prevents it from capturing rich contextual information, which can impede the detection of objects with only a small number of voxels. Additionally, efforts to mitigate this issue by increasing the maximum theoretical receptive field of each voxel, such as by modifying the voxel size, kernel size, downsampling stride, or number of layers, can result in numerous computational consumptions. To address these issues, Mao et al. proposed a voxel transformer (VoTr) [13], a voxel-based transformer backbone explicitly designed for 3D object detection. The VoTr utilized sparse voxel and submanifold voxel modules and employed self-attention mechanisms to model long-distance relationships between voxels. Moreover, it used local attention and dilated attention to expand the receptive field of attention without sacrificing computational efficiency. He et al. developed a voxel set transformer (VoxSeT) [14] to efficiently extract point cloud features for 3D object detection using set-to-set translation. The voxel-based set attention module, serving as the central component of VoxSeT, effectively overcomes the limitations of group-based and convolution-based attention modules and establishes long-distance dependencies more efficiently.

Despite their ability to utilize the transformer architecture for improved attention range, reduced information loss, and enriched contextual information, both the VoTr and VoxSeT models suffer from a limitation of not effectively focusing on foreground objects. As a result, the detection of naturally sparse and incomplete 3D objects is hindered. Given the potential increase in memory and computational costs associated with simply expanding the receptive field, as well as the possibility of feature interference from irrelevant parts beyond the region of interest, it is imperative to design a new architecture that can flexibly concentrate on foreground object areas and extract more informative features.

Recently, deformable mechanisms have been employed in 2D vision tasks due to their capability of selectively concentrating on areas with more information [15]. It has been demonstrated in references [16,17] that learning the deformable receptive field of a convolutional filter can efficiently selectively focus on more informative regions on a data-dependent basis. Xia et al. introduced a deformable attention transformer [18], which serves as a backbone model for tasks such as image classification and dense prediction. While the model chooses the positions of key-value pairs in self-attention based on data dependency to emphasize relevant regions, using a uniform reference point for all analyzed scenes hinders the detection of targets that may only have partial pixels or are positioned in the corners of the scene.

When applying the deformable mechanism to the detection of objects in 3D point clouds, previous studies have either built deformable self-attention modules for contextual feature extraction [19,20] or guided cross-domain feature aggregation [21]. However, it has not been recognized as a critical component in building a robust backbone, such as the deformable attention transformer. Two problems restrict the direct application of 2D deformable attention in detecting 3D point cloud objects. One problem arises from the sparsity of 3D point clouds, as the method used to determine reference points in the 2D deformable attention can lead to a substantial loss of information. Furthermore, numerous 3D point clouds in a scene exacerbate the quadratic computational complexity of self-attention, resulting in excessive memory usage and computational expenses.

To solve the problems above, we present a novel and efficient module called the deformable voxel set attention (DVSA) module for detecting objects in 3D point clouds. In general, the transformer-based 3D point cloud detection backbone has a larger receptive field, which enhances its representation capabilities. However, in object detection tasks, excessive attention to background areas is unnecessary and resource-consuming. Therefore, considering that the deformable mechanism can selectively focus on the target area, we tried to build a transformer backbone that selectively focuses on the foreground target area for 3D point cloud object detection. The target-region-sensitive deformable attention is constructed considering two factors when introducing the deformable mechanism. One factor is that performing operations involving the offset of all pixels in the deformable convolutional network feature map leads to a significant computational burden, and it also introduces confusion in the target area due to the offset of background points. Another factor is that, compared to only learning a few sets of query-independent offsets in 2D deformable attention, it is better to learn offsets for foreground queries to move the keys and values of the targets to important areas. Therefore, after the DVSA module divides the point clouds in the scene into foreground points and background points, it uses the newly constructed offset generation network to learn the offsets of the foreground queries. Then, the offset features carrying offset information are projected into keys and values, and the calculation of deformable attention is completed together with the input projected queries.

Specifically, after the DVSA divides all the points in the voxel grid into foreground and background points, the model learns the offsets of foreground points via an offset generation network. The offset features then serve as keys and values input to the multi-head attention, where the model derives the output features. Motivated by induced set attention [22], we reduce computational complexity by decomposing full self-attention into two consecutive cross-attention modules. This is based on the assumption that self-attention is low-rank [23] and can be approximated through low-rank projection [14]. The DVSA initially learns the offsets of foreground queries within the 3D voxel to acquire deformable features. Then, a set of inducing points is introduced to transform the features into hidden space by encoding cross-attention. Subsequently, a globally responsive convolutional feed-forward network (GRCFFN) enhances the hidden features. Lastly, output features are derived through decoding cross-attention.

With DVSA, we propose a novel approach called the deformable voxel set transformer (DVST) for 3D object detection from point clouds. The DVST employs deformable mechanisms to focus on and capture target-related features flexibly. It consists of multi-layer interconnected DVSA modules and a multi-layer perceptron (MLP). Through experiments conducted on the 3D detection benchmark KITTI and the Waymo open dataset, we demonstrate the effectiveness of the proposed model. The main contributions of our work are summarized as follows:

(1) The DVST deformable backbone was developed to detect objects in 3D point clouds. It combines the flexibility of deformable mechanisms, the powerful long-range modeling capability of transformers, and the linear computational complexity of set attention in a cohesive manner. By solely focusing on learning the offsets of foreground queries, the target semantic information is strengthened, the background features are weakened, and the detection performance is improved.

(2) DVSA, a deformable attention module for 3D point cloud learning, was designed for the first time. It utilizes deformable mechanisms to transfer the candidate keys and values of foreground queries to important regions. This enhancement provides the original self-attention with increased flexibility to capture target-related feature information.

(3) A novel offset generation module (OGM) was constructed for learning the offsets of foreground queries in the DVSA module. This data-dependent method of generating offsets enhances the model's robustness in detecting diverse scenarios. Moreover, a GRCFFN with a residual connection is proposed to facilitate global interaction and feature learning within the hidden space.

(4) One-stage and two-stage detection models were developed based on the DVST, and experiments were performed on the KITTI and Waymo open datasets, widely used benchmarks for 3D object detection. The findings indicate that the proposed backbone enhances the performance of 3D detection while ensuring computational efficiency.

2. Related Work

2.1. Three-Dimensional Object Detection from Point Clouds

Point-Based methods. Point-based detectors typically abstract geometric information from unstructured point clouds using PointNet [3] and its variants [4,24], which process the point cloud directly to generate 3D boxes that preserve precise position information.

Shi et al. introduced PointRCNN [25], a two-stage 3D object detection model that utilizes separate subnetworks for the first and second stages to generate and refine 3D proposals. To strike a balance between accuracy and efficiency, Yang et al. developed a point-based 3D single-stage object detector named 3DSSD [5]. This detector eliminates the feature propagation layer and refinement module and incorporates a novel fusion sampling strategy. Point-GNN [26] is a graph neural network comprising an auto-registration mechanism. Shi et al. constructed it for performing 3D point cloud object detection using graph representation. Given the significance of foreground points compared to background points in the detection of point clouds, two down-sampling strategies were proposed by Zhang et al. for the stratified selection of foreground points to the desired objects, resulting in the development of an efficient 3D detector named IA-SSD [6]. Ren et al. developed DGT-Det3D [10], a 3D object detection network that uses dynamic graph conversion to extract point-by-point semantic features that are beneficial in identifying distant and obstructed objects. Despite their ability to avoid the quantization errors caused by voxelization through direct feature learning, point-based methods have limited learning capabilities and efficiency.

Voxel-Based methods. Voxel-based detectors typically convert unordered point clouds into regular spatial units, usually of equal size. Yan et al. introduced sparse convolution into voxel-based 3D detectors and proposed a sparse embedded convolutional (SECOND) framework [14], improving detection accuracy and runtime speed. Lang et al. proposed PointPillars [7], which partitions point clouds into pillars, enabling faster encoding of point cloud features. To improve the detection accuracy with a coarse voxel granularity, Deng et al. developed the two-stage framework Voxel-RCNN [8], achieving comparable detection accuracy to point-based models with lower computational costs. Yin et al. presented an anchor-free detection framework called CenterPoint [27], which regresses other attributes based on detecting object centers. Yang et al. constructed an end-to-end multi-feature fusion network [28] composed of a voxel convolutional module, a local point feature module, and a detection head, obtaining a higher detection accuracy by extracting richer voxel features. Voxel-based methods can effectively balance accuracy and computational efficiency through the regular representation of point clouds. However, the introduction of inevitable quantization errors limits the learning ability of voxel-based methods on point cloud features, thereby restricting their detection performance.

Point–Voxel methods. Some approaches address the challenges posed by voxel-based and point-based methods by combining point and voxel representation. Shi et al. developed the PV-RCNN [11], which leverages 3D sparse convolution to extract voxel features and uses farthest-point sampling to sample key points. The voxel set abstraction module is then used to fuse point and voxel features, improving the detection accuracy. To fully utilize the structural information of point clouds, SA-SSD [29] incorporates an auxiliary network that transforms the convolutional features of the backbone into point-level representation. Noh et al. used hybrid voxel-point representation (HVPR) [30] for 3D object detection. Specifically, HVPR generates mixed 3D features by integrating point and voxel streams in the dual-stream encoder. Shuang et al. introduced AFE-RCNN [31], a point–voxel integrated network consisting of critical components such as the residual of dual attention proposal generation module, the multi-scale feature extraction module based on feature adaptive adjustment, and the refinement loss function module with vertex correlation. Nevertheless, achieving a balance between detection accuracy and efficiency poses a more significant challenge for these methods. Furthermore, the extensive reliance on handcrafted feature design in hybrid approaches hampers the potential for model improvement.

2.2. Transformers in 3D Object Detection

Transformer-based models are highly suitable for learning 3D point clouds with unordered and unstructured characteristics due to their capability of encoding positional information, long-range contextual information, and permutation invariance. Several 3D object detection models that incorporate transformers have recently been proposed. For instance, Mao et al. developed the VoTr model [13] to overcome the limitation of convolutional backbone networks in effectively capturing contextual information. Sheng et al. proposed CT3D [32], a two-stage 3D object detection framework that utilizes a channel-wise transformer to enhance the candidate boxes generated by the region proposal network. He et al. introduced the VoxSeT model [14], which leverages voxel-based set attention (VSA) for point cloud learning from sets. Guan et al. proposed M3DETR [33], a multi-representation and multi-scale integrated method that employs an M3 transformer to model point cloud features from multiple perspectives. Inspired by a window-based transformer, Sun et al. proposed a sparse window transformer [34] that fully exploits the sparsity of point clouds. In the weakly supervised point cloud transformer framework [35], Tang et al. adopted a self-attention mechanism to extract global features.

Moreover, Zhou et al. improved the performance of the anchor-free object detection network using the CenterFormer model [36], which utilizes the features of center candidates as query embeddings and incorporates cross-attention to fuse multi-frame features. Building upon the concept of anchor points, Liu et al. introduced the AnchorPoint model [37]. This detector encodes foreground points, which act as anchor points, as object queries. This encoding strategy allows each object query to possess a distinct and meaningful physical interpretation. Ning et al. proposed PV-SSD [38], a point–voxel and bird's eye view (BEV) representation aggregation network for single-stage 3D detection, which adaptively integrates multi-level semantic features using a windows transformer spatial-semantic aggregation module. Ren et al. designed DGT-Det3D [10], a dynamic graph transformer 3D object detection network that captures long-range dependencies and refines candidate boxes through dynamic graph transformer and proposal-aware fusion modules.

Unlike the methods mentioned above, the proposed DVST combines the advantages of a deformable mechanism, voxel-based model, and set transformer, enabling a flexible focus on foreground target regions while efficiently modeling long-range contextual information and maintaining computational efficiency.

2.3. Deformable Mechanisms in 3D Object Detection

Deformable mechanisms are commonly incorporated into 3D object detection models using point clouds to enhance the extraction capability of salient and discriminative features. Bhattacharyya et al. introduced the Deformable PV-RCNN [19] model by integrating deformable mechanisms into the PV-RCNN. This model adaptively extracts informative and discriminative features through deformable convolution operations. Subsequently, Bhattacharyya et al. developed the deformable self-attention (DSA) module in SA-Det3D [20]. This module calculates offsets for sample points obtained through farthest-point sampling, extracting context information containing representative and informative feature subsets. Recognizing that relevant object information is mainly located in adjacent positions, Chen et al. proposed a deformable cross-attention feature alignment module [21]. This module guides the aggregation positions in the feature map using deformable convolutions. To learn more effective contextual features, Tang et al. designed a deformable offset self-attention (DOSA) structure [39]. This structure first samples a voxel subset using the farthest point sampling method and then learns offsets for this voxel subset. Offset self-attention is performed on the voxel subset, which is upsampled to cover each original voxel position.

It is evident that combining the deformable mechanism with the self-attention mechanism is a more appropriate approach for learning useful features of 3D point clouds, as opposed to solely utilizing deformable convolutions to process 2D feature maps. Nonetheless, it should be noted that the DSA and DOSA modules mentioned above do not effectively employ the deformable mechanism to modify the key-value pairs of each query during the computation of the attention mechanism. In other words, the adaptive capability of the deformable mechanism for extracting target-related features has not been fully utilized. Additionally, DSA and DOSA sample the same number of points or voxels using the farthest-point sampling method for all scenes, result in a somewhat diminished model performance. To fully integrate the ability of the deformable mechanism to attend to target regions and the power of the self-attention mechanism to extract contextual information, we propose the DVSA module for 3D point cloud object detection, taking inspiration from 2D deformable attention.

3. Deformable Voxel Set Transformer

3.1. Overall Architecture

This paper presents a novel deformable voxel set transformer (DVST) for point-cloud-based 3D object detection. The DVST serves as a transformer-based 3D backbone, as depicted in Figure 1. By combining the deformable mechanism and transformer, the DVST can adaptively focus on the target region while possessing robust long-range modeling capabilities. We introduce induced set attention to reduce computational complexity, which breaks down full self-attention into two consecutive cross-attentions. Built upon the conventional voxel-based 3D detection pipeline [13], the DVST can seamlessly integrate into most voxel-based 3D detection frameworks. Specifically, the proposed 3D detector takes a voxelized point cloud as the input to extract features in the DVST backbone network, encodes them into BEV representation, applies a 2D convolutional neural network (CNN) to enhance the feature density further, and finally utilizes anchor-based detection heads to generate detection results. Following the traditional transformer paradigm, the DVST backbone consists of interconnected DVSA modules and MLPs. Moreover, we package the DVSA modules into residual blocks to ensure optimal gradient flow.

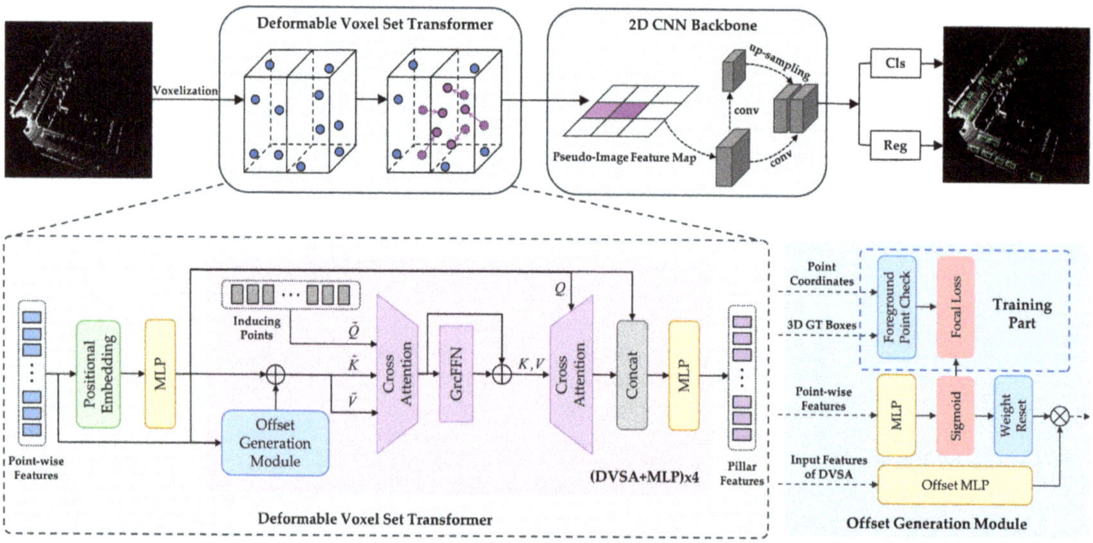

Figure 1. The overall architecture of DVST. The DVST is a transformer-based 3D backbone that can be utilized in various voxel-based 3D detection frameworks. It comprises a sequence of DVSA modules and MLPs. The DVSA is a deformable attention module explicitly designed for learning from 3D point clouds. It consists of the OGM, encoding cross-attention, GRCFFN, and decoding cross-attention. The OGM is employed to generate deformable offsets.

The DVSA module is a vital component of the DVST and is purpose-built for processing 3D point cloud data. The query features are acquired by encoding the position of the voxelized point cloud and processing it using an MLP. Subsequently, these query features are utilized as inputs for the DVSA module. Inspired by the 2D deformable attention module [18], we propose a novel OGM to generate offsets for the queries. The OGM utilizes point-wise features to determine confidence scores for each query, indicating whether it is a foreground query. Using a threshold, we set the scores of non-foreground queries to zero. Meanwhile, the OGM employs an offset MLP to generate initial offsets for all queries multiplied by the confidence scores to obtain the final offset values. Through this approach, DVSA can adaptively learn the offsets of foreground queries for different scenes, overcoming the limitation of 2D deformable attention in learning the same position offsets for all scenes.

Motivated by induced set attention [22], which posits that a low-rank projection can approximate self-attention, we use two cross-attentions guided by a set of inducing points to approximate the complete self-attention within the set. The first cross-attention maps the inducing points to the hidden space by attending to the input set, while the second cross-attention directs its focus towards the hidden features and generates the output set. The output features are then encoded as BEV features and fed into a 2D backbone network. Finally, we employ a detection head for 3D object detection.

3.2. *Deformable Voxel Set Attention*

Unlike images, point clouds are extensively distributed and possess weak semantic correlations at the scene level, although they exhibit strong structural details in localized regions. Thus, we suggest using deformation attention to model the relationships between markers, guided by the significant regions in the voxel grid. These focused regions are determined by multiple sets of deformable foreground points, which are learned from queries using the OGM. Deformable features are extracted from these points and then projected to obtain deformed keys and values. Simultaneously, a set of inducing points is assigned for each voxel in the grid, and the deformable features are encoded into the hidden space using encoding cross-attention. Next, a GRCFFN with a residual connection enhances the global interaction of hidden space features. Finally, decoding cross-attention is applied to focus on the original queries and aggregate features from deformed values. This DVSA module specifically focuses on learning the point cloud features of important areas to detect 3D objects during the set-to-set transformation process. The overall architecture of the DVSA module can be observed in Figure 2, and the design of each module component will be elaborated below.

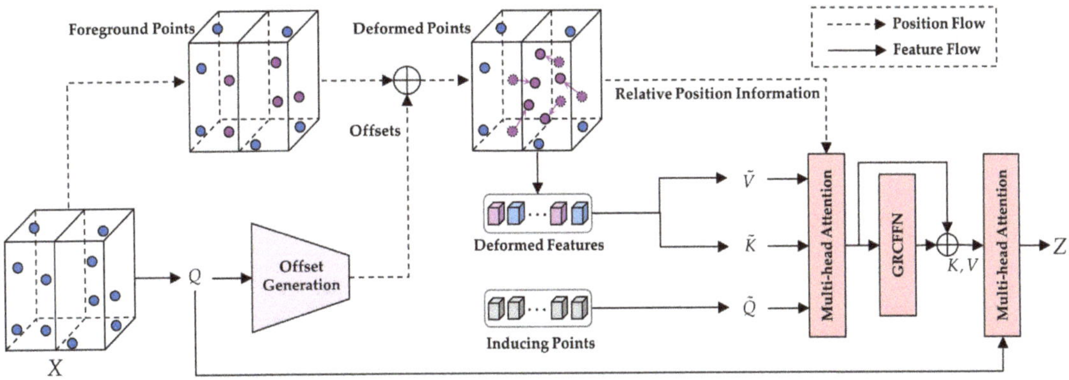

Figure 2. An illustration of DVSA. The figure presents the information flow of the DVSA. A group of foreground points is identified, and their offsets are learned from the queries through the OGM. Then, the deformed keys and values are projected from the deformable features. Subsequently, the deformable features are compressed into a latent space using a set of inducing points, and the features are refined using the GRCFFN. Finally, the output features are obtained through multi-head attention. For clarity of presentation, only two voxel grids and five foreground points are shown, although there are more points in practical implementation.

3.2.1. Positional Embedding

Given the significance of preserving the local structure of point clouds to enhance detection performance [40], we introduce a positional embedding (PE) module. This module encodes the local coordinates of point clouds within voxels into high-dimensional features, which are subsequently fed into the DVSA. Specifically, the PE module applies Fourier parameterization to obtain values $[sin(f_k \pi x), cos(f_k \pi x)]$, where x represents the normalized local coordinates and f_k is the k-th frequency with a bandwidth of L. The resulting Fourier embedding has a dimension of $3L$, which is then mapped to the input dimension of the first MLP using a trainable linear layer. The positional encoding feature not only maintains the local structure of the point cloud but also aids in generating more precise deformable offsets.

3.2.2. Offset Generation Module

The 2D deformable attention [18] determines the focal area by deforming a fixed position reference point. However, this method is unsuitable for the deformable attention of three-dimensional point clouds, as they are sparse and unstructured. To resolve this problem, we introduce an intuitive and practical approach: dividing the point cloud into foreground and background points and determining the focal area by deforming the foreground points. We developed an OGM module, as illustrated in Figure 1.

The provided point cloud $P = \{p_1, \ldots, p_n\}$ consists of n points. It undergoes processing to extract d_p-dimensional point-wise features $f_p \in R^{n \times d_p}$, which are then encoded with the PE module to generate d-dimensional DVSA input features $X \in R^{n \times d}$. The OGM initially takes f_p as the input to determine the confidence of the foreground $S \in R^n$ as follows:

$$S = G_p(f_p), \tag{1}$$

where $G_p(\cdot)$ represents a three-layer MLP with a sigmoid activation function designed for predicting the foreground confidence within the range of [0, 1]. Then, $S' \in R^n$ is obtained by resetting the confidence of non-foreground points to zero using a foreground confidence threshold T_f, as shown in Equation (2):

$$S_i' = \begin{cases} S_i & S_i > T_f \\ 0 & S_i \leq T_f \end{cases}. \tag{2}$$

The variable i represents the foreground confidence of the i-th point among n points. The default threshold for the foreground confidence is set at 0.5. Afterward, the feature X is fed into a shared subnetwork to generate the offset $\triangle p$, as depicted in Equation (3):

$$\triangle p = \theta_{offset}(X), \tag{3}$$

where $\theta_{offset}(\cdot)$ is a three-layer MLP network. The final offset value for each point in the foreground is determined by assigning the confidence of each point to the generated offset, referred to as $\triangle p'$ in Equation (4):

$$\triangle p' = \triangle p \cdot S'. \tag{4}$$

After generating the offset of foreground points, it is added to the input vector X to obtain the deformable feature $f_D \in R^{n \times d}$, as described in Equation (5):

$$f_D = X + \triangle p'. \tag{5}$$

The OGM employs focal loss [41] with default hyperparameters in the training process. This technique addresses the imbalance between foreground and background points in the training set. By multiplying the confidence of foreground points with the corresponding offset, the module generates offset sizes that vary based on the confidence levels of the foreground points. This somewhat helps to mitigate the adverse effects caused by foreground threshold edge points. Moreover, the proposed OGM uses two learnable networks to identify foreground points and determine their offsets. This approach overcomes the drawbacks of manual offset generation design and enhances the DVST's adaptive focusing capability on foreground regions.

3.2.3. Encoding Cross-Attention

As shown in Figure 2, the complete query of the deformable attention is acquired by projecting the input feature X, while the keys and values are obtained by projecting the deformable feature f_D. Following acquiring the deformable feature, we use an alternative approach to self-attention to model the global contextual relationship. To bypass the quadratic computational complexity of self-attention, we utilize VSA to learn the global features of the point clouds through set-to-set transformation. Given an input set with a size of n and a dimension of d, and by defining a k-dimensional vector $I \in R^{k\times d}$, known as inducing points, the output set $Z \in R^{n\times d}$ from the VSA [14] can be represented as:

$$H_f = \text{CrossAttention}(I, X) \in R^{k\times d}, \tag{6}$$

$$\hat{H} = \text{ConvFFN}\left(H_f\right) \in R^{k\times d}, \tag{7}$$

$$Z = \text{CrossAttention}(X, \hat{H}) \in R^{n\times d}. \tag{8}$$

The inducing points I are trainable parameters that are learned along with other parameters in the network. The first cross-attention transforms I into hidden features H_f by attending to the input set. The hidden features are then input to a convolutional feed-forward network (ConvFFN) for updating the information. The second cross-attention attends the input set X to the resulting hidden features. For clarity, we refer to the first and second cross-attentions as encoding cross-attention and decoding cross-attention, respectively. Both cross-attentions utilize standard multi-head attention. Furthermore, we propose a novel GRCFFN with residual connection to enhance the comprehensiveness and flexibility of information updating.

In the encoding cross-attention, we begin by performing a linear projection to project the deformable features f_D into the key $\widetilde{K} \in R^{n\times d}$ and the value $\widetilde{V} \in R^{n\times d}$:

$$\widetilde{K} = f_D W_{\widetilde{K}}, \widetilde{V} = f_D W_{\widetilde{V}}. \tag{9}$$

where $W_{\widetilde{K}}$ and $W_{\widetilde{V}}$ are projection matrices. Then, cross-attention is performed between the key $\widetilde{K}$ and query I to generate an attention matrix $A \in R^{n\times k\times d}$, as shown in Equation (10):

$$A = I^T \widetilde{K}. \tag{10}$$

The query $I \in R^{k\times d}$ is the given induced vector. Next, the attention matrix A is voxel-normalized using the scatter kernel function:

$$\widetilde{A} = \text{Softmax}_{scatter}(A, v), \tag{11}$$

where $v \in R^{n\times 3}$ denotes the voxel coordinates of n points. Let $\{p_i = (x_i, y_i, z_i) : i = 1, \ldots, n\}$ represent the point cloud coordinates and $\left[d_x, d_y, d_z\right]$ be the size of the three-dimensional voxel. The voxel coordinates can be calculated using Equation (12):

$$v = \left\{v_i = \left(\left\lfloor \frac{x_i}{d_x} \right\rfloor, \left\lfloor \frac{y_i}{d_y} \right\rfloor, \left\lfloor \frac{z_i}{d_z} \right\rfloor\right) : i = 1, \ldots, n\right\}, \tag{12}$$

where $\lfloor\cdot\rfloor$ is the floor function. The CUDA kernel library scatter function allows for symmetric reductions on various matrix sections, including maximum, average, and sum. In DVSA, the input set is treated as a single matrix where each row represents a point-wise feature, and each voxel associated with a point-wise feature can be indexed by voxel coordinates. Given point-wise features, their reduced voxel form after the symmetric scatter function can be represented as:

$$Y = \text{F}_{scatter}(X, v) = \{\text{F}_{scatter}(\{X_i : v_i = j\}) : j = 1, \ldots, m\}, \tag{13}$$

where m is the number of non-empty voxels. After obtaining the voxel-normalized attention matrix $\widetilde{A}$, it is multiplied by the values to produce the hidden feature H_f:

$$H_f = \widetilde{A}^T \widetilde{V}. \tag{14}$$

After that, voxel reduction of the hidden features according to the voxel index is used for feature refinement in the hidden space, as shown in Equation (15):

$$H_r = \text{Sum}_{scatter}(H, v). \tag{15}$$

3.2.4. Globally Responsive Convolutional Feed-Forward Network

After encoding deformable features into the hidden space using inducing points, the hidden features are input into a feed-forward network to achieve more flexible and complex feature refinement. In a previous study [14], the hidden features were scattered into a 3D sparse tensor, and a ConvFFN was constructed to learn the information in the corresponding tensor. While this approach successfully captures inter-voxel information interaction, it may impede the model's performance when dealing with deformable features due to limited feature diversity. To address this limitation, ConvNeXt V2 [42] proposed global response normalization (GRN), which normalizes the feature maps channel-wise, thereby enhancing feature competition and promoting feature diversity. Notably, GRN serves as a convolutional neural network layer that requires no additional parameters and has no learnable parameters. In light of this, we introduce the GRN into the ConvFFN to develop an improved version, GRCFFN, specifically designed to amplify feature diversity. The architecture of GRCFFN is illustrated in Figure 3.

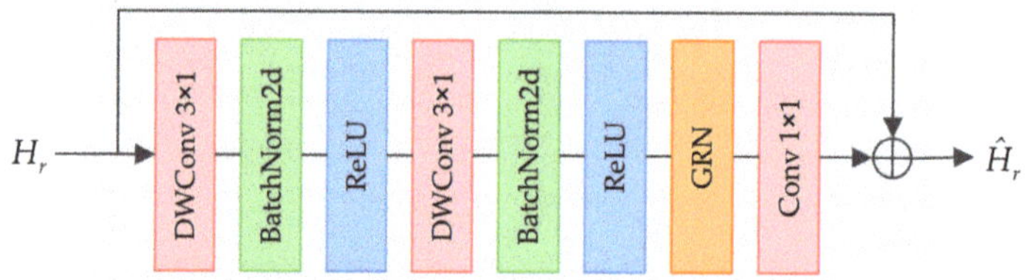

Figure 3. The network structure of GRCFFN.

Specifically, we disperse the reduced hidden features into a 3D sparse tensor based on voxel coordinates to adaptively integrate voxel features and global dependencies. Subsequently, two depth-wise convolutions (DWConv) are applied to these 3D tensors to facilitate feature interactions in the spatial domain, and a GRN layer is used to enhance feature diversity. By incorporating a residual connection, we can retain the spatial information in the input hidden features and integrate this information after feature interaction. Given the reduced hidden features $H_r \in R^{m \times k \times d}$, the features $H' \in R^{H \times W \times C}$ obtained from performing two DWConvs can be written as:

$$H' = \sigma_1(\text{DWConv}(\sigma_2(\text{DWConv}(\tau(H_r, C_r); W_1)); W_2)). \tag{16}$$

where W_1 and W_2 are convolutional weights, σ_1 and σ_2 represent non-linear activation functions, C_r denotes voxel coordinates, and τ refers to the formulation of sparse tensor.

When performing GRN, the first step is to use the global function $G(\cdot)$ to aggregate the feature H' into the vector gh:

$$G(H') := H' \in R^{H \times W \times C} \rightarrow gh \in R^C. \tag{17}$$

Next, we apply the response normalization function to the aggregate value as follows:

$$N(\|H_i'\|) := \|H_i'\| \in R \rightarrow \frac{\|H_i'\|}{\sum_{j=1,\ldots,C} \|H_j'\|} \in R, \quad (18)$$

where i is the i-th channel and $\|H_i'\|$ is the L2-norm of the i-th channel. The response normalization function computes the relative importance of each channel with the other channels. Finally, the normalized scores are used to correct the original input response:

$$H_i' = H_i' * N(G(H')_i) \in R^{H \times W}. \quad (19)$$

After completing GRN, given the convolution weights W_3, the rich hidden features $\hat{H}_r \in R^{m \times k \times d}$ from GRCFFN can be written as:

$$\hat{H}_r = H_r + \text{Conv}(H'; W_3). \quad (20)$$

3.2.5. Decoding Cross-Attention

The output set is reconstructed by the decoding cross-attention, as demonstrated in Figure 2, using the rich hidden features $\hat{H}_r$. Initially, the hidden features are broadcasted to generate features $\hat{H}_r \in R^{n \times k \times d}$ with the same length as the input set. Then, the decoding cross-attention projects the input set onto the query $Q \in R^{n \times d}$ and projects the rich hidden features containing offset information onto the key $K \in R^{n \times k \times d}$ and the value $V \in R^{n \times k \times d}$:

$$Q = XW_Q, K = \hat{H}_r W_K, V = \hat{H}_r W_V, \quad (21)$$

where W_Q, W_K, and W_V are projection matrixes. Lastly, multi-head attention is performed on the query Q, key K, and value V. The output of the DVSA can be calculated as:

$$A = [A_1, \ldots, A_n] = \left[K_1 Q_1^T, \ldots, K_n Q_n^T\right], \quad (22)$$

$$\widetilde{A} = \left[\text{Softmax}\left(\widetilde{A}_1\right), \ldots, \text{Softmax}\left(\widetilde{A}_n\right)\right], \quad (23)$$

$$Z = [Z_1, \ldots, Z_n] = \left[\widetilde{A}_1^T V_1, \ldots, \widetilde{A}_n^T V_n\right]. \quad (24)$$

In summary, the DVSA module initially learns the offset of foreground queries through the proposed OGM module. Subsequently, it transforms the deformed features with offset information to the hidden space utilizing encoding cross-attention. The hidden features are then refined by the proposed GRCFFN. Finally, decoding cross-attention is employed to derive the output features. By dynamically prioritizing significant regions during the set-to-set transformation process, the DVSA can effectively detect 3D objects.

3.3. *BEV Feature Encoding and Detection Head*

As illustrated in Figure 1, the DVST backbone is employed to extract features from the input point cloud. Subsequently, these extracted features are encoded into BEV representation, and a shallow 2D CNN backbone is applied to enhance the feature density. The reason for performing BEV encoding is that, in the context of 3D point cloud detection, models utilizing dense BEV features [8,12] generally exhibit higher recall rates than models utilizing sparse point-wise features [25,43].

The 2D CNN network utilized in this study comprises two strides, with each stride encompassing three convolutions. The point-wise features generated by the DVST are aggregated into pillars measuring 0.36 m × 0.36 m, which is followed by soft-pooling to generate BEV features. Given the point-wise output feature $O^j \in R^{k \times d}$, the pooled pillar-wise feature P^j can be represented as Equation (25):

$$P^j = \sum_{m=1}^{k} O^j_m * w^j_m, w^j_m = \frac{e^{O^j_m}}{\sum_{m=1}^{k} e^{O^j_m}}. \quad (25)$$

The pillar features are then fed into a shallow 2D CNN. The convolution features of two strides are connected and passed into the detection head to predict bounding boxes. The detection head follows a traditional anchor-based design [7,12].

Our DVST can be extended to accommodate two-stage detectors. The CT3D [32] is a prime example of a two-stage detector that boasts exceptional performance and minimally relies on manual design. In the two-stage detector, we utilize the RoI head of CT3D as our second-stage module. This module performs proposal-aware embedding of point features and aggregates contextual information through a channel-wise transformer. We validate the DVST by constructing single-stage and two-stage detectors in the experimental section.

3.4. Training Losses

We employ an end-to-end strategy in training the DVST. The overall training loss is calculated by the OGM foreground segmentation loss L_{seg}, the bounding box classification loss L_{cls}, the bounding box regression loss L_{reg}, and the bounding box orientation loss L_{dir}:

$$L_S = L_{seg} + \frac{1}{N_p}\left(L_{cls} + L_{reg} + L_{dir}\right). \quad (26)$$

where N_p denotes the number of positive samples, both L_{seg} and L_{cls} adopt focal loss, L_{reg} adopts Smooth-L1 loss, and L_{dir} adopts binary entropy loss.

When the DVST is expanded to a two-stage model, in addition to the losses mentioned above, the loss from the second stage of the bounding box regression and confidence prediction must also be included.

4. Experiments and Results

In this section, we assess our proposed DVST using two publicly available 3D point cloud detection datasets, the KITTI and Waymo open datasets. We begin by presenting the implementation details and evaluation settings of DVST, followed by a comparison with state-of-the-art 3D point cloud detection models. Lastly, we perform a comprehensive analysis of the components of DVST through ablation experiments.

4.1. Dataset and Implementation Details

KITTI dataset. In 3D point cloud object detection, the KITTI dataset [44] is widely used for evaluating model performance. The dataset classifies the 3D detection task into three difficulty levels, i.e., easy, moderate, and hard, based on object size, occlusion status, and truncation level. It comprises a total of 7481 training samples and 7518 testing samples. According to the usual protocol, the training samples are subdivided into a training set containing 3712 annotated samples and a validation set containing 3769 annotated samples. To assess the effectiveness of the DVST, we conducted experiments on an online testing server. When evaluating the test set, the model was trained using 80% of the training set data, while the remaining 20% was employed for validation purposes.

According to the KITTI official evaluation criteria, the intersection over union (IoU) threshold for the car category was set to 0.7. For the pedestrian and cyclist categories, the IoU thresholds were set to 0.5. The mean average precision (mAP) and the average orientation similarity (AOS) results on the test set were computed using 40 recall positions on the online testing server. The mAP results on the validation set were computed using 11 recall positions to ensure a fair comparison with other methods. We present two DVST-based detection architectures, including a DVST single-stage detector (DVST-SSD) for single-stage detection and a DVST two-stage detector (DVST-TSD) for two-stage detection. The single-stage model adopts the anchor settings from SECOND [12], while the two-stage model utilizes the RoI head from CT3D [32].

Waymo open dataset. The Waymo open dataset [45], serving as a large-scale 3D detection dataset, comprises 1000 sequences, consisting of 798 training sequences and 202 validation sequences. The training set contains approximately 158 k point cloud samples, while the validation set contains around 40k samples. The validation process for the Waymo open dataset employs two main approaches. The first approach categorizes objects based on difficulty levels: LEVEL_1 pertains to objects with at least five lidar points, whereas LEVEL_2 encompasses objects directly labeled as second level or objects with at least one lidar point. The second approach organizes objects according to their distance from the sensor, with three distinct ranges: 0–30 m, 30–50 m, and >50 m.

The mAP and the mean average precision with heading accuracy weighting (mAPH) are official evaluation metrics. These metrics are calculated based on an IoU threshold of 0.7 for vehicles and 0.5 for other categories. The models were trained using 20% of the training data and validated using the complete validation data set. To evaluate the effectiveness of the DVST on the Waymo open dataset, we replaced the 3D backbone networks in the PointPillars [7] and CenterPoint [27] models with the DVST while keeping the other modules unchanged. Consequently, we used two detection architectures: DVST-SSD, an anchor-based single-stage detection architecture, and DVST-TSD, an anchor-free two-stage detection architecture.

Implementation details. On the KITTI dataset, the point cloud ranges for the X, Y, and Z axes were set to [0 m, 70.4 m], [−40 m, 40 m], and [−3 m, 1 m], respectively. The voxel size was set to [0.32 m, 0.32 m, 4 m]. The feature dimensions of the four DVSAs were 16, 32, 64, and 128. The number of inducing points for each DVSA was eight in the one-stage model and four in the two-stage model. The bandwidth of the PE module was set to 64. In the two-stage models, 512 RoIs were sampled during training, and 128 RoIs were sampled during inference. During the post-processing stage, both the one-stage and two-stage models utilized NMS with an IoU threshold of 0.01 to predict bounding boxes. For the final prediction, the one-stage and two-stage models selected bounding boxes with confidence scores surpassing 0.1 and 0.5, respectively. The models were trained end-to-end for 100 epochs using the Adam optimizer on four NVIDIA 3080Ti GPUs. The batch size and weight decay were set to 2 and 0.01, respectively. The one-stage model used a learning rate of 0.001, while the two-stage model utilized a learning rate 0.0005. Additionally, the learning rate decayed with the one-cycle strategy, where the momentum damping range was set to [85%, 95%]. For other default settings, please refer to the OpenPCDet [46] toolbox employed in this study.

On the Waymo open dataset, the point cloud ranges for the X, Y, and Z axes were set as [−75.2 m, 75.2 m], [−75.2 m, 75.2 m], and [−2 m, 4 m], respectively. The voxel size was set as [0.32 m, 0.32 m, 6 m]. The feature dimensions of the four DVSAs with four inducing points were 16, 32, 64, and 128. The bandwidth of the PE module was set to 64. During the post-processing stage, the DVST-SSD and DVST-TSD predicted the bounding boxes using an IoU threshold of 0.7. The final predictions were determined by selecting the bounding boxes with confidence scores above 0.1. The learning rates for the two models were set to 0.00075 and 0.001, respectively. Furthermore, both models were trained end-to-end using the Adam optimizer on four NVIDIA GeForce RTX 3080Ti GPUs. The training process involved 30 epochs with a batch size of 1 and a weight decay of 0.01.

4.2. *Three-Dimensional Detection Results on KITTI Dataset*

4.2.1. KITTI Test Set

The test set was evaluated by submitting the detection results of the DVST-TSD to the official KITTI test server, and the performance of the DVST was compared with state-of-the-art methods. The DVST-TSD adopted the anchor setting in SECOND and used the RoI head in CT3D. Tables 1 and 2 present the results of the test set, including 3D detection mAP, BEV detection mAP, and 3D detection AOS metrics. Since some methods do not focus on the orientation evaluation of 3D objects, the results of the AOS metric are not listed in their respective literature. For these methods, the AOS metric was obtained either from the official KITTI leaderboard or was not compared in Table 2.

A total of nine specific quantitative indicators are involved in Tables 1 and 2. As shown in Tables 1 and 2, our DVST had five top-three indicators. It can be seen that among the voxel-based detection methods, our DVST had apparent advantages in the number of the top-three indicators. In addition to the point–voxel combined PV-RCNN, which had the same number of top-three-ranked indicators, the DVST had more indicators ranked in the top three than the other methods in the table. Thus, the comprehensive performance of the DVST was validated. In other words, the DVST demonstrated exceptional overall performance in 3D car detection, BEV detection, and orientation estimations due to its incorporation of deformable mechanisms, which offer flexibility, and transformers, which facilitate powerful long-range modeling. Consequently, the DVST effectively concentrated on the target area and extracted rich intrinsic contextual features, enabling it to adapt better to various detection tasks and difficulty levels.

Table 1. Performance evaluation of 3D object and BEV detection for the DVST on the KITTI test set. The results regarding mAP at an IoU threshold of 0.7, along with 40 recall positions, are reported. The results obtained using our DVST are in bold, while the top-three-ranked results are underlined.

Method	Type	Car-3D Detection (%)			Car-BEV Detection (%)		
		Easy	Moderate	Hard	Easy	Moderate	Hard
PointRCNN [25] (CVPR 2019)	Point-based	86.96	75.64	70.70	92.13	87.39	82.72
EPNet [47] (ECCV 2020)		89.81	79.28	74.59	94.22	88.47	83.69
3DSSD [5] (CVPR 2020)		88.36	79.57	74.55	92.66	89.02	85.86
Point-GNN [26] (CVPR 2020)		88.33	79.47	72.29	93.11	89.17	83.90
IA-SSD [6] (CVPR 2022)		88.34	80.13	75.04	92.79	89.33	84.35
AnchorPoint [37] (IEEE T-ITS 2023)		-	80.16	-	-	87.39	-
DGT-Det3D [10] (KBS 2023)		87.89	80.68	76.02	-	-	-
SA-SSD [29] (CVPR 2020)	Point–voxel	88.75	79.79	74.16	95.03	91.03	85.96
PV-RCNN [11] (CVPR 2020)		90.25	81.43	76.82	94.98	90.65	86.14
DVFENet [48] (Neurocomputing 2021)		86.20	79.18	74.58	90.93	87.68	84.60
HVPR [30] (CVPR 2021)		86.38	77.92	73.04	-	-	-
DSA-PV-RCNN [20] (ICCV 2021)		88.25	81.46	76.96	92.42	90.13	85.93
AFE-RCNN [31] (Remote Sensing 2022)		88.41	81.53	77.03			
FusionPillars [49] (Remote Sensing 2023)		86.96	75.74	73.03	92.15	88.00	85.53
SECOND [12] (Sensors 2018)	Voxel-based	83.13	73.66	66.20	88.07	79.37	77.95
PointPillars [7] (CVPR 2019)		82.58	74.31	68.99	90.07	86.56	82.81
Part-A2 [50] (IEEE TPAMI 2020)		87.81	78.49	73.51	91.70	87.79	84.61
TANeT [51] (AAAI 2020)		84.39	75.94	68.82	91.58	86.54	81.19
CIA-SSD [52] (AAAI 2021)		89.59	80.28	72.87	93.74	89.84	82.39
DFAF3D [39] (IMAVIS 2023)		88.59	79.37	72.21	93.14	89.45	84.22
CT3D [32] (ICCV 2021)		87.83	81.77	77.16	92.36	88.83	84.07
VoxSeT [14] (CVPR 2022)		88.53	82.06	77.46	92.70	89.07	86.29
DVST-TSD (Ours)		**88.70**	**81.66**	**77.18**	**92.72**	**88.71**	**85.97**

Table 2. Comparison of AOS performance of different methods on the KITTI test set. Our DVST results are shown in bold, and the top-three-ranking results are underlined.

Method	Type	Car-Orientation (%)		
		Easy	Moderate	Hard
PointRCNN [25] (CVPR 2019)	Point-based	95.90	91.77	86.92
EPNet [47] (ECCV 2020)		96.13	94.22	89.68
Point-GNN [26] (CVPR 2020)		38.66	37.20	36.29
IA-SSD [6] (CVPR 2022)		96.07	93.47	90.51
SA-SSD [29] (CVPR 2020)	Point–voxel	39.40	38.30	37.07
DVFENet [48] (Neurocomputing 2021)		95.33	94.44	91.55
PV-RCNN [11] (CVPR 2020)		98.15	94.57	91.85
DSA-PV-RCNN [20] (ICCV 2021)		95.84	94.52	91.93
AFE-RCNN [31] (Remote Sensing 2022)		95.84	94.63	92.07
PointPillars [7] (CVPR 2019)	Voxel-based	93.84	90.70	87.47
Part-A2 [50] (IEEE TPAMI 2020)		95.00	91.73	88.86
TANeT [51] (AAAI 2020)		93.52	90.11	84.61
CIA-SSD [52] (AAAI2021)		96.65	93.34	85.76
DFAF3D [39] (IMAVIS 2023)		96.54	93.20	90.03
CT3D [32] (ICCV 2021)		96.26	93.20	90.44
VoxSeT [14] (CVPR 2022)		96.15	95.13	90.38
DVST-TSD(Ours)		**96.02**	**94.67**	**92.03**

Specifically, for 3D car detection, the DVST outperformed most of the methods in Table 1 at the easy level with an mAP of 88.70%. Additionally, the DVST exhibited a higher mAP in 3D detection for the moderate and hard difficulty levels than all the point-based and point–voxel fusion methods in Table 1. Specifically, regarding the moderate-difficulty 3D detection, the DVST increased the mAP by at least 0.98% compared to the point-based methods and by at least 0.13% compared to the point–voxel fusion methods. For the hard-difficulty 3D detection, the proposed DVST showed mAP improvements of at least 1.16% and 0.15% compared to the point-based methods and point–voxel fusion methods, respectively. Similarly, when considering voxel-based methods, the DVST performed comparably to CT3D and VoxSeT for the medium- and hard-difficulty 3D detection mAP but outperformed them for the easy level.

Concerning car BEV detection, the point–voxel fusion method exhibited an advantage over the point-based and voxel-based methods. Notably, the DVST secured second place in the BEV detection mAP for the hard difficulty among the voxel-based methods, surpassing the third-ranked Part-A2 model by 1.36%. Regarding the car category orientation estimation, our DVST enhanced the AOS by at least 0.45% for the moderate difficulty compared to the point-based methods and by at least 0.04% compared to the point–voxel fusion methods. Among the voxel-based methods, our DVST achieved the highest AOS for the hard difficulty level, surpassing the second-ranked VoxSeT model by 1.65%. Based on the detailed analysis, it is evident that the DVST demonstrated a commendable overall performance on the KITTI test set.

4.2.2. KITTI Validation Set

We evaluated the performance of the DVST on the KITTI validation set, as presented in Table 3. For the VoxSeT, we reproduced the results based on the reference [14] utilizing the OpenPCDet toolbox [46]. Our findings are as follows: (1) For 3D car detection, the proposed DVST demonstrated competitive performance, achieving an mAP of 85.05% at moderate difficulty. The mAP was 0.23% higher than the point-based detector DGT-Det3D and at least 1.06% higher than the point–voxel detector. (2) Among the voxel-based detectors, the DVST-TSD outperformed VoTr-TSD and VoxSeT in the 3D detection of cars at the moderate difficulty level, surpassing them by 1.01% and 0.47% in terms of the mAP, respectively, despite all of them utilizing transformer-based backbone networks. (3) It is noteworthy that the DVST achieved the highest mAP of 78.99% among the voxel-based methods for the car type at the hard difficulty level.

Table 3. Evaluation of the KITTI validation set. The results are reported using the mAP with a 0.7 IoU threshold and 11 recall positions. The results of our DVST are highlighted in bold, and the top-three-ranked results are underlined.

Method	Type	Car-3D Detection (%)		
		Easy	Moderate	Hard
PointRCNN [25] (CVPR 2019)	Point-based	88.88	78.63	77.38
3DSSD [5] (CVPR 2020)		89.71	79.45	78.67
Point-GNN [26] (CVPR 2020)		87.89	78.34	77.38
AnchorPoint [37] (IEEE T-ITS 2023)		89.70	83.21	78.79
DGT-Det3D [10] (KBS 2023)		89.65	84.82	78.76
SA-SSD [29] (CVPR 2020)	Point–voxel	90.15	79.91	78.78
PV-RCNN [11] (CVPR 2020)		89.35	83.69	78.70
Deformable PV-RCNN [19] (ECCV 2020)		-	83.30	-
DVFENet [48] (Neurocomputing 2021)		89.81	79.52	78.35
HVPR [30] (CVPR 2021)		90.24	82.05	79.49
AFE-RCNN [31] (Remote Sensing 2022)		89.61	83.99	79.18
SECOND [12] (Sensors 2018)	Voxel-based	88.61	78.62	77.22
PointPillars [7] (CVPR 2019)		86.46	77.28	74.65
Part-A2 [50] (IEEE TPAMI 2020)		89.55	79.40	77.84
TANeT [51] (AAAI 2020)		87.52	76.64	73.86
CIA-SSD [52] (AAAI2021)		90.04	79.81	78.80
VoTr-TSD [13] (ICCV 2021)		89.04	84.04	78.68
CT3D [32] (ICCV 2021)		89.54	86.06	78.99
VoxSeT [14] (CVPR 2022)		89.24	84.58	78.87
DVST-TSD(Ours)		**89.30**	**85.05**	**78.99**

The mAP of the DVST for 3D car detection at moderate difficulty was lower than that of CT3D. After analysis, one primary reason for this discrepancy was the difference in the voxel size settings. In CT3D, the voxel size was set as [0.05 m, 0.05 m, 0.1 m], whereas in the DVST, it was set to [0.32 m, 0.32 m, 4 m]. Although a more minor voxel size resulted in higher precision in CT3D, it also led to increased computational complexity. We conducted experiments with a batch size of two on four NVIDIA RTX 3080Ti GPUs. The average floating-point operation of CT3D was 200.11 G with an inference time of 118.03 ms, while with the DVST, the values were only 89.61 G and 79.34 ms. Another relevant factor is the difference in the IoU threshold settings, with CT3D using 0.81 and the DVST using 0.7. Consequently, the following conclusions can be drawn: the DVST effectively utilized both deformable and attention mechanisms to extract rich object features from large voxels, and the DVST achieved a good balance between accuracy and speed in 3D object detection.

4.2.3. Performance on Pedestrian and Cyclist Classes

To assess the performance of the multi-class object detection model, we conducted experiments on the KITTI validation set using the single-stage DVST model. The DVST-SSD uses the proposed DVST as the backbone and adopts the same anchor settings as SECOND. The results of these experiments are presented in Table 4. We also compared the DVST-SSD with several baseline models, such as SECOND, PointPillars, and VoxSeT. The results in Table 4 illustrate the varying degrees of improvement in the mAP for the car, pedestrian, and cyclist categories across nine different situations of three difficulty levels. The corresponding improvements are shown in the last row of the table.

Table 4. Performance of DVST-SSD on the KITTI validation set. All results are reported in terms of mAP with 40 recall positions, using an IoU threshold of 0.7 for the car category and 0.5 for the pedestrian and cyclist categories. Results from DVST-SSD are shown in bold, and the best result in each column is underscored. The improvements are displayed in italics.

Method	Car (%)			Pedestrian (%)			Cyclist (%)		
	Easy	Moderate	Hard	Easy	Moderate	Hard	Easy	Moderate	Hard
SECOND [12]	88.61	78.62	77.22	56.55	52.98	47.73	80.58	67.15	63.10
PointPillars [7]	86.46	77.28	74.65	57.75	52.29	47.90	80.04	62.61	59.52
VoxSeT [14]	88.45	78.48	77.07	60.62	54.74	50.39	84.07	68.11	65.14
DVST-SSD (Ours)	**88.98**	**78.95**	**77.76**	**60.79**	**57.34**	**52.44**	**84.63**	**70.65**	**65.53**
Improvements	*+0.37*	*+0.33*	*+0.54*	*+0.17*	*+2.60*	*+2.05*	*+0.56*	*+2.54*	*+0.39*

The improvements demonstrate that the DVST exhibited strong detection capabilities for the car category and performed well in detecting pedestrians and cyclists, whose spatial sizes are smaller. The DVST-SSD achieved significant improvements of 2.60% and 2.05% in the mAP for the pedestrian category at the moderate and hard difficulty levels, respectively. Moreover, a substantial 2.54% increase in the mAP was observed for the cyclist category under moderate difficulty. These noteworthy enhancements suggest that the deformable attention utilized in the DVST-SSD effectively learned the target features under occlusion and truncation conditions. The overall improvement in the model performance further confirms the effectiveness of the DVST.

4.2.4. Qualitative Analysis

In this section, we present a comparison of the visualization results between VoxSeT and the DVST, as illustrated in Figure 4, where the circled parts are false-positive targets. Lidar-only detectors often suffer from a common failure mode: false-positive detections caused by more distant targets with similar shape characteristics. As shown in Figure 4a, VoxSeT and the DVST showed four and one false-positive detections, respectively. Figure 4b shows four false-positive detections for VoxSeT and two for the DVST. Figure 4c shows one false-positive detection for VoxSeT, while the DVST had none. From this comparison, we observed that the DVST used a deformable mechanism to transfer the candidate keys and values of the foreground query to essential areas to obtain more discriminative features, which can significantly reduce the number of false positives in the scene to be detected and improve the detection performance.

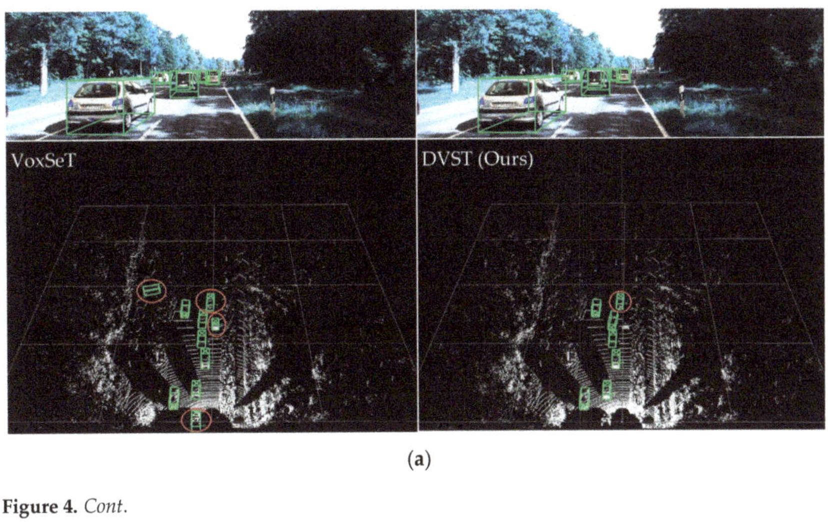

(a)

Figure 4. *Cont.*

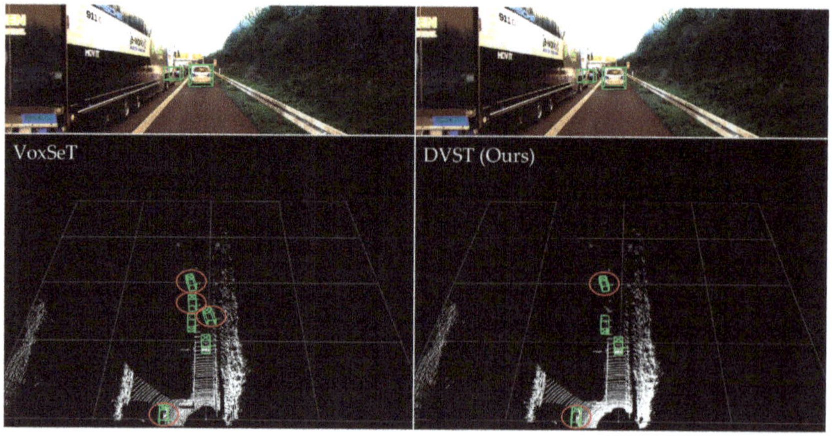

(b)

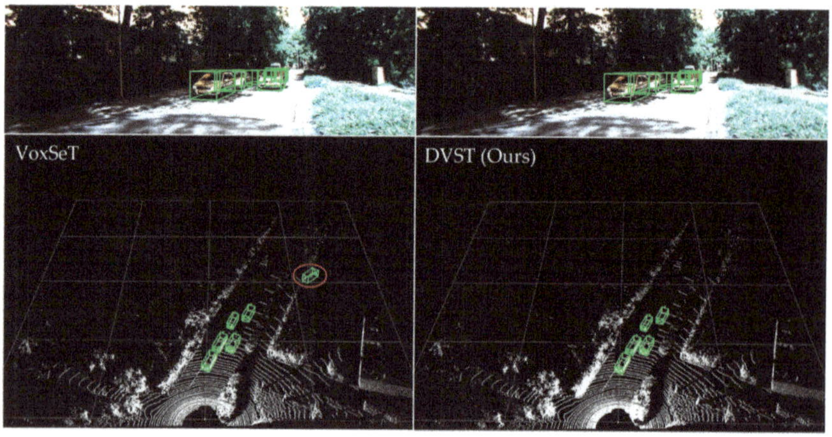

(c)

Figure 4. Qualitative analysis results of 3D object detection on the KITTI dataset. Three scenes, denoted as (**a**–**c**), are showcased. Each scene includes the ground truth box. The detection results for each scene using VoxSeT and DVST models are depicted on the left and right sides, respectively. False detection targets are circled for clarity.

4.3. Three-Dimensional Detection Results on Waymo Open Dataset

We performed experiments on the large-scale Waymo open dataset to validate the effectiveness of the proposed DVST. The experimental results for the mAP and mAPH are presented in Table 5, encompassing the three categories of vehicles, pedestrians, and cyclists, evaluated at two difficulty levels: LEVEL_1 and LEVEL_2. We replaced the backbone networks of two 3D object detection models, PointPillars and VoxSeT, with the DVST to construct the DVST-SSD and DVST-TSD, respectively, for the experiments and comparisons.

From the results in Table 5, it can be observed that compared to PointPillars, the DVST-SSD achieved improvements in the mAP and mAPH across all twelve metrics for the vehicle, pedestrian, and cyclist classes at both the LEVEL_1 and LEVEL_2 difficulty levels. Mainly, the DVST-SSD showed significant improvements in the pedestrian and cyclist scores, increasing them by 5.44%, 11.44%, 5.37%, 10.42%, 6.52%, 7.85%, 6.28%, and 7.57% across the eight metrics. Similarly, the DVST-TSD exhibited improvements compared to VoxSeT, with increases of 1.20%, 1.20%, 1.16%, 1.16%, 1.39%, 1.48%, 1.55%, 1.59%, 1.93%,

1.92%, 1.85%, and 1.83% across the twelve metrics. These improvements across the metrics for the DVST-SSD and DVST-TSD provide strong evidence of the performance of the DVST in large-scale datasets. Compared to state-of-the-art models such as PV-RCNN, AFE-RCNN, IA-SSD, and AnchorPoint, although the DVST-TSD did not achieve the best performance in the vehicle class, it achieved the best accuracy in all eight metrics for the pedestrian and cyclist classes. The results were consistent with the results of the pedestrian and cyclist classes in the KITTI validation set, further validating the efficacy of the DVST for 3D object detection.

Table 5. Comparison of 3D detection results on 202 validation sequences from the Waymo open dataset. All models were trained using 20% of the training samples. 'L1' and 'L2' represent LEVEL_1 and LEVEL_2, respectively. The best results are underlined below. Results from DVST-SSD and DVST-TSD are shown in bold, and the improvements are displayed in italics.

Method	Vehicle-L1		Vehicle-L2		Pedestrian-L1		Pedestrian-L2		Cyclist-L1		Cyclist-L2	
	mAP	mAPH	mAP	mAPH	mAP	mAPH	mAP	mAPH	mAP	mAPH	mAP	mAPH
SECOND [12]	68.03	67.44	59.57	59.04	61.14	50.33	53.00	43.56	54.66	53.31	52.67	51.37
CenterPoint [27]	70.50	69.96	62.18	61.69	73.11	61.97	65.06	55.00	65.44	63.85	62.98	61.46
Part-A2 [50]	74.66	74.12	65.82	65.32	71.71	62.24	62.46	54.06	66.53	65.18	64.05	62.75
AnchorPoint [37]	73.91	73.40	65.10	64.64	70.37	61.76	61.77	54.08	70.18	68.65	67.76	66.28
IA-SSD [6]	70.53	69.67	61.55	60.80	69.38	58.47	60.30	50.73	67.67	65.30	64.98	62.71
AFE-RCNN [31]	71.23	70.53	62.62	61.99	-	-	-	-	59.69	43.14	57.44	41.51
PV-RCNN [11]	<u>75.41</u>	<u>74.74</u>	<u>67.44</u>	<u>66.80</u>	71.98	61.24	63.70	53.95	65.88	64.25	63.39	61.82
PointPillars [7]	69.37	68.73	61.22	60.65	64.29	44.64	56.03	38.84	54.14	51.19	52.11	49.26
DVST-SSD	**69.93**	**69.39**	**61.62**	**61.14**	**69.73**	**56.08**	**61.40**	**49.26**	**60.66**	**59.04**	**58.39**	**56.83**
Improvements	*+0.56*	*+0.66*	*+0.40*	*+0.49*	*+5.44*	*+11.44*	*+5.37*	*+10.42*	*+6.52*	*+7.85*	*+6.28*	*+7.57*
VoxSeT [14]	70.09	69.59	61.63	61.18	74.04	65.56	65.78	58.13	68.72	67.42	66.18	64.93
DVST-TSD	**71.29**	**70.79**	**62.79**	**62.34**	<u>**75.43**</u>	<u>**67.04**</u>	<u>**67.33**</u>	<u>**59.72**</u>	<u>**70.65**</u>	<u>**69.34**</u>	<u>**68.03**</u>	<u>**66.76**</u>
Improvements	*+1.20*	*+1.20*	*+1.16*	*+1.16*	*+1.39*	*+1.48*	*+1.55*	*+1.59*	*+1.93*	*+1.92*	*+1.85*	*+1.83*

4.4. Ablation Studies

4.4.1. Effect of the DVST Components

We performed ablation experiments on the KITTI validation set to evaluate the various components of our method, and the results are presented in Table 6. As the DVST is built upon VoxSeT, we used VoxSeT as the baseline for comparison. The results in the first and second rows indicate that incorporating OGM improved the model's mAP scores in the car category from 84.58% to 84.71%, which suggests that including the deformable mechanism enhances the model's performance. Comparing the second and third rows, we observed a 0.09% increase in the moderate difficulty, demonstrating that our proposed GRCFFN can enrich global features in the hidden space. Furthermore, comparing the second and fourth rows reveals that adding residual connections played a critical role in boosting the model's performance. Comparing the performance of the fourth and fifth rows indicates that GRCFFN outperformed the ConvFFN used in VoxSeT.

Table 6. Ablation experiments on the KITTI validation set regarding the effects of the proposed OGM, GRCFFN, and residual connection (RC). The results are reported by the mAP with a 0.7 IoU threshold and 11 recall points. The best results are shown in bold.

Methods	OGM	GRCFFN	RC	Easy (%)	Moderate (%)	Hard (%)
VoxSeT	-	-	-	89.24	84.58	78.87
DVST-TSD	√	-	-	89.10	84.71	78.88
	√	√	-	89.19	84.80	78.84
	√	-	√	**89.34**	84.89	78.90
	√	√	√	89.30	**85.05**	**78.99**

4.4.2. Effect of the Number of Inducing Points

To analyze the impact of the number of inducing points used in the DVSA, we conducted experiments on the KITTI validation set using the DVST-SSD. As indicated in Table 7, we varied the number of inducing points between two, four, and eight, respectively, and evaluated the mAP across three difficulty levels within the car, pedestrian, and cyclist classes.

Table 7. Ablation experiments on the KITTI validation set regarding the impact of the number of induction points. The results are reported with mAP at 0.7 IoU threshold and 11 recall positions. The best results for each column are shown in bold.

Number of Inducing Points	Car (%)			Pedestrian (%)			Cyclist (%)		
	Easy	Moderate	Hard	Easy	Moderate	Hard	Easy	Moderate	Hard
2	88.34	78.33	77.04	59.08	55.17	50.32	85.69	67.16	63.79
4	88.46	78.49	77.37	**61.25**	56.80	52.27	83.78	68.22	63.39
8	**88.98**	**78.95**	**77.76**	60.79	**57.34**	**52.44**	**84.63**	**70.65**	**65.53**

The results unveiled in the table show that an increase in the number of inducing points allowed for encoding more comprehensive contextual information, enhancing the DVST's modeling capability and improving the detection. These findings are consistent with those observed in VoxSeT, which explored various latent codes. It is important to note that in the DVST-TSD, as indicated in the first and fifth rows of Table 6, the model achieved superior performance compared to VoxSeT when four inducing points were utilized, despite VoxSeT using eight inducing points. This phenomenon highlights that introducing deformable mechanisms strengthened the DVST's ability to model context and global features, leading to more accurate detection outcomes using fewer inducing points.

4.5. Efficiency Analysis

Table 8 presents the mAP, the number of parameters, the number of floating-point operations (FLOPs), and the inference time of SECOND, PointPillars, VoxSeT, and the DVST-SSD on the KITTI validation set. All the models were executed using an NVIDIA 3080Ti GPU. Analyzing the results in Table 8 reveals that the DVST-SSD model outperformed the baseline models SECOND and PointPillars, exhibiting precision improvements of 0.33%, 4.36%, and 3.50% for cars, pedestrians, and cyclists under moderate difficulty, respectively, while having fewer parameters and FLOPs. Compared to VoxSeT, the DVST-SSD achieved a superior detection accuracy while only undergoing a slight increase in parameters, FLOPs, and inference time. This model offers a reliable solution for achieving a balance between efficiency and accuracy in 3D object detection tasks.

Table 8. Performance comparison of DVST-SSD on the KITTI validation set. The best results in each column are shown in bold.

Method	Car (%)	Pedestrian (%)	Cyclist (%)	Parameters (M)	FLOPs (G)	Inference_Time (ms)
SECOND	78.62	52.98	67.15	4.61	76.86	53.10
PointPillars	77.28	52.29	62.61	4.83	63.69	**28.73**
VoxSeT	78.48	54.74	68.11	**2.93**	**53.01**	48.85
DVST-SSD	**78.95**	**57.34**	**70.65**	4.42	57.94	52.80

5. Discussion

In this paper, we used the proposed backbone network DVST to construct single-stage and two-stage models, respectively, conduct comprehensive experiments on the KITTI and Waymo datasets, and compare the proposed model with other methods on the corresponding benchmarks. In addition to the better results achieved by our approach, there are several points worthy of attention. As shown in Tables 4 and 5, the improvements in the DVST were more prominent for the pedestrian and cyclist categories than for the vehicle categories. Combined with the overall performance of the DVST in the KITTI and Waymo datasets, we believe that there were two main reasons. One reason was that the proposed deformable voxel attention module could better learn the offsets of small-sized targets, thus significantly improving the detection accuracy of small-sized targets. For the larger vehicle categories, the original method achieved better detection results, so the DVST showed a relatively small improvement. Another reason was that the trained DVST could better adapt to the data distribution of the corresponding dataset than the other models. Another point worth noting is that the accuracy of the division of foreground and background points affected the performance of the DVST. The DVST uses a three-layer MLP and foreground threshold to divide foreground and background points. It is reasonable to suspect that DVST can achieve better results with the blessing of a deeper or more targeted network. Therefore, a more accurate division of foreground and background points is one of the focuses of our follow-up research. In addition, although the DVST provides a reliable solution to achieve a balance between efficiency and accuracy of 3D object detection from point clouds, in practical applications, higher accuracy and faster speed are still the goals we pursue.

6. Conclusions

In this work, we proposed a transformer-based method called the DVST for 3D object detection from point clouds. The DVST addresses the issue of excessive attention caused by the large receptive field of the transformer backbone network by employing deformable voxel set attention as its core. It also utilizes induced set attention to ensure computational efficiency. The DVST effectively learns target-related features and concentrates on significant foreground regions by selectively deforming only the foreground queries. The clever integration of a deformable mechanism, transformer, and induced set attention in the DVST strikes a favorable balance between average precision and computational efficiency. Moreover, as a voxel-based backbone network, the DVST can be applied to any voxel-based 3D detector. Extensive experiments on the KITTI and Waymo open datasets demonstrated that DVST performs similarly to state-of-the-art lidar-only detectors. However, to meet the higher accuracy and faster speed requirements of practical application scenarios, we will explore the integration of point cloud information with images and lightweight model optimization in future research.

Author Contributions: Conceptualization, Y.N., J.C. and C.B.; methodology, Y.N.; software, Y.N.; validation, Y.N.; formal analysis, Y.N. and J.C.; investigation, Y.N.; resources, J.C. and C.B.; data curation, Y.N.; writing—original draft preparation, Y.N.; writing—review and editing, Y.N., J.C. and C.B.; visualization, Y.N.; supervision, J.C. and Q.H.; project administration, Q.H.; funding acquisition, J.C. All authors have read and agreed to the published version of the manuscript.

Funding: This research received no external funding.

Data Availability Statement: The datasets generated and analyzed during this study are available in the KITTI repository (https://www.cvlibs.net/datasets/kitti/eval_object.php?obj_benchmark=3d, accessed on 17 October 2023) and the Waymo open dataset (https://waymo.com/open/, accessed on 17 October 2023).

Acknowledgments: The authors thank Dexian Yin, Xue Huang, and Chang Zhou for their help in the paper and the anonymous reviewers for their valuable comments and suggestions.

Conflicts of Interest: The authors declare no conflict of interest.

References

1. Yang, J.; Shi, S.; Wang, Z.; Li, H.; Qi, X. ST3D++: Denoised Self-Training for Unsupervised Domain Adaptation on 3D Object Detection. *IEEE Trans. Pattern Anal. Mach. Intell.* **2022**, *45*, 6354–6371. [CrossRef] [PubMed]
2. Wang, M.; Chen, Q.; Fu, Z. LSNet: Learned Sampling Network for 3D Object Detection from Point Clouds. *Remote Sens.* **2022**, *14*, 1539. [CrossRef]

3. Qi, C.R.; Su, H.; Mo, K.; Guibas, L.J. PointNet: Deep Learning on Point Sets for 3D Classification and Segmentation. In Proceedings of the IEEE Conference on Computer Vision and Pattern Recognition, Honolulu, HI, USA, 21–26 July 2017; pp. 652–660.
4. Qi, C.R.; Yi, L.; Su, H.; Guibas, L.J. PointNet++: Deep Hierarchical Feature Learning on Point Sets in a Metric Space. In Proceedings of the 31st International Conference on Neural Information Processing Systems, Red Hook, NY, USA, 4 December 2017; pp. 5105–5114.
5. Yang, Z.; Sun, Y.; Liu, S.; Jia, J. 3DSSD: Point-Based 3D Single Stage Object Detector. In Proceedings of the IEEE/CVF Conference on Computer Vision and Pattern Recognition, Seattle, WA, USA, 13–19 June 2020; pp. 11040–11048.
6. Zhang, Y.; Hu, Q.; Xu, G.; Ma, Y.; Wan, J.; Guo, Y. Not All Points Are Equal: Learning Highly Efficient Point-Based Detectors for 3D LiDAR Point Clouds. In Proceedings of the IEEE/CVF Conference on Computer Vision and Pattern Recognition, New Orleans, LA, USA, 18–24 June 2022; pp. 18931–18940.
7. Lang, A.H.; Vora, S.; Caesar, H.; Zhou, L.; Yang, J.; Beijbom, O. PointPillars: Fast Encoders for Object Detection from Point Clouds. In Proceedings of the IEEE/CVF Conference on Computer Vision and Pattern Recognition, Long Beach, CA, USA, 15–20 June 2019; pp. 12689–12697.
8. Deng, J.; Shi, S.; Li, P.; Zhou, W.; Zhang, Y.; Li, H. Voxel R-CNN: Towards High Performance Voxel-Based 3D Object Detection. In Proceedings of the AAAI Conference on Artificial Intelligence, Virtually, 2–9 February 2021; pp. 1201–1209.
9. Wu, H.; Wen, C.; Li, W.; Li, X.; Yang, R.; Wang, C. Transformation-Equivariant 3D Object Detection for Autonomous Driving. In Proceedings of the AAAI Conference on Artificial Intelligence, Washington DC, USA, 7–14 February 2023; pp. 2795–2802.
10. Ren, S.; Pan, X.; Zhao, W.; Nie, B.; Han, B. Dynamic Graph Transformer for 3D Object Detection. *Knowl.-Based Syst.* **2023**, *259*, 110085. [CrossRef]
11. Shi, S.; Guo, C.; Jiang, L.; Wang, Z.; Shi, J.; Wang, X.; Li, H. PV-RCNN: Point-Voxel Feature Set Abstraction for 3D Object Detection. In Proceedings of the IEEE/CVF Conference on Computer Vision and Pattern Recognition, Seattle, WA, USA, 13–19 June 2020; pp. 10529–10538.
12. Yan, Y.; Mao, Y.; Li, B. SECOND: Sparsely Embedded Convolutional Detection. *Sensors* **2018**, *18*, 3337. [CrossRef] [PubMed]
13. Mao, J.; Xue, Y.; Niu, M.; Bai, H.; Feng, J.; Liang, X.; Xu, H.; Xu, C. Voxel Transformer for 3D Object Detection. In Proceedings of the IEEE/CVF International Conference on Computer Vision, Montreal, QC, Canada, 10–17 October 2021; pp. 3144–3153.
14. He, C.; Li, R.; Li, S.; Zhang, L. Voxel Set Transformer: A Set-to-Set Approach to 3D Object Detection from Point Clouds. In Proceedings of the IEEE/CVF Conference on Computer Vision and Pattern Recognition, New Orleans, LA, USA, 18–24 June 2022; pp. 8407–8417.
15. Zhu, X.; Su, W.; Lu, L.; Li, B.; Wang, X.; Dai, J. Deformable DETR: Deformable Transformers for End-to-End Object Detection. In Proceedings of the International Conference on Learning Representations. *arXiv* **2020**, arXiv:2010.04159.
16. Dai, J.; Qi, H.; Xiong, Y.; Li, Y.; Zhang, G.; Hu, H.; Wei, Y. Deformable Convolutional Networks. In Proceedings of the IEEE International Conference on Computer Vision, Venice, Italy, 22–29 October 2017; pp. 764–773.
17. Wang, W.; Dai, J.; Chen, Z.; Huang, Z.; Li, Z.; Zhu, X.; Hu, X.; Lu, T.; Lu, L.; Li, H.; et al. InternImage: Exploring Large-Scale Vision Foundation Models with Deformable Convolutions. In Proceedings of the IEEE/CVF Conference on Computer Vision and Pattern Recognition, Vancouver, BC, Canada, 17–24 June 2023; pp. 14408–14419.
18. Xia, Z.; Pan, X.; Song, S.; Li, L.E.; Huang, G. Vision Transformer with Deformable Attention. In Proceedings of the IEEE/CVF Conference on Computer Vision and Pattern Recognition, New Orleans, LA, USA, 18–24 June 2022; pp. 4784–4793.
19. Bhattacharyya, P.; Czarnecki, K. Deformable PV-RCNN: Improving 3D Object Detection with Learned Deformations. *arXiv* **2020**, arXiv:2008.08766.
20. Bhattacharyya, P.; Huang, C.; Czarnecki, K. SA-Det3D: Self-Attention Based Context-Aware 3D Object Detection. In Proceedings of the IEEE/CVF International Conference on Computer Vision, Montreal, BC, Canada, 11–17 October 2021; pp. 3022–3031.
21. Chen, Z.; Li, Z.; Zhang, S.; Fang, L.; Jiang, Q.; Zhao, F. Deformable Feature Aggregation for Dynamic Multi-Modal 3D Object Detection. In Proceedings of the European Conference on Computer Vision, Tel Aviv, Israel, 23–27 October 2022; pp. 628–644.
22. Lee, J.; Lee, Y.; Kim, J.; Kosiorek, A.R.; Choi, S.; Teh, Y.W. Set Transformer: A Framework for Attention-Based Permutation-Invariant Neural Networks. *arXiv* **2019**, arXiv:1810.00825.
23. Wang, S.; Li, B.Z.; Khabsa, M.; Fang, H.; Ma, H. Linformer: Self-Attention with Linear Complexity. *arXiv* **2020**, arXiv:2006.04768.
24. Paigwar, A.; Erkent, O.; Wolf, C.; Laugier, C. Attentional PointNet for 3D-Object Detection in Point Clouds. In Proceedings of the IEEE/CVF Conference on Computer Vision and Pattern Recognition, Long Beach, CA, USA, 16–17 June 2019; pp. 1297–1306.
25. Shi, S.; Wang, X.; Li, H. PointRCNN: 3D Object Proposal Generation and Detection from Point Cloud. In Proceedings of the IEEE/CVF Conference on Computer Vision and Pattern Recognition, Long Beach, CA, USA, 15–20 June 2019; pp. 770–779.
26. Shi, W.; Rajkumar, R. Point-GNN: Graph Neural Network for 3D Object Detection in a Point Cloud. In Proceedings of the IEEE/CVF Conference on Computer Vision and Pattern Recognition, Seattle, WA, USA, 13–19 June 2020; pp. 1708–1716.
27. Yin, T.; Zhou, X.; Krahenbuhl, P. Center-Based 3D Object Detection and Tracking. In Proceedings of the IEEE/CVF Conference on Computer Vision and Pattern Recognition, Nashville, TN, USA, 20–25 June 2021; pp. 11779–11788.
28. Yang, S.; Lu, H.; Li, J. Multifeature Fusion-Based Object Detection for Intelligent Transportation Systems. *IEEE Trans. Intell. Transp. Syst.* **2023**, *24*, 1126–1133. [CrossRef]
29. He, C.; Zeng, H.; Huang, J.; Hua, X.-S.; Zhang, L. Structure Aware Single-Stage 3D Object Detection from Point Cloud. In Proceedings of the IEEE/CVF Conference on Computer Vision and Pattern Recognition, Seattle, WA, USA, 13–19 June 2020; pp. 11870–11879.

30. Noh, J.; Lee, S.; Ham, B. HVPR: Hybrid Voxel-Point Representation for Single-Stage 3D Object Detection. In Proceedings of the IEEE/CVF Conference on Computer Vision and Pattern Recognition, Nashville, TN, USA, 20–25 June 2021; pp. 14600–14609.
31. Shuang, F.; Huang, H.; Li, Y.; Qu, R.; Li, P. AFE-RCNN: Adaptive Feature Enhancement RCNN for 3D Object Detection. *Remote Sens.* **2022**, *14*, 1176. [CrossRef]
32. Sheng, H.; Cai, S.; Liu, Y.; Deng, B.; Huang, J.; Hua, X.-S.; Zhao, M.-J. Improving 3D Object Detection with Channel-Wise Transformer. In Proceedings of the IEEE/CVF International Conference on Computer Vision, Montreal, QC, Canada, 10–17 October 2021; pp. 2723–2732.
33. Guan, T.; Wang, J.; Lan, S.; Chandra, R.; Wu, Z.; Davis, L.; Manocha, D. M3DETR: Multi-Representation, Multi-Scale, Mutual-Relation 3D Object Detection with Transformers. In Proceedings of the IEEE/CVF Winter Conference on Applications of Computer Vision, Waikoloa, HI, USA, 3–8 January 2022; pp. 2293–2303.
34. Sun, P.; Tan, M.; Wang, W.; Liu, C.; Xia, F.; Leng, Z.; Anguelov, D. SWFormer: Sparse Window Transformer for 3D Object Detection in Point Clouds. In Proceedings of the European Conference on Computer Vision, Tel Aviv, Israel, 23–27 October 2022; pp. 426–442.
35. Tang, Z.; Sun, B.; Ma, T.; Li, D.; Xu, Z. Weakly Supervised Point Clouds Transformer for 3D Object Detection. In Proceedings of the IEEE 25th International Conference on Intelligent Transportation Systems, Macau, China, 8 October 2022; pp. 3948–3955.
36. Zhou, Z.; Zhao, X.; Wang, Y.; Wang, P.; Foroosh, H. CenterFormer: Center-Based Transformer for 3D Object Detection. In Proceedings of the European Conference on Computer Vision, Tel Aviv, Israel, 23–27 October 2022; pp. 496–513.
37. Liu, H.; Ma, Y.; Wang, H.; Zhang, C.; Guo, Y. AnchorPoint: Query Design for Transformer-Based 3D Object Detection and Tracking. *IEEE Trans. Intell. Transp. Syst.* **2023**, *24*, 10988–11000. [CrossRef]
38. Ning, K.; Liu, Y.; Su, Y.; Jiang, K. Point-Voxel and Bird-Eye-View Representation Aggregation Network for Single Stage 3D Object Detection. *IEEE Trans. Intell. Transp. Syst.* **2023**, *24*, 3223–3235. [CrossRef]
39. Tang, Q.; Bai, X.; Guo, J.; Pan, B.; Jiang, W. DFAF3D: A Dual-Feature-Aware Anchor-Free Single-Stage 3D Detector for Point Clouds. *Image Vis. Comput.* **2023**, *129*, 104594. [CrossRef]
40. Pan, X.; Xia, Z.; Song, S.; Li, L.E.; Huang, G. 3D Object Detection with Pointformer. In Proceedings of the IEEE/CVF Conference on Computer Vision and Pattern Recognition, Nashville, TN, USA, 20–25 June 2021; pp. 7459–7468.
41. Lin, T.-Y.; Goyal, P.; Girshick, R.; He, K.; Dollar, P. Focal Loss for Dense Object Detection. *IEEE Trans. Pattern Anal. Mach. Intell.* **2020**, *42*, 318–327. [CrossRef] [PubMed]
42. Woo, S.; Debnath, S.; Hu, R.; Chen, X.; Liu, Z.; Kweon, I.S.; Xie, S. ConvNeXt V2: Co-Designing and Scaling ConvNets with Masked Autoencoders. *arXiv* **2023**, arXiv:2301.00808.
43. Yang, Z.; Sun, Y.; Liu, S.; Shen, X.; Jia, J. STD: Sparse-to-Dense 3D Object Detector for Point Cloud. In Proceedings of the IEEE/CVF International Conference on Computer Vision, Seoul, Republic of Korea, 27 October 2019; pp. 1951–1960.
44. Geiger, A.; Lenz, P.; Urtasun, R. Are We Ready for Autonomous Driving? The KITTI Vision Benchmark Suite. In Proceedings of the 2012 IEEE Conference on Computer Vision and Pattern Recognition, Providence, RI, USA, 16–21 June 2012; pp. 3354–3361.
45. Sun, P.; Kretzschmar, H.; Dotiwalla, X.; Chouard, A.; Patnaik, V.; Tsui, P.; Guo, J.; Zhou, Y.; Chai, Y.; Caine, B.; et al. Scalability in Perception for Autonomous Driving: Waymo Open Dataset. In Proceedings of the IEEE/CVF Conference on Computer Vision and Pattern Recognition, Seattle, WA, USA, 13–19 June 2020; pp. 2443–2451.
46. OpenPCDet Development Team. OpenPCDet: An Open-Source Toolbox for 3D Object Detection from Point Clouds. 2020. Available online: https://github.com/open-mmlab/OpenPCDet (accessed on 17 October 2023).
47. Huang, T.; Liu, Z.; Chen, X.; Bai, X. EPNet: Enhancing Point Features with Image Semantics for 3D Object Detection. In Proceedings of the European Conference on Computer Vision, Glasgow, UK, 23–28 August 2020; pp. 35–52.
48. He, Y.; Xia, G.; Luo, Y.; Su, L.; Zhang, Z.; Li, W.; Wang, P. DVFENet: Dual-Branch Voxel Feature Extraction Network for 3D Object Detection. *Neurocomputing* **2021**, *459*, 201–211. [CrossRef]
49. Zhang, J.; Xu, D.; Li, Y.; Zhao, L.; Su, R. FusionPillars: A 3D Object Detection Network with Cross-Fusion and Self-Fusion. *Remote Sens.* **2023**, *15*, 2692. [CrossRef]
50. Shi, S.; Wang, Z.; Shi, J.; Wang, X.; Li, H. From Points to Parts: 3D Object Detection from Point Cloud with Part-Aware and Part-Aggregation Network. *IEEE Trans. Pattern Anal. Mach. Intell.* **2020**, *43*, 2647–2664. [CrossRef] [PubMed]
51. Liu, Z.; Zhao, X.; Huang, T.; Hu, R.; Zhou, Y.; Bai, X. TANet: Robust 3D Object Detection from Point Clouds with Triple Attention. In Proceedings of the AAAI Conference on Artificial Intelligence, New York, NY, USA, 7–12 February 2020; pp. 11677–11684.
52. Zheng, W.; Tang, W.; Chen, S.; Jiang, L.; Fu, C.-W. CIA-SSD: Confident IoU-Aware Single-Stage Object Detector from Point Cloud. *arXiv* **2020**, arXiv:2012.03015. [CrossRef]

MDPI
St. Alban-Anlage 66
4052 Basel
Switzerland
www.mdpi.com

MDPI Books Editorial Office
E-mail: books@mdpi.com
www.mdpi.com/books

www.ingramcontent.com/pod-product-compliance
Lightning Source LLC
Chambersburg PA
CBHW061606220926
51850CB00005B/436